S0-AIT-488

ANIMAL NUTRITION

ANIMAL NUTRITION

ARON A. BONDI
Emeritus Professor of Animal Nutrition and Biochemistry,
Hebrew University of Jerusalem,
Faculty of Agriculture, Rehovot, Israel

In partial collaboration with
David Drori
Agricultural Research Centre,
Beth Dagan, Israel

A Wiley–Interscience Publication

JOHN WILEY & SONS
Chichester · New York · Brisbane · Toronto · Singapore

First published in Hebrew by the Magnes Press,
The Hebrew University of Jerusalem, 1982.

Library of Congress Cataloging in Publication Data:

Bondi, Aron A.
Animal Nutrition.

Translation of: Hazanat ba'ale-hayim.
'Wiley–Interscience publication.'
Includes index.
1. Animal nutrition. I. Drori, David. II. Title.
SF95.B615513 1987 636.08'52 85-26376
ISBN 0 471 90375 2

British Library Cataloguing in Publication Data:

Bondi, Aron A.
Animal nutrition.
1. Animal nutrition
I. Title II. Drori, David
636.08'52 SF95

ISBN 0 471 90375 2

Printed and Bound in Great Britain

This book is dedicated to the memory of our late beloved son Dr. Elchanan Esra Bondi, who devoted his main interests to the natural sciences. In 1971 he succumbed at the age of 32 to severe illness.

Contents

Part II. CHEMISTRY AND METABOLISM OF NUTRIENTS

Part IV. NUTRITIVE REQUIREMENTS FOR BODY PROCESSES AND PRODUCTIVE FUNCTIONS

Preface

The main purpose of animal production is to supply high-quality food for humans. Although animal products are not essential dietary components for humans other than the very young, meat, milk, cheese and eggs are very satisfying to the palate as well as providing satiety. However, animal proteins are of higher nutritional quality than plant proteins because they contain large amounts of the essential amino acids required by man. Also, animal products supply vitamins, particularly B_{12}, and valuable minerals. Livestock production is vital to man today and will be equally important to us in the future. The consumption of animal proteins by the affluent population of the USA is very high and provides almost 70% of the protein consumed, whereas the world population obtains only one-third of the dietary protein from animal sources. Increase of the world population and rise in the standard of living will bring about an increased demand for animal proteins.

In the economics of animal production, the cost of feedstuffs is an item of greatest importance. World economy strives to reduce the losses incurred in the transformation of feedstuffs, mostly from vegetable origin, to animal products. Success of the animal enterprises depends primarily on the proper utilization of the nutrients offered. Progress in animal nutrition has been achieved mainly by discovering the metabolic processes underlying the utilization of all the nutrients which benefit the animal. In writing this textbook, attention has been focused on those biochemical aspects of nutrition that have been consolidated by recent research. Moreover, the area of bioenergetics has received quite intensive treatment, since the provision of energy is the main reason for the food supply to farm animals.

The book presents the principles of nutrition of various species of farm animals such as cattle, sheep, goats, horses, pigs and poultry. The comparative consideration of the metabolic processes occurring in various species of animals will contribute to a better understanding of the general mode of utilization of nutrients by various species and this will be profitable even for readers interested in the nutrition of one particular species. It is not the purpose of this book to provide instructions for practical livestock feeding or descriptions of feeding stuffs. Data on nutrient requirements of the various species are not given here, since they are frequently subjected to changes according to accruing new experimental results; also, the recommended requirements are easily accessible in the currently appearing official publications of the ARC (Agricultural Research Council in Great Britain) and the NRC (National Research Council in USA).

This book is based on my textbook of animal nutrition that has been published in Hebrew in 1982 by the Magnes Press, Hebrew University, Jerusalem, Israel. The English edition has been thoroughly revised and updated. Some data on the nutrition of pigs not dealt with in the Hebrew book, have been included here.

Citations of relevant books, monographs, review articles and a limited number of significant research papers are given at the end of each chapter and a list of relevant textbooks at the end of the book.

In general, the book is designed for undergraduate and graduate students in agriculture and in veterinary sciences. It can also serve breeders and agricultural advisers with a background in chemistry, biochemistry, human nutritionists, medical scientists and physicians. The basic principles of human and animal nutrition are the same, therefore the nutrient needs of humans are occasionally compared to those of animals throughout the text.

It is my pleasure to extend sincere thanks to my colleagues who have helped in the preparation of this book. Dr David Drori has collaborated in diverse chapters. The following colleagues have collaborated in chapters dealing with fields of their particular interest as indicated in the text or assisted by reading and commenting on several sections of the book: Drs E. Alumot (Mrs), A. Arieli, I. Ascarelli, A. Bar, R. Braude (Reading, England), P. Budowski, Y. Folman, D. Levi, H. Neumark (late), I. Nir, Z. Nitsan (Mrs), B. Robinson and D. Sklan. Mrs N. Gestetner is thanked for drawing most of the figures and Mrs E. Abraham for preparing the photographs. Above all, I thank my wife, Eva, for her support, apart from many arduous hours of typing, patience and devotion, for my work.

A. BONDI.

Part I

FATE AND FUNCTION OF NUTRIENTS IN THE ANIMAL BODY

CHAPTER 1

Chemical Composition of Feedstuffs and Animals

INTRODUCTION

Man and animals require permanent food intake in order to permit normal functioning of the life processes. The food of farm animals consists mainly of plants and plant products. Solar energy enables plants to synthesize their components—substances of complex chemical structure, proteins, fats and carbohydrates—from simple substances such as carbon dioxide from the air, and water and inorganic substances from the soil. Considerable amounts of energy originating from solar radiation are stored as chemical energy within the plant components. When animals ingest food of plant origin, the energy contained therein is used by the animals for maintaining bodily functions (respiration, blood flow and nervous system function), for tissue gain in growing animals and for formation of animal products (milk, eggs, wool, etc.).

Water is the quantitatively predominant constituent of the animal body, surpassing the total amount of all other constituents. The bulk of the dry matter in animals and plants is made up mainly of three groups of organic compounds—proteins, fats and carbohydrates—and also of inorganic matter (minerals) and some minor constituents such as vitamins, nucleic acids and others. However, there are striking differences between plants and animals in the composition and quantitative relationship of the three groups of organic compounds. The dry matter of plants consists principally of carbohydrates (75–80%); the animal body contains only about 1%, even though in the animal carbohydrates fulfil vital functions in energy production. It should be noted that the rate of synthesis and breakdown of carbohydrates occurring in the animal body are very rapid (Chapter 4). Soluble carbohydrates—primarily starch—serve as an energy store in the plant, while insoluble carbohydrates (cellulose) preserve structure and mechanical stability. Protein provides the structure of soft animal tissue, and fat deposited in the body serves as energy reserve. The protein content of the animal body is generally higher than of plant matter, except protein-rich, oil-bearing seeds. The synthesis of protein in the animal body, leading to the formation of body tissues such as muscle (meat), organs, soft tissues and animal products (milk, eggs, etc.), should be considered as the main objective of farm animal nutrition.

The characteristics and external appearance of organs of animals and parts of plants are greatly influenced by their water content, which varies considerably with species and age of the animal or plant. In addition to the main groups of components mentioned, plant and animal matter contain small amounts of nucleic acids, organic acids and vitamins as shown in the following diagram:

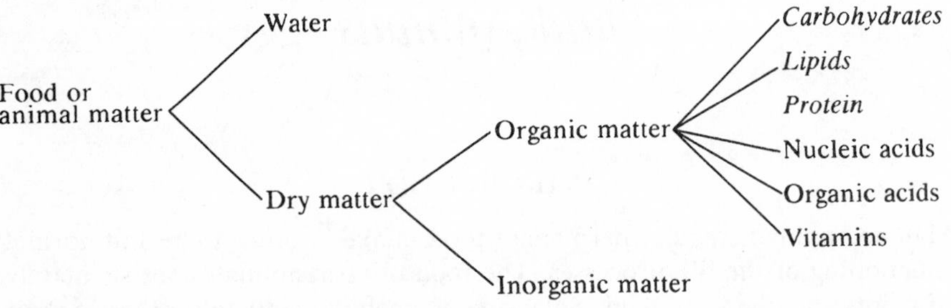

The dry matter present in plants and animals is divided into organic and inorganic matter, the inorganic matter consisting of a large number of elements present in varying amounts in the various parts of plants and animals. Certain minerals are present in considerable amounts in both animals and plants: calcium, phosphorus, magnesium, sodium, potassium, chlorine and sulphur; while other elements, such as iron, copper, cobalt, iodine, manganese, zinc, fluorine, selenium, molybdenum and others, appear only in minute amounts in living matter.

There are components of foods, such as starch or fats, which may be metabolized as a source of energy but for which the body has no specific need. On the other hand, there are nutrients with specific physiological functions, such as amino acids, minerals and vitamins. The overall amount of energy-supplying nutrients required is considerably greater than that of nutrients with specific functions.

PRINCIPLES OF ANALYSIS OF FEEDSTUFFS

Analyses of individual compounds present in samples of food or animal tissues are tedious to carry out and the results are often difficult to relate to animal utilization. Much of the existing information we have about the composition of foods and parts of animals has been obtained by a scheme of proximate analysis known as the Weende Method, named after the experimental station in Germany where it was developed some 120 years ago. According to this method, a feedstuff is divided into six fractions: water, ether extract (crude fat), crude fibre, nitrogen-free extract, crude protein, and ash. Five of these fractions are determined by chemical analyses. The sixth one—nitrogen-free extract—is determined by calculation of the differences: 100 minus the sum of the five other fractions.

Each of these fractions, except water, groups together a variety of food constituents on the basis of their common chemical characteristics (see Figure 1.1).

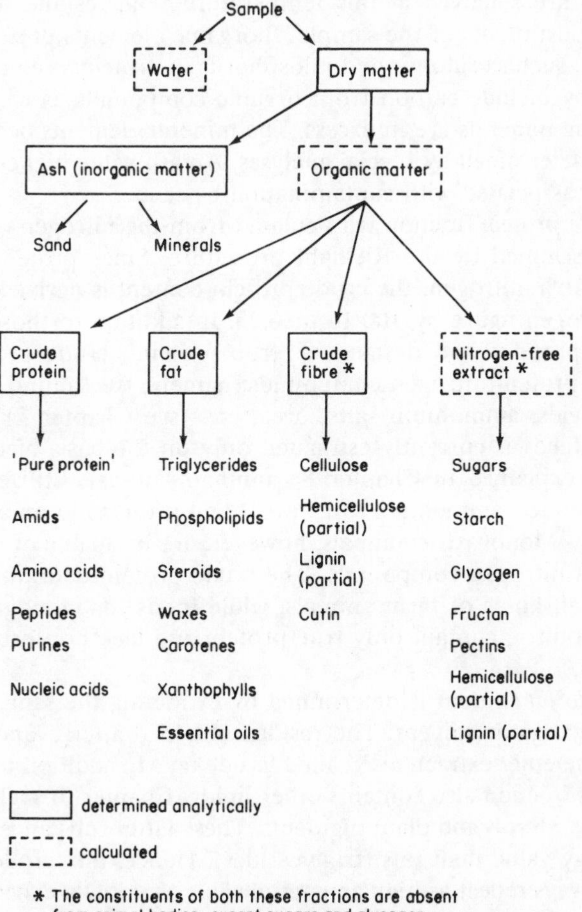

Figure 1.1. Components of different fractions in the proximate analysis of feedstuffs (Weende Method).

The *water content* of natural feedstuffs and animal tissues varies quite widely, and a knowledge of the water content is of the utmost importance, particularly when comparing analytical data for different feeds. After analysis, nutrient composition can be expressed on a dry basis. The most simple means of determining dry matter is to place the test material in an oven at 105 °C and dry it until all the free water has evaporated. Water content is usually measured by the difference in weight of a sample before and after drying. This method is satisfactory for most feeds, but some feedstuffs, particularly silage contain volatile compounds such as volatile fatty acids or ammonia. In order to avoid loss of volatiles during the drying of such feedstuffs, their moisture content should be determined by oven drying at 70 °C, freeze-drying or by toluene distillation.

The *ash* content is determined by ignition of a sample at 500 °C. Organic compounds are removed at this temperature. The residue represents the inorganic constituents of the sample. Inorganic elements present in organic compounds, such as sulphur and phosphorus in proteins, remain in the ash. The ash may include carbon from organic compounds as carbonate when base-forming minerals are in excess. The mineral elements occurring in ash have to be determined by special analyses. An ash value higher than normal is generally associated with contamination by sand.

The *crude protein* fraction is calculated from the nitrogen content of the sample determined by the Kjeldahl procedure. Since proteins contain an average of 16% nitrogen, the crude protein content is derived by multiplying the nitrogen figure by 100/16 or 6.25. In addition to those compounds defined as proteins and designated 'true protein', crude protein also includes non-protein nitrogen compounds; namely, free amino acids, amides of amino acids, ammonium salts, urea, etc. (see Chapter 7). The protein content of feeds is currently estimated only on the basis of crude protein figures. As explained in Chapter 8, ruminant animals utilize non-protein nitrogen (amides, ammonium salts, urea) and proteins to an approximately equal extent. Monogastric animals, however, are incapable of utilizing most non-protein nitrogen compounds. The crude protein data of feeds are applicable to all kinds of farm animals, while foods for monogastric species, including poultry, contain only true protein and lack non-protein nitrogen compounds.

The *crude fat* content is determined by extracting the sample with ether or a similar organic solvent. The residue obtained after evaporation of the solvent is the ether extract, also called 'crude fat'. In addition to fats (triglycerides), this fraction also contains other lipids (Chapter 6) such as phospholipids, waxes, sterols and plant pigments. These latter substances are of much lower energy value than fats (triglycerides). Hence, the 'crude fat' content does not always reflect as high an energy value as initially expected (Chapter 14).

The Weende method divides food carbohydrates into two fractions: sparingly soluble *crude fibre* and well-soluble *nitrogen-free extract*. The former is determined by subjecting the ether-extracted sample to successive treatments with boiling dilute acid and base; the insoluble residue remaining is the crude fibre containing all the cellulose present in the sample and part of its hemicellulose and lignin. Nitrogen-free extract is made up primarily of readily available carbohydrates, such as sugars and starch. Apart from the soluble carbohydrates, a considerable proportion of lignin and hemicellulose is contained in the nitrogen-free extract, particularly in such feedstuffs as forage, depending upon the species and stage of growth of the plants. Differentiation of carbohydrates into two groups of different solubility according to the Weende procedure is based on the assumption that there is a direct relationship between solubility of a carbohydrate and its nutritional efficiency. In fact crude fibre is a carbohydrate fraction not utilizable by monogastric animals which utilize nitrogen-free extract as a main energy source in their food. As far as ruminants are concerned, however, crude fibre values may be a misleading index for low availability of this fraction since crude fibre is well utilized by ruminants and in certain feeds at an even higher rate than nitrogen-free extract. The results of analyses of feeds according to the Weende procedure do not reflect their exact feeding value. We therefore need to supplement these analytical values with other data, such as the coefficient of digestibility (Chapter 13) or the energy content (Chapter 15) of the respective feedstuffs or with more detailed chemical analyses carried out using modern techniques (Chapter 13). In spite of these limitations, the Weende analysis is still generally used in tables of food composition and is the basis for the chemical description of animal feeds. These figures are the buyers' assurance that a commercial feedstuff contains at least those minimal amounts of the higher-cost items, particularly protein, and not more than the specified quantities of the less valuable items, the crude fibre and ash.

Energy may be defined as the capacity to work. There are various forms of energy such as chemical, thermal, electrical and radiant energy, all of which are interconvertible by suitable means; for example, solar energy is transformed into chemical energy stored in plant matter. The nutrients: carbohydrates, fats and proteins are the fuels that the animal cell is capable of converting into various forms of energy for mechanical work of muscles, for synthesis of macromolecules from simpler ones, for providing heat and for other functions.

Since all forms of energy can be converted quantitatively into heat, the energy of feedstuffs or of parts of the animal body is measured as heat. The energy value of a given sample is determined by burning it and measuring the heat produced (see Chapter 15). The basic unit of heat energy is the calorie (cal), defined as the amount of heat required to raise the temperature of 1 gram of water 1 °C, measured from 14.5 °C to 15.5 °C. Since this unit is very small, the kilocalorie (kcal = 1000 cal) or the megacalorie (megacal or Mcal = 1 000 000 cal) is commonly used. In some countries the joule (J) has recently been accepted as the unit of energy, with, consequently, the kilojoule (kJ) and megajoule (mJ). The kilojoule is defined as the energy expended when 1 kg is moved 1 m by a force of 1 newton (N). The conversion factor for kilocalories to kilojoules is 4.2, or 1 kJ = 0.239 kcal.

CLASSIFICATION OF ANIMAL FEEDSTUFFS

Roughage and concentrate feeds are the main kinds of feedstuffs.

Roughage is mainly used for ruminant animals. Such feeds are bulky, having a low weight per unit volume and a high content of cell wall material (usually between 25% and 30% crude fibre in the dry matter). They comprise plants fed green as pasture or harvested and given indoors, and products obtained from green crops by conservation, such as hay, dried grass or silage. Straw, chaff, hulls and corn cobs are extremely fibrous roughage of lower value. Roots and tubers are of high moisture content (75–92%) like green crops, but differ from the latter by their low crude fibre content.

Concentrates are the almost exclusive food of monogastric farm animals; they are included in the diet of ruminant animals to supplement forage to various extents. The term 'concentrate' is used for foods low in crude fibre (less than 10% on a dry matter basis), low in moisture and highly digestible. There are concentrates fed primarily for energy and low in protein, such as cereal grains and milling by-products (e.g. wheat bran) which are high in starch, and also fats, molasses, etc.

Protein concentrates containing >20% crude protein are important nutrients for rapidly growing animals and high producing adults. Protein concentrates are of plant or animal origin. The most important protein concentrates of plant origin are oilseed cakes and meals which are the residues after the greater part of the oil from oilseeds (such as soybeans, cotton seeds, etc.) has been removed for human consumption. Meat-, fish- and dairy by-products are expensive and are given in small amounts to certain species of animals.

In order to improve animal productivity, *feed additives*—materials other than the feeds themselves—are added to the diets, such as vitamin and mineral supplements, antibiotics, etc.

Composition of Some Representative Animal Feeds

The composition of some typical forage plants and feedstuffs is given in Table 1.1. These data are presented for the purpose of comparison with those given in Table 1.2 for the animal body, and for indicating the main differences between types of feeds belonging to various classes of feeding stuffs. Comprehensive tables on the composition of animal feeds have been published by the National Research Council (1982), Nehring *et al.* (1972) and by Schneider and Flatt (1975).

The results of analyses of lucerne cut at three different stages of growth indicate a sharp increase in dry matter and crude fibre content and a drop in protein content (calculated on the basis of dry matter). Similar changes

Table 1.1 Composition of some feedstuffs (%)

Feedstuff	Water	Protein	Fat	Ash	Crude fibre	Nitrogen-free extract	Calcium	Phosphorus
Lucerne (very young)	83.0	4.3	0.5	2.0	4.0	6.2	0.41	0.06
Lucerne (blossoming)	77.3	5.2	0.8	2.6	5.9	8.2	0.58	0.07
Lucerne (in full bloom)	71.0	5.5	1.0	3.0	8.7	10.8	0.51	0.07
Lucerne hay	10.8	16.5	2.4	7.5	25.5	37.3	1.32	0.21
Vetch and oats hay	12.0	13.1	1.9	8.5	25.8	38.7	0.66	0.18
Wheat straw	9.9	3.0	2.0	11.0	35.1	38.9	0.31	0.06
Barley, grains	10.3	10.8	1.8	5.5	4.7	66.9	0.05	0.38
Wheat bran	11.9	14.3	4.6	4.6	10.3	54.3	0.15	1.05
Soybean oil meal	10.8	44.5	0.6	5.6	4.5	34.0	0.18	0.66
Soybeans	9.1	37.9	17.4	4.9	5.3	25.4	0.26	0.62
Fish meal	9.0	62.0	2.5	23.8	–	–	4.10	2.70

occur in other plants during their growth. Hay may be stored for several years due to its low water content (less than 15%). The very high content of nitrogen-free extract (largely starch) in barley is typical of most feed grains. Wheat bran is characterized by a high content of crude fibre compared to the wheat grain as a whole, it is also richer in protein, fat, vitamins and mineral matter than the whole grain. A comparison of the composition of soybean meal and whole soybeans points to the differences which arise in the wake of oil extraction from the soybeans. Commercial fish meals, used mostly in swine and poultry rations, are of relatively constant composition containing more than 60% protein.

COMPOSITION OF THE ANIMAL BODY

Much attention has been paid to the chemical composition of animal bodies, not only for studying processes of food utilization, but also for assessment of meat quality. Species, strain, age, sex and state of nutrition all influence body composition. The percentage composition of various animal species is given in Table 1.2.

Table 1.2 Percentage composition of the animal body* (taken from Maynard *et al.*, *Animal Nutrition* 7th edn, 1979, p. 10)

					Dry, fat-free	
Species	Water	Protein	Fat	Ash	Protein	Ash
Calf, newborn	74	19	3	4.1	82.2	17.8
Calf, fat	68	18	10	4.0	81.6	18.4
Steer, thin	64	19	12	5.1	79.1	20.9
Steer, fat	43	13	41	3.3	79.5	20.5
Sheep, thin	74	16	5	4.4	78.2	21.8
Sheep, fat	40	11	46	2.8	79.3	20.7
Pig, 8 kg	73	17	6	3.4	83.3	16.7
Pig, 30 kg	60	13	24	2.5	84.3	15.7
Pig, 100 kg	49	12	36	2.6	82.4	17.6
Hen	56	21	19	3.2	86.8	13.2
Horse	61	17	17	4.5	79.2	20.8
Man	59	18	18	4.3	80.7	19.3

* Less contents of digestive tract.

The data given in Table 1.2 show a sharp decline in water content during the course of an animal's development. A cattle embryo contains 95% water, the water content of a newborn calf is 75–80%, that of a 5-month-old calf 66–72% and of a mature animal 50–70%. It is most significant that fatty tissues are practically free of water, contrary to proteinaceous ones which contain some 75% water. Growth of animals is accompanied by a sharp drop in water content, a parallel increase in fat content and a moderate decrease in protein and ash percentages as well (see Table 1.2 and Figure 1.2). Data on the composition of the gain of growing animals are given in Chapter 18.

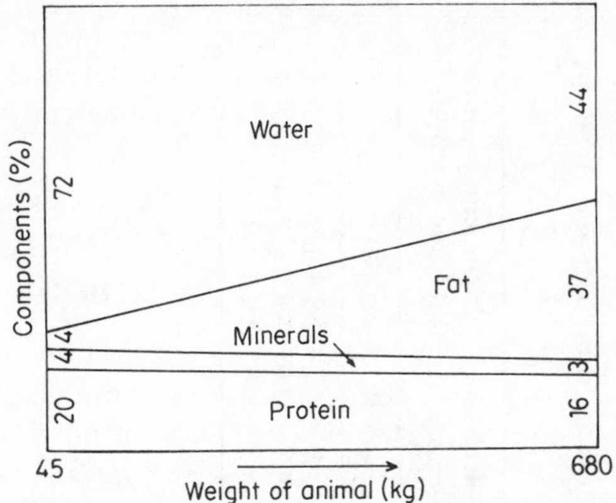

Figure 1.2. Changes in the body composition of a calf during growth (after Kolb and Gürtler, 1971). *Nutritive Physiology of Farm Animals* (in German). *Reproduced by permission of Gustav Fischer Verlag.*

Fluctuations in the composition of the body of animals of the same species are very considerable (due to strain, age and sex of animals and to nutritional conditions). Fat is the most variable of gross body constituents, whereas the composition of the fat-free body is relatively constant with respect to water, protein and ash (see Table 1.3). The protein/ash ratio is likewise highly stable, as seen in the values of their contents in fat-free dry matter.

The distribution of constituents within the various body tissues and organs is by no means uniform. Water, which has numerous functions (Chapter 9), is distributed throughout all parts of the body. The water content of various parts of the body of a cow is as follows: blood plasma, 90–92%; heart, kidneys and lungs, 80%; muscles, 72–75%; bones, 45%; and tooth enamel, only 5%.

Proteins are the main constituent of the dry matter of organs and soft tissues such as the muscles, liver, heart, lung and kidneys. Muscles, which account for almost half the body weight of the body of cattle, contain 75–80% of protein in their dry matter. Protein is also present in connective tissues and tendons, as well as in skin, hair, feathers, wool, hooves and even in bones.

A large amount of *fat* may be found in the adipose tissue, i.e. in fat depots which occur under the skin as well as around the kidneys and around the intestines and other organs. Certain quantities of fat are also present in various organs such as liver, kidneys, heart and lungs and smaller quantities even in muscles and bones. Minute quantities of *carbohydrates* in the form of glucose and glycogen are found in the liver, muscles and blood.

Table 1.3 Chemical composition of bodies of pigs, sheep and cattle (taken from National Academy of Sciences, 1968)

	Composition of ingesta-free body (%)				Mean composition of fat-free body (%)			Mean composition of fat-free dry body (%)	
Species	Water	Fat	Protein	Ash	Water	Protein	Ash	Protein	Ash
Pigs	30.7–80.8	1.1–61.5	8.3–19.6	1.3–5.6	77.0 ± 2.69	19.2 ± 2.34	3.9 ± 0.79	83.4 ± 2.33	17.0 ± 1.69
Sheep	39.6–73.8	4.9–46.6	10.7–19.5	1.7–5.8	74.9 ± 1.02	20.3 ± 0.84	4.8 ± 0.67	81.1 ± 2.40	18.9 ± 2.39
Cattle	39.8–77.6	1.8–44.6	12.4–20.6	3.0–6.8	72.9 ± 2.01	21.6 ± 1.53	5.3 ± 0.95	80.3 ± 1.69	19.7 ± 1.69

The different parts of the body contain varying amounts of *minerals* according to their functions. The body of a steer (less the content of the digestive tract) contains the following percentages of minerals: calcium, 1.33%; phosphorus, 0.74%; sodium, 0.16%; chlorine, 0.11%; potassium, 0.19%; magnesium, 0.04%; sulphur, 0.15%. In addition to these elements, there are many others which are present in smaller amounts, some of which are known to be essential for animals (Chapter 10).

PRINCIPLES OF ESTIMATING BODY COMPOSITION

Body composition may be determined by slaughtering the animals and analysing the carcass. Analyses of large animals are difficult and tedious. Thus the determination of body composition in living animals rather than in dead animals is advantageous also because such determinations can be carried out with the same animal at various stages of feeding experiments.

Extent of fatness or leanness of animals can be concluded from measurements of their specific weight under the assumption that the density of fat would be 0.900 and that of lean beef 1.300. The technique is to use the principle of Archimedes and weigh the animal in air and under water, thus obtaining the specific gravity. Measurements of body volume by displacing air rather than water are quite possible.

Many procedures for estimating body composition are based on the determination of the water content of the body. This is possible since the fat and water contents of the body are inversely related (see Figure 1.3) and the relation between protein and ash content of the fat- and water-free body substance is quite constant (see above).

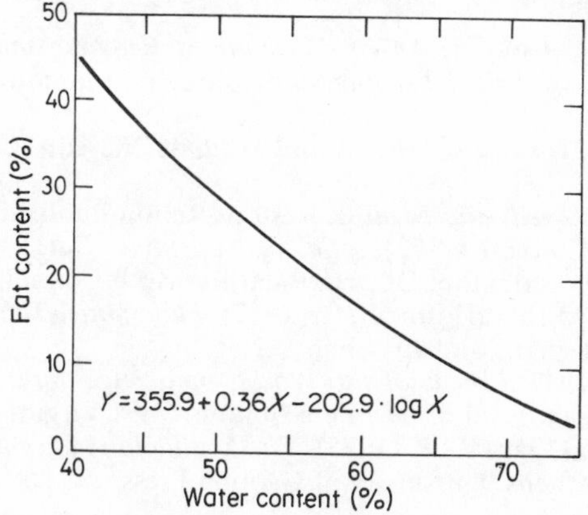

x = water content (%); y = fat content (%)

Figure 1.3. Correlation between fat and water content of the body of cattle (after Kirchgessner, 1978), *Animal Nutrition* (in German). *Reproduced by permission of DLG Verlag.*

The water content of the body may be estimated by measuring the dilution of a water-soluble tracer substance of which a known amount is injected intravenously into the bloodstream. Appropriate tracers are highly water soluble, rapidly penetrate the water of the body, do not react with any other body constituent and, lastly, are eliminated by the body at a constant rate. Antipyrine or isotopes of deuterium or tritium, injected as heavy water (H^3HO or ^2H_2O), may serve as tracers. After equilibrium has been reached, the dilution of the marker is determined quantitatively. Body fat is free of water while in contrast all water present in the tissues may be found together with proteins. So we may infer both the fat and the protein content from the body's water content. For example, the following regression equations are given for calculating the fat and protein content of the body of a sheep from that of tritium in the whole body.

$$\text{Fat (percentage of live weight)} = 1.26x - 97.3$$
$$\text{Protein (percentage of live weight)} = 0.145x + 4.8$$

where $x = H^3HO$ as a percentage of live weight.

Another means of determining the protein content of the body relies on the fact that the potassium concentration in cells is remarkably constant per gram of protein. The body's potassium content may be assessed directly through measuring the gamma-rays emitted by the radioactive isotope of potassium, ^{40}K.

BIBLIOGRAPHY

Composition of feedstuffs

Agricultural Research Council (1976). *The Nutrient Requirements of Farm Livestock, No. 4, Composition of British Feeding Stuffs.* Slough: Commonwealth Agricultural Bureau.

Church, D. S. (1983). *Livestock Feeds and Feeding,* 2nd edn. Corvallis: O. and B. Books.

Cullison, A. (1975). *Feeds and Feeding.* Reston: Reston Publishing Company.

National Research Council (1982). *Atlas of Nutritional Data on United States and Canadian Feeds.* Washington, DC: National Research Council.

Nehring, K., Beyer, M., and Hoffman, B. (1972). *Futtermittel-Tabellenwerk,* 2nd edn. Berlin: Ver. Deutscher Landwirtschaftsverlag.

Schneider, B. H. (1947). *Feeds of the World, their Digestibility and Composition.* Morgentown: Agricultural Experiment Station, West Virginia University.

Schneider, B. H., and Flatt, W. P. (1975). *The Evaluation of Feeds through Digestibility Experiments.* Athens: University of Georgia Press.

Composition of animals

National Academy of Sciences (1968). *Body Composition in Animals and Man,* Publication 1968, National Research Council.

Cuthbertson, A. (1978). Carcass evaluation of cattle, sheep and pigs. *World Rev. Nutr. Diet.,* **28**, 211–235.

Hedrick, H. B. (1983). Methods of estimating live animal and carcass composition. *J. Anim. Sci.*, **57**, 1316–1327.

Pearson, A. M. (1965). *Body Composition in Newer Methods of Nutritional Biochemistry* (A. A. Albanese, ed.), Vol. 2, pp. 2–40. New York and London: Academic Press.

Reid, I. T., and White, O. D. (1976). Diet and Body composition: cattle, sheep, swine and poultry. *Proc. Maryland Nutr. Conf.*, pp. 55–67. Maryland: Department of Animal Science, University of Maryland.

Symposium on 'Comparative aspects of body composition of farm and laboratory animals'. (1986). *Proc. Nutr. Soc.*, **45**, 45–130.

CHAPTER 2

*Functions and Fate of Nutrients in the Animal Body**

Animals obtain food as complex raw materials. Nutrients consist of large molecules, and before they can be utilized they are digested, i.e. broken down into small basic nutrient units (e.g. proteins to amino acids or starch to glucose). This breakdown occurs in the lumen of the gastrointestinal tract and is accomplished by a combination of mechanical and enzymatic processes (see Chapter 3). These simple compounds are absorbed through the walls of the gastrointestinal tract and are transported via the circulatory system to the various organs (liver, kidneys, muscle, etc.). In the organs the nutrients serve their ultimate purpose: participation in the metabolic activities of the tissues and cells upon which the life of the animal depends, as well as the formation of products such as meat, milk, eggs, wool, etc.

The simple compounds derived from food merge indistinguishably in the body with the respective compounds derived from the breakdown of complex molecules present in the body and together they form metabolic pools. The participation of dietary and body constituents in the metabolic pool is indicated in Figure 2.1. The metabolic transformation of organic body constituents is called endogenous metabolism compared with the exogenous metabolism of compounds of dietary origin.

The amount of a substance in the pool is termed pool size and is expressed in moles or weight units. For example, the pool size of muscle protein is about 3.7 kg in a pig with a body weight of 50 kg.

Food constituents taken up by the cells are constantly being transformed

* In collaboration with P. Budowski.

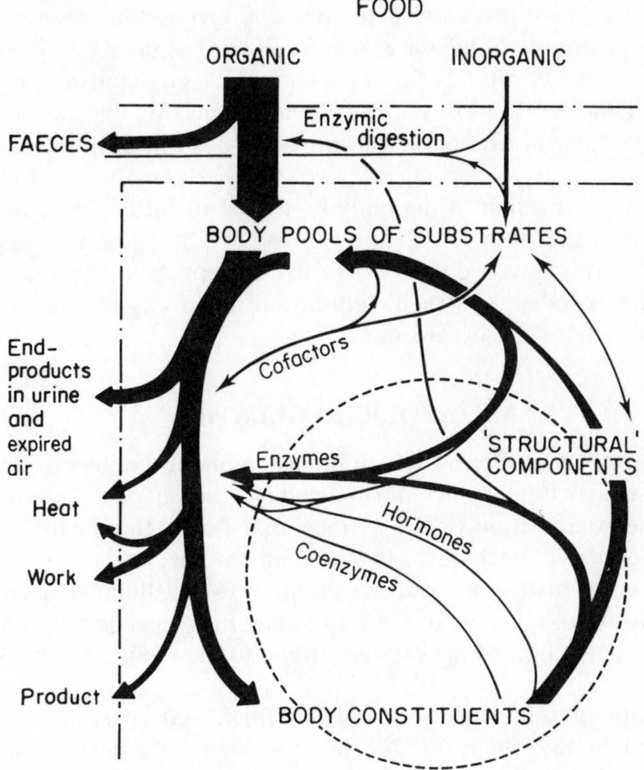

Figure 2.1. Participation of dietary and body constituents in the metabolic pool (from Phillis, J. W., *Veterinary Physiology*, 1976, p. 544). *Reproduced by permission of W. B. Saunders Co.*

for biosynthetic (anabolic) processes, or being degraded (catabolic processes). The continuous entry and disappearance of molecules into and from the pool result in a continuous change in the pool. This is termed 'turnover' of the pool. The rates of turnover depend on the particular compound, the tissues and the physiological and nutritional state of the animal. The quantitative approach to the dynamic state of body constituents was made possible with the advent of isotope tracer techniques.

The amount of substance flowing into and out of the pool per unit time is called the flux, or entry rate. Flux rate is the product of the turnover rate and the pool size. The term entry rate is based on the assumption that the rate of entry equals the rate of disappearance.

In biosynthetic processes metabolites are transformed into body constituents and animal products (milk, eggs, wool, etc.) or replace those that are continuously broken down. Energy production is achieved by degradation of food constituents or their metabolic conversion products, some of which serve as metabolic fuels (glucose, fatty acids, etc.). The energy liberated by oxidative processes is temporarily invested in energy-rich compounds, the prototype of which is adenosine triphosphate (ATP, see later). These furnish energy directly for the various energy-requiring processes in the body: for the synthesis of chemical compounds, for the translocation of tissue constituents in the body (transport and absorption), and for muscle work.

Energy sources present in the body in greater quantities than needed are stored in cells or tissues as fat (Chapter 6) and to a limited extent as glycogen (Chapter 4). Fat or glycogen in certain tissues represent reserves of energy available when needed, and their withdrawal in no way interferes with the metabolic activity of the organism.

METABOLIC REGULATION

Cells take up nutrients from their fluid environment, i.e. blood plasma and other extracellular fluids. Since normal cell function depends upon the constancy of the composition of the extracellular fluids, there exists in higher animals an effective mechanism to maintain the medium constant, i.e. the concentrations of many compounds present in extracellular fluids are maintained within narrow limits, in spite of sometimes considerable changes in food supply. These limits may vary according to the requirements of specific tissues.

The concept of self-regulation was put forth 100 years ago by Claude Bernard, and in the 1920s W. N. Cannon coined the term homeostasis. Homeostasis is defined as 'the various physiological arrangements which serve to restore the normal state, once it has been disturbed'. Temperature homeostasis is well known, but there are many examples of biochemical homeostasis, such as the buffering properties of the body fluids to the presence of excess acid or alkali, or the maintenance of the glucose concentration in plasma (see Chapter 4). As long as normal conditions are maintained in the internal environment, the cells of the body will function properly. Since metabolic pathways consist of a series of enzyme-catalysed reactions which convert a substrate into a product, the main metabolic control systems are concerned with the modification of enzymatic activities. The following four factors are mainly responsible for controlling metabolic pathways: (1) substrate availability, (2) product removal, (3) cofactor availability, and (4) feedback regulation.

Substrate Availability. Any enzymatic reaction can be regulated by the availability of the substrate. A reduction in substrate concentration will decrease the activity of an enzyme whereas an increase in substrate concentration stimulates the respective reaction. An example for control by substrate availability is the concentration of long-chain free fatty acids in plasma which determines the rate of oxidation in tissues such as muscles (Chapters 4 and 6).

Product Removal. If a substrate is converted to a product the removal of the product will enhance the enzymatic reaction. For example, the rate of conversion of pyruvate to lactate

$$(CH_3.CO.COOH \longrightarrow CH_3.CHOH.COOH)$$

occurring in muscle and catalysed by lactate dehydrogenase is promoted by diffusion of lactic acid from the muscle to the blood.

Cofactor Availability. A regulating mechanism based on availability of a cofactor is somewhat similar to control by substrate availability. NADPH produced in the pentose-phosphate pathway of glucose oxidation is necessary for the synthesis of long-chain fatty acids occurring, for example, in the mammary gland. These fatty acids are transformed into milk fat (Chapter 21).

Feedback Regulation of Enzyme Activity. This is a widespread mechanism of metabolic control. If a cell is supplied with a product it normally synthesizes, it will cut off its own production by inhibiting synthesis of the enzymes involved. When the synthesis of a product is the result of a series of enzyme actions, accumulation of the end-product inhibits the first enzymatic reaction in the chain (Figure 2.2) thereby preventing further production of the end-product; for example, cholesterol inhibits the preliminary steps of the biosynthetic pathway in which it is formed (see Chapter 6).

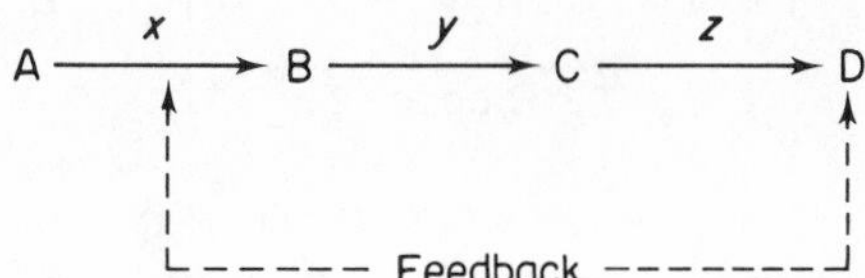

Figure 2.2. Feedback inhibition. Accumulation of product D inhibits enzyme *x* required for reaction A → B.

In addition to the chemical mechanisms described, metabolic regulation also occurs by transfer of metabolites from one cell compartment to another, particularly from mitochondria to cytosol and vice versa. (The mitochondria are sausage-shaped bodies dispersed in the cytoplasm.) Many of the enzymes involved in oxidation processes are attached to the inner membrane surfaces of the mitochondria. Cellular compartmentation keeps anabolic and catabolic pathways separated, with highly controlled translocation of metabolites, some of which act as signals to switch off the opposing pathway. For example, the aerobic oxidation of glucose, which yields a comparatively large amount of ATP, occurs inside mitochondria, and the ATP formed is transferred to cytosol where it serves for biosynthetic processes, i.e. for the synthesis of long-chain fatty acids.

METABOLIC FUNCTIONS OF HIGH-ENERGY PHOSPHATE COMPOUNDS

The energy liberated by catabolic reactions is not used directly by cells, but is applied instead to the formation of high-energy phosphate compounds . The most important energy-rich compound is ATP (for structural formula, see Figure 2.3). ATP is a nucleotide composed of the base adenine, the sugar ribose (see Chapter 4) and three phosphate radicals. The last two phosphate radicals are connected with the remainder of the molecule by high-energy phosphate bonds which are represented by the symbol ≈. This ubiquitous molecule is the energy storehouse of the body. Each of the high-energy phosphate bonds contains about 52 kJ per mol of ATP. Not all organic phosphates are of the high-energy type; many phosphate bonds, such as occur in glucose 6-phosphate, are low-energy phosphate bonds. Their formation requires 8–12 kJ/mol and the same amount of energy is liberated on hydrolysis:

$$\text{Glucose 6-phosphate} \underset{-H_2O}{\overset{+H_2O}{\rightleftharpoons}} \text{Glucose} + H_3PO_4$$

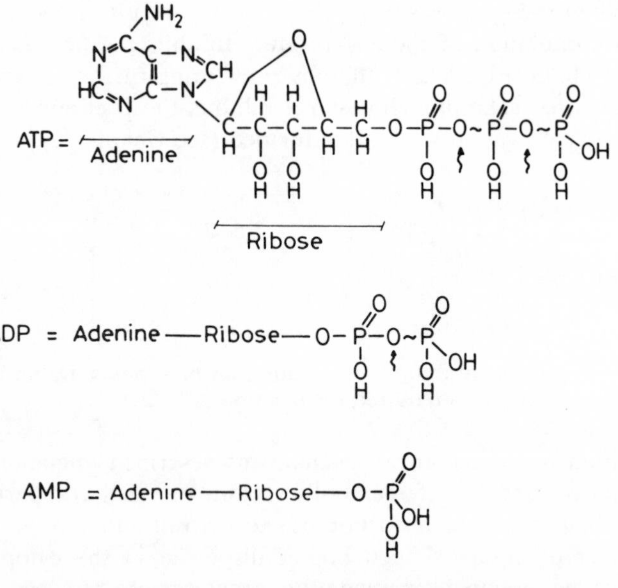

ATP = Adenine — Ribose — O–P(=O)(OH)–O~P(=O)(OH)–O~P(=O)(OH)–OH

ADP = Adenine — Ribose — O–P(=O)(OH)–O~P(=O)(OH)–OH

AMP = Adenine — Ribose — O–P(=O)(OH)–OH

Figure 2.3. Structure of ATP, ADP and AMP.

The high-energy phosphate bond in ATP is very labile and can be split instantly on demand when energy is required to promote biosynthetic reactions, or for mechanical work such as muscular contraction. When ATP releases its energy (52 kJ per mol) a phosphate radical is split away and ADP is formed:

$$ATP \rightleftharpoons ADP + P_i \; (P_i = \text{inorganic phosphate}).$$

Loss of another phosphate, to form AMP, liberates more energy:

$$ATP \rightleftharpoons AMP + P\text{-}P \; (P\text{-}P = \text{pyrophosphate}).$$

Energy derived from biological oxidation causes the ADP and phosphate to recombine to form new ATP, the entire process continuing over and over again. The role of ATP in the trapping and utilization of energy is illustrated diagrammatically in Figure 2.4.

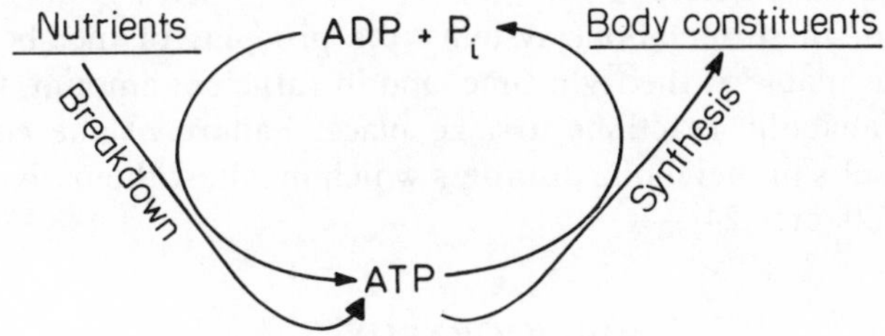

Figure 2.4. Interaction of exergonic and endergonic reactions—formation of ATP and utilization of energy invested in ATP.

Animals derive their energy from the oxidation of nutrients. In its simplest form the oxidation of food can be regarded as

$$\text{Food} + O_2 + ADP + P_i \longrightarrow CO_2 + H_2O + ATP$$

In the case of glucose, the reaction may be formulated as follows:

$$C_6H_{12}O_6 + 6\,O_2 + 38\,ADP + 38\,P_i \longrightarrow 6\,CO_2 + 6\,H_2O + 38\,ATP$$

38 ATP molecules are formed for each molecule of glucose oxidized to carbon dioxide and water. Thus, 1976 kJ (38 × 52) are stored in the form of ATP, while 2870 kJ are released by complete oxidation of each gram molecule of glucose in an oxygen atmosphere. The process of biological oxidation is not very efficient, since 69% (1976 : 2870) of the energy is conserved in the form of high-energy bonds, the rest being evolved as heat which cannot be used by the cells to perform additional functions and is in this sense wasted.

The decisive step in the action of ATP is its transfer from the inside of the mitochondria to the cytosol to become available there in reactions that require energy. In turn, ADP is continually transferred from the cytosol back into the mitochondria for conversion into ATP.

METABOLIC REGULATION BY HORMONES

Hormones exert metabolic control in response to physiological, nutritional and psychological factors. The hormones are secreted by endocrine glands, and are transported in the blood to target organs where they cause physiological actions. Synthesis and catabolism of carbohydrates, proteins and fats are influenced by many hormones. Certain hormones increase the supply of substrate for synthetic processes while others act on specific enzymatic reactions which regulate the flow of intermediates along biochemical pathways. Transport of substrate across the cell membrane is an important action of several hormones. Some complex metabolic processes are stimulated by one hormone and inhibited by another. Fine adjustments in the rate or direction of the reactions are possible.

As a result of these metabolic systems, the products of metabolic reactions become available at the right time, and in sufficient amount, for other catabolic and anabolic reactions to take place. Failure of the control of metabolism results in metabolic disorders which manifest themselves in certain diseases (Chapter 24).

BIBLIOGRAPHY

Books

Banks, P., Bartley, W., and Birt, L. M. (1976). *The Biochemistry of the Tissues*, 2nd edn. Chichester: John Wiley.

Denton, R. M., and Pogson, C. I. (1976). *Metabolic Regulation.* London: Chapman and Hall.

Newsholm, E. A., and Start, C. (1973). *Regulation in Metabolism.* London: John Wiley.

Riis, P. M. (1983). *Dynamic Biochemistry of Animal Production. Amsterdam: Elsevier.*

Articles

Armstrong, D. G. (1969). Cell bioenergetics and energy metabolism. In *Handbook of Animal Nutrition* (W. Lenkeit *et al.*, eds), Vol. 1, pp. 385–414. Hamburg and Berlin: Paul Parey.

Matthews, D. E., and Bier, D. M. (1983). Stable isotope methods for nutrition investigation. *Annu. Rev. Nutr.*, **3**, 309–339.

Milligan, L. P. (1971). Energetic efficiency and metabolic transformations. *Fed. Proc.*, **30**, 1454–1458.

CHAPTER 3

*The Digestive Tract and its Functions**

The common functions of the digestive tract are the intake of food, its digestion and absorption and the excretion of unabsorbed dietary components and waste materials. Digestion is the preparation of food for absorption, i.e. reduction of food particles in size and solubility by mechanical and chemical means. Most food constituents are of large molecular weight, such as proteins or starch which have to be broken down to low molecular weight compounds (amino acids, glucose, etc.) to allow for absorption. Mechanical breakdown of feedstuffs is performed by chewing in the mouth and contractions of the muscles of the gastrointestinal walls: Chemical breakdown is mainly effected by enzymes secreted in various digestive juices and by microorganisms. Absorption is the passage of digested nutrients through the mucous membranes of the alimentary canal into blood and lymph. The various parts of the digestive tract of typical farm animals are shown in Figure 3.1.

The digestive or gastrointestinal tract can be considered as a long hollow muscular tube extending from mouth to anus and lined with mucous membranes. The gastrointestinal tract of mammals and avian species includes the mouth, stomach, and small and large intestine. Ruminants differ from other mammals in having a greatly enlarged stomach divided into four compartments, the largest of which, the rumen, acts as a large fermentation incubator; in the ruminant, contrary to most other types of animals, most of the ingested food is fermented actively by microbes before it is exposed to the typical processes of digestion occurring in the true stomach and intestine. The great capacity of the rumen microorganisms to digest cellulose is

---

* In collaboration with the late Dr H. Neumark.

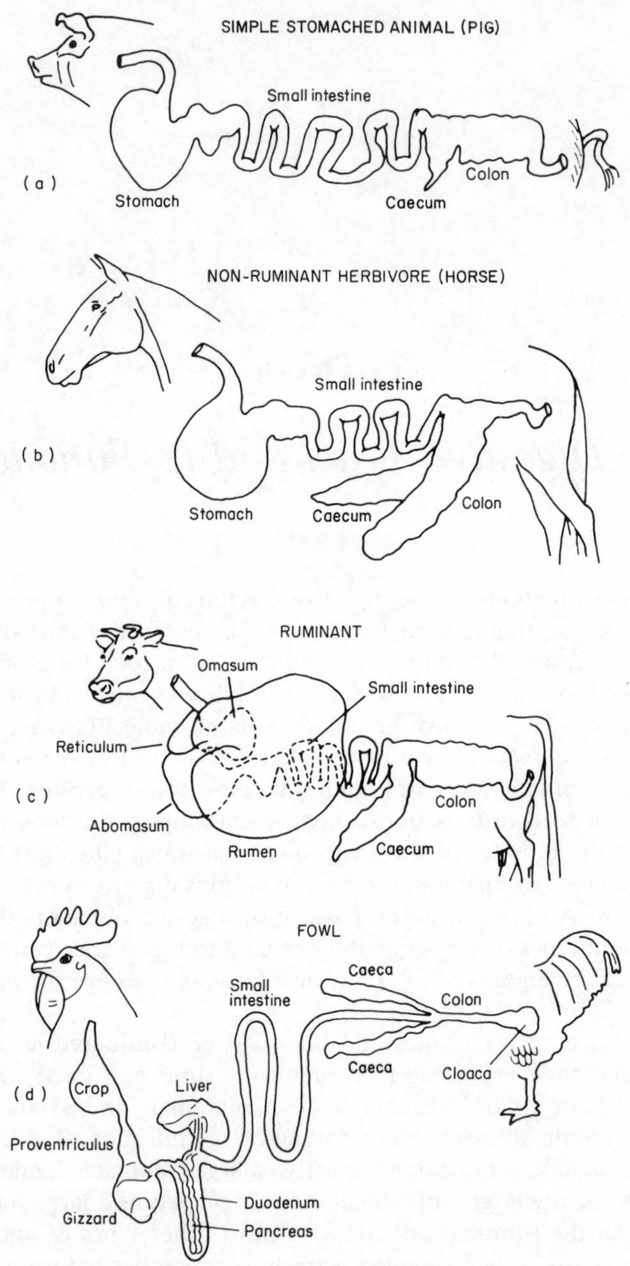

Figure 3.1. The digestive tract of farm animals: simple-stomached animal (pig); non-ruminant herbivora (horse); ruminant (cow); fowl (from Moran, 1982, pp. 2 and 4). *Reproduced by permission of the author.*

essential for ruminants to enable them to utilize cellulose which mammalian enzymes cannot hydrolyse. Some herbivorous non-ruminant animals such as horses or rabbits have a sacculated stomach with quite intensive microbial activity. Moreover, some microbial breakdown of carbohydrates occurs in the large intestine of various species of mammals and birds. Most species derive only little benefit from the limited microbial activity taking place in caecum and colon, because of the very limited absorbing capacity through the walls of the lower intestine. This is different in horses and ruminants; the main site of cellulose breakdown in horses is their enlarged colon in addition to the stomach. Horses and ruminants also differ from most other species of mammals by the capacity to absorb nutrients through the epithelium of the large intestine.

Structural and functional variations in the digestive tract are clearly related to diets and feeding habits. Table 3.1 lists species variations in the dimensions and capacity of various segments of the gastrointestinal tract.

Table 3.1 Capacity (litres) of the digestive organs of some different species

	Man	Pig	Horse	Sheep	Cattle
Body weight, kg	75	109	450	80	575
Reticulo-rumen				17	125
Omasum				1	20
Abomasum				2	15
Total stomach	1	8	8	20	160
Small intestine	4	9	27	6	65
Caecum		1	14	1	10
Large intestine	1	9	41	3	25
Total digestive tract	6	27	90	30	260

ORGANS OF DIGESTION, THEIR STRUCTURE AND FUNCTIONS

Functions of the Mouth and the Stomach in Monogastric Species

The Mouth

The mouth serves for food intake, for mechanical breakdown of the food and for mixing it with saliva. There are large differences among animal species in the relative importance of the mouth for breaking down the food mechanically and chewing it. Different animal species differ particularly in the adaptation of teeth for these purposes. Omnivorous species such as humans and pigs and also horses use the incisor teeth to bite off pieces of fibrous food materials and the molar teeth for mastication. Ruminants have no upper incisors; these are replaced by a tough dental pad which provides a surface against which the lower incisors can put pressure to bite off fibrous food material. Thorough mastication and grinding of forages occur in ruminants when regurgitated material reaches the mouth a second time rather than at the time the forage is consumed. Poultry have no teeth and they swallow food whole or after some reduction of its size by the beak. Grinding is done by the action of grit in the bird's gizzard.

Saliva excreted by the salivary glands contains 99% water and is a solution of salts and mucoproteins (see Chapter 7). Because of the presence of mucoproteins, saliva is a highly viscous fluid and acts as a lubricant. Thus saliva aids in forming the food into a bolus which may be swallowed easily, thereby increasing the surface area of the food to allow the digestive enzymes to act on it when passing through the gastrointestinal tract.

Saliva of monogastric animals is slightly acid and that of ruminants alkaline. Saliva is highly buffered, particularly that of ruminants, due to the presence of bicarbonate and phosphate ions. It serves for neutralizing volatile fatty acids formed in the rumen. Man secretes about 1.5 litres of saliva daily, a relatively small amount compared with sheep which secrete 5–10 litres per day, pigs with 15 litres daily, and cattle which produce 130–180 litres. The amount of saliva secreted is also influenced by the amount and quality of the diet ingested, since salivation may be enhanced by mechanical stimulation of the salivary glands.

Pigs are the only farm animals whose saliva contains the enzyme ptyalin with amylase activity but it is doubtful whether much starch is digested in the mouth since the food mass passes quickly to the stomach (see later). The presence of lysozyme in the saliva of many species explains the bactericidal action of saliva. This enzyme breaks down polysaccharides present in the cell wall of bacteria, thereby killing them.

The Stomach

The stomach provides a reservoir of diet for controlled release to the small intestine and to initiate digestion, mainly of protein. The stomach is not essential for food utilization. Some species of fish, such as carp, have no stomach; they are continuous feeders. The shape of the stomach of the various species varies as does its relative size. Viewed from its exterior the stomach is divided into four regions; the oesophageal, cardiac, fundic and pyloric regions (see Figure 3.2). The oesophagus is a muscular tube, lined by stratified epithelium, and acts as a passage for food and liquid. The oesophagus is the only region of the stomach lacking secretory cells, whereas in the other three regions gastric glands are embedded in the underlying connective tissue which is lined by mucous membrane. The three regions of the stomach (apart from the oesophagus) differ in the composition of the lining epithelium and by the types of embedded glands. There are also species variations in the sizes of the four regions of the stomach. The stomach of the horse is relatively small, but the oesophagus occupies a major part of it. A quite intensive microbial activity takes place there, particularly digestion of starch (see Chapter 4). However, the oesophageal region is very small in the relatively larger stomach of the pig (capacity of the whole stomach 8 litres), and the cardiac region occupies the major part of the pig stomach.

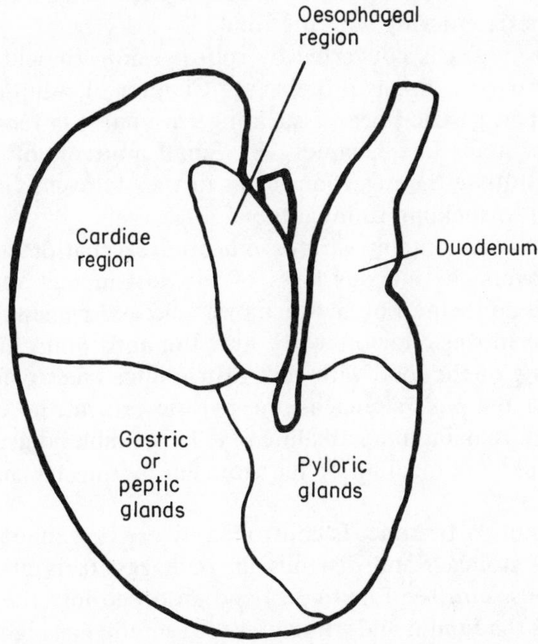

Figure 3.2. The main regions of the stomach of the pig (from Moran, 1982, p. 49). *Reproduced by permission of the author.*

The types of glands present in the various regions of the stomach and their secretions will not be detailed here; the secretory cells secrete collectively hydrochloric acid, enzymes and mucus. Hydrochloric acid is secreted by the cardiac glands; the following mechanism has been proposed for its formation. Hydrogen ions originate from body water by the hydration of carbon dioxide (mediated by the enzyme carbonic anhydrase) and the subsequent dissociation of the carbonic acid formed:

$$H_2O + CO_2 \xrightleftharpoons{\text{Carbonic anhydrase}} H_2CO_3 \rightleftharpoons H^+ + HCO_3^-$$

The hydrogen ions are secreted into the lumen of the stomach by an energy-dependent process requiring ATP. Chloride ions originate from sodium chloride in the blood plasma. The release of chloride ions into the lumen is accompanied by discharge of bicarbonate ions into the bloodstream.

The concentration of hydrochloric acid in the gastric juice of most farm animals is about 0.1 M; it fluctuates somewhat with the diet, but is sufficient to lower the pH in the stomach contents to 2.0. Hydrochloric acid fulfils important functions in the stomach. Proteins are denaturated by hydrochloric acid and so become more susceptible to the action of proteolytic enzymes (see later). The enzyme pepsin acts only in an acid medium. Hydrochloric acid dissolves minerals ingested with the food such as $Ca_3(PO_4)_2$, and kills pathogenic bacteria entering with the food.

Secreted pepsinogen is converted by self-digestion in acid medium into pepsin, which partially digests protein (see Chapter 7). Another proteolytic enzyme present in gastric juice of suckling ruminants is rennin which coagulates milk proteins (see Chapter 18). Small amounts of lipase present in gastric juice initiate the digestion of fat in the stomach. Gastric lipase is more important in suckling ruminants.

The gastric mucus contains various mucopolysaccharides and mucoproteins. Mucus covers the mucous layer of the gastric wall and so prevents corrosion of the epithelium by hydrochloric acid and pepsin. Mucoproteins also contain the intrinsic factor which aids the absorption of vitamin B_{12}. Thorough mixing of the food with acid gastric juice takes place only in the lowest region of the pig's stomach—the pyloric region; the content of the anterior stomach remains at an alkaline pH. This enables digestion of starch in the slightly alkaline medium of the stomach by salivary amylase and by bacteria.

The secretion of gastric juice is controlled by the hormone gastrin. When food enters the stomach and distends its wall, gastrin is produced in the glands of the anthrum (see Figure 3.2) and absorbed into the bloodstream, where it reaches the fundus and stimulates the secretion of highly acid gastric juice. Accumulation of acid in the stomach inhibits the production of gastrin and hence acid.

The digestive tract of fowl, particularly the stomach, differs considerably from that of monogastric mammalians, such as the pig (see Figure 3.1). In the upper part of the fowl's oesophagus copious amounts of mucus are secreted, lubricating the food bolus and facilitating swallowing of the food. Very small amounts of saliva secreted into the mouth of birds are of minor importance. A part of the oesophagus is expanded, forming the crop, where the food is stored and soaked. Some microbiological activity occurs in the crop and results in the formation of organic acids (see later). In the fowl, differently from monogastric animals, stomach digestion takes place in two compartments—the proventriculus and the ventriculus or gizzard. The proventriculus is the site of production of gastric juice containing hydrochloric acid and pepsinogen (see above). The acid concentration in gastric juice and also the pepsin output (per kg live weight per day) is higher in fowl than in mammals as in pigs. After a short stay in the proventriculus, the digesta mixed with gastric juice pass to the gizzard where crushing and grinding of the food is achieved by numerous contractions of the gizzard's powerful muscles, aided by non-calcareous insoluble grit ingested with the food. Enzymatic processes cannot take place in the gizzard because of the low water content of its contents—no more then 44%. On the other hand, the fine dispersion of the digesta leaving the gizzard aids the enzymatic activities occurring in the intestine. In other words, the three organs (crop, proventriculus and gizzard) combine the functions of teeth and stomach in monogastric mammals.

STRUCTURE AND FUNCTION OF THE RUMINANT STOMACH

Ruminants differ from other mammals in having a greatly enlarged stomach divided into four chambers: rumen, reticulum, omasum and abomasum (Figure 3.3). The first two compartments are considered together as the reticulorumen; the first three compartments are called together the forestomach, and the last the true stomach. Unlike the monogastric stomach, the ruminant stomach not only digests food, but also absorbs its metabolites. The first three compartments are lined with non-glandular keratinized stratified epithelium. The entire surface area of the rumen and reticulum is covered with keratinized projections called papillae. These papillae are very small in young animals; their number and size increase with ingestion of solid food due to mechanical and chemical stimulation, the latter by metabolites originating from the food. The inner surface area of the omasum is increased by many thin folds called laminae which enable the absorption of water and volatile fatty acids. In addition, the omasum aids in reducing the particle size of ingested food.

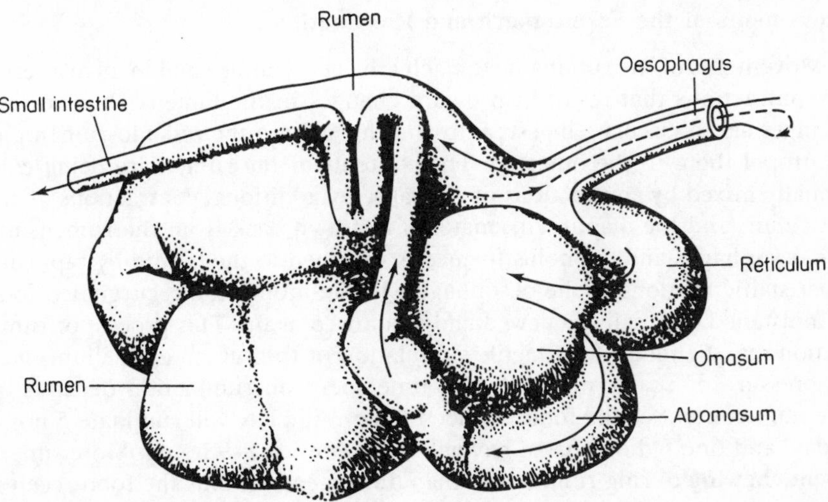

Figure 3.3. Structure of the ruminant stomach (From Annison and Lewis, 1962, p. 15). *Reproduced by permission of the author*

The structure and function of the fourth stomach, the abomasum, are very similar to those of the stomach in monogastric animals. The abomasum is divided in two areas, the proximal fundic and the smaller distal pyloric regions. The fundic area contains most of the glands secreting hydrochloric acid, pepsin, mucus and gastrin, and, in the young calf, rennin. The fluid entering the abomasum has a pH of 6 and that leaving it has a pH of about 2.5. Microorganisms reaching the abomasum are killed due to the acidity of the stomach juice, and digestion of the cells by peptic enzymes begins.

Development of the Forestomach During Growth of Ruminants

In the young ruminant the forestomach is relatively underdeveloped and non-functional because the suckling animal, like monogastric species, depends primarily on the abomasum and intestine. As the calf or lamb starts to consume solid food, the three chambers of the forestomach develop rapidly. Thus in the newborn calf the reticulorumen occupies about 30% of the total volume of the stomach, while the abomasum occupies about 70%. In the adult, the reticulorumen occupies about 85% of the total capacity of the stomach and the omasum and abomasum together occupy the remainder.

In the young ruminant, the suckled milk by-passes the reticulorumen, thus escaping bacterial fermentation, and passes directly into the omasum through the oesophageal groove. This is a tubelike fold of tissue beginning at the lower end of the oesophagus and ending at the reticulo-omasal orifice. This groove is non-functional in older animals unless they continue to take liquid diets.

Movements of the Forestomach and Rumination

Movements of the ruminant stomach are very complex and involve a series of contractions that result in pressure changes in the lumen. The pressure changes act to circulate ingesta into and throughout the reticulo-rumen and to propel them to the omasum. The contents of the rumen are being continually mixed by contractions of its walls. By additional contractions of the reticulum and the diaphragm, material is drawn back from the rumen into the oesophagus and the bolus formed is returned to the mouth by rapid antiperistaltic motions of the oesophagus. In the mouth the regurgitated food is moistened with saliva, chewed and swallowed again. This process of rumination may be induced by tactile stimulation of the rumen epithelium, such as pressure of coarse material. The time spent on rumination depends on the fibre content of the food. A sheep fed on fine hay will ruminate 5 hours a day, and one fed on coarse hay will ruminate for 8–9 hours. More time is spent chewing during rumination than during eating, thus the food receives a more thorough mastication during rumination than during eating—a feature peculiar to ruminants. The material ruminated in the mouth is returned to the rumen.

Microbial Breakdown of Food Occurring in the Reticulorumen

The chemical breakdown of food in the reticulorumen is achieved by enzymes secreted by microorganisms, not by the animal itself. The importance of microbial digestion occurring in the rumen is indicated by the fact that 70–85% of the digestible dry matter is degraded by the microorganisms with the production of volatile fatty acids (see Chapter 5), carbon dioxide, methane, ammonia, and microbial cells. Major carbohydrate components of ruminant diets such as cellulose and starch (see Chapter 5) and also many proteins (see Chapter 8) are degraded by microbial activities. Since many of the microorganisms can synthesize proteins (see Chapter 8) and B vitamins, the ruminant can be maintained on diets free of otherwise essential B vitamins and amino acids (see Chapter 12).

The gas mixture in the rumen is composed of 30–40% methane, 20–65% carbon dioxide, 5% hydrogen and small amounts of nitrogen and oxygen. Carbon dioxide is produced by fermentation of sugars, by degradation of amino acids (see Chapter 7) and by neutralization of volatile fatty acids with bicarbonate present in saliva. Methane is formed by microbial reduction of carbon dioxide (see Chapter 5). The gas production in the rumen is most rapid immediately after a meal, and in the cow may exceed 30 litres/hour following feeding, but this decreases to about 10 litres after about 4 hours. The gases formed in the rumen are eliminated by eructation (belching) and to a significant extent through absorption across the rumen wall and exhalation via the lungs. Failure to eructate results in bloat which is a potentially fatal condition (see Chapter 24). The volatile fatty acids are mainly absorbed through the rumen wall (see later). The microbial cells, together with undegraded food components, pass to the abomasum and intestines where they are digested by the host animal.

The reticulorumen provides a continuous culture system for anaerobic microorganisms. In order to maintain normal function for microbial growth, constancy of conditions in the rumen is achieved as follows. There is a relatively constant influx of food providing a regular supply of substrate for the microorganisms. Rumen pH is maintained at 6–7 by the high buffering capacity of saliva (see above) and removal of fermentation products such as volatile fatty acids and ammonia by absorption through the rumen wall into the blood. Saliva is the major factor in the maintenance of a relatively constant fluid volume; the rumen contents contain 85–95% of water. The osmotic pressure stays within the necessary limits by the flux of ions between rumen content and blood. Finally, the temperature of the rumen remains at 38–42 °C by the temperature-regulating mechanism of the animal.

Rumen Microorganisms

Microbial activity in the reticulorumen is achieved largely by strictly anaerobic bacteria and protozoa and to a much lesser extent by certain fungi.

Rumen content of adult animals under normal feeding conditions contains about 10^{11} bacteria per ml and protozoa a much smaller number (10^6 per ml). One group of bacteria use as substrate nutrients such as cellulose, starch or glucose, and a second group ferments metabolic products formed by the first group, such as lactic acid. The total number of bacteria and the relative ratio of various species vary according to an animal's diet; for example, diets rich in concentrate feeds stimulate high total bacterial counts and promote the development of lactobacilli. Most protozoa present in adult animals are ciliates belonging to two families: oligotrichs and holotrichs. Oligotrichs, but not holotrichs, engulf food and cell particles and utilize cellulose. Protozoa fulfil important functions in protein metabolism. Although they are present in much smaller numbers than bacteria, they are larger in size and the total mass of protozoa may thus equal that of bacteria. Since the metabolic activity of microorganisms is inversely related to the size of the organism, bacteria are responsible for a much greater share of the metabolic work than protozoa.

Many types of rumen bacteria attach to the surfaces of feed particles, some even to the rumen epithelium. Such attachment of bacteria to surfaces is advantageous for the host animal, since free unattached bacteria will tend to wash out at too fast a rate before their microbial activity can be fully exploited (see Chapter 8).

Growth and division of microorganisms is accompanied by death and autolysis of others, so that in the rumen there are always living, dead and damaged cells. All of these cells formed by transformation of dietary nutrients pass with the digesta to the abomasum and intestine where they are digested by the host animal together with undegraded food residues.

The rumen of a newly born animal is empty, but soon acquires a population of bacteria, mainly lactobacilli, by oral contact with older animals, from milk regurgitated from the abomasum or from other feed or water supplies. At 9–13 weeks of age the microbial rumen population is very similar to that of adult cattle.

DIGESTION AND ABSORPTION IN THE SMALL INTESTINE

The small intestine serves in all types of farm animals as a major site for digestion and absorption of required nutrients. The small intestine is a hollow muscular tube lying in between the pyloric and ileocaecal sphincters. The small intestine is arbitrarily divided into the duodenum, jejunum and ileum. As the food reaches the duodenum, the intestinal wall begins a complicated series of contractions—to and fro movements that mix the food with digestive juices, bring it in contact with the absorptive mucosa (see later) and push the chyme distally. The following digestive juices are added to the content of the intestine: the pancreatic secretion, the bile, and intestinal secretions such as the duodenal juice and the succus entericus.

Duodenal glands produce an alkaline secretion (mucus and electrolytes). This secretion acts as a lubricant and also protects the duodenal wall against hydrochloric acid entering from the stomach.

A second secretion from the intestinal mucosa, the succus entericus, is produced in the crypts of Lieberkühn, the tubular depressions between the villi (see Figure 3.4). Contrary to previous views, the secretions of intestinal glands do not contain the enzymes responsible for the hydrolysis of disaccharides (see Chapter 4) and peptides (see Chapter 7). However, the respective enzymes act on the surface of the intestinal mucosa or intracellularly inside the mucosal cells (see later). Bile synthesized by the liver cells passes to the duodenum through the bile duct; it contains sodium chloride and bicarbonate, sodium and potassium salts of bile acids, chiefly taurocholic and glycocholic acids (see Chapter 6), and the bile pigments biliverdin and bilirubin which are responsible for the colour of bile, faeces and urine. The bile pigments are breakdown products of haemoglobin. In all types of farm animals, except the horse, bile is stored in the gallbladder until required. Bile salts provide an alkaline pH in the small intestine and fulfil an important function in digesting and absorbing fats (see Chapter 6).

Pancreatic juice is a bicarbonate-rich solution containing many extracellular enzymes acting in the intestinal lumen, such as proteases (see Chapter 7) and amylase (see Chapter 4). Pancreatic juice is secreted by the pancreas, a gland which opens into the duodenum through the pancreatic duct. Pancreatic secretions are controlled by hormones produced in the anterior region of the small intestinal mucosa in response to the digesta reaching the intestine. Such hormones are secretin and cholecystokinin. Acid chyme from the stomach stimulates the liberation of secretin; this hormone elicits the release from the pancreatic cells of a watery solution, rich in bicarbonate, but poor in enzymes. Cholecystokinin, a hormone released under the stimulus of partially digested proteins, stimulates the secretion of pancreatic juice rich in enzymes.

Pancreatic juice and bile are secreted by small ducts into the duodenum of simple-stomached animals within a short distance of the pylorus, but in ruminants somewhat more distal. Thus the intestinal contents of the latter animals remain acid throughout the upper part of the small intestine and the action of pepsin continues to act in the weakly acid medium of the upper intestine. The ruminant intestine does not deal with the rather sudden inflows of food and water that pass the duodenum in monogastric animals. Apparently, intestinal digestion in ruminants benefits in efficiency from the continuous flow of digesta of quite constant composition (after having undergone the action of microorganisms in the rumen).

ABSORPTION OF DIGESTA FROM THE GUT

The small intestine is the most important site of absorption of required nutrients in monogastric mammals and in avians. In addition, in all herbivores the large intestine is adapted for absorption, at least to a limited extent (see later) and in ruminants, in addition to the absorption from the intestine, some nutrients are absorbed through the walls of the forestomach (see above). The small intestine and particularly the duodenum fit this purpose because the inner surface area is greatly increased by folding of the epithelium and the presence of villi, tiny fingerlike projections extending into the intestinal lumen (see Figures 3.4. and 3.5). The large surface area is further increased by the brush borders, revealed by the electron microscope to be microvilli which extend from the entire free surface of the epithelial cells of the villi (see PLATE I).

Each villus contains an arteriole and a venule together with a drainage tube of the lymphatic system, a lacteal. Thus, nutrients pass across the epithelial cell and enters either blood capillaries or the lymphatic system (in birds only the blood). The nutrients are carried through the portal vein to the liver or through the lymphatic system to the heart. Villi undergo movements for facilitating contact with digested nutrients, particularly pamping movements which are under nervous control and are augmented by the hormone villikinin.

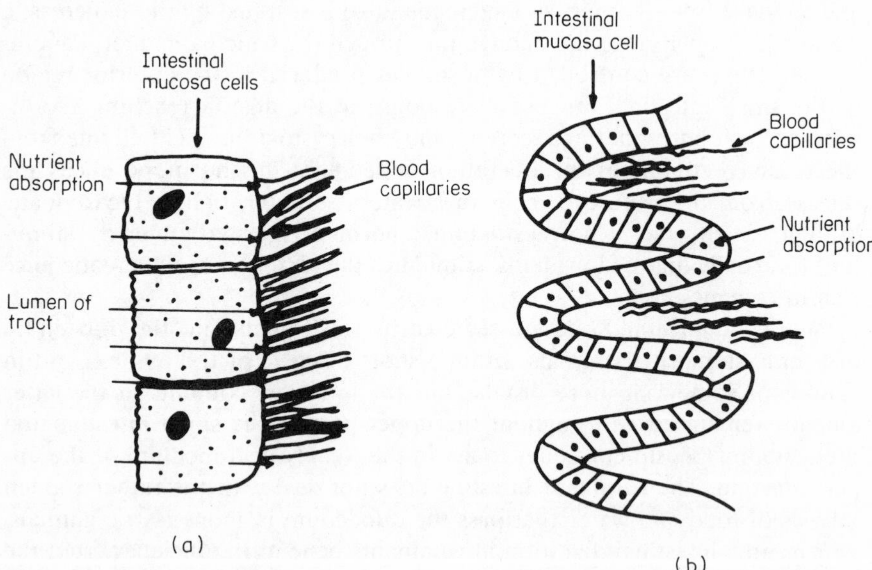

Figure 3.4. (a) Diagram of epithelial cells lining the the intestinal tract and villi which increase the absorption surface (from Church and Pond, *Basic Animal Nutrition*, 1974, p. 23). (b) Arrangement of intestinal mucosa cells (from Church and Pond, 1974, p. 23). Corvallis: O. & B. Books. *Reproduced by permission of Professor D. C. Church.*

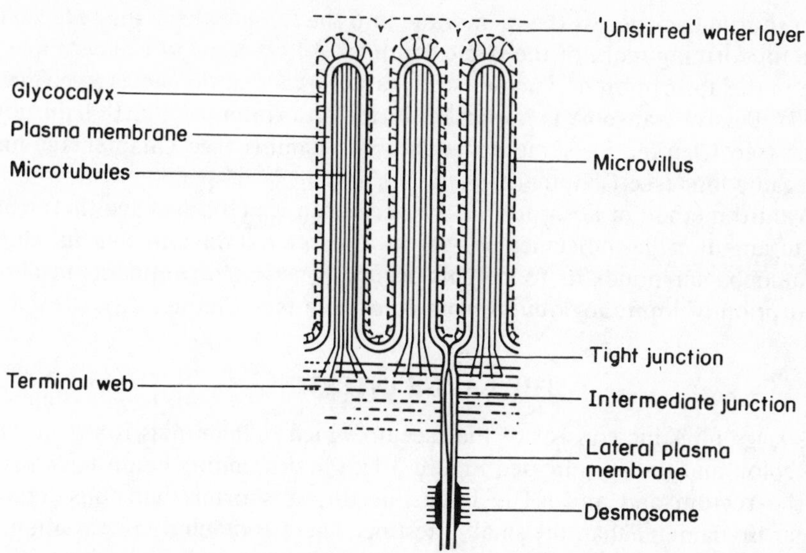

Figure 3.5 Apical portion of absorptive cell (from Madge, 1975, p. 72). The microvilli covering the internal surface of the intestine are coated with glycocalix* and are considered to be proportional to digestive and absorptive capacity. The figure illustrates also the arrangement of absorbed substances into the cells through the junctions (see also Plate I). *Reproduced by permission of the author.*

* The glycocalix is a network of aglycoprotein fibres (see Chapter 7) that are superimposed on the microvilli.

The passage of individual nutrients from the intestinal lumen into the enterocytes (absorbing cells) takes place by passive transport, active transport with use of specific carriers, or pinocytosis.

Passive absorption involves simple diffusion of a substance from a region of high concentration outside the cell to a region of low concentration inside. The rate of transfer is directly proportional to the concentration gradient across the membrane. The chemical concentration gradient, or in case of electrolytes the electrochemical gradient, is the only driving force of the transport and additional energy is not required.

Active transport involves the movement of material in the opposite direction, i.e. from a region of low concentration outside the cell to one of high concentration inside the cell, a movement against a concentration gradient or an electrochemical potential difference. This process requires energy and implies direct coupling to metabolic energy-yielding reactions. Active transport involves the use of a specific carrier for facilitating the transfer. The carrier has two specific binding sites and the organic nutrient is attached to one of those while the other site picks up sodium. The ternary complex moves across the intestinal membrane and the organic nutrient and the sodium are discharged inside the cells. The empty carrier and the sodium then shuttle back across the membrane into the lumen where they are available for forming more of the triple complex. Active transport plays a major role in the absorption of glucose (see Chapter 4) and amino acids (see Chapter 7). Passive transport is important in the absorption of short-chain fatty acids (see Chapter 5), some water-soluble vitamins (see Chapter 12) and inorganic ions (see Chapter 10).

A third method of absorption is *pinocytosis* in which cells have the capacity to engulf large molecules or ions in a manner similar to that in which an amoeba surrounds its food. This occurs in newborn ruminants to allow absorption of immunoglobulin from colostrum (see Chapter 18).

THE LARGE INTESTINE

The large intestine consists of the caecum which in mammals is a blind sac, the colon and the rectum (see Figure 3.1); the descending colon terminates as the rectum and anus. The large intestine is shorter, but considerably larger in diameter than the small intestine. There is much more variation in the large intestine from one species to another than in the small intestine. The horse has the biggest and most complex large intestine of any of the domestic animals. The caeca of birds are paired blind-ended tubes which arise at the junction of the small and large intestines. The large intestine of birds communicates with the exterior through the cloaca, an orifice which serves the treble purpose of excreting faeces, urine and eggs.

Most of the absorption of digested nutrients takes place in the upper gastrointestinal tract of all species. Food constituents resistant to digestive enzymes or those escaping rumen fermentation in ruminants, enter the large intestine. The secretions of the large intestine consist of an enzyme-free watery fluid containing sodium bicarbonate and mucus, which lubricates food residues passing the large intestine, and its inner surface. Digestion in the large intestine occurs as a result of microbial activity exerted by a population similar to that growing in the rumen (see above). Microbial activity, particularly breakdown of cellulose (see Chapters 4 and 5), is extensive in the large intestine of the horse, somewhat lower in that of ruminants and pigs and very limited in chicks and humans. Volatile fatty acids produced in the lumen of the large intestine, water and electrolytes are absorbed through the epithelium of the large intestine. The large intestine is very efficient in water absorption. The nutritional significance of B vitamins and proteins synthesized in the large intestine is doubtful because of the limited absorption of these nutrients through the wall of the large intestine.

EXCRETIONS

Faeces

Waste material or faeces voided from the large intestine via the anus consist of water, undigested residues of food material, unabsorbed residues of gastrointestinal secretions, sloughed epithelial cells from the gastrointestinal tract, inorganic salts and bacteria. Faecal components not of direct dietary origin, such as unused digestive enzymes or desquamated mucosal cells, are called metabolic components* and will be dealt with in Chapters 7 and 13. A large part of the electrolytes present in the faeces are of metabolic origin (see Chapter 10). The colour of faeces is due to plant pigments and stercobilinogen produced by bacterial reduction of bile pigments. The odour is from products of bacterial decomposition—primary indole and skatole—which are derived from deamination and decarboxylation of tryptophan in the large intestine (see Chapter 7). The amount and composition of faeces excreted are largely dependent on the species of animal and type of diet consumed, as can be seen from Table 3.2. Faeces from monogastric species are primarily a mass of bacterial cells, a substantial part of faeces of ruminants consuming mainly roughages is made up greatly from cell-wall components.

Table 3.2 Faeces of various species: their amounts excreted daily, water content and pH

Species of animals	Body weight (kg)	Faeces daily excreted (kg)	Water content (%)	pH†
Horse	550–650	6–25	73–78	6.7
Cattle	550	15–45	75–86	6.8
Pig	100	0.5–3	65–75	7.7
Sheep	45–55	0.9–3	57–75	7.6
Goat	40–45	0.5–1.2	57–75	6.5
Hen‡	2–3	0.10–0.15	70–75	7.5

† Average value fluctuating according to composition of the diet.

‡ Faeces and urine secreted together by the cloaca.

Urine

Waste products from the blood are voided by the kidneys. It is the function of the kidneys to keep the composition of blood plasma constant by removing water, nitrogenous and sulphurous compounds and electrolytes as the need arises. The amount of urine that will be excreted per day is extremely variable as it is affected by the water intake, drinking, the food intake of nitrogen and electrolytes, and the rate of respiration as influenced by environmental and body temperature. Urea is the principal end-product of protein metabolism in mammals (see Chapter 7). The amount of urea excreted is primarily a function of nitrogen intake (see Chapter 7). In birds the urine is rich in uric acid. Uric acid is formed in mammals by oxidation of nucleic acids; it is oxidized in tissues of ruminants and excreted in their urine as allantoin.

* Nowadays often called endogenous components; the original definition of endogenous N is given in Chapter 7.

Table 3.3 Urine of various species: the amount excreted daily, chemical composition and pH (after Nehring)

Species	Daily amount (litre)	pH	Composition (g/litre)							
			Dry matter	Total nitrogen	Urea	Urea nitrogen	Hippuric acid	Hippuric acid nitrogen	Uric acid	Uric acid nitrogen
Horse	3–15	6.8–8.4	80–130	15	31	14.2	25	1.9		
Cattle	6–25	7.0–8.7	50–80	3–10	18.5	8.5	16.5	1.3		
Pig	1.5–6	6.4	30–60	2–4	5.5	2.5	5.0	0.4		
Sheep	0.5–2		100–150	13–25	22.1	10.3	32.4	2.5		
Hen		5.0		0.20*					0.15*	0.05*
Human	1–3	6.1			20	9.2	2.0	0.2	1.4	0.5

* g/day

Urea Uric acid Allantoin Creatinine

Other nitrogenous compounds present in the urine of farm animals in small amounts are ammonia, creatinine—which is related to body muscle mass and less influenced by diet—and conjugates of glycine with benzoic acid (i.e. hippuric acid, $C_6H_5.CO.NH.CH_2.COOH$) and with phenylacetic acid (i.e. phenaceturic acid, $C_6H_5.CH_2.CO.NH.CH_2.COOH$). These latter two acids both appear to be derived from microbial metabolism of phenylalanine and are excreted in much larger amounts by ruminants and horses than by other species.

Aromatic compounds of low solubility such as phenols may be detoxified by conversion into highly soluble esters with sulphuric acid ($RO.SO_3K$) which are excreted in the urine apart from other detoxified benzene derivatives such as hippuric acid and phenaceturic acid. Sulphuric acid is formed by breakdown and oxidation of sulphur-containing amino acids occurring in tissues (see Chapter 8). The excretion of sulphur in the urine is often proportional to the nitrogen/sulphur ratio in body proteins (about 10 : 1).

Urine is usually the principal route of excretion of the mineral elements chlorine, phosphorus, potassium and sodium. The pH of urine of ruminants fed on roughage is slightly basic (between 7.4 and 8.4) because of the large amounts of potassium found in green plants. However, the pH of urine from ruminants or other farm animals fed a high-protein diet is slightly acidic (5–7) due to the presence of HSO_4^- and HPO_4^{2-} in the urine.

The colour of urine is due principally to urochrome, a metabolite of bile pigments. The amount and composition of urine secreted by various species of farm animals are given in Table 3.3.

BIBLIOGRAPHY

Books

Annison, E. T., and Lewis, D. (1962). *Metabolism in the Rumen.* London: Methuen.

Boorman, K. N., and Freeman, B. M. (1976). Digestion in the fowl. *Poultry Science Symposium*, Series No. 11. Edinburgh: British Poultry Science Ltd.

Hungate, R. E. (1966). *The Rumen and its Microbes.* New York: Academic Press.

Kidder, D. E., and Manners, M. J. (1978). *Digestion in the Pig.* Bristol: Scientechnia.

Madge, D. S. (1975). *The Mammalian Alimentary System. A Functional Approach,* Special Topics in Biology Series, 9. London: Arnold.

Moran, E. T. (1982). *Comparative Nutrition of Fowl and Swine.* Guelph: Ontario Agricultural College, University of Guelph.

Swenson, M. J. (ed.) (1984). *Duke's Physiology of Domestic Animals*, 10th edn., Chapters 15–27. Ithaca and London: Cornell University Press.

Symposia on ruminant physiology, digestion and metabolism

Second Symposium, Ames, Jowa 1965. R. W. Doughtery (ed.). London: Butterworths.

Third Symposium, Cambridge 1970. A. T. Phillipson (ed.). Newcastle upon Tyne: Oriel Press

Fourth Symposium, Sydney, Australia 1974. I. W. McDonald and A. C. I. Warner (eds). Armidale: The University of New England Publish. Unit.

Fifth Symposium, Clermont-Ferrand, France 1979. Y. Ruckbush and P. Thivand (eds). Lancaster: MTP Press.

Articles

Symposium on nutritional implications of microbial action in the non-ruminant alimentary tract (1984). *Proc. Nutr. Soc.*, **43**, 1–86.

Symposium on digestive and absorptive processes in poultry (1985). *J. Nutr.*, **115**, 663–697.

Baldwin, R. L., and Allison, M. J. (1983) Rumen metabolism, *J. Anim. Sci.*, **57**, Suppl. 2, 461–477.

Buttery, P. I. (1977). Biochemical basis of rumen fermentation. In *Recent Advances in Animal Nutrition* (D. Lewis and W. Haresign eds), pp. 8–14. London: Butterworths.

Czerkawski, I. W. (1978). Reassessment of efficiency of sythesis of microbial matter in the rumen, *J. Dairy Sci.*, **61**, 1261–1273.

Hoover, W. H. (1978) Digestion and absorption in the hindgut of ruminants, *J. Anim. Sci.*, **46**, 1789–1799.

Owens, F. N. *et al.* (1979). Economy of rumen digestion. *Fed. Proc.*, **38**, 2701–2719.

Part II

CHEMISTRY AND METABOLISM OF NUTRIENTS

CHAPTER 4

Carbohydrates and their Metabolism in Non-ruminants

CHEMICAL STRUCTURE OF CARBOHYDRATES AND THEIR DISTRIBUTION IN FEEDS AND IN THE ANIMAL BODY

The name carbohydrate derives from the fact that these compounds contain carbon combined with hydrogen and oxygen, with the last two elements present in the same ratio as in water. This definition is not strictly correct since not all carbohydrates contain hydrogen and oxygen in this ratio and some contain nitrogen, sulphur or phosphorus in addition to the atoms carbon, hydrogen and oxygen. Carbohydrates are polyhydroxy-aldehydes or -ketones or substances that yield these on hydrolysis.

Carbohydrates constitute the greatest proportion of food consumed by animals, except carnivores. They are the most important components of plants, making up 75% of the dry matter present in feeds of plant origin. Comparatively small amounts of glycogen and glucose are present in the animal body, where they fulfil essential functions in energy metabolism. They are also present in membranes of bacteria. In plants, the insoluble carbohydrates, particularly cellulose, are responsible for structure stability and mechanical firmness; the more soluble carbohydrates such as starch serve as energy reserves. Solar energy is utilized by plants for converting atmospheric carbon dioxide and water to glucose by photosynthesis:

$$6\,CO_2 + 6\,H_2O + 2870\text{ kJ} \longrightarrow \underset{\text{(glucose)}}{C_6H_{12}O_6} + 6\,O_2$$

This transformation of energy by photosynthesis is essential for food production and the oxygen formed as a by-product is vital for animal life.

The carbohydrates are divided into three groups: monosaccharides or simple sugars, oligosaccharides (containing 2–8 sugar units), and polysaccharides (containing a large number of sugar units). Polysaccharides are subdivided into homopolysaccharides, containing one type of monosaccharide, and heteropolysaccharides built from different ones. Upon hydrolysis by acid or enzymes, oligosaccharides and polysaccharides are broken down into various intermediate products and finally into monosaccharides. Carbohydrates of nutritional interest are given in Table 4.1.

Table 4.1 Carbohydrates important in nutrition

1. *Monosaccharides*
   - *Pentoses* ($C_5H_{10}O_5$)
     - xylose, arabinose, ribose
   - *Hexoses* ($C_6H_{12}O_6$)
     - glucose, fructose, galactose, mannose

2. *Disaccharides*
   - saccharose, (composed of glucose and fructose)
   - lactose (milk sugar, composed of glucose and galactose)
   - maltose (composed of two units glucose)

3. *Homopolysaccharides* (containing only a single type of building unit, frequently glucose)
   - Pentosans $(C_5H_8O_4)_n$
     - arabans, xylans
   - Hexosans $(C_6H_{10}O_5)_n$
     - starch, glycogen, cellulose, fructosans

4. *Heteropolysaccharides* (containing several different types of building units such as monosaccharides together with derivatives of the latter)
   - hemicelluloses, pectins, sulphated polysaccharides and aminopolysaccharides

Monosaccharides

Monosaccharides or monosugars are divided into aldoses and ketoses according to whether they contain an aldehyde or ketone group. The presence of these reducing groups classifies these monosaccharides as reducing sugars. Monosaccharides are characterized by the number of carbon atoms present in the molecule, e.g. trioses ($C_3H_6O_3$), tetroses ($C_4H_8O_4$), pentoses ($C_5H_{10}O_5$), hexoses ($C_6H_{12}O_6$) and heptoses ($C_7H_{14}O_7$). Free sugars are water soluble and have a sweet taste. Small amounts of free hexoses—glucose and fructose—occur in green plants. Forage plants contain 1–3% glucose + fructose in the dry matter. Hexoses linked to polysaccharides (starch, cellulose, hemicellulose) are the most important constituents of plant feeds. Thus glucose is the major end-product of digestion of starch in non-ruminants (see later). It is the primary form of energy to be utilized in those animals and circulated in their blood (see later in this chapter). Some derivatives of glucose play an important role in metabolic reactions in living organisms or as constituents of biologically interesting compounds such as phosphoric acid esters, amino sugars in which one hydroxyl group is replaced by an amino group, sugar acids and glycosides (see later).

Fructose is the only important ketohexose in nature. It occurs free in green plants, in semen and in blood of fetal and newborn animals, and is a component of disaccharides (sucrose) and polysaccharides (fructosans) in green plants and tubers. Galactose is a component of lactose (milk sugar) and of galactolipids present in nervous tissues and green plants (see Chapter 6). The pentose sugar ribose is present in all living cells as a component of ribonucleic acid, certain vitamins (see Chapter 12) and energy-rich compounds (ADP, ATP). Xylose and arabinose are present in polymerized form in hemicellulose (see later).

Disaccharides

Disaccharides consist of two monosaccharide molecules which combine together with the elimination of one water molecule:

$$2\,C_6H_{12}O_6 \longrightarrow C_{12}H_{22}O_{11} + H_2O$$

Saccharose (or sucrose) consists of one molecule of α-D- glucose and one molecule of β-D-fructose. It occurs in sugar beet and sugar cane, whose by-products serve as animal feeds, and in somewhat lower concentrations in fodder beets and turnips. Saccharose occurs in larger amounts than the monosaccharides, in the order of 2–8% in the dry matter of herbage; the concentration of sugars in grasses is higher than in legumes. The amount of sugars in plants increases during the day, apparently as a result of photosynthesis, and decreases again during the night. Success in conserving herbage as silage often depends on the presence of readily fermentable carbohydrates (sugars, starch or fructosan) in the plants.

Lactose or milk sugar consists of one molecule of β-D-glucose and one molecule of β-D-galactose. Lactose occurs only in milk (see Chapter 21).

Maltose and cellobiose are the fundamental repeating units of starch and cellulose, respectively. Maltose consists of two molecules of α-D-glucose joined together in an α-1,4 linkage and cellobiose of two molecules of β-D-glucose joined together in a β-1,4 linkage. Maltose is produced as an intermediate product in the breakdown of starch by digestive enzymes. It is also the fundamental repeating unit of glycogen (see Figure 4.1).

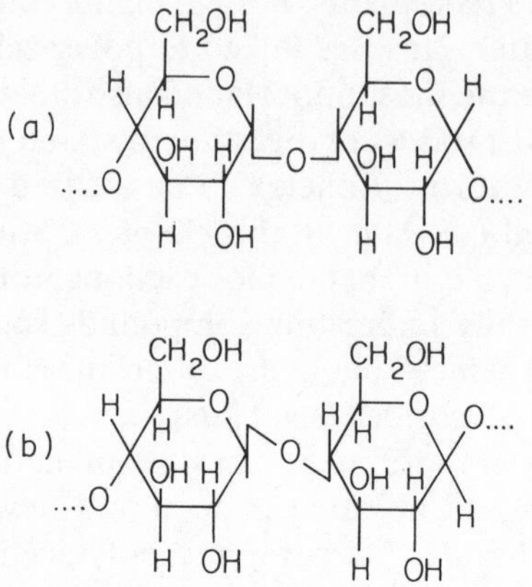

Figure 4.1. Repeating maltose unit of starch (a) and repeating cellobiose unit of cellulose (b).

This difference in the configuration of the glycosidic link of maltose and cellobiose is responsible for the very different susceptibilities or starch and cellulose to the action of digestive enzymes (see later).

Disaccharides (saccharose, lactose or maltose) cannot be absorbed into the circulation from the gut and must first undergo breakdown to monosugars by suitable digestive enzymes (see later).

Polysaccharides

These are polymers of monosaccharides joined through glycosidic links.

Starch

The major dietary polysaccharide ingested by non-ruminants is starch. It is the principal reserve carbohydrate in plants, and is found in grains, seeds (which may contain up to 70%), cereal by-products and tubers (with about 30% starch in the dry matter). The green parts of plants, leaves and stems contain small amounts of starch. Most kinds of starch consist of a mixture of two types of polymers—amylose and amylopectin. Amylose is a linear molecule in which 250–300 glucose units are joined by α-1→4 linkages. Amylopectin is a highly branched molecule with branching occurring every 24–30 glucose units, with α-1→6 links at the branch points and α-1→4 links within the chains of each branch (see Figures 4.2 and 4.5). The ratio of amylose to amylopectin varies in different starches, but it is usually in the range of one amylose to three amylopectins. Starch exists in plants as small granules which are characteristic for each species. A source of starch can be identified by microscopic examination of its granule shape and arrangement.

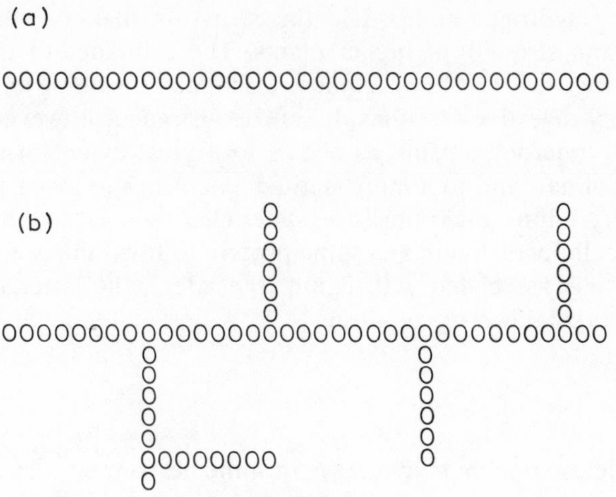

```
(a)

ooooooooooooooooooooooooooooooooooo



(b)            o                o
               o                o
               o                o
               o                o
               o                o
               o                o
ooooooooooooooooooooooooooooooooooo
        o               o
        o               o
        o               o
        o               o
        o               o
        oooooooo        o
        o
```

Figure 4.2. Structure of starch: segments of amylose (a) and of amylopectin (b).

In its natural state, tuber starch such as that in potatoes exists in an insoluble granular form which resists digestion. Such starch-containing foods as potatoes must be cooked before they can be utilized by chickens and pigs. The cooking of foods markedly aids digestion of starch by breaking down and solubilizing the starch granules.

Glycogen

The small amount of carbohydrate reserve in the animal body occurs in liver and muscle as glycogen which resembles amylopectin in structure. Branching occurs every 10–12 glucose units. The glycogen reserve stored in the animal body is very limited and not sufficient to sustain a starving animal for more than 24 hours. The livers of various animal species contain the following percentages of glycogen: calves 2–5, adult cattle 1.5–4, hens 3–4, geese 4–6. Muscles contain 0.5–1% of glycogen. About 40% of the total amount of glycogen present in the body occurs in the liver and 45% in the muscles. It may be noted that the glycogen content in muscles of resting and well-nourished horses goes up to 4%.

Cellulose

This is the most abundant carbohydrate in the world. It is the major structural component of plant cell walls and amounts to about 20–40% of the dry matter of green plants. Chemically cellulose is a linear polymer of β-1,4-linked D-glucose units; the number of polymerized glucose molecules may range from 900 to 2000. Cellulose is present in plant tissues as fibres composed of crystalline microfibrils consisting of cellulose chains held together by strong hydrogen bonds. It is this structure that contributes most of the rigidity and strength of higher plants. The β-linkage of the glucose molecules (see Figure 4.1) makes cellulose essentially insoluble and resistant to breakdown by digestive enzymes. It can be degraded, however, through fermentation by microorganisms, as occurs to a great extent in the rumen of polygastric animals and to a more limited extent in the lower tract of all animals. The crystalline microfibrils are encircled by a largely amorphous matrix of hemicelluloses, lignin and some protein to form the cell walls. Cellulose is intimately associated with lignin (see later), the latter interfering with microbial breakdown of cellulose.

Fructosans

These are fructose polymers that serve in some herbage species and tubers as storage carbohydrates instead of starch.

Heteropolysaccharides

Hemicelluloses

These are a mixture of linear and highly branched polysaccharides containing various sugar residues, e.g. xylose, arabinose, glucose, galactose and uronic acid. Hemicelluloses occur in cell walls of forage plants. They are associated with lignins and are laid down around cellulose fibres. The hemicelluloses are less resistant to chemical degradation than cellulose being hydrolysed by relatively mild acid while celluloses require highly concentrated acid. Non-ruminants digest hemicellulose better than cellulose. In the ruminant, most cellulose is digested in the rumen, but some hemicellulose escapes the rumen to be fermented in the lower tract (Chapter 5).

Pectin

This is found in the spaces between cell walls and intracellular layers. It may function as an intracellular cement, while hemicellulose occurs in the secondary wall thickening. The structure of pectin is more uniform than that of hemicellulose. It consists essentially of a polysaccharide chain substituted with araban and galactan side chains (see Figure 4.3). Many of the carboxyl groups in the galacturonic acid units are esterified with methyl groups or neutralized with calcium or magnesium. Pectins are only digested by microbial action and are thus well utilized by ruminants (see Chapter 8). Their water holding capacity (through gel formation) increases with methoxyl content; pectins prepared from citrus peels are used to reduce diarrhoea in calves.

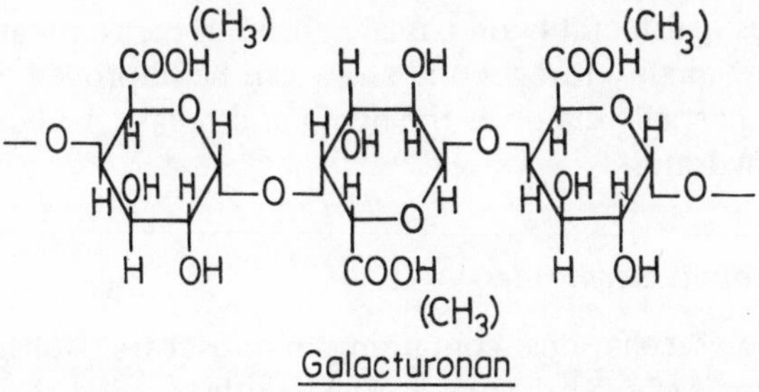

Galacturonan

Figure 4.3. Repeating unit of pectin: polygalacturonic acid (partly substituted by methoxyl groups).

Lignin

Lignin is a high molecular weight polymer of phenylpropane derivatives such as *p*-coumaryl, *p*-coniferyl and sinapyl alcohols (see Figure 4.4). It is not a carbohydrate, but because of its association with cellulose and hemicellulose it is usually discussed with carbohydrates. Lignin stiffens the cellulose fibres, providing strong structural support for the plant. It constitutes

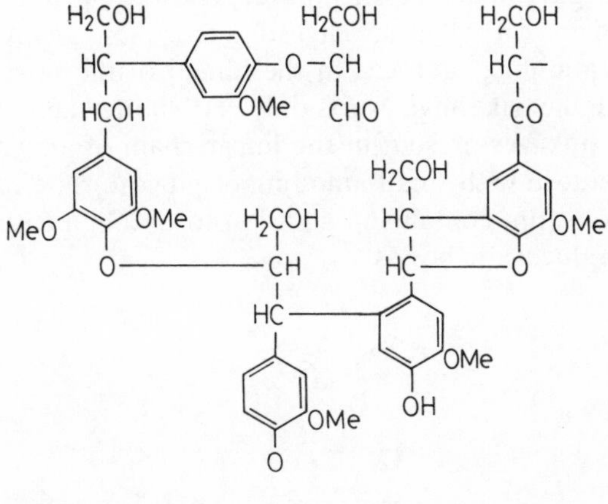

Figure 4.4. Structure of lignin (composed of phenylpropanoid units).

about 5–10% of the dry matter of annual plants. Its percentage presence in plants and its composition such as the proportion of the 3-phenylpropanol units, methoxyl and nitrogen content, vary with the plant species and maturity. The methoxyl content (OCH_3 or OMe) of wood lignin is about 15%, that of lignin present in grasses is about 8% and of lignin in legumes only 5%. Lignin is very stable due to its condensed structure and it is rarely attacked by microorganisms. Cellulose and hemicellulose present in forages and agricultural waste are digested by ruminants only at a low rate because of their close association with lignin (see above). Only the methoxyl content of lignin decreases considerably on passage through the ruminant tract. The digestibility of low-quality forages and straw can be improved by treatment with alkali which partially dissolves the lignin and breaks the bonds between lignin and carbohydrates.

Substituted Heteropolysaccharides

Some heteropolysaccharides containing monosugars, substituted with amino, sulphate or carboxyl groups are widely distributed throughout the animal body and fulfil important biological functions as constituents of the organic matter of bones, connective tissues or substances with blood group specificity. For carbohydrates containing nitrogen or conjugated with proteins see Chapter 7, and for carbohydrates connected with lipids see Chapter 6.

DIGESTION AND ABSORPTION OF CARBOHYDRATES

Starch, the only polysaccharide which is highly utilizable by monogastric animals, and dietary disaccharides have to be broken down by digestion into monosugars, the molecules suitable for passage through the intestinal mucosal cells. Digestion and absorption of starch occur in the upper intestinal tract. The major carbohydrate-digesting enzymes, α-amylase, is secreted with the pancreatic juice and acts in the lumen of the intestine; α-amylase secreted in pig and chick saliva exerts only very limited activity. This enzyme splits the α-1,4 linkages present in the linear chain of amylose, resulting in a mixture of maltose with small amounts of glucose. The main constituent of starch, amylopectin, contains in addition to α-1,4 glucosidic bonds some branched α-1,6 glucosidic bonds.

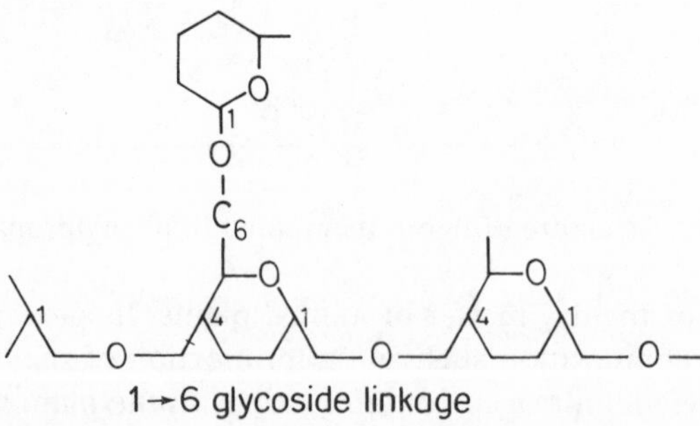

1→6 glycoside linkage

Figure 4.5. α-1,6 glucoside linkage.

Since α-amylase is unable to hydrolyse the α-1→6 branching points, the end-products of amylopectin hydrolysis by this enzyme include, in addition to maltose, branched oligosaccharides containing about 10 molecules of glucose, the so-called 'limit dextrins'. The branched oligosaccharides are finally decomposed by an additional enzyme, oligo-1,6-glucosidase, and maltose and glucose are liberated. No further hydrolysis of disaccharides takes place in the intestinal lumen, neither of maltose originating from starch nor of dietary disaccharides such as saccharose or lactose. On the other hand, different disaccharidases responsible for the final hydrolysis of the disaccharides are located in the outer brush border of the microvillous membrane. For example, α-glucosidase splits saccharose and maltose, both of which disaccharides contain an α-glucosidic bond, and β-glucosidase splits lactose which has a β-glucosidic bond. The following scheme illustrates the mode of action of carbohydrases:

$$\text{Starch} \xrightarrow[\text{Oligo-1,6-glucosidase}]{\alpha\text{-Amylase} +} \text{maltose} \xrightarrow{\alpha\text{-Glucosidase}} \text{glucose}$$

$$\text{Saccharose} \xrightarrow{\alpha\text{-Glucosidase}} \text{glucose} + \text{fructose}$$

$$\text{Lactose} \xrightarrow{\beta\text{-Glucosidase}} \text{glucose} + \text{galactose}$$

Intracellular amylases localized within the outer layer of the microvillous membrane complete the hydrolysis of starch started by pancreatic amylase in the intestinal lumen. Disaccharidases are arranged on the external surface of the brush border membrane in very close proximity to the mechanism for the transfer of monosaccharides. Those that are released by hydrolytic action are transported by mobile carriers, coupled with sodium ions, across the microvillous membrane into the absorptive cell. The mechanism for intestinal absorption of sugars involves active transport (Chapter 3). The rate of absorption of sugars decreases in the following order: galactose > glucose > fructose > pentose. The absorbed sugars are carried by the portal blood to the liver.

A striking adaptation in carbohydrase activity occurs with changes in the diet; an eight- to ten-fold increase in amylase activity is found in rats on a diet high in starch compared to a diet high in protein. The activity of disaccharidases is paralleled by changes in the diet in growing mammals. In young animals, the activity of lactase which hydrolyses milk lactose remains high until weaning and then declines (see Chapter 18) while the activity of maltase, associated with a solid diet, rises just before weaning and thereafter remains high. Feeding considerable amounts of sucrose to young mammals results in serious diarrhoea because of the insufficiency of sucrase (similar to maltase) in these animals. The diarrhoea is caused by the osmotic effect of undigested disaccharides attracting water into the lumen of the intestine.

Humans with a genetic deficiency of lactase in their intestinal tissues are unable to utilize lactose properly. Affected individuals are unable to tolerate milk products.

Nutritive Importance of Non-digestible Carbohydrates in the Diet of Monogastric Animals

Cellulose and hemicellulose escaping the small intestine of poultry or pigs are substrates for fermentation which occurs in the caecum. This microbial action resembles the ruminal fermentation of polysaccharides (see Chapter 5) but it is quantitatively limited. The volatile fatty acids formed in the caecum are absorbed into the portal blood (see Chapters 5 and 22).

The presence of non-digestible carbohydrates, which are usually determined as crude fibre (see Chapter 1) in the diet of monogastric animals, is nutritionally interesting. The capacity of non-digested carbohydrates to absorb water increases the bulkiness of the chyme passing through the tract. This encourages the peristaltic action by which food residues are driven forward through the intestine, and mechanical digestion is enhanced (see Chapter 3). The presence of a limited amount of fibre in the diet of monogastrics, and even of human beings, is desirable because of the water-holding capacity of the fibre and its influence on the composition of the intestinal flora. The fibre occurring in vegetables, bread and particularly in wheat bran relieves constipation by decreasing the rate of passage of the food through the gut. The use of fibre-deficient diets apparently plays an important causative role in some serious illnesses prevalent in western civilization such as appendicitis, ischaemic heart disease and cancer of the colon and rectum. The concept of crude fibre, developed in animal nutrition, has in recent years gained great importance in human nutrition.

METABOLISM AND METABOLIC FUNCTIONS OF CARBOHYDRATES INSIDE THE TISSUES OF MONOGASTRIC SPECIES

Production of Metabolites from Digested Carbohydrates in Monogastric Animals Compared with Ruminants

The metabolism of carbohydrates is of great importance to all farm animals since carbohydrates in the body are the essential source of energy and starting materials for the biosynthesis of fats and non-essential amino acids. The digestion products of carbohydrates given to monogastric animals differ considerably from those obtained in multigastric animals. The major product of carbohydrate digestion in monogastric animals is glucose, which originates mainly from starch which is the quantitatively dominant constituent of pig and poultry diets.

In ruminants, however, insoluble and soluble carbohydrates are fermented to volatile fatty acids which replace glucose as the most important source of energy and starting material for biosynthesis of body constituents and animal products. Direct absorption of glucose which escapes fermentation occurs in ruminants only to a very limited extent. The volatile fatty acids are absorbed from the rumen into the circulation and transported to the organs. Enzymatic systems necessary for the utilization of volatile fatty acids are well developed in the tissues of ruminant animals (Chapter 5).

Metabolic Functions of Carbohydrates in Monogastric Animals

Glucose is the main source of energy for monogastric animals and is the starting material for biosynthetic processes. The central transporting medium for glucose is the blood. The blood glucose concentration is normally maintained within narrow limits; in simple-stomached mammals and also in humans it fluctuates between 70 and 100 mg per 100 ml; in ruminants the concentration is lower (40–70 mg per 100 ml) and in poultry higher (130–260 mg per 100 ml). It is the result of two opposing processes: (1) entry of glucose into the blood from the intestine (originating from food), liver and other organs, and (2) withdrawal of glucose from the blood into various tissues (liver, muscles, kidney, adipose tissue, brain) and its utilization in these tissues for oxidation and biosynthetic purposes.

The level of blood sugar is maintained within the rather narrow range by conversion of circulating blood glucose into glycogen (glycogenesis, mainly in the liver) and by reconversion of glycogen to glucose (glycogenolysis):

$$\text{Glucose} \underset{\text{Fasting}}{\overset{\text{Abundant feeding}}{\rightleftarrows}} \text{Glycogen}$$

The blood glucose level, which is elevated after abundant carbohydrate intake (hyperglycaemia), returns to the normal level by storage of glucose as glycogen. On the other hand, low levels of glucose (hypoglycaemia) found after fasting or because of increased requirements of glucose for energy purposes (e.g. for muscle activity), are prevented by release of glucose from glycogen.

Sources of Blood Glucose

Blood glucose derives from the following sources: absorbed glucose originating from food; glucose synthesized from various precursors, mainly in the liver; and glucose liberated from glycogen stored mainly in the liver.

(1) Absorption of glucose resulting from digestion of oligo- and polysaccharides is discussed in Chapter 3.

(2) A second continuing source of glucose results from its biosynthetic formation in body tissues particularly in the liver from non-carbohydrate metabolites, the glucogenic substances such as amino acids, lactic acid, propionic acid and glycerol (gluconeogenesis). The quantity of glucose synthesized in the tissues of omnivorous non-ruminant animals is considerable, but never reaches the amount originating from food. Carnivorous animals obtain only very small amounts of glucose from food and the greatest part

of their glucose requirement is provided by biosynthesis, mainly from amino acids. Glucose is the second important source of energy for ruminants (after volatile fatty acids) and is formed mainly biosynthetically in these animals (see Chapter 5).

Gluconeogenesis is the reverse process of glycolysis. Pyruvate, the most important intermediate compound in the transformation of glucogenic substances to glucose, is transformed by the enzyme pyruvate kinase (PK) to oxaloacetate which is itself transformed by the enzyme phosphoenolpyruvate carboxykinase (PEPCK) to phosphoenolpyruvate. Two moles of this C_3 compound are condensed to one mole of hexose. The activity of the enzyme PEPCK, which occurs in most species in the cytoplasma, is altered by nutritional changes; for example, it increases considerably after feeding carbohydrate-free diets to chickens.

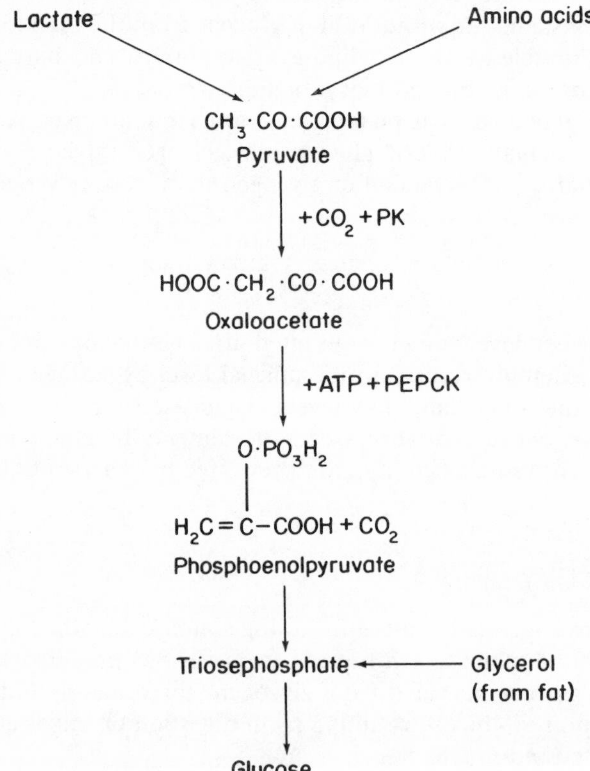

Figure 4.6. Gluconeogenesis: transformation of pyruvate to glucose—entry of precursors into the pathways of gluconeogenesis.

(3) Glycogen stored in the liver serves as a reserve for glucose when the latter is needed in metabolic processes (see later). The glucose is released from glycogen by enzymatic breakdown called glycogenolysis:

Glycogen ⟶ glucose 1-phosphate ⟶ glucose 6-phosphate ⟶ glucose

The glycogen stores themselves are derived from absorbed glucose or from glucose formed by gluconeogenesis.

The Fate of Glucose Removed from Blood

Glucose removed from the blood into cells of organs, particularly the liver, may be utilized in the following ways: (1) for the synthesis of glycogen; (2) conversion into fat; (3) conversion into amino acids; (4) as a source of energy.

Synthesis and Storage of Glycogen. As pointed out above, glycogen is an important store of readily available energy in the liver, muscles and other tissues, and plays an important role in maintaining blood glucose levels. Glycogen synthetase, which acts as follows

$$\text{UDP-glucose} + \text{glycogen}_{n-1} \longrightarrow \text{glycogen}_n + \text{UDP}$$

is stimulated after a high carbohydrate meal in order to enable storage of additional glucose (the immediate precursor of glycogen is the nucleotide: diphosphate uridine diphosphate glucose).

Conversion of Glucose into Fat. The amount of glucose that can be stored as glycogen is limited. Therefore, when ingestion of carbohydrate exceeds the amounts required for both energy production and storage as glycogen, glucose is converted into fat. Fat synthesis occurs in liver or in adipose tissues (see Chapter 6). The partition of this process between liver and adipose tissue differs in different animal species. Biosynthesis of fat in poultry occurs mainly in the liver, but in the adipose tissue in pigs and rats.

Fat formation from glucose involves breakdown of glucose to pyruvate (by glycolysis, see formula, Figure 4.7); the pyruvate formed is available

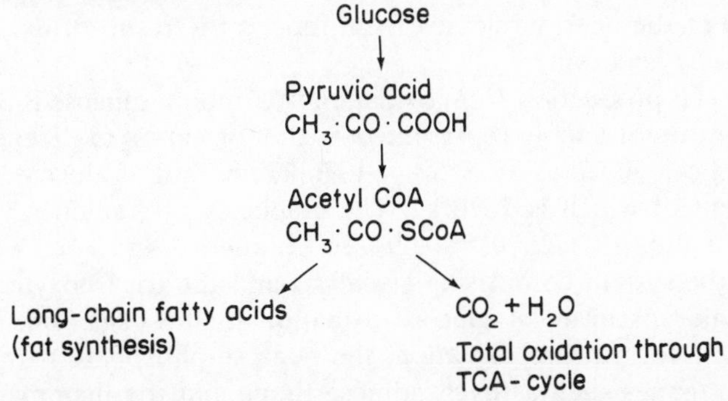

Figure 4.7. Metabolism of glucose—conversion to fat and total oxidation.

for transformation to acetyl-CoA which in turn is converted into long-chain fatty acids or serves other synthetic purposes or enters the tricarboxylic acid cycle for production of energy (see Chapter 2 and later). On the other hand, the reverse process of conversion of long-chain fatty acids into glucose does not occur in animal tissues, since there exists no enzyme enabling carboxylation of acetyl-CoA to pyruvate. The fatty acids formed may be stored in large amounts as triglycerides or used immediately for energy production, particularly in muscles (see Chapter 22). The glycerol moiety in the triglycerides is also derived largely from glucose.

Synthesis of Amino Acids. Non-essential amino acids synthesized in the body derive their carbon skeletons from intermediates produced by breakdown of glucose (see Chapter 7).

Full Oxidation to Carbon Dioxide and Water Yielding ATP as Energy Source. Glucose is the main fuel in the body or monogastric animals. It enters pathways leading to oxidation to carbon dioxide and water, yielding ATP as energy source. Glucose is the only source of energy in red blood cells, and quantitatively it is a very important fuel for the brain, nervous tissue and muscle. The latter tissues are capable of deriving their energy needs partly from other nutrients such as fatty acids or ketone bodies (see Chapter 6).

The importance of pyruvate formed as intermediate in the conversion of glucose into fat and in its oxidation has been stressed above. The principal steps in the oxidation of pyruvate for gaining energy are as follows: pyruvate is decarboxylated to form acetyl-CoA. The oxidation of acetyl-CoA is effected through the tricarboxylic acid cycle. The cycle involves condensation of a C_2 compound (acetyl-CoA) with a C_4 compound (oxaloacetic acid) yielding a C_6 compound (citric acid), with subsequent losses of carbon dioxide at two steps to yield a C_4 compound leading to the reappearance of oxaloacetate. The cycle is outlined in Figure 4.8. The tricarboxylic acid cycle is the final pathway not only in the oxidation of carbohydrates but also in the oxidation of fats and proteins. In a quantitative sense this cycle is also the most important phase in the oxidation of foodstuffs, since approximately 90% of the energy released from food is the result of oxidation by the tricarboxylic acid cycle.

The total ATP production from oxidation of 1 mol of glucose is 38 mol of ATP. The capture of energy represented by the formation of 38 energy-rich bonds may be calculated as $38 \times 52 = 1976$ kJ per mol of glucose, the total energy content of which is 2870 kJ. The efficiency of free-energy capture by the body is thus $1976/2870 = 0.69$ (see Chapters 2 and 14).

Although the system comprising glycolysis and the tricarboxylic acid cycle is the major pathway of glucose oxidation in the organism, an alternative pathway for glucose oxidation, the pentose-phosphate pathway, occurs in some tissues such as liver, adipose tissue and the mammary gland, which have a specific demand for reduced coenzyme NADPH (particularly

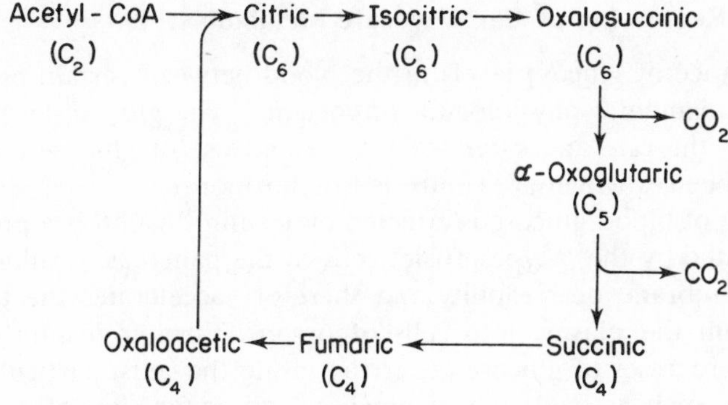

Figure 4.8. Scheme of tricarboxylic acid cycle (after Krebs).

needed for synthesizing fatty acids). The reduced nicotine adenine dinucleotide phosphate is produced by the pentose-phosphate pathway according to the formula:

$$\text{Glucose 6-phosphate} + 12\,\text{NADP}^+ \longrightarrow 6\,\text{CO}_2 + \text{PO}_4{}^{3-} + 12\,(\text{NADPH} + \text{H}^+)$$

The yield of energy-rich bonds produced by glucose oxidation in the pentose-phosphate pathway equals approximately the amount of ATP obtained by glucose oxidation through the tricarboxylic acid cycle.

Reutilization of Glucose via the Cori Cycle

Repeated and thus economic utilization of glucose in the body occurs via a cyclic pathway called the Cori cycle named after the discoverers of this process (see Figure 4.9). Muscle glycogen stores serve as an anaerobic energy supply for working muscles when the oxygen supply is not sufficient for total oxidation of glucose via the tricarboxylic acid cycle. Under anaerobic conditions glucose, mobilized from muscle glycogen, is converted to lactate by glycolysis. Lactate cannot be metabolized in the muscle, but passes readily out of the muscle cell into the blood. It is taken up by the liver for resynthesis into glucose and thence into glycogen. Glucose may then be mobilized from liver glycogen and passes via the circulation to the muscles where glycogen is resynthesized as outlined in Figure 4.9.

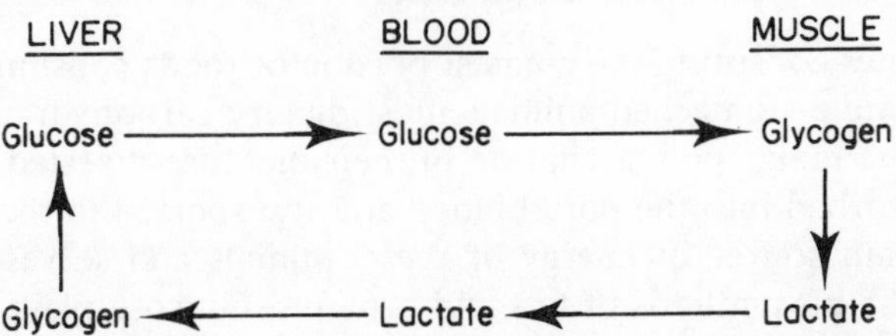

Figure 4.9. Cori or lactic acid cycle (muscle glycogen contributes to blood glucose by this pathway).

Hormonal Regulation of Carbohydrate Metabolism

Maintenance of glucose levels in the blood between certain narrow limits is of fundamental physiological importance. The glucose level in blood depends on the rate and extent of release and use of glucose by different tissues—processes which are controlled by hormones.

Lowering of blood glucose is effected by insulin. Insulin is a protein hormone secreted by the β-type of islet cells in the pancreas. Insulin increases the cell membrane permeability and therefore accelerates the transfer of glucose from the plasma into cells of organs. Further insulin stimulates metabolic processes of glucose occurring inside the cells, particularly those of the liver such as oxidation of glucose, and conversion of glucose into glycogen and into fat. The first step of all these processes, the phosphorylation of glucose by the enzyme glucokinase,

$$\text{Glucose} + \text{ATP} \longrightarrow \text{glucose 6-phosphate} + \text{ADP}$$

is enhanced by insulin. As a result insulin counteracts increased blood glucose levels and thus the insulin secretion rate is responsive to rises in blood glucose level. The increased insulin level causes the blood glucose concentration to fall to its normal level.

The action of the following hormones is opposite to that of insulin: adrenaline produced in the adrenal medulla, the protein hormone glucagon, secreted from the α-cells of the pancreas, adrenocortical hormone and growth hormone secreted by the anterior pituitary. The output of the latter hormones is increased by decreased blood glucose levels and so the ratio of insulin to these other hormones decreases. A decrease in the ratio insulin/glucagon favours the breakdown of glycogen to yield blood glucose. Adrenaline not only stimulates glycogen breakdown, but at the same time inhibits synthesis of glycogen from glucose in the liver; as a result of both processes, glucose becomes maximally available in the blood. The complicated mechanism involved in the activation of phosphorylase, the enzyme responsible for glycogen breakdown, by adrenaline is well elucidated, but for details the reader is referred to biochemical texts.

SUMMARY

Carbohydrates constitute the greatest portion of foods consumed by farm animals. In simple-stomached animals most dietary carbohydrates, particularly starch, the major polysaccharide in their diet, are digested to glucose. Glucose is absorbed into the portal blood and transported to the liver. Glucose is the main source of energy of these animals and serves as starting material for the biosynthesis of fats and proteins. Glucose plays a vital role in the metabolic processes of animal cells, particularly those of nerves, muscles and liver. Adequate blood concentrations of glucose are maintained due to hormonal control. Between meals, when not enough glucose is available from food, glucose is released from glycogen stores in the liver. In order to replenish glycogen stores, glucose is produced in the liver biosynthetically (gluconeogenesis), mainly from breakdown products of protein and fats.

BIBLIOGRAPHY

Books

Berdanier, C. D. (1976). *Carbohydrate Metabolism, Regulation and Physiological Role.* New York: John Wiley.

Dickens, F., Randle, P. J., and Whelan, W. J. (1968).*Carbohydrate Metabolism and its Disorders,* Vols. 1 and 2. New York: Academic Press.

Inglett, G. E., and Falkehag, S. I. (1975). *Dietary Fibers: Chemistry and Nutrition.* New York: Academic Press.

Shreeve, W. W. (1974). *Physiological Chemistry of Carbohydrates in Mammals.* Philadelphia: W. B. Saunders.

Sipple, H. H., and McNutt, K. W. (1974). *Sugars in Nutrition.* New York: Academic Press.

Articles

Bailey, R. W. (1973). Structural carbohydrates. In *Chemistry and Biochemistry of Herbage* (G. W. Butler and R. W. Bailey, eds), Vol. 1, pp. 157–212. London: Academic Press.

Mendeloff, A. J. (1975). Dietary fiber. *Nutr. Rev.*, **33**, 321–326.

Moran, E. T. (1985). Digestion and absorption of carbohydrates in fowl. *J. Nutr.*, **115**, 665–674.

Smith, D. (1973). The non-structural carbohydrates. In *Chemistry and Biochemistry of Herbage* (G. W. Butler and R. W. Bailey, eds), Vol. 1, pp. 106–156. London: Academic Press.

Vohra, P. (1975). New sources of foodstuffs for poultry, carbohydrates in particular. *World Rev. Nutr. Diet,* **22,** 95–152.

CHAPTER 5

Carbohydrate Metabolism in Ruminants

There are fundamental differences between ruminants and monogastric animals concerning the mode of digestion and metabolic pathways of carbohydrates and types of end-products formed by the two groups of animals. The bulk of carbohydrates in ruminant feeds are polymers: cellulose, hemicellulose, starch, fructan and pectin (see Chapter 4). Fodder plants contain on a dry basis 20–30% of cellulose, 14–20% of hemicellulose and up to 10% of pectins—apart from 2–12% of lignin. The major part of carbohydrates, solubles (sugars and starch) and less solubles (cellulose, hemicellulose), are fermented by ruminants to volatile fatty acids when the food passes through the rumen. This is due to the action of microbial enzymes. The cellulose in roughage, practically unavailable to human beings and most species of monogastric animals, may be well utilized by ruminants due to the ability of microorganisms to break it down. The mixture of volatile fatty acids present in rumen liquor is composed mainly of acetic, propionic and butyric acids but contains also small amounts of formic, isobutyric, valeric, isovaleric and caproic acids. These small amounts of branched-chain fatty acids originate from bacterial breakdown of protein (see Chapter 7).

$CH_3 \cdot COOH$ Acetic acid

$CH_3 \cdot CH_2 \cdot COOH$ Propionic acid

$CH_3 \cdot CH_2 \cdot CH_2 \cdot COOH$ Butyric acid

$(CH_3)_2CH \cdot COOH$ Isobutyric acid

$CH_3 \cdot CH(CH_3) \cdot CH \cdot COOH$ Isovaleric acid

Some gases such as methane, carbon dioxide and hydrogen are by-products of the ruminal fermentation; they are removed from the body by belching. Minor amounts of ingested carbohydrates and a part of polymer carbohydrates present in microorganisms escape fermentation in the forestomach and pass with the digesta to the lower tract; their fate there will be dealt with later.

Volatile fatty acids (acetic, propionic and butyric acids) formed in the rumen are absorbed through the rumen walls into the circulation and are transported to the tissues. The metabolic processes of these compounds take place in the animal tissues: catabolic processes which supply energy and biosynthetic processes such as biosynthesis of fats from acetic and butyric acid or of glucose from propionic acid. Enzymatic systems needed for the metabolism of volatile fatty acids act in the ruminant tissues and this can be seen as an adaptation of the animals enabling them to utilize the end-products of microbial activity. The volatile fatty acids fulfil in the ruminant most of the functions that in monogastric animals are fulfilled by glucose, especially energy release. Nevertheless, of great importance is the participation of certain, even limited, amounts of glucose in the metabolic processes occurring in ruminant tissues. Since most of the glucose ingested with the food is broken down in the rumen, ruminants produce their required glucose biosynthetically, largely from propionic acid (see later) and from other simple compounds. The quantitative limitation of glucose in the ruminant body manifests itself likewise in the blood glucose level: 40–70 mg% compared with 100 mg% in the blood of monogastric animals. The blood glucose level in suckling preruminant calves and lambs reaches 100–120 mg% since glucose is the main source of energy in very young ruminants before weaning and the metabolism of carbohydrates takes place in their tissues like in those of monogastric animals (see Chapter 18).

MICROBIAL BREAKDOWN OF CARBOHYDRATES IN THE RUMEN

The breakdown of carbohydrates to volatile fatty acids occurring in the rumen may be divided into two stages (see Figures 5.1–5.3). (1) The extracellular hydrolysis of complex carbohydrates (cellulose, hemicellulose, pectin) to short-chain oligosaccharides mainly disaccharides (cellobiose, maltose, xylobiose) and to simple sugars (see Figure 5.1). (2) Oligosaccharides and simple sugars are unstable towards rumen microorganisms, are taken up by them and they are for the most part broken down rapidly to volatile fatty acids by the action of intracellular enzymes. First, the monosaccharide is converted to pyruvate (see Figure 5.2); starting with pyruvate, the metabolic pathways leading to each of the volatile fatty acids (acetic, propionic and butyric acid) differ. These pathways, presented in Figure 5.3, will be discussed later.

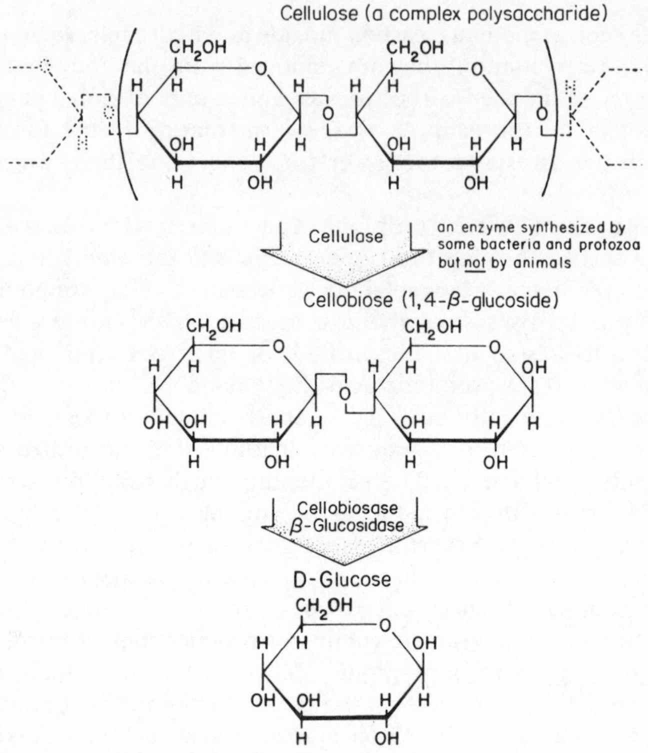

Cellulose (a complex polysaccharide)

Cellulose — an enzyme synthesized by some bacteria and protozoa but not by animals

Cellobiose (1,4-β-glucoside)

Cellobiosase β-Glucosidase

D-Glucose

Figure 5.1. Microbial breakdown of cellulose to glucose. (After Scott, M. L. *et al.*, *Nutrition of the Chicken*, 2nd Edn, 1976, p. 15.) *Reproduced by permission of Scott & Associates*

The Fermentative Breakdown of Polysaccharides

Cellulose. As noted before, a great advantage of ruminants is their ability to utilize cellulose-rich forages with the aid of rumen microorganisms, whereas the digestive juices of monogastric animals lack cellulase. Cellulose is decomposed by microbial cellulases to cellobiosase (see Figure 5.1) which is then converted again microbially (by cellobiase) to glucose or through the action of α-phosphorylase to glucose 1-phosphate.

Hemicellulose (see Chapter 4). Hydrolysis of hemicellulose proceeds primarily through the release of the disaccharide xylobiose, which is decomposed by an intracellular enzyme, xylodiase, to xylose. In addition, uronic acid is obtained as an intermediate product of the breakdown of hemicellulose.

Pectin. Uronic acids are also produced in the rumen from pectins which are first hydrolysed to pectic acid and methanol (see Chapter 4). The pectic acid is then attacked by polygalacturonases to release galacturonic acid.

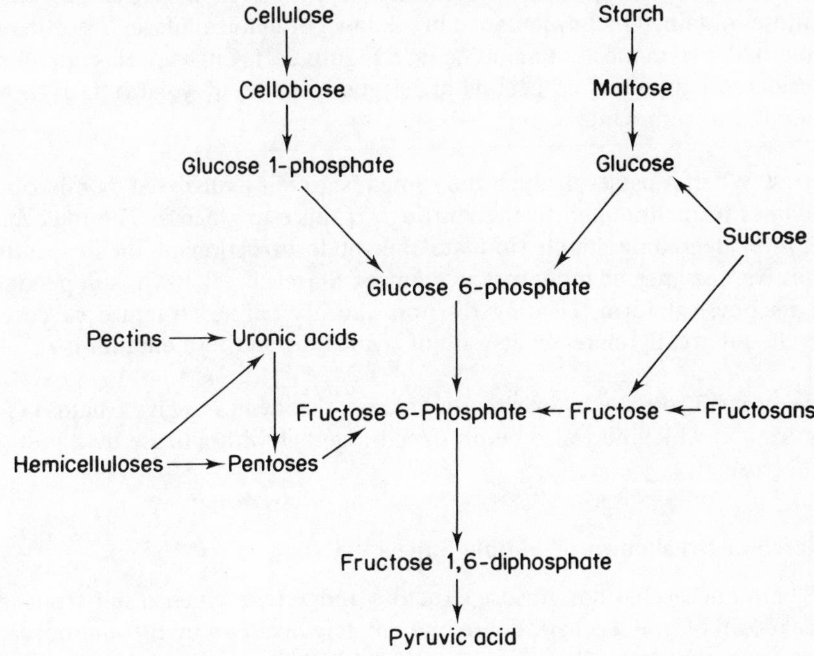

Figure 5.2. Conversion of carbohydrates to pyruvate in the rumen. (After McDonald, P., *et al.*, *Animal Nutrition*, 3rd Edn, 1981, p. 133.) *Reproduced by permission of Longman Group Ltd.*

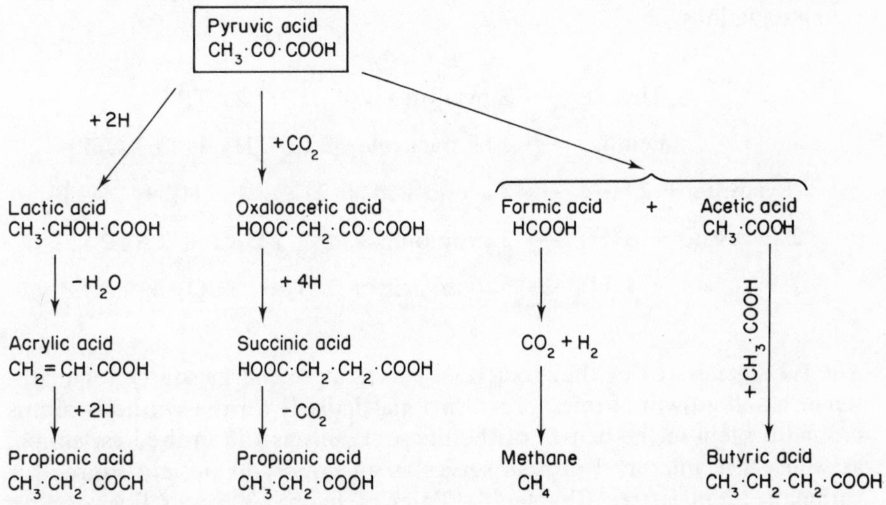

Figure 5.3. Conversion of pyruvate to volatile fatty acids (VFA) in the rumen.

The further fermentation of galacturonic acid involves formation of xylose. Pentoses obtained by fermentative breakdown of hemicellulose or pectin are converted enzymatically to glucose (see Figure 5.2). Citrus peels containing considerable amounts of pectins are a good source of volatile fatty acids when fed to ruminants.

Starch. When it passes through the rumen, starch is hydrolysed by microbial amylases to maltose and further through maltase to glucose. The microbial enzymes degrading starch simulate the mode of action of the respective digestive enzymes of monogastric animals. Starch breakdown is dependent on its physical form. Heating disrupts the crystalline structure of cereal starch and greatly increases its rate of hydrolysis and fermentation.

Fructosans. These are hydrolysed by microbial enzymes to give fructose (see Figure 5.2) which may also be produced together with glucose by digestion of sucrose.

Microbial Breakdown of Simple Sugars

The monosaccharides glucose, fructose and xylose which result from the breakdown of polysaccharides are immediately taken up by the microorganisms from the rumen liquor and are degraded intracellularly (xylose after conversion to glucose; see Figure 5.2) and this is the reason why it is difficult to detect monosaccharides produced during the first stage of carbohydrate digestion. The energy released at each step of monosaccharide breakdown, i.e. conversion of sugar to pyruvic acid and the subsequent formation of volatile fatty acids, is generated as ATP according to the following summarizing equations:

$$\begin{aligned}
\text{Hexose} &\longrightarrow 2\text{ pyruvate} + 4\,(\text{H}) + 2\,\text{ATP}\\
\text{Pentose} &\longrightarrow 1.67\text{ pyruvate} + 1.67\,(\text{H}) + 1.67\,\text{ATP}\\
2\text{ Pyruvate} + 2\,\text{H}_2\text{O} &\longrightarrow 2\text{ acetic acid} + 2\,\text{CO}_2 + 2\,\text{H}_2 + 2\,\text{ATP}\\
2\text{ Pyruvate} + 8\,(\text{H}) &\longrightarrow 2\text{ propionic acid} + 2\,\text{H}_2\text{O} + 2\,\text{ATP}\\
2\text{ Pyruvate} + 4\,(\text{H}) &\longrightarrow \text{butyric acid} + 2\,\text{H}_2 + 2\,\text{CO}_2 + 2\,\text{ATP}
\end{aligned}$$

The ATP generated by these reactions serves as an energy source for maintenance and growth of microorganisms, particularly for the synthesis of microbial protein to the benefit of the microorganisms and to the host animal as well, since microbial protein serves as an important protein source for ruminant animals (see Chapter 8). Pairs of hydrogen atoms liberated on conversion of monosaccharides to pyruvic acid provide protons (H^+) and electrons which are picked up by NAD reducing it to NADH:

$$C_6H_{12}O_6 + 2\,NAD^+ \longrightarrow 2\,CH_3.CO.COOH + 2\,NADH + 2\,H^+$$

NADH transfers protons and electrons used for reduction reactions occurring in bacteria, such as the production of methane (see later), release of hydrogen, reduction of pyruvic acid, reduction of unsaturated fatty acids (see Chapter 7), reduction of sulphate to sulphide (see Chapter 10) and of nitrate to nitrite and ammonia (see Chapters 7 and 8).

Mechanisms for the Conversion of Pyruvic Acid to Volatile Fatty Acids

Summarizing equations of the second step of the fermentation of sugars leading from pyruvic acid to individual volatile fatty acids are given above. These pathways are summarized in Figure 5.3.

Production of Acetate. Oxidation of pyruvic acid to acetic acid occurs via acetyl-CoA or acetyl phosphate as intermediate stages. Groups of microorganisms differ in the nature of the fate of pairs of electrons removed in these reactions, or they are transferred to protons which are then liberated as molecular hydrogen or transferred to carbon dioxide and formate is produced.

$$\text{Pyruvate} + \text{CoA} + H^+ \longrightarrow \text{acetyl-CoA} + CO_2 + H_2$$

$$\text{Pyruvate} + P_i \longrightarrow \text{acetyl-P} + CO_2 + H_2$$

$$\text{Pyruvate} + \text{CoA} \longrightarrow \text{acetyl-CoA} + HCOO^-$$

$$\text{Pyruvate} + P_i \longrightarrow \text{acetyl-P} + HCOO^-$$

P_i = inorganic phosphate

Production of Propionate. Two pathways of propionic acid production in the rumen are known. The first involves carbon dioxide fixation to phosphoenolpyruvate to form oxaloacetate and reduction through malate and fumarate to succinate and finally liberation of carbon dioxide from succinate:

$$\underset{\text{Phosphoenolpyruvic acid}}{H_2C{=}C(OPO_3H_2)\cdot COOH} + CO_2 + ADP \longrightarrow \underset{\text{Oxaloacetic acid}}{HOOC\cdot CH_2\cdot CO\cdot COOH} + ATP$$

$$\underset{\text{Oxaloacetic acid}}{HOOC\cdot CH_2\cdot CO\cdot COOH} \xrightarrow{NADH_2} \underset{\text{Malic acid}}{HOOC\cdot CH_2\cdot CHOH\cdot COOH} \xrightarrow{-H_2O} \underset{\text{Fumaric acid}}{HOOC\cdot CH{=}CH\cdot COOH}$$

$$\xrightarrow{+H_2} \underset{\text{Succinic acid}}{HOOC\cdot CH_2\cdot CH_2\cdot COOH} \xrightarrow{-CO_2} \underset{\text{Propionic acid}}{CH_3\cdot CH_2\cdot COOH}$$

The transformation of succinic acid to propionic acid by microbial enzymes occurs through some intermediate steps which will not be described here (see Chapters 10 and 12).

The second pathway, called the reductive or acrylate pathway, is the more important in the rumen of animals fed rations high in grains. Pyruvate is reduced directly to propionate, via the coenzyme A derivatives of lactate and acrylate:

$$\underset{\text{Pyruvic acid}}{CH_3.CO.COOH} \xrightarrow{+H_2} \underset{\text{lactic acid}}{CH_3.CHOH.COOH} \longrightarrow$$

$$\underset{\text{lactyl-CoA}}{CH_3.CHOH.CO.CoA} \xrightarrow{-H_2O} \underset{\text{acryl-CoA}}{CH_2{=}CH.CO.CoA} \xrightarrow{+H_2}$$

$$\underset{\text{propionyl-CoA}}{CH_3.CH_2COCoA} \longrightarrow \underset{\text{propionic acid}}{CH_3.CH_2.COOH}$$

Lactate serving as substrate for propionate formation occurs in significant amounts in the rumen under conditions of high-grain feeding

Production of Butyrate. Two pathways are available for butyrate synthesis from acetate in ruminal organisms. The first pathway is the reversal of β-oxidation according to the summarizing formula (see biochemical texts):

$$2\,CH_3.COOH + 2\,ATP + 2\,NADH_2 \longrightarrow CH_3.CH_2.CH_2.COOH + 2\,ADP + 2\,P_i + 2\,NAD$$

The biosynthesis of butyrate from acetate may serve also the purpose of oxidizing reduced NADH to NAD which may allow further fermentation of glucose to proceed (see above):

$$\text{Glucose} + 2\,NAD + 2\,ADP + 2\,P_i \longrightarrow 2\text{ pyruvate} + 2\,NADH_2 + 2\,ATP$$

A second pathway for the conversion of acetate involving malonyl-CoA (like the synthesis of long-chain fatty acids; see Chapter 6 and biochemical texts) occurs according to the following scheme:

$$\text{Acetyl-CoA} + CO_2 + ATP \longrightarrow \text{malonyl-CoA} + ADP + P_i$$

$$\text{Malonyl-CoA} + \text{acetyl-CoA} \longrightarrow \text{acetoacetyl-CoA} + \text{CoA}$$

The Absorption of Volatile Fatty Acids through the Rumen Wall and their Metabolism within the Rumen Epithelial Cells

The major part (about 75%) of the volatile fatty acids are absorbed directly from the reticulorumen into the blood, about 20% from abomasum and omasum, and only 5% pass with the digesta to the small intestine to be absorbed there into the blood. The absorption of the volatile fatty acids is passive, i.e. the extent and rate depend on the difference between their concentration in the rumen liquid and that in the epithelial cells or blood. The rates at which the volatile fatty acids are absorbed from the rumen increase as the pH of the rumen fluid decreases. This has led to the concept that the undissociated acids pass through the rumen wall more rapidly than the anions.

Butyric acid and some of the propionic acid produced in the rumen are converted to β-hydroxybutyric acid and lactic acid, respectively, in their passage across the rumen wall; this occurs due to the presence of the necessary enzymes in the epithelial cells.

$$\underset{\text{Propionic acid}}{CH_3.CH_2.COOH} \longrightarrow \underset{\text{lactic acid}}{CH_3.CHOH.COOH}$$

$$\underset{\text{Butyric acid}}{CH_3.CH_2.CH_2.COOH} \longrightarrow \underset{\beta\text{-hydroxybutyric acid}}{CH_3.CHOH.CH_2.COOH}$$

The fact that the rumen epithelium actively metabolizes butyric acid, influences positively the rate of its absorption. β-Hydroxybutyric acid may then be used as a source of energy by a number of tissues, such as skeletal muscles or liver, where the remaining steps of energy production take place (see Chapter 22).

Methane Production

Methane formed in the rumen appears to be a necessary by-product of anaerobic sugar fermentation. Considering the high energy value of methane, its expulsion is an unfortunate but inevitable loss of energy from the carbohydrates fed to ruminants (see Chapter 14). Rumen methane is produced mainly by reduction of carbon dioxide with hydrogen gas, a reaction performed by all species of rumen bacteria according to the following equation:

$$4\,H_2 + HCO_3^- + H^+ \longrightarrow CH_4 + 3\,H_2O$$

Carbon dioxide originates from the conversion of pyruvate to acetate (see above). The breakdown of sugars to pyruvic acid and the subsequent conversion of pyruvic acid to acetate are the sources of hydrogen. Smaller amounts of hydrogen and carbon dioxide are formed from formic acid by means of the enzyme formic dehydrogenase:

$$HCOOH \longrightarrow CO_2 + H_2$$

Formic acid is produced in the rumen by the conversion of pyruvic acid to acetate. The mixture of volatile fatty acids in the rumen contains about 1% formic acid. Hydrogen gas is very rapidly used by methane-producing bacteria and it functions also as the hydrogen donor in other reductive microbial processes (see above). Hydrogen is therefore maintained at very low partial pressure in the rumen, in order to assure favourable conditions for the growth of microorganisms. The following three additional processes of methane formation are performed by certain species of rumen bacteria. (1) Formate is readily decomposed to methane and carbon dioxide:

$$HCOO^- + H_2O \longrightarrow CH_4 + HCO_3^-$$

(2, 3) Further precursors of methane of minor importance are acetate and methanol (formed by pectin degradation; see Chapter 4).

$$CH_3COO^- + H_2O \longrightarrow CH_4 + HCO_3^-$$

$$4\,CH_3OH \longrightarrow 3\,CH_4 + HCO_3^-$$

Production of methane by microorganisms is a complicated biochemical process in which folic acid and vitamin B_{12} participate (see Chapter 12).

As methane cannot be metabolized by the animal, its production is connected with considerable loss of energy; therefore efforts have been made to find means of lowering methane production and thus diverting the energy to compounds metabolizable by the animal. For this purpose it has been suggested that unsaturated fatty acids are added to the ration; these then become saturated by the influence of the microorganisms. Hydrogen is needed for this and this is a good way to reduce the amount of methane formed in the rumen and to induce a desired rise in propionate. There are other feed additives which reduce methane production: chloroform, chloral hydrate and copper salts; however, these chemicals were found not to be specific inhibitors of methane production, but to repress growth of microorganisms in general. Overwhelming production of gas (methane, hydrogen and carbon dioxide) beyond the limits of eructation leads to bloat. Exaggerated amounts of free sugars are present, for example, in young lush pasture or in sugar beet or their offal. Bloat is a potentially fatal condition (see Chapter 24).

Intestinal Digestion of Carbohydrates by Ruminants

Fermentation in the rumen is by far the most important means whereby ruminants digest and utilize carbohydrates. However, minor portions of ingested carbohydrates (starch, cellulose and hemicellulose) and the cellular polysaccharides of microorganisms themselves escape fermentation in the rumen and pass with the digesta to the lower tract; feed and bacterial carbohydrates not degraded in the rumen account together for 10–20% of the carbohydrates ingested. Bacterial polysaccharides are lost from the microorganisms before the bacteria are killed by the acid conditions in the abomasum and undergo digestion in the intestine. The amounts of cellulose or starch resistant to breakdown in the rumen are influenced by the nature of the feedstuffs and their processing. It is remarkable that 20% of the starch in ground maize is digested in the small intestine by enzymes, as in monogastric animals. Digestion of starch in the intestine enables a more efficient utilization of energy than in its ruminal breakdown since losses as methane or heat which occur in the rumen by fermentation are avoided.

Residual cellulose, hemicellulose and starch passing the ileocaecal valva are fermented in the caecum to volatile fatty acids, carbon dioxide and methane by pathway similar to those occurring in the rumen. Volatile fatty acids formed in the caecum of ruminants or horses are absorbed into the circulation and utilized in tissues in the same way as those produced in the rumen.

NUTRITIONAL IMPORTANCE OF VOLATILE FATTY ACIDS FORMED IN THE RUMEN

Acetic, propionic and butyric acids, the main end-products of the metabolism of carbohydrates in the rumen, are the major sources of energy and fulfil important biosynthetic functions in ruminants.

Efficient energy trapped in the high-energy compound ATP is obtained in two stages of the metabolism of volatile fatty acids: (1) at the microbial transformation of glucose to VFA, and (2) by oxidation of absorbed volatile fatty acids which occurs inside the tissues. In addition, the volatile fatty acids absorbed into the body fulfil biosynthetic functions: acetic and butyric acid are starting materials for the biosynthesis of body and milk fat, and propionic acid for glucose (see later in this chapter and Chapter 21). Knowledge of the amounts of volatile fatty acids present in the rumen is of paramount importance with regard to the different energetic efficiencies of each individual volatile fatty acid and its biosynthetic functions. The amounts of volatile fatty acids produced in the rumen during 24 hours are very high: 3–4 kg in the reticulorumen of a cow and 300–400 g in that of a sheep. About an additional 10% of these amounts is produced in the large intestine and also utilized by the host animal (see above). The amounts of volatile fatty acids formed in the rumen and the relative proportions of the three principal acids formed—acetic, propionic and butyric acids—vary widely and are affected by many dietary factors such as composition of the diet, particularly the ratio between concentrate feeds and roughage, the physical form of feedstuffs (e.g. particle size), level of intake and frequency of feeding. The quantities and relative proportions of the volatile fatty acids present in the rumen depend on the type of fermentation, which in turn is determined by the microbial population which develops as a direct result of nutrition. Feeding starch-rich diets such as diets high in grains causes an increased propionate production, whereas the same absolute amounts of acetate are produced in high grain-fed ruminants as in forage-fed animals. Acetic acid is the predominant volatile fatty acid produced under all feeding conditions, but it is the ratio of acetic acid to propionic acid that varies with dietary treatments. Feeding starch-high diets causes a rise in the number of amylolytic (starch-digesting) microbes and therefore a rise in the proportion of propionic acid to other acids (acetic + butyric) present in the rumen content. Such nutritional treatments which cause a rise in the proportion of propionic acid to acetic acid result in a more efficient utilization of the feeding stuffs for meat production (see Chapter 18). In order to attain this purpose it was recently suggested to use feed additives of microbial origin such as monensin (see Chapter 18). On the other hand, diets rich in roughage (high in cellulose) lead to an increased ratio of acetic acid to propionic acid, which, in the case of lactating cows, is accompanied by an increase in the fat content of the milk. Grinding and pelleting of diets, particularly of those high in concentrates, causes a further increase in the proportion of propionic acid.

The pattern of volatile fatty acids in the rumen provides a relatively simple way of characterizing the fermentation and efficiency of rations. The generally accepted method to express levels of the volatile fatty acids is as their molar proportions (in per cent) on the basis of their gas-chromatographic determination. Such data are presented in Table 5.1.

Apart from the three main volatile fatty acids, the rumen liquor contains small amounts of acids with 5–6 C atoms such as valeric, isovaleric and caproic acids (see above), as well as formic acid and lactic acid. These latter two acids undergo further metabolic conversions in the rumen; formic acid is transformed into carbon dioxide and hydrogen (see above) and lactic acid may be transformed into propionic and butyric acid.

To determine the total daily output of volatile fatty acids, the rumen volume and turnover of the respective acids, i.e. interconversion of one acid into another, must be known. The volume of the rumen fluid varies because of the influx of water with the saliva and passage of water to the blood. Measurements of volatile fatty acid production have been made using isotope dilution techniques in which continuous infusions of labelled acids were made and the specific radioactivity of the rumen volatile fatty acids measured. These data show that appreciable amounts of butyrate are derived from acetate as a result of the condensation of 2 mol of acetate (see above). However, production of volatile fatty acids has been shown to be highly related to their concentration. The following equation was derived to estimate production:

$$Y = 0.068X - 1.75$$

where Y is the total production (mol/day), and X the total concentration (mmol/l), of volatile fatty acids in rumen liquor. The existence of such a relationship justifies the common use of volatile fatty acid concentration as a parameter for expressing the efficiency of diets, as mentioned.

Table 5.1 Volatile fatty acids (VFA) in the rumen liquor of milk cows fed on various diets (after analyses carried out by Mr A. Halevi, Rehovot)

Ration	Individual VFA (molar %)					
	Acetic	Propionic	Butyric	Valeric	Isovaleric	Caproic
90% concentrates + 10% straw	50.1	35.1	9.3	1.3	3.7	0.5
90% concentrates + 10% hay	45.7	39.5	8.8	1.2	4.1	0.7
70% concentrates + 30% hay	56.0	29.6	9.9	1.2	3.0	0.3
40% concentrates + 60% hay	66.1	20.2	9.9	1.3	2.1	0.3

Rumen concentrations of volatile fatty acids are regulated by a balance between production and absorption. The concentrations rise after a meal and consequently the pH drops. The peak of fermentation is about 4 hours after feeding on a hay diet, but occurs sooner with diets containing much concentrates. To ensure optimal growth of microorganisms, the rumen pH is maintained between 6.0 and 7.0 by the high buffering capacity of saliva and the removal of volatile fatty acids by absorption. However, this mechanism for buffering the pH in the rumen is not sufficient after feeding larger amounts of readily digestible carbohydrates (sugar or starch). Then, greater amounts of propionic and butyric acids and significant amounts of lactic acid (which is a stronger acid than propionic acid) are formed and the pH drops to 4.5–5. The lower pH inhibits the growth of cellulolytic bacteria and thus the digestibility of roughage is depressed when it is fed together with large amounts of concentrate feeds. Serious cases of overproduction of acids will damage the rumen mucosa (see Chapter 24).

METABOLIC PATHWAYS OF VOLATILE FATTY ACIDS OCCURRING INSIDE ANIMAL TISSUES

As stressed before, volatile fatty acids absorbed from the reticulorumen pass to the tissues where they undergo oxidation and provide energy for the animal or serve for the biosynthesis of fat or glucose. The proportional amount of each acid used for each of these purposes varies with the individual acid. Whereas 50% of the acetic acid present in the tissues of milk cows is oxidized, two-thirds of the butyric acid and only one-quarter of the propionic acid undergo oxidation. The metabolism of propionic and butyric acids occurs exclusively in the liver; however, 60% of acetic acid is metabolized in peripheral tissues (muscles and adipose tissue) and only 20% in the liver. In lactating animals acetic acid serves for the synthesis of milk fat in the udder (see Chapter 21).

The principles of the processes of oxidation of the three main volatile fatty acids and the yields of ATP obtained are given below.

Propionic Acid

There exist two pathways for the oxidation of propionic acid: (1) oxidation after conversion to glucose, and (2) direct oxidation of propionic acid.

(1) The mechanism of the conversion of propionic acid to glucose by gluconeogenesis (Figure 5.4) is also utilized when glucose serves other purposes than energy supply by its oxidation, such as in the formation of glycogen or lactose. The enzyme which catalyses the conversion of methylmalonyl-CoA to succinyl-CoA requires coenzyme B_{12}. Vitamin B_{12}, which contains cobalt, is synthesized by rumen microorganisms; since this vitamin is essential for gluconeogenesis, ruminants are particularly affected by cobalt deficiency (see Chapter 10). The reactions leading from succinic acid to oxaloacetic acid are

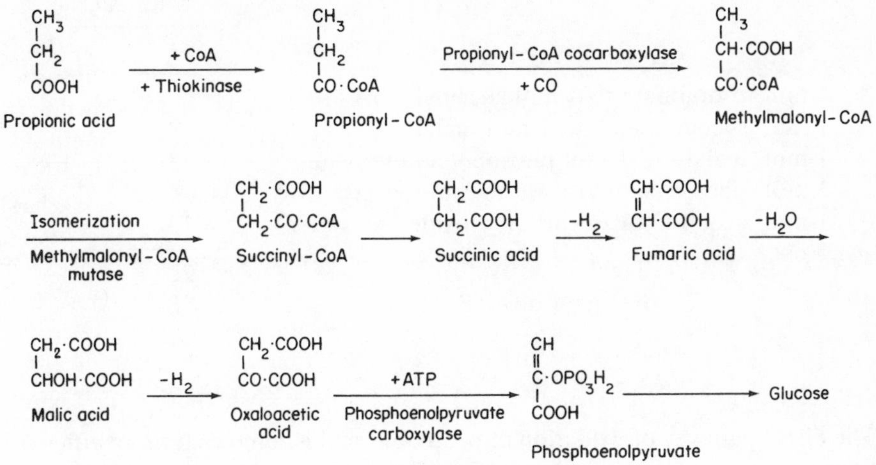

CH_3–CH_2–COOH (Propionic acid) —(+ CoA, + Thiokinase)→ CH_3–CH_2–CO·CoA (Propionyl-CoA) —(Propionyl-CoA cocarboxylase, + CO)→ CH_3–CH·COOH–CO·CoA (Methylmalonyl-CoA)

—(Isomerization, Methylmalonyl-CoA mutase)→ CH_2·COOH–CH_2·CO·CoA (Succinyl-CoA) → CH_2·COOH–CH_2·COOH (Succinic acid) —($-H_2$)→ CH·COOH=CH·COOH (Fumaric acid) —($-H_2O$)→

CH_2·COOH–CHOH·COOH (Malic acid) —($-H_2$)→ CH_2·COOH–CO·COOH (Oxaloacetic acid) —(+ATP, Phosphoenolpyruvate carboxylase)→ CH=C·OPO_3H_2–COOH (Phosphoenolpyruvate) → Glucose

Figure 5.4. Conversion of propionic acid to glucose (gluconeogenesis).

identical with the respective intermediate stages of the tricarboxylic acid cycle (see Chapter 4). The liver of ruminants contains highly active enzymes needed for the conversion of oxaloacetic acid to phosphoenolpyruvic acid and of the latter to glucose. The pathway leading from phosphoenolpyruvic acid to glucose is the reverse of glycolysis. The yield of ATP obtained by the oxidation of propionic acid via glucose is 17 mol of ATP per mol of propionic acid, as shown in the following balance sheet:

	Mol ATP +	Mol ATP –
2 mol propionate to 2 mol succinyl-CoA		6
2 mol succinyl-CoA to 2 mol malate	6	
2 mol malate to 2 mol phosphoenolpyruvate	6	2
2 mol phosphoenolpyruvate to 1 mol glucose		8
1 mol glucose to carbon dioxide and water	38	
Total	50	16
Net gain of ATP	34	

(2) The pathway for the direct oxidation of propionic acid is the same as for its oxidation via glucose as far as phosphoenolpyruvate. The further breakdown occurs via pyruvate, acetyl-CoA and the tricarboxylic acid cycle, as outlined in Chapter 4. The balance sheet for this process is as follows:

	Mol ATP +	Mol ATP −
1 mol propionate to 1 mol succinyl-CoA		3
1 mol succinyl-CoA to 1 mol malate	3	
1 mol malate to 1 mol phosphoenolpyruvate	3	1
1 mol phospoenolpyruvate to 1 mol acetyl-CoA	4	
1 mol acetyl-CoA to carbon dioxide and water	12	
Total	22	4
Net gain of ATP	18	

The direct pathway of oxidation of propionic acid is somewhat more efficient than via glucose (18 ATP versus 17 ATP).

Butyric Acid as an Energy Source

Butyric acid formed in the rumen is almost quantitatively converted to β-hydroxybutyrate in its passage through the ruminal wall (see above). The further reactions involved in energy production, and proceeding mainly in muscle tissue, are dehydrogenation of hydroxybutyrate to acetoacetate and splitting of the latter compound to 2 mol of acetyl-CoA, which is finally fully oxidized via the tricarboxylic acid cycle (see Chapter 4). The balance sheet for the oxidation of butyric acid is the following:

	Mol ATP +	Mol ATP −
1 mol butyrate to 1 mol hydroxybutyrate	5	5
1 mol β-hydroxybutyrate to 2 mol acetyl-CoA	3	2
2 mol acetyl-CoA to carbon dioxide and water	24	
Total	32	7
Net gain of ATP	25	

Acetic Acid as an Energy Source

Acetic acid is first converted to acetyl-CoA by use of two high-energy phosphate bonds as follows:

$$CH_3.COOH + CoA + ATP \longrightarrow CH_3.CO.CoA + AMP + P\text{–}P + H_2O$$

Acetyl-CoA is then oxidized via the tricarboxylic acid cycle yielding 12 ATP per mol of acetate; therefore the net yield is 10 ATP per mol of acetate oxidized.

GLUCOSE METABOLISM IN RUMINANTS

The greatest part of the carbohydrates ingested by ruminants is fermented in the reticulorumen; the volatile fatty acids formed serve in these animals as the main source of energy, unlike monogastric species in which it is the glucose originating from food that fulfils this function. Nevertheless, glucose participates also in ruminants in important metabolic processes (see later). However, the small amount of glucose absorbed from the intestine of ruminants (see above) is insufficient and the prevailing amount of glucose used by these animals is produced in their tissues by biosynthesis (gluconeogenesis) (see above), mainly in the liver and to a small extent in the kidneys. Of the glucose present in ruminants 40–60% originates from propionic acid (see above), about 20% from protein (amino acids absorbed from the digestive tract) and the remainder from branched volatile fatty acids, lactic acid and glycerol. Gluconeogenesis is of prime importance in ruminants. The various enzymes needed for gluconeogenesis are abundantly present in the liver, particularly those needed for the conversion of propionic acid to glucose (see Figure 5.3). These enzymatic activities are very low in the livers of suckling animals, which process their food like monogastric animals. The enzymatic activities increase during weaning.

Metabolic Functions of Glucose in Ruminants

Glucose fulfils the following functions in the ruminant body:

(1) Although it is a minor source of energy in the whole ruminant body, glucose is the main source of energy in nervous tissue, particularly in the brain, and also in red blood corpuscles.

(2) Glucose is required for the metabolism of muscles and for the production of glycogen, serving as an energy store in muscles and liver.

(3) The requirement for biosynthetically formed glucose increases during lactation (see Chapter 21) and in late pregnancy (see Chapter 19). Glucose is the main precursor of lactose and glycerol (as a component of milk fat) and serves as a supply of nutrients for the fetus. Indeed the amounts of glucose required by lactating and late-pregnant ruminants increase considerably beyond that required by non-producing animals. The amounts of glucose utilized by one animal during 24 hours are: ewe, neither lactating, nor pregnant: 100 g; pregnant ewe: 180 g; lactating ewe: 320 g; dry cow: 500 g; and high producing cow: 4–6 kg.

(4) Apart from its function as a precursor of glycerol, glucose is necessary for the formation of NADPH, which in turn is required for the synthesis of long-chain fatty acids by reduction of acetate (see Chapters 6 and 21). NADPH originates from glucose oxidation via the pentose-phosphate pathway.

It is important to note that in ruminants the pathway most common in monogastric species for the conversion of glucose into long-chain fatty acids is almost non-existent. Ruminants are adapted to the very economic utilization of glucose, and their tissues lack the enzymes which in monogastric animals allow the conversion of glucose into long-chain fatty acids.

Control of Glucose Metabolism

As mentioned above, glucose provides a relatively small part of the energy requirement of ruminants which under normal conditions is mainly covered by acetate, about 70% on the average, and when energy requirements increase long-chain fatty acids are used as an additional energy source. Despite this, ruminants utilize almost as much glucose on a body-weight basis as do other species; this may be concluded from the similar turnover rates.

The rate of gluconeogenesis in ruminants increases after feed ingestion when the supply of propionic acid and other glucose precursors is increased, and this rate decreases after feed restriction. This is in contrast with monogastric species where hepatic glucose production increases markedly during fasting. Glucose entry rate in fasting ruminants is about 60–65% of that in fasting non-ruminants, but that in fed ruminants falls by only 10–20% below the entry rate in fed monogastric species.

Insulin and glucagon are regulators of glucose disposal and production also in ruminants and they control glucose homeostasis, particularly in the blood. The role of both hormones in ruminants is less important than in species absorbing large amounts of glucose. Insulin reduces glucose production from propionic acid and other glucose precursors and enhances its utilization by peripheral tissues. Glucagon counteracts the effects of insulin; it promotes gluconeogenesis from glucose precursors and the release of glucose from liver glycogen. The secretion of insulin and of glucagon responds to circulating metabolites (glucose, volatile fatty acids) and to food intake. The relationship between both hormones seems to be of greater importance in maintaining glucose homeostasis than is the absolute plasma concentration.

In addition to insulin and glucagon, which influence glucose metabolism in inverse directions, growth hormone seems to be involved in the maintenance of glucose homeostasis and acts similarly to glucagon. Whether the effects of growth hormone on glucose availability are direct (promoting gluconeogenesis) or indirect (sparing glucose utilization) has not yet been ascertained.

BIBLIOGRAPHY

Symposium on Carbohydrate Metabolism in the Ruminant (1970). In *Symposium on Ruminant Physiology, Digestion and Metabolism.* (A. T. Phillipson, ed.) Third Symposium, pp. 406–455. Newcastle upon Tyne: Oriel Press.

Symposium on carbohydrate metabolism in the ruminant (1979). *Proc. Nutr. Soc.*, **38**, 269–314.

Armstrong, D. G., and Smithard, R. R. (1979). The fate of carbohydrates in the small and large intestines of the ruminant. *Proc. Nutr. Soc.*, **38**, 283–294.

Baldwin, R. L., and Allison, M. J. (1983). Rumen metabolism. *J. Anim. Sci.*, **57**, Suppl. 2, 461–477.

Ballard, F. I., Hanson, R. W., and Kronfeld, D. S. (1969). Gluconeogenesis and lipogenesis in tissues from ruminant and non-ruminant animals. *Fed. Proc.*, **28**, 218–231.

Bergman, E. N. (1973). Glucose metabolism in ruminants as related to hyperglycemia and ketosis. *Cornell Vet.*, **63**, 341–382.

Black, A. L. (1969). Carbohydrate metabolism. In *Handbook of Nutrition* (W. Lenkeit, K. Breirem and E. Crasemann, eds), Vol. 1, pp. 415–449. Hamburg and Berlin: Paul Parey.

Bryant, M. P. (1979). Microbial methane production–theoretical aspects. *J. Anim. Sci.*, **48**, 193–201.

Czerkawski, J. W. (1984). Microbial fermentation in the rumen (model system in nutritional research). *Proc Nutr. Soc.*, **43**, 101–108.

Lindsay, D. B. (1978). Gluconeogenesis in ruminants. *Biochem. Soc. Trans.*, **6**, 1152–1156.

Russel, J. B., and Hespell, R. B. (1981). Microbial rumen fermentation. *J. Dairy Sci.*, **64**, 1153–1169.

Sutton, J. D. (1980). Digestion and end-product formation in the rumen from production rations. In *Digestive Physiology and Metabolism in Ruminants* (Y. Ruckebush and P. Thivend, eds), pp. 271–308. Lancaster: MTP Press.

Young, J. W. (1977). Gluconeogenesis in cattle. Significance and methodology. *J. Dairy Sci.*, **60**, 1–13.

CHAPTER 6

*Lipids and their Significance in the Nutrition of Monogastric and Ruminant Animals**

Lipids are organic compounds found in plant and animal tissues, and are oily or greasy substances soluble in organic solvents such as benzene, ether or chloroform, but only sparingly soluble in water. Substances of the latter type are designated as 'hydrophobic', in contrast to materials that are soluble in water or wettable by water, and hence 'hydrophilic'. In routine feed analysis, all kinds of lipids are determined together as the ether extract (see Chapter 1). There are different classes of lipids which are listed in Table 6.1.

Table 6.1 Classification of lipids

Saponifiable		Non-saponifiable
Simple	Compound	
Fats Waxes	Phospholipids Sphingolipids Glycolipids Lipoproteins	Steroids Carotinoids Fat-soluble vitamins

CHEMICAL STRUCTURE AND PROPERTIES OF LIPIDS

Simple Saponifiable Lipids

Fats (Triglycerides)

From the standpoint of the amounts present in the animal body and its

* In collaboration with P. Budowski.

food, fats make up the largest fraction of lipids (about 98% of the lipids present in most concentrate feedstuffs). Fats are important sources of stored energy in plants and animals as well, and are characterized by their high energy value. One gram of fat yields 9.3 kcal (= 39.1 kJ) of heat when completely combusted compared with about 3.7 kcal (= 15.5 kJ) for 1 g of carbohydrate (see Chapter 14). Fats are esters of fatty acids formed with the trihydric alcohol glycerol, and are called triglycerides. The carbon atoms of glycerol are identified by numbers 1, 2 and 3, or by Greek letters α, β and α'.

$$
\begin{array}{ll}
1\,(\alpha) & CH_2.O.CO.R_1 \\
 & \quad | \\
2\,(\beta) & CH.O.CO.R_2 \\
 & \quad | \\
3\,(\gamma) & CH_2.O.CO.R_3
\end{array}
$$

Fatty acids are long-chain organic acids, having 4–24 carbon atoms (mostly 16 or 18) and a single carboxyl group. The same or different fatty acids may be in all three positions. The chain length and degree of unsaturation of the fatty acids making up the triglyceride determine its physical and chemical properties. Triglycerides of saturated fatty acids containing at least ten carbon atoms are solid at room temperature, whereas those with less than ten carbon atoms are liquids. In oils the percentage of unsaturated long-chain fatty acids surpasses that of saturated fatty acids.

The acids found in fats usually have an even number of carbon atoms, which is to be expected in view of their mode of biosynthesis (see later). The fatty acids most frequently occurring in fats are:

$C_{16}H_{32}O_2$	palmitic acid
$C_{18}H_{36}O_2$	stearic acid
$C_{18}H_{34}O_2$	oleic acid

Oleic acid contains one double bond in the middle of the chain:

$$CH_3.(CH_2)_7.CH{=}CH.(CH_2)_7.COOH$$

Oleic acid containing the one double bond is a monoenoic acid, unsaturated acids with two double bonds are called dienoic acids, and those with many double bonds polyenoic or polyunsaturated acids (for their importance as essential fatty acids, see later).

Chemical Properties of Fats (Triglycerides). Fats can be hydrolysed by boiling with alkali; then glycerol and water-soluble alkali salts of fatty acids (soaps) are formed. This hydrolysis is thus called saponification:

$$
\begin{array}{lcll}
CH_2.O.COR & & CH_2OH & \\
| & & | & \\
CH.O.COR & + 3\,KOH \longrightarrow & CH.OH & + 3\,R.COOK \\
| & & | & \quad \text{soap} \\
CH_2.O.COR & & CH_2OH & \\
\text{Fat} & & \text{glycerol} &
\end{array}
$$

Triglycerides also undergo hydrolysis under the influence of enzymes called lipases which are secreted into the small intestine by the pancreas. Lipases are also derived from bacteria and moulds (see later). Triglycerides containing largely unsaturated fatty acids, which are thus liquid at room temperature, can be converted into solid fats by hydrogenation of their double bonds. Oleic acid, for example, yields stearic acid:

$$
\begin{array}{ccccc}
CH_3 & & & & \\
| & & & & \\
(CH_2)_7 & & & & \\
| & & & & CH_3 \\
CH & & H & & | \\
\| & + & | & \longrightarrow & (CH_2)_{16} \\
CH & & H & & | \\
| & & & & COOH \\
(CH_2)_7 & & & & \text{stearic acid} \\
| & & & & \\
COOH & & & & \\
\text{Oleic acid} & & & &
\end{array}
$$

This process is widely employed for converting liquid oils into margarine. Rumen microorganisms enzymatically hydrogenate unsaturated fatty acids ingested with the diet (see later).

The fatty acid composition of fats of animal and plant origin is given in Tables 6.2 and 6.3. The composition of fats present in the animal body depends on the species of animal, and varies in the different tissues; also it is influenced to some extent by the diet. Fats of plants, fish and birds are more highly unsaturated than those of mammals. For example, subcutaneous fat contains more saturated fatty acids than the fat from the liver of the same mammalian species, which is richer in unsaturated fatty acids. Fats in the ruminant body, particularly tallow, are highly saturated. This is due to considerable hydrogenation of unsaturated dietary fats occurring in the rumen. Ruminant milk fat contains a considerable percentage of short-chain fatty acids (2–8 C atoms). Therefore, milk fat is softer than the body fats of these animals. Ruminant lipids contain small amounts of odd-length and branched-chain fatty acids which are products of microbial metabolism (see Chapter 8). Table 6.3 shows the high content of unsaturated fatty acids present in oils extracted from seeds.

Table 6.2 Composition (%) of fat in some animal species

	Saturated fatty acids		Unsaturated fatty acids			Consistency
	C_{14} and less	C_{16} + C_{18}	Oleic acid	Linoleic + linolenic acid	C_{20} and more	
Fish oil		11–15		20	50	Liquid
Chicken fat		25	40	25–30		Soft
Pork fat		35	50	5–7		Soft
Butter	15	30–40	35	0.50		Soft
Beef fat		50	35	0.5		Solid

Table 6.3 Composition (%) of fat in seeds

	Saturated fatty acids		Unsaturated fatty acids			Consistency
	C_{14} and less	C_{16} + C_{18}	Oleic acid	Linoleic acid	Linolenic acid	
Linseed oil		6–16	13–36	10–25	30–50	Liquid
Soybean oil		7–10	23–30	50–60	5–9	Liquid
Sunflower seed oil		6–15	20–50	30–60		Liquid
Coconut oil	80	11	5–7			Solid
Cottonseed oil		24–29	15–20	49–57		Liquid

Essential Fatty Acids

The formulae of three polyunsaturated fatty acids originally considered as essential to the rat are given here:

Linoleic acid	$C_{18}H_{32}O_2$
Linolenic acid	$C_{18}H_{30}O_2$
Arachidonic acid	$C_{20}H_{32}O_2$

These acids contain the characteristic group $-CH{=}CH{-}CH_2{-}CH{=}CH-$, which is two double bonds separated by a methylene group.

Complete removal of fat from the diet of rats results in poor growth and skin lesions. These symptoms can be reversed by small amounts of essential fatty acids. Arachidonic acid is formed in the liver of mammalians and birds from linoleic acid and is therefore required in the diet only if linoleic acid is not available. Like other polyunsaturated fatty acids, arachidonic acid forms part of the structural lipids of cell membranes and is a source material for the biosynthesis of prostaglandins and thromboxanes, hormone-like substances which regulate many different cell functions, particularly in reproductive organs of man and animals. Linolenic aid is essential for some species of fish, but its role in mammals has not been clearly established. Because linoleic acid and its higher homologues are normally present in high concentration in phosphoglycerides (see later) of the central and peripherous nervous system, they are thought to play a vital role there.

Essential fatty acids are required by chicks, calves and goats. Only linoleic acid is an essential nutrient for chicken. The importance of linoleic acid for chick growth and for maximum egg size will be discussed in Chapter 20. There is seldom a deficiency of essential fatty acids in practical diets. Excessive amounts are not desirable because of the susceptibility of polyunsaturated acids to oxidation (see later).

Waxes

These are simple lipids consisting of a long-chain fatty acid esterified with monohydric alcohol of high molecular weight. They are not readily digested by animals and have no nutritive value. They occur as protective coatings in plants and animals; wool and feathers are protected against water by the hydrophobic nature of the wax coating.

Compound Saponifiable Lipids

Compound lipids are esters of fatty acids containing hydrophilic polar groups (nitrogen base, sugars) in addition to hydrophobic residues of fatty acids and alcohols. This group of lipids, also called polar lipids, includes phospholipids, sphingolipids, glycolipids and lipoproteins.

Phospholipids (phosphatides) are esters of glycerol in which two of the hydroxyl groups of glycerol are esterified by long-chain fatty acids. The third is esterified by phosphoric acid. The most commonly occurring phospholipids in animals and plants are the lecithins in which the phosphoric acid is also esterified by the nitrogen base choline. The general formula for lecithin is:

$$\begin{array}{l} HOCH_2.CH_2.N(CH_3)_3 \\ \qquad\qquad\quad | \\ \qquad\qquad\quad OH \end{array}$$

Choline

$$\begin{array}{lll} & CH_2.O.CO.R & \\ & | & \\ R'.CO.O. & CH & \;\; O \\ & | & \;\; \| \\ & CH_2.O. & P.OCH_2.CH_2.N(CH_3)_3 \\ & & | \\ & & O^- \end{array}$$

Lecithin (1,2-diacylglycero-3-phosphorylcholine)

Due to the presence of the hydrophilic phosphate group and the hydrophobic fatty acid chains within the same molecule, the phospholipids exert emulsifying properties and fulfil important functions in lipid transport in blood and as components of animal cell membranes. Phospholipids are abundant in heart, kidney and nervous tissues. The phospholipids in these tissues are higher in unsaturated fatty acids than the triglycerides of adipose tissue. The lipids of soybean seeds also contain lecithin. Lecithin isolated from soybeans is used as an emulsifying agent in milk substitutes for calves (Chapter 18).

Sphingolipids contain instead of glycerol the amino alcohol sphingosine to which are added a fatty acid, phosphate and choline. Like the lecithins they are surface active and are components of membranes, particularly in nervous tissue.

Glycolipids are the main lipid component of forages. In these compounds two of the hydroxy groups of glycerol are esterified by fatty acids, largely linoleic acid. One or two moles of galactose are attached to the C-3 of the glycerol:

Monogalactosyl diglyceride	Digalactosyl diglyceride
Galactosyl.O.CH_2	Galactosyl-galactosyl.O.CH_2
|	|
Acyl.O.CH	Acyl.O.CH
|	|
Acyl.O.CH_2	Acyl.O.CH_2

The lipid content of leaf tissues ranges between 3% and 10% of the dry weight; as mentioned, the leaf lipids are predominantly (about 60%) galactolipids whereas triglycerides, which are the main lipids of concentrate feeds and seeds, form only a negligible fraction of leaf lipids. Other lipid fractions present in leaf lipids are waxes, chlorophyll and other pigments and essential oils; all of them lack nutritive value.

Lipoproteins are lipids associated with specific proteins. They play a basic role in the transport of lipids via the blood from the small intestine to the tissues (see later). The lipoproteins of blood plasma are classified on the basis of their density which in turn is a reflection of their lipid content.

Non-saponifiable Lipids

Non-saponifiable lipids, which do not contain fatty acids and cannot form soap, include steroids, terpenes, carotinoids and fat-soluble vitamins. *Steroids* are a large group of physiologically important compounds in plants and animals. All are derived from the cyclopentanophenanthrene ring:

A B C D

Perhydrocyclopentanophenanthrene

CH_3 CH_3
$CH-CH_2-CH_2-CH_2-CH$
CH_3
HO 3 5 6 17

Cholesterol

The most abundant steroids are the sterols, which are steroid alcohols. Cholesterol, the most important animal sterol, is a component of the cell membrane. For a discussion on a possible causal relationship between nutritional conditions and cholesterol levels in blood and tissue, see later.

Terpenes are plant components which have characteristic odours and flavours. They are substances that yield isoprene

$$H_2C{=}\overset{\displaystyle CH_3}{\dot{C}}{-}CH{=}CH_2$$

upon degradation. They represent only a small part of the ether extract of some forages and provide no energy to the animal.

Carotinoids, plant pigments and fat-soluble vitamins are also non-saponifiable lipids; these will be discussed in Chapter 11.

DIGESTION AND ABSORPTION OF FATS

Fat digestion and absorption differ principally from the respective processes of digestion and absorption of carbohydrates and proteins, since fats are non-polar and are not miscible with water. The primary object of lipid digestion is to arrange the lipid in a form that is water miscible and can be absorbed through the microvilli of the small intestine, which are covered by an aqueous layer. The sequence of events is the same in all animals: lipolysis,* micellar solubilization of the products of lipolysis, uptake of the solubilized products by the intestinal mucosa, resysnthesis of triglycerides in the mucosal cells, and secretion of triglycerides into the blood. But there are important differences in the mechanism involved between ruminants and non-ruminants.

In Monogastric Animals

Mechanical separation of lipids from the other nutrients occurs in the stomach. Under the influence of the peristaltic action of the stomach and duodenum, a coarse fat emulsion enters the duodenum which is the site of the major processes of fat digestion and absorption. Further emulsification of fat occurs in the small intestine after contact with bile salts due to the detergent properties of the latter. Thus the lipid particle size is reduced to 500–1000 Å. Common bile acids such as cholic, taurocholic and glycocholic acids, are detergents, the sterol nucleus being lipid-soluble and the ionized conjugates of glycine and taurine being water-soluble. The smaller size of the particles formed under the influence of bile acids allows for greater surface exposure to pancreatic lipases. Pancreatic lipase acts only at an oil–water interface, which explains why emulsification is required for the digestion of fats.

---

* = Hydrolysis of lipids.

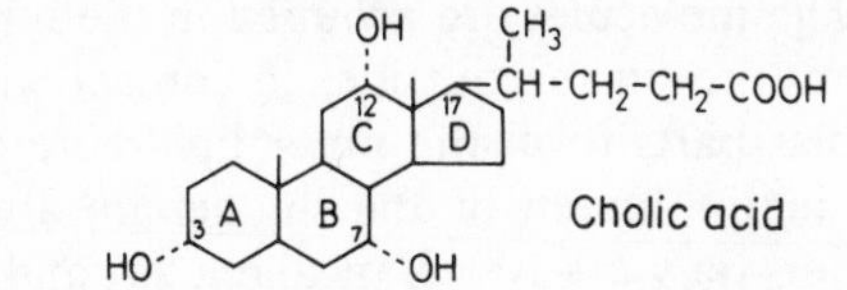

Cholic acid

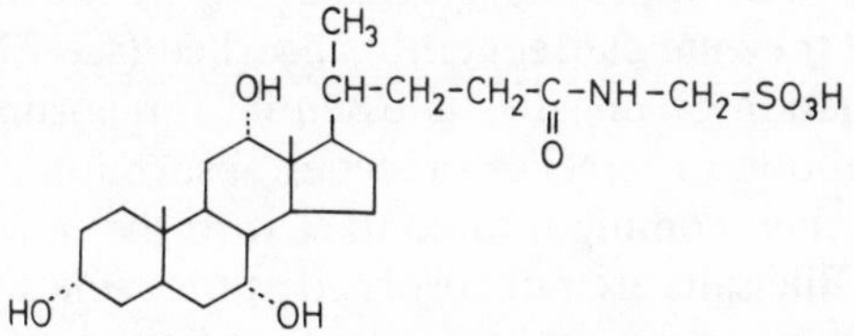

Taurocholic acid

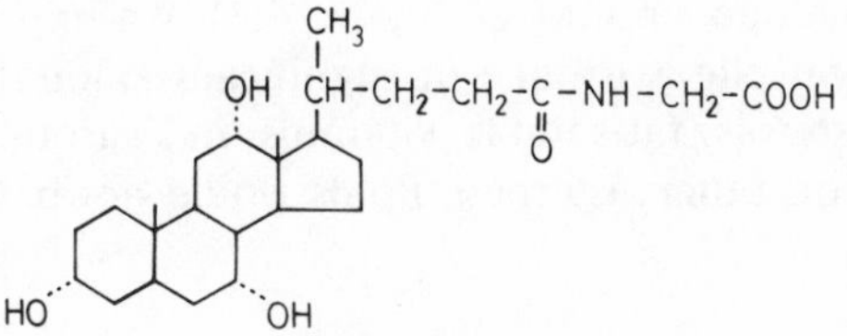

Glycocholic acid

Breakdown of triglycerides is achieved by pancreatic lipase in the presence of bile salts. The enzyme does not completely hydrolyse the triglyceride and the breakdown stops at the 2-acylmonoglyceride stage. Moreover, two molecules of free fatty acid are liberated from every molecule of triglyceride.

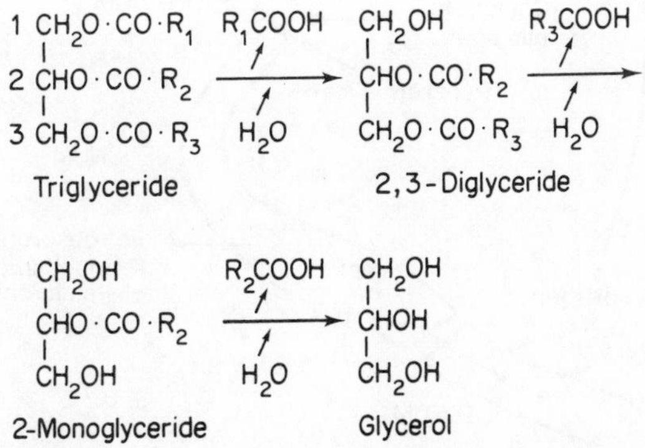

$$\begin{array}{l} 1\ CH_2O\cdot CO\cdot R_1 \\ 2\ CHO\cdot CO\cdot R_2 \\ 3\ CH_2O\cdot CO\cdot R_3 \end{array} \xrightarrow[H_2O]{R_1COOH} \begin{array}{l} CH_2OH \\ CHO\cdot CO\cdot R_2 \\ CH_2O\cdot CO\cdot R_3 \end{array} \xrightarrow[H_2O]{R_3COOH}$$

Triglyceride 2,3-Diglyceride

$$\begin{array}{l} CH_2OH \\ CHO\cdot CO\cdot R_2 \\ CH_2OH \end{array} \xrightarrow[H_2O]{R_2COOH} \begin{array}{l} CH_2OH \\ CHOH \\ CH_2OH \end{array}$$

2-Monoglyceride Glycerol

Monoglycerides, fatty acids and bile acids—each of which has polar and non-polar groups—have the property of aggregating together and forming micelles suitable for absorption.

Micelles are water-soluble aggregates of lipid molecules containing polar and non-polar groups. The molecules are grouped in the micelles in such a way that the polar groups are on the outside, in contact with the aqueous phase, while the non-polar parts form the inner lipid core of the micelles. The micelles produced in the lumen of the duodenum are very fine dispersions of lipids in water, only 50–100 Å in diameter, and carry the lipid digestion products (fatty acids, monoglycerides) to the mucosal cells of the small intestine where they are subsequently absorbed (see Figure 6.2).

Although the formation of micelles is essential for normal fat digestion and absorption, disruption of micelles precedes absorption of their components. The micelles, when coming into contact with the microvillous membrane, are disrupted. Bile salts are not absorbed at the same site as fatty acids and monoglycerides. Absorption of the products of lipolysis occurs primarily in the duodenum and upper jejunum, whereas bile salts are absorbed in the ileum and are recirculated in the portal blood to the liver and hence to the bile for re-entry to the duodenum (see Figure 6.1). An important property of the lipid–bile salt micelle is to take up significant amounts of non-polar compounds such as sterols, fat-soluble vitamins and carotinoids within its lipid non-polar interior; otherwise these lipids would not be absorbed.

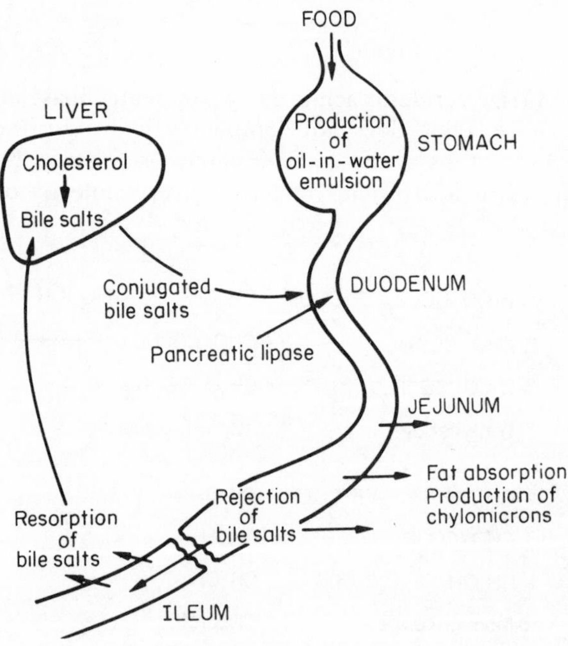

Figure 6.1. Scheme for digestion, absorption and transport of fat (after Gurr and James, 1981, p. 107). *Reproduced by permission of Chapman and Hall and the authors.*

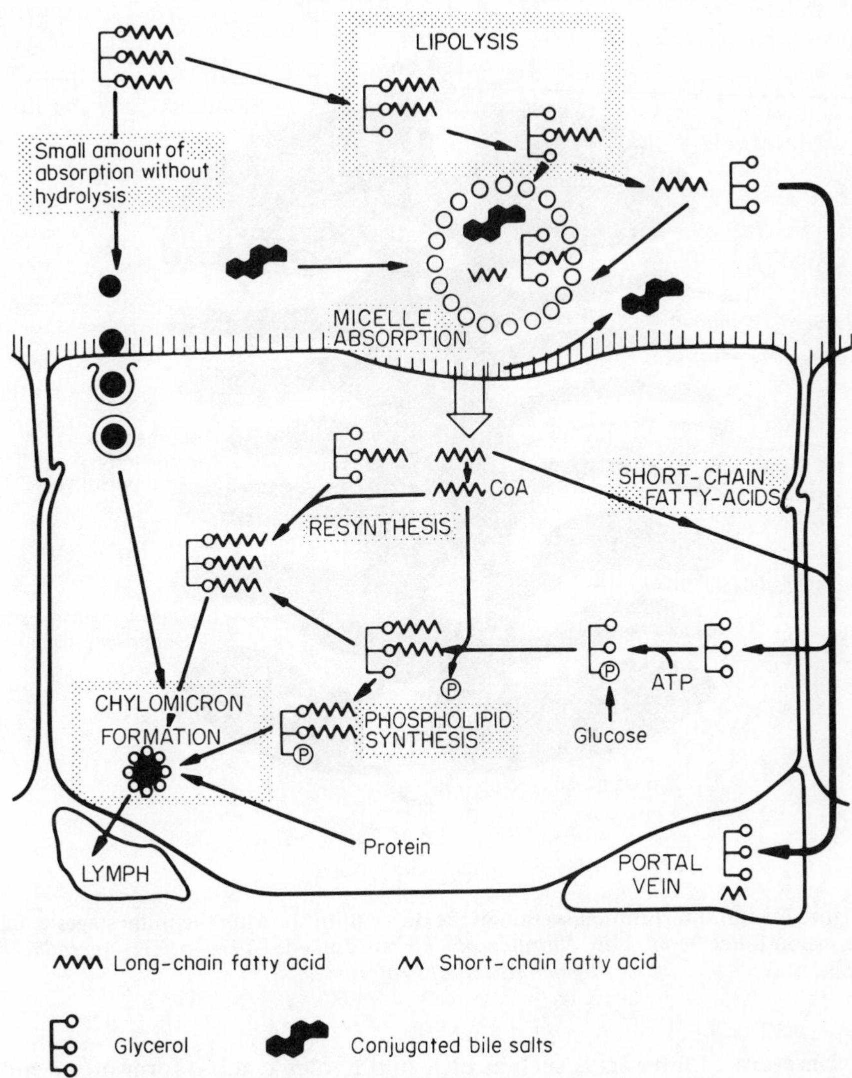

Figure 6.2. Schematic diagram for digestion and absorption of fat (after Gurr and James, 1981, p. 110). *Reproduced by permission of Chapman and Hall and the authors.*

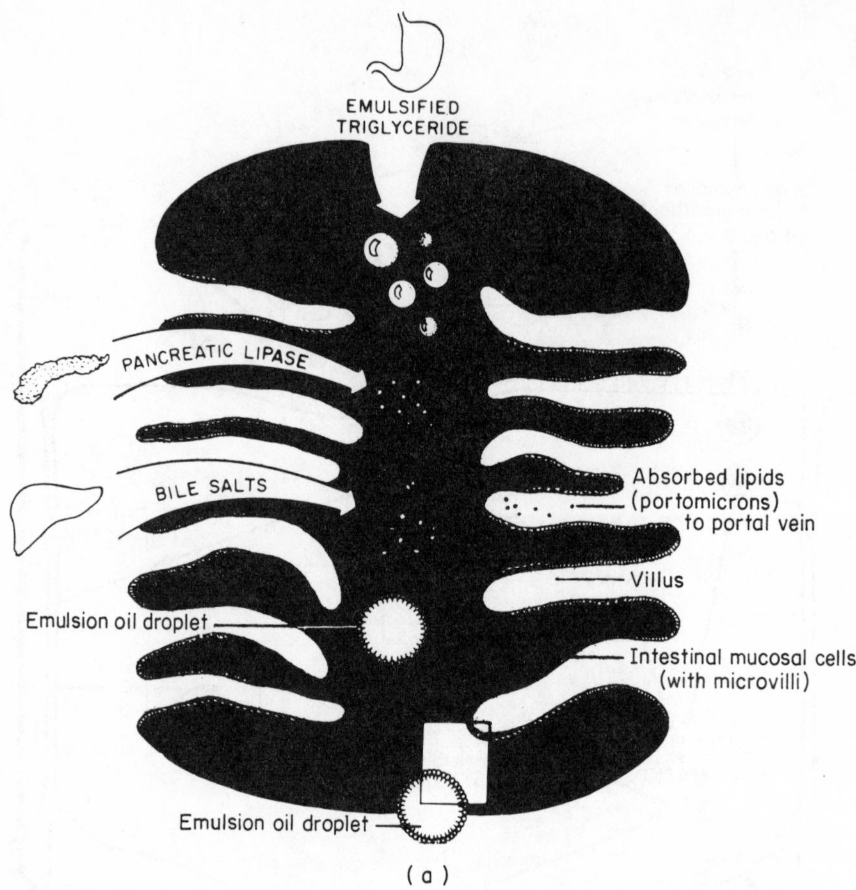

Figure 6.3. (a) Interluminal section of the duodenum showing the inital stages of fat digestion (after Scott *et al., Nutrition of Chicken,* 2nd edn, 1976, p. 27). *Reproduced by permission of Professor Scott.*

Unsaturated fatty acids such as oleic and linolenic acids, form mixed micelles with bile salts, whereas the solubility of long-chain saturated fatty acids in bile acid micelles is very low. Monoglycerides and unsaturated fatty acids act synergistically to promote the incorporation of saturated fatty acids into micelles. This effect is shown in young chicks which utilize saturated fats much better when fed together with unsaturated fat.

Schemas representing the processes of digestion and absorption of fats occurring in the intestine are given in Figures 6.1 and 6.2, and a diagram of the processes involved in hydrolysis and absorption of fats is shown in Figure 6.3a and b.

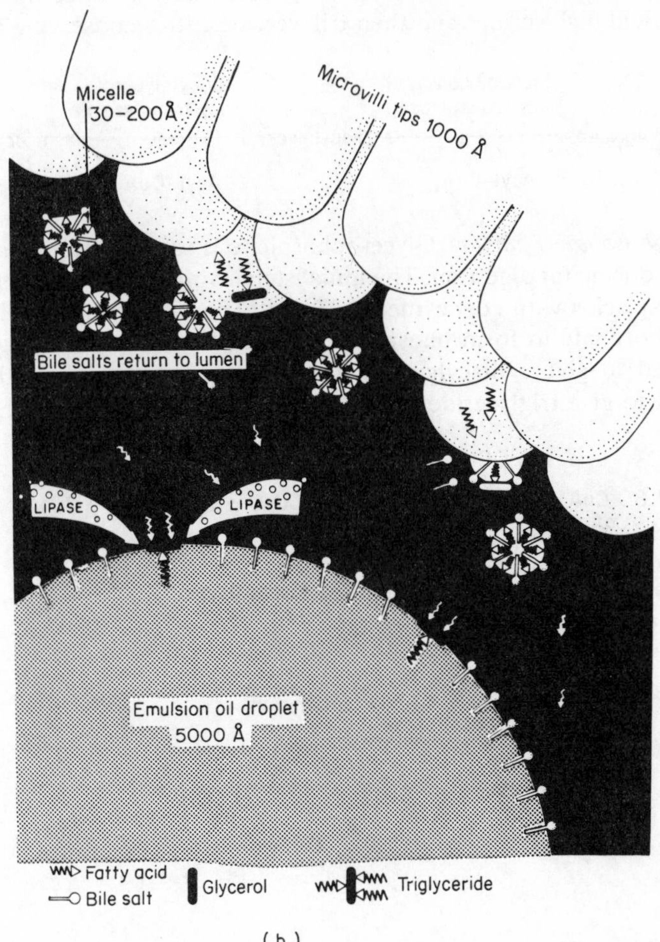

Figure 6.3. (b) Enlarged section of (a) showing the absorption of micelles (after Scott *et al.*, 1976 p. 28). *Reproduced by permission of Professor Scott.*

The major route of fat absorption in mammals and birds is via micelles; in addition, limited amounts of triglycerides may be absorbed intact as a fine emulsion, and glycerol, liberated by lipolysis, is absorbed like glucose by simple diffusion with a carrier. The main products of fat digestion, monoglycerides and long-chain fatty acids enter (after disruption of the micelles) the absorptive intestinal mucosal cells by diffusion (passive transport; see Chapter 3). Resynthesis of triglycerides from monoglycerides proceeds inside the mucosal cells via the monoglyceride-pathway or the glycerol 3-phosphate pathway. In both pathways the free fatty acids are first converted to their coenzyme A derivatives.

Monoglyceride Pathway. Monoglycerides are directly acylated by fatty acyl-CoA to yield diglycerides and then triglycerides, thus conserving energy.

$$\text{Monoacylglycerol} \xrightarrow[\text{Acyl-CoA}]{\text{Monoacylglycerol acyltransferase}} \text{Diacylglycerol} \xrightarrow[\text{Acyl-CoA}]{\text{Diacylglycerol acyltransferase}} \text{Triacylglycerol}$$

Glycerophosphate Pathway. Glycerophosphate is formed from free glycerol liberated during fat digestion. The initial step involves the activation of long-chain fatty acids with coenzyme A. The fatty acyl-CoA then interacts with glycerophosphate to form phosphatidic acid. The phosphatidic acid is then hydrolysed to give a diacylglyceride which then reacts with a third fatty acyl-CoA to give triglyceride.

$$\underset{\text{Acyl-CoA}}{2R{\cdot}CO{\cdot}CoA} + \underset{\text{Glycerol phosphate}}{\begin{array}{c} HOCH_2 \\ | \\ HOCH \\ | \\ H_2CO{\cdot}PO_3H_2 \end{array}} \longrightarrow \underset{\text{Phosphatidic acid}}{\begin{array}{c} R{\cdot}CO{\cdot}O-CH_2 \\ | \\ R{\cdot}CO{\cdot}O-CH \\ | \\ H_2CO{\cdot}PO_3H_2 \end{array}} + 2CoA$$

$$\begin{array}{c} R{\cdot}CO{\cdot}O-CH_2 \\ | \\ R{\cdot}CO{\cdot}O-CH \\ | \\ H_2CO{\cdot}PO_3H_2 \end{array} \xrightarrow{\text{Phosphatase}} \underset{\text{Diacylglycerol}}{\begin{array}{c} R{\cdot}CO{\cdot}O-CH_2 \\ | \\ R{\cdot}CO{\cdot}O-CH \\ | \\ H_2COH \end{array}} + H_3PO_4$$

$$\begin{array}{c} R{\cdot}CO{\cdot}O-CH_2 \\ | \\ R{\cdot}CO{\cdot}O-CH \\ | \\ H_2COH \end{array} + R{\cdot}CO{\cdot}CoA \longrightarrow \underset{\text{Triacylglycerol}}{\begin{array}{c} R{\cdot}CO{\cdot}O-CH_2 \\ | \\ R{\cdot}CO{\cdot}O-CH \\ | \\ R{\cdot}CO{\cdot}O-CH_2 \end{array}} + CoA$$

Triglycerides resynthesized in the mucosa are transported to various tissues where they are deposited as an energy reserve or enter a wide range of metabolic processes (see later).

In Ruminants

The mechanism of digestion and absorption of lipids in suckling calves and lambs, where the rumen is non-functional, is the same as that in non-ruminants (see Chapters 3 and 18).

The lipids in diets of ruminants are mainly present in an esterified form as mono- and digalactoglycerides in forages and as triglycerides in concentrate feeds. The lipids present in these diets contain a high proportion of polyunsaturated fatty acids in galactolipids of forages and in triglycerides of cereal grains. In the rumen extensive hydrolysis of esterified dietary lipids, triglycerides, galactolipids and phospholipids occurs by the action of microbial lipases, releasing free fatty acids and allowing galactose and glycerol to be fermented to volatile fatty acids (see Chapter 5). The polyunsaturated fatty acids (linoleic and linolenic acid) released from ester combination are hydrogenated by bacteria, yielding first a monoenoic acid and, finally, stearic acid. Less than 10% of the polyunsaturated fatty acids usually escape hydrogenation, so that the digestion products reaching the major site of absorption in the upper small intestine are mainly saturated fatty acids (i.e. palmitic and stearic acids), with minor amounts of monounsaturated fatty acids (including *trans*-isomers), very small amounts of polyunsaturated fatty acids and microbial lipids. Both linoleic and linolenic acid present in feedstuffs of plant origin have all-*cis* double bonds; but before they are hydrogenated, one double bond in each of both acids is converted to the more stable *trans* form and the latter acids are absorbed by the animal.

Despite the fact that the greatest part of dietary essential fatty acids is destroyed by biohydrogenation, ruminants do not appear to suffer from essential fatty acid deficiency. The small amount of unchanged essential fatty acids passing through the rumen is sufficient for the needs of these animals.

Rumen microorganisms are also capable of synthesizing some odd-chain fatty acids from propionate (see Chapter 5) and branched-chain fatty acids from the carbon skeleton of the amino acids valine, leucine and isoleucine (see Chapter 8). These acids are eventually incorporated in the milk and body fat of ruminants.

As outlined above, most lipids in the ruminant enter the duodenum as free fatty acids with a very high proportion of saturated fatty acids. Free fatty acids are present in the small intestine of the ruminant in the form of thin layers on the surface of the feed particles. Monoglyceride, which promotes the micellar solubilization of lipids in simple-stomached animals, is removed by hydrolysis in the rumen and is therefore lacking in ruminants. The free fatty acids present in the small intestine of the latter animals are transferred from digesta particles to micellar phase. Micelle formation of saturated fatty acids occurs here with bile salts and lysolecithin. Lysolecithin which replaces monoglyceride as emulsifying agent is formed from biliary and dietary phospholipid by the action of pancreatic lipase.* Resynthesis of triglycerides in the mucosal cells proceeds via the glycerophosphate pathway since no monoglyceride is absorbed from the intestine of adult ruminants. Lipid discharge from the mucosal cells is similar to what occurs in non-

* Lysolecithin is formed by partial hydrolysis of lecithin with removal only of one fatty acid residue.

ruminants. The poor absorbability of carotene by ruminants may be due to defective micellar solubilization of this compound (see Chapter 11).

TRANSPORT OF LIPIDS IN THE BODY

Lipids are transported from the mucosal cells of the intestine to various tissues by the circulation, mainly as lipoproteins and to a lesser extent as free fatty acids. The protein moiety of lipoproteins imparts water-soluble properties to lipids and allows their transport in the blood. The lipoproteins can be separated by a density-gradient procedure into different fractions. The lipoproteins present in blood range from those of very low density to those of high density. The density increases as the proportion of protein in the lipoprotein increases and the lipid decreases. Before leaving the mucosal cell, the mixed lipids present there—mainly triglycerides and smaller amount of phospholipids and cholesterol ester—become coated with a thin layer of protein. These protein-coated particles are called chylomicrons; chylomicrons contain the lightest lipoproteins. In mammals the chylomicrons pass first into the lymph and then enter the systemic circulation, whereas in birds most of the lipids enter the portal system directly as very-low-density lipoproteins to be carried to the liver. In the lymph of simple-stomached mammals, chylomicrons are the prevailing lipid fraction, but in ruminant lymph very-low-density lipoproteins are the prime carrier of lipids. Because of the continuous nature of digestion in ruminants (see Chapter 3), the intestinal lymph has a permanently milky appearance, but in simple-stomached mammals this only occurs after ingestion of a high-fat meal. The concentration of blood lipids is very high in the laying hen; this is necessary for the deposition of the large amounts of fat needed for yolk formation (see Chapter 20).

In addition to lipoproteins and chylomicrons, the plasma contains low levels of unesterified fatty acids complexed with albumin; however, the turnover rates are extremely rapid and thus the free fatty acids in the blood are of great metabolic importance (see later).

DEPOSITION OF FATS IN TISSUES

Apart from intestinal absorption of dietary lipids, the lipids circulating in the blood are derived from mobilization of lipids stored in tissues or from synthesis. The lipids are very rapidly removed from the blood by adipose tissue, liver and other tissues. Uptake of triglycerides by these tissues is followed by hydrolysis to free fatty acids and glycerol; this process is catalysed by the lipoprotein lipase present in the capillary walls of the tissues. The free fatty acids can be disposed of in the tissues in a variety of ways: (1) they may be completely oxidized to carbon dioxide and water for release of energy; (2) they may be esterified to reform triglycerides which may be released again into circulation or deposited in tissues for storage; (3) a small proportion is transported in the blood complexed with albumin (see above). A scheme for the transport and storage of lipids is presented in Figure 6.4.

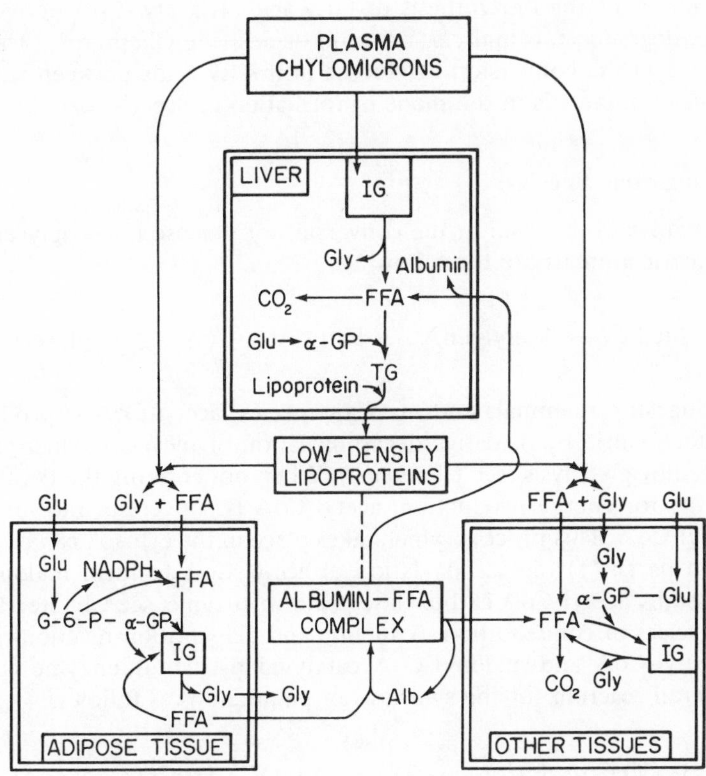

Figure 6.4. Transport, storage and metabolism of lipids (after *Dukes' Physiology of Domestic Animals*, 9th edn, 1977, p. 337). *Reproduced by permission of Cornell University Press.*

TG = Triglycerides	α-GP = Glycerol phosphate
FFA = Free fatty acids	G-6-P = Glucose 6-phosphate
Gly = Glycerol	Alb = Albumin
Glu = Glucose	

Adipose tissue is the most notable storage site of fats in animals. It consists of aggregates of spherical cells, termed adipocytes; large lipid droplets fill most of the cells. Adipose tissue is mainly found under the skin and also around internal organs (heart, kidney). Contrary to the previously accepted view that adipose tissue is a metabolically inert storage tissue E. Wertheimer and B. Shapiro in Jerusalem provided in 1948 evidence that adipose tissue is a highly active dynamic tissue inside which fat is continually being synthesized and degraded.

BIOSYNTHESIS OF FATS

Liver, adipose tissue and lactating mammary gland (see Chapter 21) are the primary sites of biosynthesis of fatty acids and triglycerides. The main starting material for the biosynthesis of fatty acids is acetyl-CoA derived from glucose, degraded fats and certain amino acids (see Chapter 7). The major difference in the conversion of glucose into fatty acids between ruminants and non-ruminants is in the mode of formation of acetyl-CoA.

In Monogastric Species

The pathways involved in the conversion of glucose into triglycerides by monogastric animals are the following:

$$\text{Glucose} \longrightarrow \text{Acetyl-CoA} \longrightarrow \text{Fatty Acyl-CoA} \longrightarrow \text{Triglyceride}$$

In monogastric mammals and in avian species, acetyl-CoA is produced in the mitochondria by oxidative decarboxylation of pyruvate which, in turn, is formed by glycolysis (see Chapter 4). When not entering the tricarboxylic acid cycle for energy production, acetyl-CoA is converted into long-chain fatty acyl-CoA. This process, which takes place in the cytosol, can be divided into two parts: (1) the carboxylation of acetyl-CoA to form malonyl-CoA (catalysed by acetyl-CoA carboxylase, a biotin enzyme; see Chapter 12); and (2) the series of condensation, reduction and dehydration reactions starting with acetyl-CoA and malonyl-CoA catalysed by a multienzyme complex. The overall reaction for the synthesis of palmitate is as follows:

$$(1)\quad \underset{\text{Acetyl-CoA}}{CH_3.CO.CoA} + CO_2 + ATP \longrightarrow \underset{\text{malonyl-CoA}}{HOOC.CH_2.CO.CoA} + ADP + P_i$$

$$(2)\quad 7\,HOOC.CH_2.CO.CoA + CH_3.CO.CoA + 14\,NADPH + 14\,H^+ \longrightarrow$$

$$CH_3(CH_2)_{14}COOH + 7\,CO_2 + 8\,CoA + 14\,NADP^+ + 6\,H_2O$$

As mentioned above, the conversion of pyruvate to acetyl-CoA occurs within the mitochondria, whereas the synthesis of fatty acetyl-CoA occurs in the cytosol. Therefore acetyl-CoA must be transported to the cytosol. The mitochondrial membrane is impermeable to CoA derivatives; the most likely mechanism of transfer of acetyl-CoA from the mitochondria is the intramitochondrial formation of citrate by the condensation of acetyl-CoA with oxaloacetate. Citrate permeates the mitochondrial membrane. In the cytosol, citrate is cleaved by citrate lyase to oxaloacetate and acetyl-CoA which becomes available for the fatty acid synthetase system:

$$\text{Citrate} + \text{CoA} \xrightarrow{\text{Citrate lyase}} \text{acetyl-CoA} + \text{oxaloacetate}$$

ATP → ADP

In the cytoplasm oxaloacetate is rapidly reduced with NADH to form malate:

$$\begin{array}{l} COOH \\ | \\ CO \\ | \\ CH_2 \\ | \\ COOH \\ \text{Oxaloacetate} \end{array} + NAD + H^+ \longrightarrow \begin{array}{l} COOH \\ | \\ CHOH + NAD^+ \\ | \\ CH_2 \\ | \\ COOH \\ \text{malate} \end{array}$$

Malate is oxidatively decarboxylated to pyruvate which re-enters the mitochondria, and by this reaction NADPH is also provided:

$$\begin{array}{l} COOH \\ | \\ CH_2 \\ | \\ CHOH + NADP^+ \\ | \\ COOH \end{array} \xrightarrow[\text{enzyme}]{\text{Malic}} \begin{array}{l} CO_2 \\ + \\ CH_3 \\ | \\ C{=}O \\ | \\ COOH \end{array} + NADPH + H^+$$

Citrate transport not only transfers acetyl-CoA to the cytoplasm but also provides NADPH which supplies about 50% of the reducing power needed for fatty acid synthesis. The other 50% of the NADPH required for this purpose is supplied by glucose oxidation through the pentose-phosphate pathway.

The last step in triglyceride biosynthesis, the esterification of long-chain fatty acyl-CoA with glycerol, occurs in the same way as the re-esterification of fatty acids in mucosal cells as outlined above.

In Ruminant Animals

Young ruminants possess the ability to convert glucose into fatty acids. When the rumen becomes functional, however, this ability is lost, and acetate, a major product of rumen fermentation (see Chapter 5), becomes the major carbon source utilized in fatty acid synthesis. Acetic acid diffuses into the blood from the rumen and is converted in the tissues by cytosolic acetyl-CoA synthetase into acetyl-CoA, the energy being supplied by hydrolysis of ATP to AMP (as shown in Chapter 5).

This pathway of fatty acid synthesis occurring in adipose tissue of ruminants is markedly different from that in non-ruminating animals. Lack of utilization of blood glucose for fatty acid synthesis in ruminants is due to the very low activity of citrate lyase and malic enzyme in their tissues. These animals have thus evolved a mechanism for conserving glucose for more vital functions since their blood glucose is mainly produced by gluconeogenesis, unlike monogastric species which receive most of their blood glucose from the diet.

Conversion of acetyl-CoA into long-chain fatty acids occurs in ruminants in the same way as in non-ruminants. The reducing equivalents (NADPH) arise from both the pentose-phosphate cycle and the isocitrate cycle (conversion of isocitrate to α-ketoglutarate).

Fatty Acid Interconversions

Tissue lipids are derived from dietary lipids as well as from lipogenesis. This synthetic process results in the immediate formation of saturated even-chain fatty acids. In order to provide the specific fatty acid composition of various tissue lipids, interconversion of fatty acids occur by successive addition or removal of C_2 units. Furthermore, fatty acids with one double bond are formed by dehydrogenation by the action of specific enzymes; for example, formation of oleic acid from stearic acid. In chicks, lipogenesis favours the production of unsaturated acids, contrary to most mammals in whose tissues saturated fatty acids are predominant. Synthesis of linoleic or linolenic acid, the essential fatty acids, by animal tissues is not possible since animals have no enzymes capable of introducing a double bound beyond C_{-9} (see above).

Sites of Biosynthesis of Lipids in Various Species

The relative importance of liver and adipose tissue varies with species. Fatty acid synthesis in the pig and in non-lactating ruminants occurs almost exclusively in the adipose tissue. In contrast, this synthesis in man and avian species occurs predominantly in the liver, with adipose tissue acting merely as a storage site for fats. In rats, mice and rabbits, both liver and adipose tissue are important sites of fatty acid synthesis.

Influence of Dietary Factors on the Biosynthesis of Fats

The synthesis of fatty acids is controlled by dietary factors and hormones which influence the rate of production of enzymes connected with this synthesis, particularly of acetyl-CoA carboxylase. This enzyme is inhibited by fatty acyl-CoA and by free fatty acids, which would be abundant with a high fat diet, and stimulated by tricarboxylic acids formed from fed carbohydrates. Thus intake of lipids and carbohydrates provides control of fatty acid synthesis through feed-back and feed-forward mechanisms. Saturated fat has a greater effect in suppressing fatty acid synthesis than unsaturated fat. The rate of lipogenesis is reduced under the following conditions when the energy requirement is mainly covered by fat of exogenous and endogenous origin: (1) in fasting animals which obtain the greatest part of their energy demands from degradation of body fat (see Chapter 17); (2) in suckling piglets; and (3) in chick embryos before hatching.

The type of dietary carbohydrate also influences fatty acid synthesis. Substituting sucrose or fructose for starch or glucose in the diet results in increased lipogenesis, particularly in fowl. The activation of acetyl-CoA carboxylase by citrate may be attributed to its aggregation to the active polymer form of the enzyme, whereas fatty acyl-CoA causes disaggregation and deactivation of the enzyme.

The positive effect of carbohydrates on fatty acid biosynthesis may also be related to the circulating level of insulin. When a high carbohydrates diet is consumed, not only does the blood glucose level rise, but insulin output increases. This allows a greater portion of glucose to enter adipose cells, where it can be converted to fatty acids and stored as fat. Monogastric species consuming a carbohydrate-rich diet produce a high level of fat.

Not only the composition of the diet but also the pattern of food intake has an effect on the rate of lipogenesis (Leveille, 1970). Meal feeding of rats or chicks rather than nibbling (continuous feeding) results in an appreciable increase in fatty acid synthesis. This may be due to alteration of the pattern of circulating substrates and hormones in response to meal feeding. Therefore animals trained to eat their entire day's diet in a single meal become fatter than those consuming the same amount of food in several meals throughout the day.

Influence of Nutrition on Body Fat Deposition

Each monogastric species and each different tissue within a given species has a certain constant fatty acid composition. Triglycerides derived from liponeogenesis (i.e. from carbohydrates and proteins) and some interconversions of fatty acids (see above) will fit the specific composition of a given tissue. The consumption of significant amounts of lipids, however, changes the fatty acid composition of tissue lipids in hogs and chicks in varying degrees. The adipose fat is more responsive to the influence of dietary fatty acid composition than the fat in organs. It is noted that rations containing cottonseed oil produce lard graded as hard, and those rich in highly unsaturated fats, e.g. corn or soybean oil, produce soft lard. As mentioned above, fat deposited in adipose tissue is not static. When unsaturated fats have been fed to growing pigs, the subcutaneous fat can be hardened by feeding the pigs rations of low fat content for two months.

In contrast to non-ruminants, the fatty acid composition of adipose tissue in ruminant animals is not markedly affected by the fatty acid composition of the diet. The extensive hydrogenation of unsaturated fat by rumen microorganisms results in the preformed fatty acids taken up by adipose tissue being predominantly saturated, regardless of the diet. Thus ruminant fats are characteristically hard (see above).

CATABOLISM OF FATS

During periods of fast, or excessive energy demands, fat is mobilized from storage mainly in adipose tissue and oxidized for energy production. Triglycerides must be hydrolysed to their constituents before further catabolism can proceed. The breakdown of fat stored in adipose tissue into glycerol and fatty acids is catalysed by intracellular hormone-sensitive lipase (different from the lipoprotein lipase mentioned above, which is located in capillaries, which hydrolyses circulating triglycerides). The hormone-sensitive lipase is converted from an inactive form by a protein kinase which in turn is activated by cyclic AMP and the latter by hormones (epinephrine, glucagon). The production of these hormones is stimulated by fasting and stress. On the other hand, insulin output increased by high carbohydrate feeding inhibits mobilization of free fatty acids and enhances re-esterification of free fatty acids and thereby synthesis of fats (see above). Thus the rate of fat metabolism and the control between accumulation and breakdown of fat stores depend mainly on the control of the rate of lipolysis.

Fatty acids and glycerol liberated from stores diffuse into the blood where the fatty acids become bound to albumin. The albumin-bound fatty acids are transported to other tissues where they can be used as fuel. Most tissues that can carry out aerobic oxidation (muscles, liver, heart, brain, etc.) have some capacity to use fatty acids as an energy source. The major function of fatty acids is production of energy. In fasting animals dependent on body reserves for energy, fat is the predominant source of energy (see Chapter 17).

The transport of fatty acids in blood is most important. Their concentration in plasma is very low (100–200 μmol/l in fed sheep or only a few per cent of the total plasma lipid). However, free fatty acids in blood are physiologically important, due to their high turnover rate. The concentration and composition of free fatty acids in plasma vary with their mobilization from stores, their uptake and utilization by tissues and their absorption from the gastrointestinal tract. They are intermediates in energy metabolism and their levels in plasma may be used as a biochemical parameter for energy supply (Giesecke, 1983). In conditions such as stress, starvation or prolonged muscular activity, the plasma fatty acid level in monogastric species is raised about five-fold and provides an alternative fuel to glucose, which can be conserved for use by other tissues. The comparatively high concentration of plasma fatty acids in lactating and pregnant ruminants deserves interest for assessing their nutritive status (see Chapters 19 and 21).

β-Oxidation of Fatty Acids

Fatty acids are completely oxidized to carbon dioxide and water. The major pathway of their oxidation is the well-known β-oxidation (see biochemical texts). Fatty acid oxidation starts with activation of the fatty acid with coenzyme A in the presence of ATP and acyl-CoA synthetase to give an acyl-CoA:

$$\text{RCOOH} + \text{CoA} + \text{ATP} \longrightarrow \text{acyl-CoA} + \text{AMP} + \text{PP}$$

The β-oxidation proceeds by stepwise shortening of the carbon chain of acyl-CoA through removal of two carbon atoms as acetyl-CoA at a time. Each step of removing the two carbon acetyl-CoA produces 5 mol of ATP. The process continues until the carbon chain has been converted completely to acetyl-CoA. Any such C_2 compound enters the tricarboxylic acid cycle to be oxidized to carbon dioxide and water, each mole of acetyl-CoA so oxidized providing 12 mol of ATP. The large energy production arising from the oxidation of the 16-carbon palmitate is as follows:

			Mol ATP	
			+	−
Palmitic acid	⟶	palmityl-CoA		2
Palmityl-CoA	⟶	8 acetyl-CoA (7 cleavages × 5)	35	
8 acetyl-CoA	⟶	16 H_2O + 16 CO_2 (8 × 12)	96	
		Net gain of ATP per mol of palmitate	129	

The total energy content of 1 mol of palmitic acid is 9586 kJ; the capture of energy by the formation of 129 energy-rich bonds is 6708 kJ per mol (129 × 52). The efficiency of free energy capture is thus

$$\frac{6708}{9586} = 69\%$$

and equals approximately the efficiency of glucose oxidation (see Chapters 2 and 14).

Oxidation of Glycerol

A small part of the energy derived from fat, about 20%, may be provided by glycerol, which is liberated from stores together with fatty acids. Glycerol as a gluconeogenic substance is converted to glucose which is oxidized to produce energy (see Chapters 4 and 5).

Formation of Ketone Bodies

During periods of great physiological demand for glucose, as in cattle at the peak of lactation and in sheep in late pregnancy, animals cannot fully utilize fatty acids for oxidation and energy production. In such conditions degradation of carbohydrates is reduced and thus the level of oxaloacetate is inadequate to condense with acetyl-CoA. Consequently, condensation between pairs of acetyl groups takes place:

$$\underset{\text{Acetyl-CoA}}{2\,CH_3.CO.CoA} \longrightarrow \underset{\text{acetoacetic acid}}{CH_3.CO.CH_2.COOH} + 2\,CoA$$

Acetoacetic acid may be reduced with NADH to β-hydroxybutyric acid ($CH_3.CHOH.CH_2.COOH$) or decarboxylated to form acetone. These three compounds are called ketone bodies. Small amounts of ketone bodies are always formed in ruminants (see Chapter 5) and can be used as energy source for peripheral tissues, particularly skeletal muscles. Blood levels and urinary excretion of ketone bodies rise in animals affected by ketosis such as high producing cows or ewes carrying more than one foetus (see Chapter 24).

SUPPLEMENTATION OF ANIMAL DIETS WITH FATS

In the diets of animals, carbohydrates can substitute for fats as a source of energy and for conversion into body fat. Every practical diet for farm animals contains small amounts of fat necessary as a vehicle for transport and for the absorption of fat-soluble vitamins (see Chapter 11) and as source of the essential fatty acids (see above).

Nowadays, supplementation of diets for farm animals, particularly poultry, with moderate amounts of fats has been widely adopted. A wide range of fats (tallow, lard, soybean oil, rapeseed oil, by-products of edible oil refining) are used for increasing the energy concentration of diets for poultry (2–5%) and pigs (5–10%), thereby improving the efficiency of the conversion of food into animal products and to influence the composition of the latter in desired directions.

'Protected' Fats

For preruminants, high levels of fat are used in milk substitutes (15–30%). Ruminants on dry feeds are less tolerant of high fat levels (not more than 6–8%) than are monogastrics. Around 1970, in Australia, techniques were developed that allow the feeding of considerable amounts of 'protected' fats. Protecting fats is a procedure of encapsulating small droplets of oils in a thin layer of formaldehyde-treated protein. In this way the droplets avoid attack by microorganisms during passage through the rumen, but the fat is released by the acidity of the abomasum and thus becomes available for digestion and absorption from the small intestine as in non-ruminants. Feeding of unsaturated fats in a protected form results in a prompt rise in the degree of unsaturation of plasma lipids, and of milk and body fat. This has been promoted in the context of prevention of human heart disease and atherosclerosis. However, the connection between human disease and intake of saturated fat is controversial, the shelf-life of unsaturated milk is poor, and the use of protected fat in feeds is not yet economically feasible.

DETERIORATION OF FATS AND ITS PREVENTION

Fats and oils are subject to degradation of two types—hydrolytic and oxidative—with subsequent development of rancidity. Hydrolytic degradation—of fats into fatty acids and glycerol—is the result of the action of lipases occurring in moulds and bacteria. The fact that fat has undergone hydrolytic rancidity can be offensive and unacceptable to human beings, but it does not interfere with its nutritional value for animals.

Oxidative rancidity, however, apart from producing abnormal flavours and odours, results in a decrease in the energy value and destruction of essential fatty acids and some other biologically important compounds. The methylene group flanked by double bonds on each side, which characterizes polyunsaturated fatty acids, is directly attacked by oxygen. The oxygen removes a methylenic hydrogen yielding a free radical (see formula). The double bond in the free radical shifts to the conjugated compound. This structure is then attacked by molecular oxygen and the free radical is converted to a fatty acid peroxide and thence to a fatty acid hydroperoxide. The free radicals such as ROO·, RO·, OH· (indicated by an asterisk in the formula) react by eliminating hydrogen or by a variety of oxidation reactions to damage enzymes, structural membranes and other lipids, especially the fat-soluble vitamins A and E (see Chapter 11).

$$-CH_2-CH{=}CH- \xrightarrow{-H\cdot} -\overset{*}{C}H-CH{=}CH- \xrightarrow{+O_2}$$

Radical of peroxide: $-CH(-O-\overset{*}{O})-CH{=}CH-$; $+ O_2$; $-\overset{*}{C}H-CH{=}CH-$

$-CH_2-CH{=}CH-$ $\xrightarrow{-H\cdot}$; $+H\cdot$

Hydroperoxide: $-CH(-O-OH)-CH{=}CH-$

The oxidation of polyunsaturated fatty acids is catalysed by traces of heavy metals and their compounds, and by the plant enzyme lipoxidase which occurs in some seeds and in green parts of plants.

The oxidation of unsaturated fats can be delayed by the presence of antioxidants. Antioxidants, designated by the general formula AH, supply a hydrogen atom to a peroxide radical:

$$\underset{\text{Peroxide radical}}{ROO^{*} + AH} \longrightarrow \underset{\text{hydroperoxide}}{ROOH + A^{*}}$$

Consequently, the free radicals of fatty acids are reconverted into the original fatty acid and a free radical of the antioxidant (A^{*}) is formed. A^{*} does not react with oxygen, or only at a very slow rate. Thus the peroxidation is blocked.

Synthetic compounds manifesting the properties of antioxidants are phenols (butylated hydroxyanisole, butylated hydroxytoluene) or cyclic amines (ethoxyquin, DPPD); the tocopherols (vitamin E; see Chapter 11) are efficient natural antioxidants.

HINDERED PHENOLIC ANTIOXIDANTS

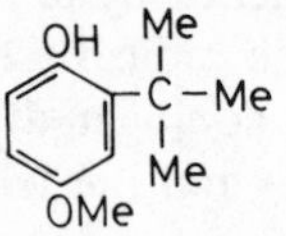

BHA ('butylated hydroxyanisole')
2-*tert.*-butyl-4-methoxyphenol

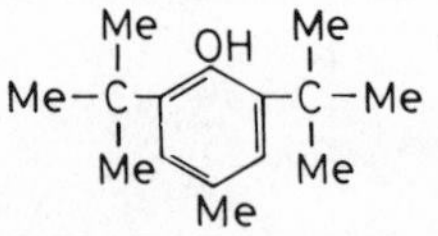

BHT ('butylated hydroxytoluene')

di-*tert.*-butyl-*p*-cresol (2,6-di-*tert.* butyl-4-methylphenol)

AROMATIC AMINE ANTIOXIDANTS

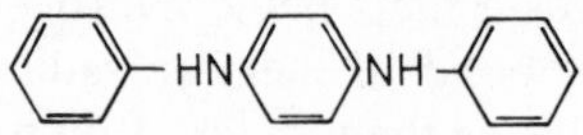

N, *N*'-diphenyl-*p*-phenylenediamine (DPPD)

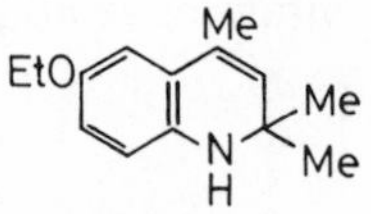

Santoquin
2,2,4-Trimethyl-6-ethoxy-1,2-dihydroqinoline

The safe use of these synthetic antioxidants in animal nutrition has been ascertained, apart from DPPD which has been prohibited due to the adverse effect of this compound on reproduction in rodents. Antioxidants are always added to fats designated for feeding purposes in order to protect these fats from oxidation. Furthermore, it is recommended to add antioxidants to fish meal and to lucerne meal (in order to preserve carotene; see Chapter 11).

If the products of lipid oxidation are allowed to form in the diet, they destroy sensitive compounds, especially vitamins, not only in the diet but also in animal tissues, since the respective peroxides are absorbed to a small extent from the gastrointestinal tract. α-tocopherol (vitamin E) is an efficient inhibitor of oxidation of polyunsaturated fatty acids in the tissues. Thus the requirements for vitamin E are related to the content of polyunsaturated fatty acids in the diet.

FATTY LIVERS

Fat normally constitutes about 5% of the wet weight of the liver, but the fat content in the liver can rise to 30% or even 50%. Fatty livers are a symptom of deranged fat metabolism; it may be associated with an over-production of fat in the liver and an increased transport of fat to the liver, on the one hand, or to an under-utilization of fat by the liver and a defective release of it from the liver, on the other.

Fatty goose livers are deliberately produced by force-feeding large quantities of cereal grains. Force-fed geese may increase their liver weight six-fold while adding only two-thirds to their body weight.

'Fatty liver syndrome' is a metabolic disorder in laying hens characterized by an enlarged and fragile liver, and accompanied by reduced egg production. The apparent mechanism for the accumulation of triglycerides in the liver is that the overconsumption of high-carbohydrate grains stimulates fat synthesis in the liver beyond the maximal ability of the liver to discharge fat into the plasma. This increase in neutral lipids is accompanied by a decrease in phospholipids.

A 'fatty liver and kidney syndrome' observed in broiler chickens receiving diets based on cereals has been attributed to deficiency of biotin which is involved in fat metabolism (see Chapter 12).

Fatty livers may also be caused by exposure to toxic chemicals such as carbon tetrachloride, which leads to excessive production of fibrous tissue in the liver, and by lack of choline in the diet; the mechanism of this action of choline will be explained in Chapter 12.

CHOLESTEROL IN ANIMAL PRODUCTS

(Influence of intake of lipids on heart conditions in human beings)

Cholesterol is widespread in animal tissue as an essential constituent of all membranes. Further, it is a precursor of steroid hormones, vitamin D and bile acids. Esters of cholesterol predominate in plasma and adrenals, whereas nearly all cholesterol present in brain and nerve tissue is in free form; milk and meat contain small amounts and egg yolk contains fairly high amounts (see Chapter 20).

Cholesterol in animal tissues originates mainly from biosynthesis and to a lesser extent from dietary cholesterol absorbed from the intestine. The liver is the major site of cholesterol synthesis, but synthesis also occurs in the intestinal mucosa, arterial wall and other tissues. Acetyl-CoA is the major precursor for biosynthetic cholesterol. The rate of the synthesis occurring in the liver is inversely proportional to the amount of cholesterol present in the body derived from biosynthesis and from dietary input as well. A reaction catalysed by hydroxymethylglutaryl-CoA reductase is the rate-limiting step among the various reactions leading from acetyl-CoA to cholesterol (see Chapter 2). Many other factors influence cholesterol synthesis in animals; for example, a high level of fat in the diet increases this synthesis and this may be explained as follows. High inputs of saturated fat furnish acetyl-CoA in excess of that required for energy production and body fat synthesis, and the excess acetyl-CoA will be available for cholesterol formation. Thus, the body cholesterol pool will enlarge and the plasma cholesterol level increase. However, if the saturated fats are replaced with polyunsaturated fats, the plasma cholesterol level falls because of a shift of cholesterol from the plasma into the tissues. This is because the presence of polyunsaturated fats in the plasma lipoproteins leads to a decrease in their capacity to carry cholesterol.

The influence of dietary factors on plasma cholesterol levels, well established by experiments with laboratory rodents, seems to be valid in all monogastric species and deserves particular interest because of the correlation of blood cholesterol level and the incidence of atherosclerotic diseases in human beings. An elevated blood cholesterol level is one of the risk factors indicating a susceptibility to atherosclerotic heart disease. Cholesterol is the principal constituent of the plaques which form in the walls of blood vessels causing them to narrow and become rigid, thus reducing the blood flow.

In order to avoid dietary risk factors in human nutrition, it has been recommended (1) to decrease the intake of saturated fat, (2) to increase the intake of polyunsaturated fat by substituting saturated fat with polyunsaturated fat, and (3) to lower cholesterol consumption. For this reason Australian authors have proposed to increase the polyunsaturated fat content of body and milk fat of ruminants by including 'protected' fats in the diet of these animals (see above).

In an attempt to decrease the cholesterol content of eggs, substances which impair the absorption of cholesterol from the intestine, such as plant sterols or pectin, were added to the mash of laying hens, but practical results have not yet been achieved.

BIBLIOGRAPHY

Books

Furth, A. J. (1980). *Lipids and Polysaccharides in Biology.* London: Edward Arnold.

Gurr, M. I., and James, A. T. (1981). *Lipid Biochemistry,* 3rd edn. London: Chapman and Hall.

Schultz, H. W., Day, E. A., and Sinnhuber, R. O. (1960). *Symposium on Foods. Lipids and their Oxidation.* Westport: Avi Publishing Co.

Snyder, F. (1977). *Lipid Metabolism in the Mammals.* New York: Plenum.

Wiseman, J. (1984). *Fats in Animal Nutrition.* London: Butterworths.

Articles

Symposium on lipid metabolism and its control (1975). *Proc. Nutr. Soc.,* **34**, 203–293.

Symposium on comparative aspects of fatty acid metabolism (1983). *Proc. Nutr. Soc.* **42**, 263–360.

Properties of lipids and their metabolism in monogastric species

Balnave, D. (1970). Essential fatty acids in poultry nutrition. *World's Poult. Sci. J.*, **26**, 442–460.

Balnave, D., Cumming, R. B., and Sutherland, T. M. (1977). A biochemical explanation of the fatty livers and kidney syndrome of broilers. *Br. J. Nutr.*, **38**, 319–328.

Bauman, D. E. (1976). Intermediary metabolism of adipose tissue. *Fed. Proc.*, **35**, 2308–2313.

Butter, G. J. (1975). Lipid metabolism in the fowl under normal and abnormal conditions. *Proc. Nutr. Soc.*, **34**, 29–34.

Freeman, C. P. (1976). Digestion and absorption of fat. In *Digestion in the Fowl* (K. N. Boorman and B. H. Freeman, eds), pp. 117–142. Edinburgh: British Poultry Science Ltd.

Freeman, C. P. (1984). Digestion, absorption and transport of fats—nonruminants. In *Fats in Animal Nutrition* (J. Wiseman, ed.). London: Butterworths.

Garrett, R. L., and Young, R. J.. (1975). Effect of micelle formation on the absorption of neutral fat and fatty acids by the chicken. *J. Nutr.*, **105**, 827–838.

Giesecke, D. (1983). Plasma free fatty acids. In *Dynamic Biochemistry of Animal Production* (P. M. Riis, ed.), pp. 197–214. Amsterdam: Elsevier.

Griminger, P. (1976). Lipid metabolism. In *Avian Physiology* (P. Sturkie, ed.), pp. 232–252. New York: Springer-Verlag.

Hurwitz, S., Bar, A., Katz, M., Sklan, D., and Budowski, P. (1973). Absorption and secretion of fatty acids and bile acids in the intestine of the laying fowl. *J. Nutr.*, **103**, 543–547.

Ivy, C. A., and Nesheim, C. M. (1973). Factors affecting the liver fat content of laying hens. *Poult. Sci.*, **52**, 281–291.

Krogdahl, A. (1985). Digestion and absorption of lipids in poultry. *J. Nutr.*, **115**, 675–685.

Leveille, G. A. (1970). Adipose tissue metabolism: influence of periodicity of eating and diet composition. *Fed. Proc.*, **29**, 1294–1301.

Meijering, A. (1979). Fatty liver syndrome in laying hens. *World's Poult. Sci. J.*, **35**, 79–94.

O'Hea, E. K., and Leveille, G. A. (1969). Significance of adipose tissue and liver as sites of fatty acid synthesis in the pig and the efficiency of utilization of various substrates for lipogenesis. *J. Nutr.*, **99**, 338–344.

O'Hea. E. K., and Leveille, G. A. (1969). Lipid biosynthesis and transport in the domestic chick. *Comp. Biochem. Physiol.*, **30**, 149–159.

Pearce, J. (1974). The interrelationship of carbohydrate and lipid metabolism. *World's Poult. Sci. J.*, **30**, 115–128.

Robinson, A. M., and Williamson, D. H. (1981). Physiological roles of ketone bodies as substrates and signals in mammalian tissues. *Physiol. Rev.*, **60**, 143–187.

Sklan, D., and Hurwitz, S. (1980). Intestinal uptake of fatty acids complexed to proteins in the chick intestine. *J. Nutr.*, **110**, 270–274.

Stallones, R. A. (1983). Ischaemic heart disease and lipids in blood and diet. *Annu. Rev. Nutr.*, **3**, 155–185.

Fat metabolism in ruminants

Book

Christie, W. W. (1981). *Lipid Metabolism in Ruminant Animals.* Oxford: Pergamon Press.

Articles

Symposium on Fat metabolism in ruminants (1970). In *Symposia on Ruminant Physiology, Digestion and Metabolism.* Third symposium, pp. 489–544 (see Bibliography to Chapter 3).

Symposium on Lipid metabolism (1974). In *Symposia on Ruminant Physiology, Digestion and Metabolism.* Fourth symposium, pp. 465–523.

Bauman, D. E., and Davis, C. L. (1974). Regulation of lipid metabolism in ruminants. In *Symposia on Ruminant Physiology, Digestion and Metabolism.* Fourth symposium, pp. 496–509.

Bell, A. W. (1981). Lipid metabolism in liver and selected tissues and in the whole body of ruminant animals. *Prog. Lipid Res.*, **18**, 117–164.

Christie, W. W. (1980). The composition, structure and function of lipids in the tissues of ruminant animals. *Prog. Lipid Res.*, **17**, 111–206.

Clapperton, J. L., and Steele, W. (1983). Fat supplementation in animal production—ruminants. *Proc. Nutr. Soc.*, **42**, 343–350.

Harfoot, C. G. (1980). Lipid metabolism in the ruminant. *Prog. Lipid Res.*, **17**, 21–54.

Harrison, F. A., and Leat, W. M. F. (1975). Digestion and absorption of lipids in non-ruminant and ruminant animals: a comparison. *Proc. Nutr. Soc.*, **34**, 203–210.

McDonald, W., and Scott, T. W. (1977). Foods of ruminant origin with elevated content of polyunsaturated fatty acids. *World Rev. Nutr. Diet.*, **26**, 144–207.

Moore, J. H., and Christie, W. W. (1984). Digestion absorption and transport of fats in ruminants. In *Fats in Animal Nutrition* (J. Wiseman, ed.). London: Butterworths.

Noble, R. C. (1980). Digestion, absorption and transport of lipids in ruminant animals. *Prog. Lipid Res.*, **17**, 55–92.

Scott, T. W., and Cook, L. J. (1974). Effect of dietary fat on lipid metabolism in ruminants. Symposia on Ruminant Physiology, Digestion and Metabolism. Fourth Symposium (I. W. McDonald and A. C. I. Warner, eds), pp. 510–523. Armidale, N.S.W., Australia: Univ. of New England.

Vernon, R. G. (1982). Lipid metabolism in the adipose tissue of ruminant animals. *Prog. Lipid Res.*, **19**, 23–106.

Wiggers, K. D. *et al.* (1977). Type and amount of dietary fat affect relative concentrations of cholesterol in blood and other tissues of calves. *Lipids*, **12**, 586–590.

CHAPTER 7

Proteins and their Metabolism in Monogastric Animals

The word protein stems from the Greek word *proteios* meaning prime or primary. This is appropriate since protein is the most fundamental component of animal tissues and a continuous supply of it is needed throughout life. All cells contain protein and a rapid cell turnover occurs. Consequently, it is essential to provide replacement protein from the diet to meet turnover requirements in all kinds of animals. In addition to protein needed for repair, protein is necessary for growth and formation of animal products (meat, milk, eggs). Most protein-containing feeds are of plant origin. Thus the conversion of plant proteins into animal proteins is the main purpose of animal production.

STRUCTURE AND AMINO ACID COMPOSITION OF PROTEINS

Proteins are complex organic compounds ranging in molecular weight from about 5000 to one million. The monomer units of proteins are amino acids. The amino acids contain one carboxyl group and at least one amino group ($-NH_2$) in the α-position according to the general formula:

$$\begin{array}{c} COOH \\ | \\ H-C-NH_2 \\ | \\ R \end{array}$$

The nature of the R group varies with different amino acids. The formulae of 20 amino acids occurring in proteins are given here.

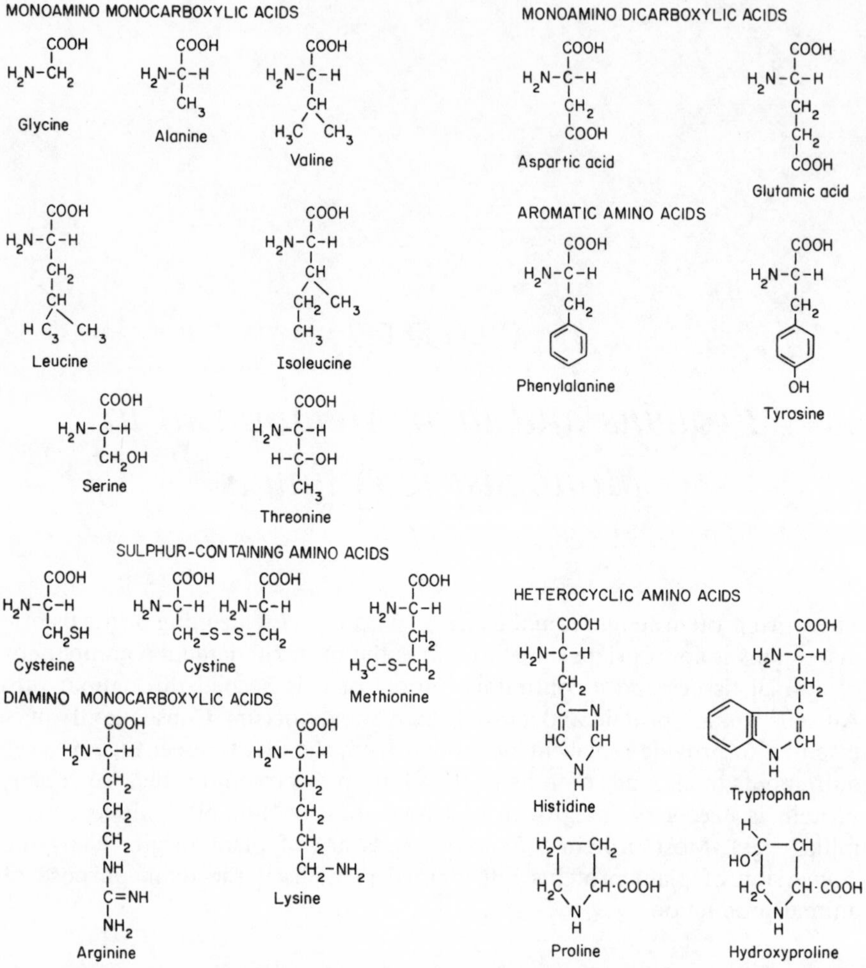

MONOAMINO MONOCARBOXYLIC ACIDS

Glycine

Alanine

Valine

Leucine

Isoleucine

Serine

Threonine

MONOAMINO DICARBOXYLIC ACIDS

Aspartic acid

Glutamic acid

AROMATIC AMINO ACIDS

Phenylalanine

Tyrosine

SULPHUR-CONTAINING AMINO ACIDS

Cysteine

Cystine

Methionine

DIAMINO MONOCARBOXYLIC ACIDS

Arginine

Lysine

HETEROCYCLIC AMINO ACIDS

Histidine

Tryptophan

Proline

Hydroxyproline

All amino acids except glycine have an asymmetrical carbon atom. Only those amino acids belonging to the L-series are commonly found in proteins, and with a few exceptions, these are the only ones that can be used by animals for metabolic purposes. Synthetic amino acids are usually found as a racemic mixture of L- and D-isomers, a fact which must be considered when synthetic amino acids are used as feed additives (see later).

L-Amino acid

D - Amino acid

The amino acids are linked together by peptide bonds

$$-\underset{\displaystyle \mathrm{H}}{\underset{|}{\mathrm{N}}}-\underset{\displaystyle \mathrm{O}}{\underset{\|}{\mathrm{C}}}-$$

in which the nitrogen of one amino group in one molecule is joined to the carboxyl group of a second amino acid through the elimination of water.

$$\underset{\text{Alanine}}{\mathrm{H_2N{-}CH(CH_3){-}C(=O){-}OH}} + \underset{\text{Glycine}}{\mathrm{H{-}NH{-}CH_2{-}COOH}} \rightarrow \underset{\text{Alanylglycine}}{\mathrm{H_2N{-}CH(CH_3){-}C(=O){-}NH{-}CH_2{-}COOH}} + \mathrm{H_2O}$$

Proteins are made up of one or more chains of amino acids. These chains are termed poylpeptides because the amino acids are linked by peptide linkages as follows:

$$\mathrm{H_2N{-}CH(R_1){-}CO{-}NH{-}CH(R_2){-}CO{-}NH{-}CH(R_3){-}CO{-}NH{-}CH(R_4){-}CO{-}NH{-}CH(R_5){-}CO{-}NH{-}CH(R_6){-}CO{-}NH{-}CH(R_7){-}CO{-}NH{-}CH(R_8){-}COOH}$$

The kind of amino acids present in a protein molecule, their relative amounts and sequential arrangements are unique for each specific protein. Composition, molecular size and spatial arrangement of the various proteins have a striking influence on the physical and chemical properties of these proteins and make them suitable for their particular functions. Proteins are structural components in organs, enzymes, transporting agents, hormones, buffers, etc. Amino acids are liberated from proteins by hydrolytic breakdown. This is effected in the laboratory by heating with strong acids and in living organisms by enzymatic actions.

$$\mathrm{R_1{-}CH(NH_2){-}C(=O){-}NH{-}CH(R_2){-}COOH} + \mathrm{H_2O} \rightarrow \mathrm{R_1{-}CH(NH_2){-}C(=O){-}OH} + \mathrm{H_2N{-}CH(R_2){-}COOH}$$

Since proteins are large molecules which cannot cross the intestinal wall, they are hydrolysed to amino acids. Amino acids are liberated from dietary proteins by digestive enzymes acting in the gastrointestinal tract. The amino acids liberated are absorbed into the mucosa, pass into the bloodstream and are removed to the tissues where they are built up into the proteins of the body. Thus, animals do not require dietary proteins per se, but rather the amino acids which are the breakdown products of proteins (see later). The nutritive value of proteins supplied in the diet depends on the percentage of amino acids that are digested and made available to animals.

CLASSIFICATION OF PROTEINS

Very few individual proteins have been isolated from animal organs or from animal feeds. Mixtures of proteins present in these sources have been investigated and the respective proteins can be classified as follows: (1) simple proteins which yield only amino acids on hydrolysis, and (2) conjugated proteins which are simple proteins combined with non-protein compounds. The non-protein entity is referred to as the prosthetic group.

Simple Proteins

There are two groups of simple proteins, globular and fibrous, distinguished by their structural configuration.

Globular Proteins

These are relatively soluble and are quite compact due to the considerable amount of folding of a long polypeptide chain. Biologically active proteins, such as enzymes, protein hormones and oxygen-carrying proteins belong to this group. Albumins are water-soluble and form a significant part of seed proteins, serum proteins and egg proteins (egg albumin; see Chapter 20). Globulins are sparingly soluble in water, but their solubility is increased by the addition of neutral salts, such as sodium chloride. Examples are serum and muscle globulins and globulins present in leguminous seeds. Glutelins (soluble only in dilute alkali) and prolamins (soluble in alcohol) characterize the insoluble and hydrophobic behaviour of wheat and corn proteins; both these proteins are resistant to the hydrolytic action of rumen microorganisms.

Leaf proteins are less well characterized than those of seeds. The major components of leaf and stem proteins are soluble protoplasmic proteins which are further subdivided into cytoplasmic and chloroplastic proteins. The much less soluble cell wall proteins form a component minor in quantity. The protoplasmic proteins represent the enzymatic machinery of plant metabolism and are of higher nutritional quality than the storage proteins of plant seeds. Utilization of proteins extracted from green plants such as lucerne has been proposed for human and animal nutrition.

Fibrous Proteins

These are composed of elongated peptide chains joined by several types of cross-linkages to form a stable and rather insoluble structure. These proteins are responsible for the mechanical properties of many animal tissues and organs in the form of collagen, elastin and keratin.

Collagen is a principal component of connective tissue, its content increasing with the age of an animal, making the meat of older animals less tender (see Chapter 18). *Elastin* is an elastic fibrous protein found in tendons and arteries.

Collagen and elastin contain high percentages of the amino acids glycine and proline. Collagen is further characterized by the presence of hydroxyproline which is unique to this protein. Considerable quantities of collagen and elastin are present in abattoir by-products used as animal feeds. Both these proteins are of low biological value because of their quite uniform amino acids composition. Collagen is resistant to digestive enzymes, but is converted to easily digestible gelatins by boiling in water or diluted acids.

Keratins are found in epithelial tissues, hair, wool, feathers, horns and hooves. They have a high cystine content, and the many disulphide bonds linking the peptide chains contribute to their mechanical strength and resistance towards digestion. Only properly hydrolysed feathers containing about 85% protein can substitute for limited amounts of other protein sources in poultry diets. (For contractile proteins taking part in muscle contraction see Chapter 22.)

Conjugated Proteins

Glycoproteins are protein–carbohydrate complexes. The chondroitin sulphates occurring in cartilage, tendon and skin contain sulphated polysaccharides. Mucoproteins containing hexoses, pentoses, amino sugars and other sugar derivatives are present in mucus secretions such as saliva, and in egg-white constituents (ovalbumin, ovomucid; see Chapter 20).

Lipoproteins are compounds containing lipids. They occur in egg yolk, nervous tissues (myelin) and membranes (such as those of erythrocytes). Soluble lipoproteins are a major form of transport for lipids in the blood (see Chapter 6).

Chromoproteins are proteins combined with a coloured compound, such as haemoglobin, myoglobin in the muscles, and chlorophyll.

PROTEIN AND NON-PROTEIN NITROGEN COMPOUNDS IN ANIMAL FEEDS

As pointed out in Chapter 1, the protein content of food and samples of animal tissues is obtained by multiplying the nitrogen content by the factor 6.25. This calculation of the crude protein content is not very exact, since (1) the nitrogen content of some proteins deviates somewhat from the average value of 16% (for example, it is 15.7% in milk protein and 17.15% in cereal proteins), and (2) many foods contain simple nitrogen compounds in addition to protein.

Plants and animal tissues contain various nitrogen-containing compounds which are not proteins, i.e. they are not amino acids joined by peptide bonds. The respective compounds are called non-protein nitrogen. Nucleic acids are non-protein nitrogen compounds that are present in all animal and plant tissues. The following non-protein nitrogen compounds are found in animal tissues and in their excreta: urea, uric acid, creatine and hippuric acid (see Chapter 3). Non-protein nitrogen compounds that occur in various feeds of plant origin are free amino acids, amides and alkaloids. Roughage (green forages, hay and silage) contains considerable amounts of non-protein nitrogen compounds; mature seeds and plant concentrate feeds contain only very small amounts. These compounds form 10–15% of total nitrogen present in green forages; this percentage is even higher in roots and tubers. Free amino acids and amides are the major components of the non-protein nitrogen fraction present in green forages (see Table 7.1). The concentration of free amino acids is high in young, rapidly growing plants and declines with maturity. The non-protein nitrogen content is particularly high in silage and makes up as much as 50% of the total nitrogen. This is due in part to the early harvesting of silage crops and in part to fermentation processes. Plant proteases in ensilaged forage hydrolyse the proteins to amino acids, and further breakdown of amino acids leads to amines and ammonia.

Table 7.1 Non-protein nitrogen compounds in some feedstuffs (according to Kirchgessner, M., *Animal Nutrition*, Third Ed., p. 72, 1978. Frankfurt: DLQ)

	Green pasture grass	Green lucerne	Maize grains
Total Nitrogen (% dry matter)	3.0	2.8	1.4
Nitrogen compounds (% of Nitrogen)			
Peptides and free amino acids	13.9	18.5	1.16
Ammonia	1.0	0.6	0.07
Amides	2.9	2.6	
Choline	0.5	0.1	0.12
Betaine	0.6	1.1	0.01
Purines	2.2	1.3	0.05
Nitrates	2.4	0.3	
Other non-protein nitrogen compounds	6.4	3.5	0.59

Free amino acids are well utilized by all kinds of animals, but only ruminants are capable of utilizing other non-protein nitrogen compounds. Since non-ruminants are incapable of utilizing most non-protein nitrogen compounds and their regular feeds do not contain them, separate determinations of protein and non-protein nitrogen are not necessary in routine analyses of feedstuffs (see Chapter 1).

PROTEIN DIGESTION AND ABSORPTION IN NON-RUMINANTS

It has already been mentioned that because they are high-molecular-weight compounds, proteins can only be absorbed into the mucosal cells of the small intestine after being broken down to low-molecular-weight compounds, primarily amino acids. Digestion of proteins takes place in the stomach and upper small intestine by enzymes from three sources: the mucosa of the stomach, the mucosa of the intestine, and the pancreas, which produces the most efficient enzymes. A summary of the enzymes involved in the digestion of protein, i.e. their source of secretion and substrate requirements, is given in Table 7.2.

Table 7.2 The proteolytic enzymes, their production sites and specificity of action

Enzyme	Site of production	Splits peptide bonds adjacent to:	pH of optimal activity
Pepsin	Mucosa of stomach	Tryptophan, phenylalanine, tyrosine, methionine, leucine	1.8–2.0
Trypsin	Pancreas	Arginine, lysine	8–9
Chymotrypsin	Pancreas	Aromatic amino acids and methionine	8–9
Elastase	Pancreas	Aliphatic amino acids	8–9
Carboxy-peptidase A	Pancreas	Aromatic amino acids*	7.2
Carboxy-peptidase B	Pancreas	Arginine, lysine*	8.0
Amino-peptidase	Mucosa of intestine	Amino acids with free NH_2 groups*	7.4

* On the end of the peptide chain.

All these protein-disintegrating enzymes are initially secreted as inactive precursors, called zymogens. Conversion of zymogens into active enzymes involves the removal (also by enzymatic process) of a small peptide, which shields and blocks the active centre of the enzyme in the precursor state. Pepsinogen, the inactive precursor of pepsin which is secreted from the mucosa of the stomach, has a molecular weight of 42 000, whereas the active enzyme pepsin has a molecular weight of 34 500.

$$\text{Pepsinogen} \xrightarrow[+\ \text{HCl}]{\text{Pepsin}} \text{pepsin} + \text{peptide}$$

The hydrogen ion concentration prevailing in the stomach permits the autocatalytic conversion of pepsinogen to pepsin with the splitting off of smaller peptide chains. Enzymes secreted by the pancreas are activated within the duodenal lumen as follows: the activation of trypsinogen by intestinal enterokinase produces trypsin which hydrolyses bonds in the other zymogens to form the active enzymes. A scheme dealing with the formation of enzymes from zymogens is given in Figure 7.1.

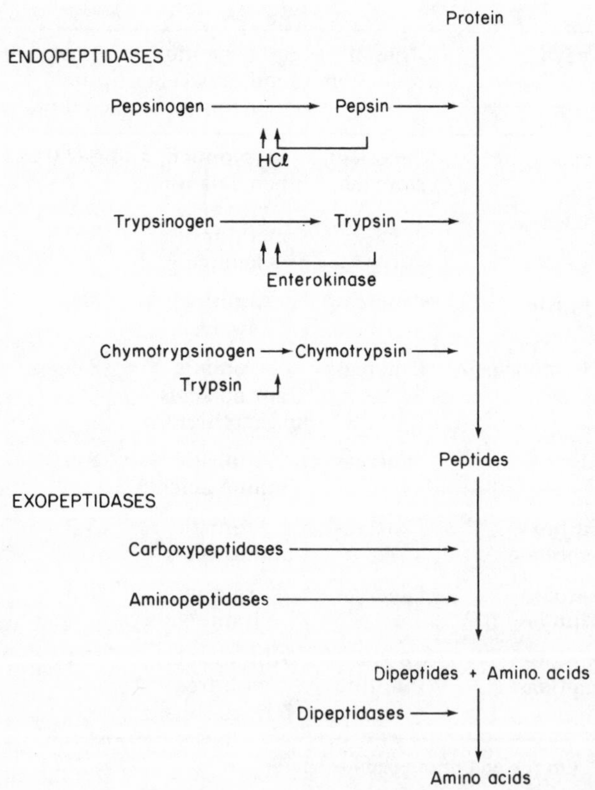

Figure 7.1. The proteolytic enzymes of the alimentary tract—modes of their formation and action.

Pepsin is known to hydrolyse several different bonds, particularly between the aromatic amino acids (see Table 7.2). Enzymes of the pancreatic juice show specificity toward bonds adjacent to lysine or arginine (e.g. trypsin), to aromatic amino acids (e.g. chymotrypsin) or to neutral aliphatic amino acids (e.g. elastase). Two types of proteolytic enzymes act in the digestive tract. The enzymes discussed above are endoenzymes, which cleave large peptides into smaller ones by acting in the middle of a large peptide chain. The second type of enzymes, the exopeptidases, prefer to act on smaller peptides and attack terminal peptide linkages to yield free amino acids. The exopeptidases may be further divided into aminopeptidases and carboxypeptidases. Aminopeptidases attack peptides which have a free terminal $-NH_2$ group, and the subsequent hydrolysis occurring in the adjacent bond leads to the liberation of the terminal amino acid. Analogously, carboxypeptidases break off peptides having a free terminal $-COOH$ group.

Sites of Protein Digestion in the Animal

The main objective of protein digestion, i.e. complete hydrolysis of dietary proteins to absorbable amino acids, is achieved by the protein- and peptide-cleaving enzymes, which are specific for certain linkages, acting in a definite sequence.

Protein digestion starts in the stomach with denaturation by hydrochloric acid followed by digestion with pepsin. Native dietary proteins show a considerable degree of resistance to the action of digestive enzymes because of the uniquely ordered steric structure of the protein. Denaturation of protein is the transition from the highly ordered protein structure to a less ordered one, rendering the protein more susceptible to enzymatic hydrolysis. Such denaturation is effected by heat treatment of foods, or by the action of hydrochloric acid as such, which is secreted into the lumen of the stomach. Denaturation by hydrochloric acid renders the pepsin-sensitive bonds present in protein foods more susceptible to the action of pepsin. The dietary proteins are then broken down by pepsin and hydrochloric acid into large polypeptides, with only small amounts of free amino acids being liberated. Protein molecules resistant to the action of the stomach, together with large peptide fragments resulting from peptic digestion, enter the duodenum to be further hydrolysed in alkaline medium by pancreatic enzymes. These enzymes release small peptides (chain length: 2–6 amino acids) and considerable amounts of free amino acids. The small peptides become hydrolysed, their breakdown occurring to a limited extent within the intestinal lumen, by the action of peptidases secreted from the pancreas and peptidases that are present in desquamated mucosal cells. Most of the small peptides formed in the intestine are hydrolysed by peptidases associated with the outer (digestive–absorptive) surface, the brush border of the epithelial cells, rather than by peptidases acting inside the mucosal cells into which the peptides are absorbed. The brush border membrane thus fulfils a double function, in absorbing amino acids and peptides and in enzymatic hydrolysis of small peptides to free amino acids (see Figure 7.2).

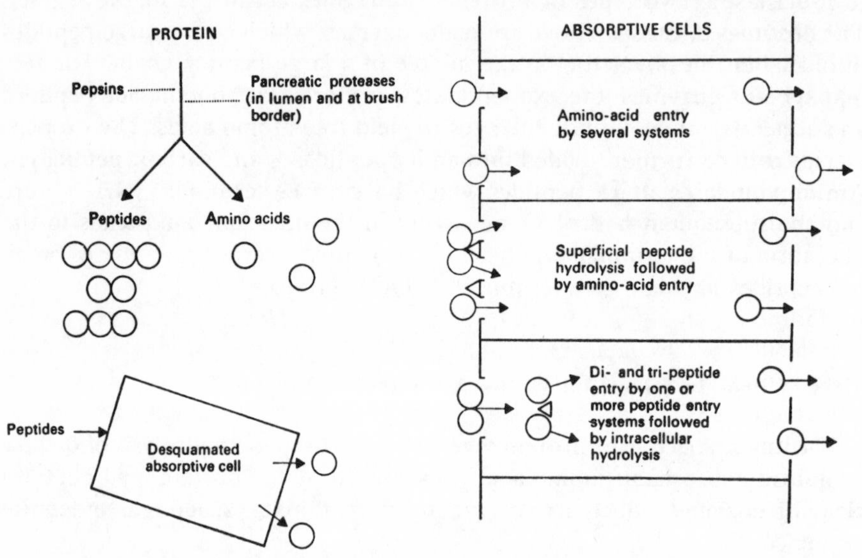

Figure 7.2. Scheme of protein absorption (after Matthews, 1972, *Proc. Nutr. Soc.* **31**, p. 172). *Reproduced by permission of Cambridge University Press.*

The preparatory step of protein digestion occurring in the stomach is not essential; for example, carp lack a stomach but are able to utilize protein very well because digestion occurs solely in the intestine.

Rennin, an enzyme of weak proteolytic activity similar to that of pepsin, is secreted into the abomasum of young ruminants and causes clotting of milk protein. Rennin liberates a glycopeptide from κ-casein. The remaining molecule, paracasein, together with calcium ions form a clot or curd which slows the passage of proteins in the intestinal tract, allowing additional time for enzyme activity and so improving the utilization of milk proteins.

Protein Absorption

A scheme of protein digestion and absorption is given in Figure 7.2. The amino acids are absorbed from the small intestine by an active transport mechanism (see Chapter 3) which in most cases is sodium-dependent. Vitamin B_6 (pyridoxine) may also enhance normal amino acid transport across the cell. Several transport systems have been described for amino acid transfer and these can be classified into three main groups for neutral, basic and acid amino acids. These different transport systems reflect the existence of different carriers, each responsible for the transport of members of one of the groups. The members of each group compete among themselves, but not with those of other groups, for absorption (see Chapter 3).

As a result of the processes of protein digestion and absorption only free amino acids pass into the portal vein and ultimately into the liver. The rate of absorption of amino acids into the portal blood depends upon the amino acid composition of the digest present in the intestine. The absorption of amino acids and simple peptides by the intestinal epithelium is quite a rapid process and there is a post-prandial increase in concentrations of amino acids in the portal circulation.

The intestinal epithelium acts as selective filter in growing and adult farm animals, and it is impermeable to protein. Only newborn mammals are able to absorb proteins ingested in the colostrum (immunoglobulins) which confer passive immunity upon suckling animals (see Chapter 18).

Some physiological importance may be attributed to peptides of medium molecular size which are produced in the duodenal lumen as intermediate products of protein digestion. These peptides stimulate the release of hormones from small intestinal mucosa, e.g. cholecystokinin-pancreozymin which induces the secretion of pancreatic enzymes and other hormones controlling intestinal motility (see Chapter 3).

Effect of Nutritional Conditions on Activity of Proteolytic Enzymes

Effect of Protein Level in the Diet

Secretion of proteolytic enzymes by the pancreas appears to be regulated by the presence of protein in the intestinal lumen. For example, it has been demonstrated in pigs that the activity of chymotrypsin varies greatly according to the protein content of the diet (an increase in activity of 250% in the case of an increase of protein level in the diet from 10 to 30%). This phenomenon may be based upon a regulation by feedback control of the proteolytic enzymes. The dietary proteins probably act by binding the enzymes in the gut content. With increase of protein intake, the amount of free enzyme present in the intestinal lumen will drop, thus enhancing the synthesis and secretion of pancreatic enzymes.

Significance of Protease Inhibitors

The effect of protease inhibitors present in soybeans and other leguminous seeds on the production of pancreatic enzymes is even more marked than the respective action induced by protein surplus. Small amounts of certain proteins present in soybeans inhibit the activity of trypsin and chymotrypsin, and the use of raw soybeans as feed for young rats and chicks leads to growth impairment. Since the protease inhibitors are heat sensitive, soybeans must be heated in order to support growth in these animals. Feeding of unheated soybeans or of isolated soybean trypsin inhibitors causes increased secretion of pancreatic enzymes and hypertrophy of the pancreas. These effects are due to the tenacious binding of trypsin to the inhibitor inside the intestine. Growth depression caused by feeding the trypsin inhibitor may be attributed to the loss of amino acids present in enzymes being secreted by a hyperactive pancreas and finally excreted in the faeces. A second factor responsible for the growth-retarding effect of raw soybeans is the resistance of native soybean protein to digestion by trypsin unless first denaturated by heat (see above).

Overheating of Protein

Mild heat treatment of many proteins enhances their susceptibility to digestion, whereas excessive heating impairs the value of proteins by the Maillard or browning reaction. In this reaction, aldehyde groups of reducing sugars react with free amino groups of the peptide chain, particularly the ϵ-amino group of lysine, and aminosugar complexes are formed making the lysine nutritionally unavailable. To avoid overheating, the process of toasting soybean meal must be carefully controlled. Properly heated soybean meal forms a very valuable food for all farm animals, since apart from protease inhibitors other inhibitory substances present in soybeans are also destroyed by mild heating and, as mentioned earlier, the availability of amino acids is increased by heat denaturation of proteins.

NUTRITIVE IMPORTANCE OF AMINO ACIDS

The nutritive effects of proteins depend on those amino acids that are released from them by digestive processes. For nutritional purposes, amino acids may be divided into two groups: essential or non-dispensable amino acids, and non-essential or dispensable amino acids (see Table 7.3).

Essential amino acids cannot be synthesized in the body at a rate adequate to meet physiological needs and they must therefore be supplied in the diet. Although the non-essential amino acids are needed for protein synthesis in the body, they are not needed as dietary components since the body can synthesize them by transferring amino groups to certain intermediates of carbohydrate metabolism (see Chapter 4) or by conversion of some essential amino acids into certain non-essential acids, such as methionine into cysteine or phenylalanine into tyrosine (see later). The reverse reactions, however, never take place in the body. In principle, the dispensable amino acids can be synthesized in the cell by using non-specific sources of amino nitrogen (glutamic acid, diammonium citrate), although a mixture of dispensable amino acids provided to animals together with the essential amino acids is more efficient than the use of a mixture of essential amino acids supplemented with one single dispensable amino acid or with ammonium salt.

This phenomenon is interesting with regard to the definition of dispensable amino acids, but is of less importance nutritionally because of the satisfactory pattern of dispensable amino acids in every practical diet. However, the definition of an indispensable amino acid as one that must be supplied preformed in the diet is not exactly correct. Most of the essential amino acids can be replaced by their corresponding α-keto or α-hydroxy analogues except those of lysine and threonine. In fact, the hydroxy analogue of methionine can be used as a dietary supplement to livestock rations (see later). It appears that the animal's requirement is for the preformed carbon skeleton of the essential amino acid which the cell cannot synthesize, but to which the amino nitrogen can be attached. However, the ability to add ammonia to keto analogues of threonine or lysine is lacking.

The non-natural D-antipode of methionine is used by animals as efficiently as the L-isomer, since it is transformed in the L-isomer by enzymatic processes occurring inside the tissues. Thus, synthetic methionine, which is a mixture of equal quantities of both antipodes, can be recommended as supplement for feed mixtures. The ability to transform the D-antipode into the L-antipode does not apply to other essential amino acids and particularly not to threonine and lysine. L-lysine prepared by microbial synthesis is pure L-lysine and is thus suitable for feeding purposes.

Table 7.3 shows the classification of essential and non-essential amino acids for young rats. The list of dispensable and indispensable amino acids varies somewhat with species, age and function of animals. Arginine was found to be a special case; it can be synthesized by the rat, but not at a sufficiently rapid rate to meet the demands for maximal growth. Adult rats, however, are not affected by absence of arginine in the diet. The amino acids

Table 7.3 Nutritional classification of amino acids with regard to their growth effect on rats

Essential (indispensable)	Non-essential (dispensable)
Lysine	Glycine
Tryptophan	Alanine
Histidine	Serine
Phenylalanine	Tyrosine
Leucine	Aspartic acid
Isoleucine	Glutamic acid
Threonine	Proline
Methionine	Cystine
Valine	Hydroxyproline
Arginine*	

* Arginine is semi-essential (see text).

indispensable for growing rats are also indispensable for adult humans. The amino acid requirements of the dog, cat and pig are fairly similar to those of the rat. Unlike the situation in mammals arginine is an essential amino acid for chicks (see Table 7.4).

Table 7.4 Nutritional classification of amino acids with regard to their growth effect on chicks

Essential	Semi-essential	Non-essential
Lysine	Glutamic acid	Alanine
Tryptophan	Glycine or serine	Tyrosine
Histidine	Proline	Aspartic acid
Phenylalanine		Cystine
Leucine		Hydroxyproline
Isoleucine		
Threonine		
Methionine		
Valine		
Arginine		

Glycine plays a special role in birds, which excrete uric acid as the end-product of nitrogen metabolism (see later), since the formation of uric acid requires glycine. However, the ability to synthesize glycine in the growing chick cannot meet the demand for maximum growth rate. As is seen in Tables 7.3 and 7.4, the amino acids glycine, serine, glutamic acid and proline are semi-essential amino acids for growing chicks, as is arginine for growing rats. Serine and glycine are interchangeable since serine is easily converted into glycine inside the tissues. The semi-essential amino acids are not necessary for survival of the animals, but rather for optimal growth. Therefore the supply of semi-essential amino acids in practical diets for chicks is as important as that of the essential amino acids.

Ruminant animals are practically independent of the supply of essential amino acids in the diet since ruminal microorganisms synthesize essential amino acids required by the host animal (see Chapter 8).

The nutritive value of protein feeds depends mainly on their amino acid composition (see Table 7.5). Since most feedstuffs used for monogastric animals are highly digestible, the amounts of amino acids determined analytically are generally fully available. It is desirable that the essential amino acid composition of the diet should approximate the requirements of animals and the ratio of essential to non-essential amino acids should be about 1 : 1. Since the amino acid requirements of growing animals are closely related to tissue amino acid composition, the amino acid contents of most animal protein feeds such as fish meal resemble the amino acid composition required for growth. The amino acid composition of high-quality protein

Table 7.5 Amino acid composition (%) of some feedstuffs (According to NRC, *Nutrient Requirements of Poultry.* National Academy of Science, Washington, 1977)

	Fish meal	Skim milk (dried)	Cotton-seed meal	Soybean meal	Maize
Arginine	3.79	1.12	4.59	3.68	0.50
Glycine	4.19	0.27	1.70	2.29	0.37
Histidine	1.46	0.84	1.10	1.32	0.20
Isoleucine	2.85	2.15	1.33	2.57	0.37
Leucine	4.50	3.23	2.41	3.82	1.10
Lysine	4.83	2.40	1.71	3.18	0.24
Methionine	1.78	0.93	0.52	0.72	0.20
Cystine	0.56	0.44	0.64	0.73	0.15
Phenylalanine	2.48	1.58	2.22	2.11	0.47
Tyrosine	1.98	1.13	1.02	2.01	0.45
Threonine	2.50	1.60	1.32	1.91	0.39
Tryptophan	0.68	0.44	0.47	0.67	0.09
Valine	3.23	2.30	1.89	2.72	0.52
Protein	60.5	33.5	41.4	48.5	8.8

feeds of plant origin does not differ greatly from that of animal proteins. Such plant proteins can be marginal with respect to one or two essential amino acids. Therefore the required amino acid level can be obtained by appropriate combination of two or more plant protein feeds. For example, cereal proteins are deficient in lysine, whereas soybean meal is commonly added to corn to satisfy the lysine requirement. The main difference between animal and plant feeds is the presence of vitamin B_{12} in animal feeds and its total absence in plant feeds (see Chapter 12). Diets for monogastric animals deficient in essential amino acids can be enriched by the addition of synthetic amino acids such as methionine or lysine (see Chapter 18).

PROTEIN METABOLISM IN THE BODY

The processes of tissue protein synthesis from amino acids and breakdown of tissue proteins into amino acids take place simultaneously. The various reactions involving amino acids in the body are shown in Figure 7.3. The amino acids act in the body as a common pool, irrespective of their origin from three main sources. The latter are indicated with arrows directed from the periphery to the centre: (1) amino acids liberated from dietary proteins by digestion are absorbed through the intestinal wall into the circulating blood and are transported to the organs; (2) amino acids liberated by breakdown of tissue proteins also pass into the circulatory system where they mix with amino acids originating from the diet and together with them are transferred to the organs; (3) non-essential amino acids synthesized in the tissues also belong to the common pool of amino acids.

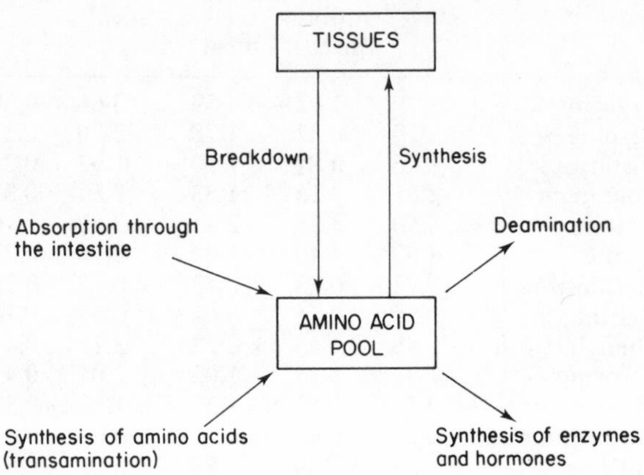

Figure 7.3. Scheme of metabolism of amino acids in the animal body.

Amino acids present in the body pool undergo the following biochemical reactions (indicated in Figure 7.3 by arrows directed from the centre of the periphery): (1) biosynthesis of proteins present in tissues and animal products (milk, eggs); (2) biosynthesis of hormones, enzymes and other biologically important nitrogen compounds such as nucleic acids, creatine and choline (see later); (3) breakdown of amino acids that have fulfilled their functions in the body or from which amounts exceeding the requirements were ingested. If the amino acid supply exceeds the potential for maximum protein production, surplus amino acids are deaminated (see later). The oxo-acids formed by deamination are fully oxidized and either provide energy for the animal or act as intermediate compounds in the transformation of proteins into carbohydrates or fats. The ammonia released by deamination is excreted as urea or uric acid (see later). The free amino acid pool in the body represents only 0.2–2.0% of the total amount of amino acids which are mainly protein bound (a small amount of bound amino acid is present in peptides). The free amino acid pool does not represent a reserve of protein synthesis precursors and depletes rapidly if the influx of amino acids into the pool suddenly ceases, by scarcity of protein in the diet, for example.

Synthesis, Degradation and Turnover of Proteins

The biosynthesis of proteins comprises a series of complex processes involving DNA, which carries the genetic information determining the exact structure of the protein being synthesized. DNA controls the formation of three types of RNA which determine the sequence of amino acids and the molecular size of the protein being formed. The mechanics of protein synthesis are covered in detail in textbooks of biochemistry. The energy requirements of protein biosynthesis, i.e. those needed for formation of a peptide bond, are explained in Chapter 14. Proteins, like other body constituents, especially fat, are continually undergoing degradation and renewal even in an adult animal. The amount of protein within a cell is governed by the difference between the rate of synthesis and the rate of degradation of proteins. The rate at which the constituents of a tissue are replaced is called the turnover rate (see Chapter 2). Protein turnover is quantitatively large and may amount to 5–10 times the daily intake of protein. About 80% of the amino acids undergoing synthesis in adult tissues originates from decomposed body proteins and only about 20% from the diet.

Rates of protein synthesis and breakdown in the live animal have been measured by administration of amino acids labelled with the stable isotopes ^{14}C or ^{15}N. Three different approaches are used. In large animals, when it is not possible to take samples of tissue, the rate of protein synthesis and breakdown can be obtained by measuring the rate of disappearance of the administered label from the free amino acid pool. Alternatively, when tissue samples can be taken, incorporation of label into the protein of that tissue can be determined and the rate of synthesis estimated. Finally, it is possible to wait until all the isotope is incorporated into the body protein and then to observe the subsequent rate of loss from the tissues, thus determining the rate of protein breakdown in addition to synthesis.

Protein turnover in different species varies considerably. This turnover, when related to 1 kg of body weight, decreases with increasing body weight as is seen in Table 7.6. The protein turnover is much higher in growing animals than in adult ones. The influence of body weight and age on protein turnover parallels other metabolic processes (see Chapter 17).

Table 7.6 Whole body protein turnover in various species (adult animals) (from Garlick, 1980)

Species	Body weight (kg)	Protein turnover (g/kg body wt per day)
Mouse	0.042	43.4
Rat	0.510	20.5
Rabbit	3.6	18.0
Dog	10.2	12.1
Sheep	67	5.3
Man	77	5.7
Cow	628	3.7

Protein turnover is controlled by the nutritional and hormonal status of the animal; growth hormone and insulin stimulate protein synthesis, while the glucocorticoids cause protein degradation. The turnover rates of protein vary considerably between tissues, even within the same species. One possible reason for this is the presence in some tissues of inherently stable proteins which result in low average turnover rates in these tissues. The intestinal mucosa is the most rapidly regenerating tissue in the body, the renewal of mucosal protein taking 1–3 days in most species. The following values for the average half-life of soluble tissue proteins of rats have been found: liver, 0.9 days; kidney, 1.7 days; heart, 4.1 days; brain, 4.6 days; and muscle, 10.7 days.

The contribution of muscle and liver to the total protein synthesis in pigs and rats is compared in Table 7.7. The liver, despite its rapid rate of turnover, contains relatively little protein and so contributes only 10% of whole body protein synthesis in both pigs and rats. By contrast, skeletal muscle is the largest organ in the body, but has a slow rate of synthesis. Nevertheless, muscle contributes 19% of the whole body protein synthesis in the rat and as much as 42% in the pig.

Table 7.7 Protein synthesis in liver and muscle of pigs and rats as a proportion of synthesis in the whole body (after Swick and Benevenga, 1977)

	Tissue weight (% of whole body)	Protein content (g/tissue)	Protein synthesis (g/day)	Protein synthesis (% of whole body)
Pig				
Liver	2	258	60.1	10
Muscle	45	6830	287	42
Rat				
Liver	4.1	0.62	0.42	10
Muscle	36	6.1	0.79	19

The rate of protein synthesis in muscle is considerably influenced by the nutritive status of the animal (see Figure 7.4), increasing after feeding and decreasing during fasting. During fasting, degradation exceeds synthesis and there is a flow of amino acids (especially alanine and glutamine) to the liver where they are converted into glucose.

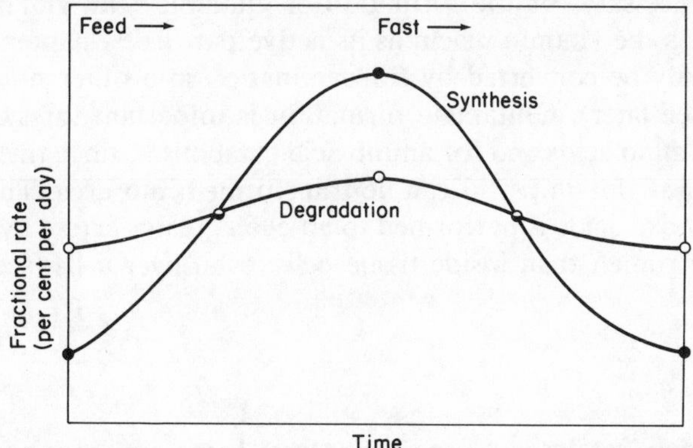

Figure 7.4. Rate of protein synthesis (measured by infusion of rats with ^{14}C-tyrosine) and degradation of skeletal muscle protein as influenced by the nutritive status of the animal (after Swick and Benevenga, 1977, *J. Dairy Sci.*, **60**, p. 513, by *permission of the Journal of Dairy Science*).

METABOLISM OF AMINO ACIDS

Biosynthesis of Amino Acids

Of the 19 amino acids commonly found in the animal body, about nine can be synthesized there, particularly in liver cells. The carbon skeleton of the non-essential amino acids can be formed from carbohydrates, fats or essential amino acids. The α-amino group present in biosynthesized amino acids originates either from ammonium ions or from the amino groups of other amino acids, according to the following two pathways.

(1) Formation of Amino Acids by Fixation of Ammonia

Glutamic acid is formed by reductive amination of α-ketoglutarate, an intermediate product of sugar breakdown:

$$\begin{array}{c} COOH \\ | \\ CO \\ | \\ CH_2 \\ | \\ CH_2 \\ | \\ COOH \\ \alpha\text{-Ketoglutaric acid} \end{array} \quad NH_3 + NADH \longrightarrow \begin{array}{c} COOH \\ | \\ CHNH_2 \\ | \\ CH_2 \\ | \\ CH_2 \\ | \\ COOH \\ \text{glutamic acid} \end{array} + NAD + H_2O$$

The enzyme necessary for the formation of glutamic acid, glutamate synthetase, contains the vitamin niacin as its active part (see Chapter 12). Glutamate can easily be converted by transamination into other non-essential amino acids (see later). Glutamate formation is important for synthesis of non-essential amino acids and for amino acid catabolism, since the ammonia incorporated into glutamic acid can be transformed into urea. The binding of ammonia by oxo acids is performed to an even greater extent by microorganisms of the rumen than inside tissue cells, even liver cells (see Chapter 8).

(2) Transamination

Transamination, catalysed by enzymes termed transaminases occurring in most animal tissues, particularly the liver, enables transformation of one amino acid into another. It is a reversible reaction between amino and α-oxo acids leading to an exchange of the amino and ketonic groups. Pyridoxal phosphate, the coenzyme form of vitamin B_6 (see Chapter 12), catalyses such reactions as, for example, that between oxaloacetic acid and glutamic acid leading to aspartic acid and ketoglutaric acid:

$$
\begin{array}{lllll}
\text{COOH} & \text{COOH} & & \text{COOH} & & \text{COOH} \\
| & | & & | & & | \\
\text{CH}_2 & \text{CH}_2 & & \text{CH}_2 & & \text{CH}_2 \\
| & | & & | & & | \\
\text{CO} & \text{CH}_2 & \longrightarrow & \text{CHNH}_2 & + & \text{CH}_2 \\
| & | & & | & & | \\
\text{COOH} & \text{CHNH}_2 & & \text{COOH} & & \text{CO} \\
 & | & & & & | \\
 & \text{COOH} & & & & \text{COOH}
\end{array}
$$

Oxaloacetic acid — glutamic acid — aspartic acid — α-ketoglutaric acid

In transamination (Figure 7.5), pyridoxal phosphate acts as an acceptor for an amino group. The resulting pyridoxamine transfers the amino group to a new α-keto acid resulting in the regeneration of pyridoxal phosphate and the formation of a new amino acid.

Other dispensable amino acids formed by transamination from breakdown products of carbohydrate metabolism are the following: alanine from pyruvic acid, serine from 3-phosphoglyceric acid, and glycine is easily obtained from serine. It is interesting to note that dicarboxylic acids (glutamic and aspartic acids) are transaminated to alanine after both these amino acids have been absorbed into intestinal mucosal cells (pyruvate + glutamate $\rightarrow$ alanine + α-ketoglutarate). Transaminases occur in most animal tissues. Certain transaminases diffuse from tissues into the blood plasma when the tissue

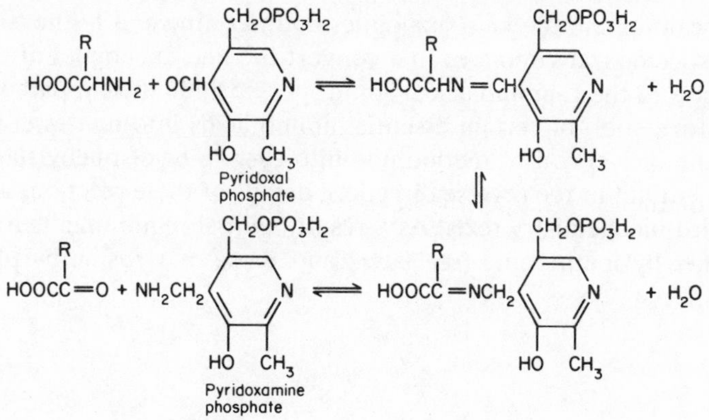

$$\mathrm{HOOC\overset{R}{\overset{|}{C}}HNH_2 + OCH\text{-}(pyridoxal\ phosphate;\ CH_2OPO_3H_2,\ HO,\ CH_3,\ N)} \rightleftharpoons \mathrm{HOOC\overset{R}{\overset{|}{C}}HN{=}CH\text{-}(ring)} + H_2O$$

Pyridoxal phosphate

$$\Updownarrow$$

$$\mathrm{HOOC\overset{R}{\overset{|}{C}}{=}O + NH_2CH_2\text{-}(pyridoxamine\ phosphate;\ CH_2OPO_3H_2,\ HO,\ CH_3,\ N)} \rightleftharpoons \mathrm{HOOC\overset{R}{\overset{|}{C}}{=}NCH_2\text{-}(ring)} + H_2O$$

Pyridoxamine phosphate

Figure 7.5. Mechanism of transamination.

suffers injury. This is the basis of a clinical test for heart muscle damage in cardiac infarction when the tissue suffers injury. Very high plasma levels are also found in disorders involving destruction of liver tissue, such as poisoning with carbon tetrachloride.

Transformation of Some Non-dispensable into Dispensable Amino Acids

As mentioned above, monogastric animals cannot synthesize carbon skeletons present in non-dispensable amino acids. However, synthetic oxo or hydroxy acids corresponding to non-dispensable amino acids ingested by monogastric animals are transformed to the respective amino acids, apparently by transamination. Hydroxymethionine, which is cheaper and more stable than methionine, has been recommended as a food additive for monogastric and ruminant animals (see Chapter 8) to meet their requirements for methionine. The conversion of hydroxymethionine to methionine occurs according to the following pathway:

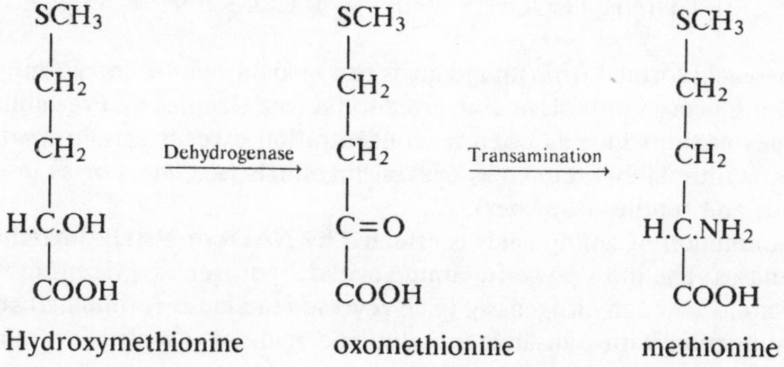

$$\begin{array}{ccccc} SCH_3 & & SCH_3 & & SCH_3 \\ | & & | & & | \\ CH_2 & & CH_2 & & CH_2 \\ | & & | & & | \\ CH_2 & \xrightarrow{\text{Dehydrogenase}} & CH_2 & \xrightarrow{\text{Transamination}} & CH_2 \\ | & & | & & | \\ H.C.OH & & C{=}O & & H.C.NH_2 \\ | & & | & & | \\ COOH & & COOH & & COOH \\ \text{Hydroxymethionine} & & \text{oxomethionine} & & \text{methionine} \end{array}$$

Likewise, the conversion of the optical D-antipode of amino acids into the biologically efficient L-antipode also occurs by transamination (see above). Animals cannot use the keto analogues of threonine and lysine, since they lack the specific transaminases that convert the keto analogues of these two amino acids to the L-amino acids. Some specific and efficient pathways exist for transformation of certain essential amino acids into non-essential ones, such as the conversion of methionine into cystine or of phenylalanine into tyrosine, but not in the reverse direction; details of these reactions are found in standard biochemistry texts. As a result, animal requirements for cystine can be met by methionine (see later) and those of tyrosine by phenylalanine.

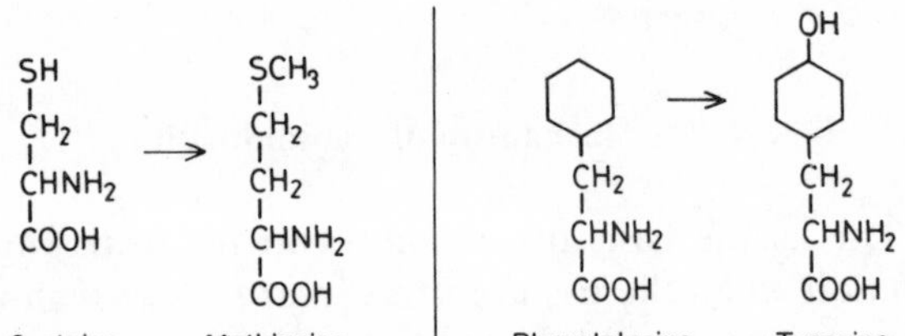

SH–CH_2–$CHNH_2$–COOH → SCH_3–CH_2–CH_2–$CHNH_2$–COOH

Cysteine → Methionine

C_6H_5–CH_2–$CHNH_2$–COOH → OH–C_6H_4–CH_2–$CHNH_2$–COOH

Phenylalanine → Tyrosine

Pathways of Amino Acid Degradation

Amino acids not used for synthetic purposes undergo catabolic reactions such as deamination or decarboxylation.

Deamination

Oxidative deamination of amino acids in tissues results in the liberation of ammonia and the conversion of the amino acid to the corresponding keto acid:

$$R.CH(NH_2).COOH + \tfrac{1}{2}O_2 \longrightarrow R.CO.COOH + NH_3$$

The oxo acid formed from amino acids can be oxidized for energy supply or used for glucose synthesis or converted to fat (see Chapter 6). Free ammonia becomes noxious in cells when its concentration exceeds certain limits and thus it is ultimately excreted as urea in mammals (see later) or as uric acid in birds and reptiles (see later).

Deamination of amino acids is effected by NAD- or NADP-linked dehydrogenases. The most powerful amino acid dehydrogenase present in tissues is glutamic acid dehydrogenase. (The reverse reaction is formulated above.) In conjunction with transaminase, glutamic acid dehydrogenase is taken to be responsible for the liberation of ammonia from other amino acids; all the amino nitrogen from amino acids which undergo transamination can be transferred to glutamate which is easily deaminated:

$$\alpha\text{-Ketoglutaric acid} + \text{amino acid} \longrightarrow \text{glutamic acid} + \alpha\text{-keto acid}$$

$$\text{Glutamic acid} + NAD^+ + H_2O \longrightarrow \alpha\text{-keto acid} + NADH + H^+ + NH_3$$

$$\text{Amino acid} + NAD^+ + H_2O \longrightarrow \alpha\text{-keto acid} + NADH + H^+ + NH_3$$

Such deaminative breakdown of branched-chain amino acids such as leucine, isoleucine and valine deserves close attention since they make up about 40% of the minimum daily requirement of indispensable amino acids in monogastric mammals. Apart from their role in protein biosynthesis, especially in muscles, they are important as metabolic fuels during stressful situations of caloric deprivation (starvation) and increased caloric needs (in working muscles; see Chapter 22). Unlike the metabolic pathways of the other essential amino acids occurring in the liver, the branched-chain amino acids, like the non-essential amino acids, are mainly degraded in muscles. The first step in the degradation of the branched-chain amino acids is the transfer of the amino group (transamination) yielding analogous branched α-keto acids (see above) followed by oxidative decarboxylation of these α-keto acids. The decarboxylation step is catalysed by a high-molecular-weight, multienzyme complex involving the same cofactors that are responsible for the conversion of pyruvic acid to acetic acid (see Chapters 4 and 6):

$$R.CO.COOH + CoAH + NAD^+ \longrightarrow R.CO.CoA + CO_2 + NADH + H^+$$

Disposal of the carbon skeleton of amino acids is the result of various reactions which are specific for each amino acid;, for example, leucine is degraded according to the following mechanism:

$$(CH_3)_2CH{-}CH_2{-}CHNH_2{-}COOH \longrightarrow (CH_3)_2CH{-}CH_2{-}CO{-}COOH \xrightarrow[+\ NAD]{+\ CoA} (CH_3)_2CH{-}CH_2{-}CO{-}CoA + CO_2 \xrightarrow{+\ FAD} (CH_3)_2C{=}CH{-}CO{-}CoA \xrightarrow[+\ H_2O]{ATP\ +\ CO_2}$$

Leucine — Ketoleucine — Isovaleryl-CoA

$$(CH_3)(HOOC{-}CH_2)C{=}CH{-}CO{-}CoA \xrightarrow{+\ H_2O} (CH_3)(HOOC{-}CH_2)C(OH){-}CH_2{-}CO{-}CoA \xrightarrow{Fission} CH_3{-}CO{-}CH_2{-}COOH + CH_3{-}CO{-}CoA$$

Acetoacetic acid

Pathways of the respective breakdown products of amino acids lead to glucose or to ketone bodies; thus amino acids are termed glucogenic and ketogenic, respectively. Valine, in common with most other amino acids, is glucogenic, whereas leucine is the only amino acid leading exclusively to ketone bodies. Isoleucine, like lysine, phenylalanine and tyrosine, is partly glucogenic and partly ketogenic. Ketone bodies are ultimately converted to fats or oxidized with liberation of energy (see Chapter 6). Glucogenic amino acids yield 1 mol of glucose per 2 mol of amino acids. Amino acids, whether from dietary origin or from breakdown of body proteins, are important sources of glucose in monogastric species when there is a shortage of carbohydrates in their diet, and always in ruminants which cannot directly utilize glucose ingested with food (see Chapter 5).

Decarboxylation

Breakdown of amino acids by decarboxylation occurs in animal tissues and in many microorganisms, particularly intestinal ones (see Chapters 3 and 8).

$$\begin{array}{c} R \\ | \\ CHNH_2 \\ | \\ COOH \end{array} \longrightarrow \begin{array}{c} R \\ | \\ CH_2NH_2 \\ + \\ CO_2 \end{array}$$

The significance of decarboxylation in animal tissues lies in the formation of important biological compounds such as histamine and serotonin (required for certain functions of nervous tissue), *p*-aminobutyric acid (in the brain), taurine (constituent of bile acids; see Chapter 6), and others.

Interorgan Amino Acid Exchange

Alanine and glutamine are quantitatively the most important amino acids for transport between the tissues. They are also the principal amino acids released from muscles into the blood, each accounting for 30–40% of the total α-amino nitrogen released from muscle. In the muscle, alanine is mainly formed by transamination of the branched-chain amino acids with pyruvic acid, and glutamine is formed by transfer of ammonia to glutamic acid:

$$
\begin{array}{l}
COOH + NH_3 + ATP \longrightarrow CONH_2 + ADP + P_i \\
\quad | \qquad\qquad\qquad\qquad\quad | \\
CH_2 \qquad\qquad\qquad\quad\; CH_2 \\
\quad | \qquad\qquad\qquad\qquad\quad | \\
CH_2 \qquad\qquad\qquad\quad\; CH_2 \\
\quad | \qquad\qquad\qquad\qquad\quad | \\
CHNH_2 \qquad\qquad\quad\;\; CHNH_2 \\
\quad | \qquad\qquad\qquad\qquad\quad | \\
COOH \qquad\qquad\qquad\; COOH \\
\text{Glutamic acid} \qquad\quad\;\; \text{glutamine}
\end{array}
$$

Glutamine released into the bloodstream is taken up mainly by intestinal tissue and the kidneys; it is an important oxidative fuel in the gut. Alanine originating mainly from muscle, is taken up by the liver where it is reconverted to glucose (see Figure 7.6). The unique role of alanine in interorgan amino acid exchange led Fehlig to the formulation of the glucose–alanine cycle which, in addition to the Cori cycle for lactate (see Chapter 4), provides the liver with precursors for glucose synthesis. Alanine also serves as an efficient transport vehicle for transferring nitrogen originating from amino acids, especially branched-chain ones, from muscle to liver for conversion into urea. The existence of the glucose–alanine cycle was established by Fehlig in humans and rats and is probably also valid in monogastric farm animals.

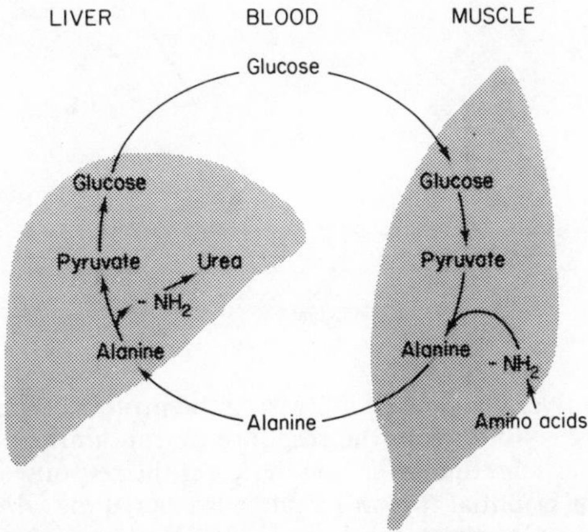

Figure 7.6. Glucose-alanine cycle (after Felig, 1975). *Reproduced with permission, from the* Annual Review of Biochemistry, *Volume 44, and from the author.* © 1975 by Annual Reviews Inc.

Metabolic Response to Amino Acids Ingested in Amounts Inadequate for the Requirements

Adaptation of Enzymatic Activities

Ingestion of excess amino acids is wasteful and in some cases even detrimental. However, the undesirable effects of a slight excess of certain amino acids in the circulatory system or in the tissues may be eliminated or alleviated by adaptation of enzymatic activities or hormonal effects (see later). Activity of deaminases or transaminases is enhanced to facilitate the breakdown of amino acids present in excess of the amounts needed for protein synthesis. For example, when the intake of threonine, an essential amino acid, is progressively increased, the activity of threonine dehydratase shows a sudden increase (see Figure 7.7) and the threonine above the critical level is destroyed. In the case of excess of non-essential amino acids, the levels of enzymes (such as glutamic acid transferase) responsible for their metabolism do not show this sudden increase, but increase progressively with intake of the amino acids. On the other hand, when the dietary level of an amino acid is low, the enzymes of its degradative pathway are present at low levels. This adaptation helps to conserve amino acids ingested at low levels.

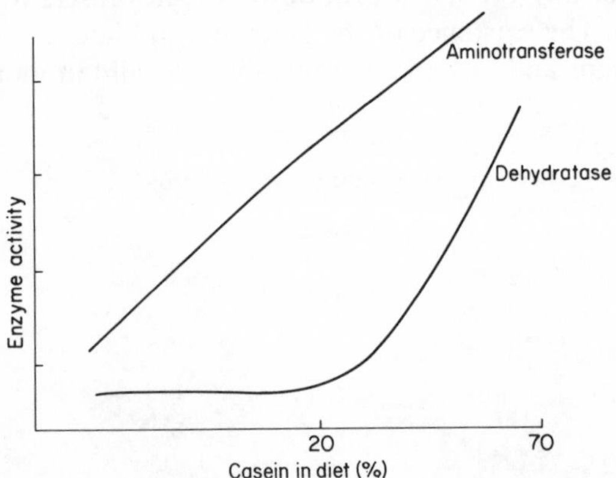

Figure 7.7. Effect on liver enzymes of increasing the protein content of a diet fed to growing rats. The upper line shows the response to transaminase metabolizing the non-essential glutamic acid, the lower line indicates the response of dehydratase, as enzyme for degrading essential threonine (after Harper, 1968. *Amer. J. Clin. Nutr.*, **21**, p. 358; adapted by Goodhart and Shils, 1978, *Modern Nutrition, etc.*, 6th edn, p. 68). *Reproduced by permission of Lea & Febiger.*

Hormonal Regulation of Amino Acid Metabolism

Insulin released after ingestion of a carbohydrate-high meal stimulates uptake of amino acids by cells of tissues such as muscle in which essential amino acids are degraded; insulin also stimulates incorporation of amino acids into proteins. Dietary supply of energy and protein increases also the rate of secretion of another hormone, thyroxine, which stimulates amino acid oxidation and protein turnover. The hormonal regulation represents an important factor in the adaptation of amino acid metabolism to variations in entry rates of nutrients from the digestive tract and to variations in tissue requirements for amino acids (see above).

CONDITIONS FOR ASSURING CORRECT SUPPLY OF AMINO ACIDS TO NON-RUMINANTS

Inadequate protein nutrition is the most common of all nutrient deficiencies. Signs of protein deficiency include reduced food intake and food utilization, reduced growth rate, infertility, reduced serum protein concentration, accumulation of fat in the liver and carcass, and reduced synthesis of certain enzymes and hormones. Deficiencies of individual essential amino acids generally produce the same symptoms (because a single amino acid deficiency prevents protein synthesis) in addition to specific lesions in some cases. Diets for non-ruminant animals must contain a reasonable ratio of essential to non-essential amino acids. In addition to the minimum daily required intake of specific essential amino acids, these must be supplied in certain specific proportions; for example, a certain protein present in tissues can be synthesized only to the extent that the essential amino acid in shortest supply (i.e. the limiting amino acid) will allow. The other amino acids present beyond their specific proportions are not used for protein synthesis and are oxidized to give energy (see above); and the nitrogen released is excreted in the urine.

To assure the profitable use of synthetic amino acids as feed additives, it is essential to know both the absolute requirements for the respective amino acids and also the relative proportions in which they should be given to the animals. Some relationships between pairs of certain amino acids have been observed by Lewis (1972) in experiments with growing chickens. The efficiency of one amino acid depends on the level of another in the diet. Figure 7.8 shows that such a relationship exists between the requirements for methionine plus cystine and the dietary lysine level. The actual requirement for methionine plus cystine is 0.8% and that for lysine is 1.0%. When the supply of lysine is only equivalent to half of the actual requirement (0.5% of the diet) then the chicks are able to utilize only 0.4% methionine plus cystine.

In establishing the methionine requirement of chicks, it is essential to consider the cystine level in the light of the fact that methionine can replace cystine, but not vice versa (see above). The methionine requirement is 0.45%; that for methionine plus cystine in 0.8%. If the dietary cystine is inadequate, methionine can satisfy the need for it. The experimental results represented in Figure 7.9 show that diets containing 0.45–0.80% methionine can be completed by the addition of cystine or methionine until the level of both sulphur-containing amino acids is raised to 0.8%. However, the efficiency of diets containing 0.8% methionine is optimal and cannot be further improved by addition of more methionine or cystine.

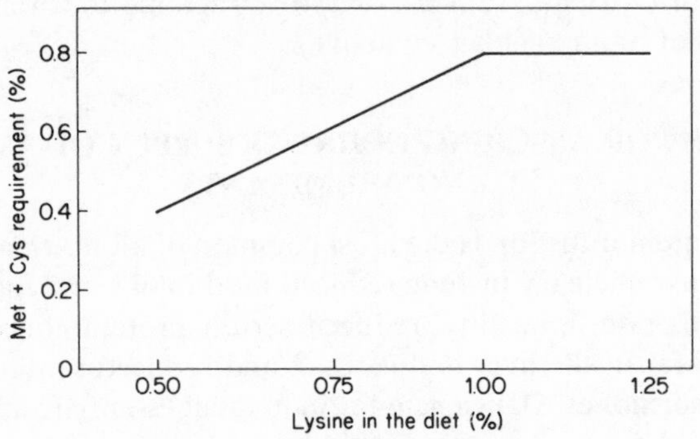

Figure 7.8. Requirement of chickens for methionine plus cystine in relation to dietary lysine level (after Lewis, 1972), *Fed. Proc.*, **31**, p. 1165). *Reproduced by permission of the Federation of American Societies for Experimental Biology.*

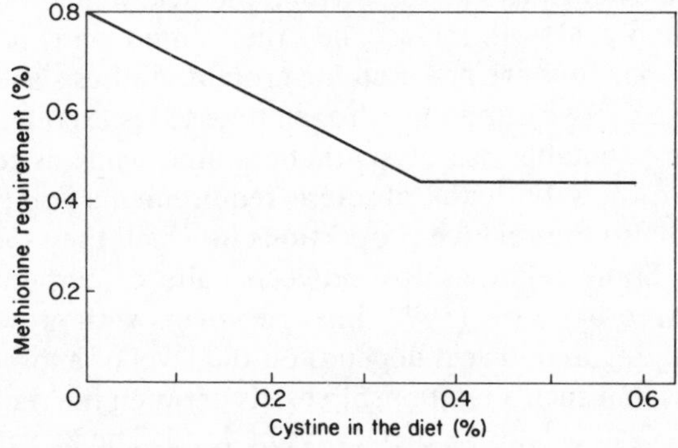

Figure 7.9. Methionine requirement of chickens in relation to dietary cystine level (after Lewis, 1972, *Fed. Proc.*, **31**, p. 1165). *Reproduced by permission of the Federation of American Societies for Experimental Biology.*

Another example of the relationship between two amino acids is the dependence of the arginine requirement on the lysine content of the diet. When there is surplus lysine in the diet there is a subsequent depression in food intake and reduction in growth rate, which can be corrected by the inclusion of arginine in the diet. This could be accounted for by an increased urinary loss of arginine or enhanced enzymatic breakdown of arginine in the body, apparently induced by surplus lysine. It has been found that the level of arginase, the enzyme responsible for arginine degradation in chicks, is increased in the kidneys when chicks are fed excess lysine.

Amino Acid Imbalance

Apart from the adverse effect of surplus lysine on utilization of arginine, many other cases of such an amino acid imbalance have been observed, i.e. changes in proportion of amino acids which depress food intake and growth. These amino acid imbalances result particularly from the addition of one or more essential amino acids to a low-protein diet, which serves as a control diet. These adverse effects can be prevented by addition of relatively small amounts of the most limiting amino acids. Table 7.8 represents the results of a feeding experiment with rats in which such disproportions were created; the adverse effects could be corrected by addition of limiting amino acids. Lysine and threonine are the limiting amino acids in rice. When the lysine content of a rice diet is held constant and the threonine content is increased stepwise, a point is reached at which the growth of rats fed on the threonine-supplemented rice is retarded unless the lysine content of the diet is also increased. The same phenomenon is seen in the reverse situation of threonine being held constant and lysine increased. These growth retardations are caused by imbalances produced by the presence of small surpluses of the second limiting amino acid in proportion to the first limiting amino acid.

Table 7.8 Effect of lysine and threonine supplements on the growth of rats fed on a diet containing 90% rice (after Harper and Rogers, 1965)

Supplements		Weight gain (g per 5 weeks)
L-lysine.HCl	DL-threonine	
–	–	57
0.1	–	78
0.1	0.1	112
0.1	0.2	138
0.1	0.3	114
0.2	0.3	151
0.2	0.1	136
0.25	0.1	52
0.3	0.1	131

A similar case of amino acid imbalance is the following. When noog protein was supplemented with lysine rather than threonine, rats fed the lysine-supplemented diet gained less weight than those receiving no supplement. Weight gains increased considerably when both limiting amino acids were added simultaneously (see Table 7.9).

Table 7.9 Effect of lysine and threonine supplements on the growth of rats feed noog protein (*Guizottia abessynica*; after Rogers, 1976)

	Diet	Weight gain (g in 10 days)
Control	Noog meal (10% protein in diet)	52
Control	Noog meal (10% protein in diet) + 0.3% threonine	56
Imbalance	Noog meal (10% protein in diet) + 0.5% lysine	39
Corrected	Noog meal (10% protein in diet) + 0.3% threonine + 0.5% lysine	68

The adverse effect of amino acid imbalance may be explained by increased incorporation of the amino acid supplied in excess into tissue proteins, particularly liver protein. This has been shown experimentally by ingestion of surplus amino acids labelled with ^{14}C or ^{15}N. Stimulation of protein synthesis causes a decrease in the amount of the specific amino acid that was originally added in excess passing into general circulation. The pattern of free amino acids in the blood becomes severely imbalanced, and resemble those produced by ingestion of diets severely deficient in the respective amino acids. These changes are monitored by an appetite-regulating centre; food intake is depressed (see Chapter 23) and consequently growth is retarded.

Amino Acid Antagonism

This refers to growth depression caused by ingestion of a surplus of one amino acid which can be overcome by adding another, structurally similar amino acid. The adverse effect of the excess amino acid may be caused by the blocking of an enzymatic reaction involved in protein biosynthesis or by competition for a transport site. Thus amino acid antagonism is distinguished from imbalance in that the effects of antagonism are not prevented by a supplement of the growth-limiting amino acid to the basic diet, but rather by a supplement of an amino acid that is structurally similar to the one added in excess. The antagonism between leucine on the one hand and isoleucine and valine on the other has been shown in an experiment with rats fed a diet containing 9% casein supplemented with methionine (see Table 7.10).

Addition of 3% leucine to this diet causes a growth depression which is not prevented by a supplement of threonine, the most limiting amino acid in this diet, and is largely prevented by additional isoleucine and valine.

Table 7.10 Effects of supplements of various amino acids given to rats on the growth retarding action of leucine (after Harper and Rogers, 1965)

Casein (%)	L-leucine (%)	Amino acid supplements	Weight gain (g in 2 weeks)
9	–	–	33
9	3.0	–	9
9	–	0.9% DL-threonine	44
9	3.0	0.9% DL-threonine	8
9	–	0.5% DL-isoleucine	31
9	3.0	0.5% DL-isoleucine	23
9	3.0	0.5% DL-isoleucine + 1.2% DL-valine	30

The antagonistic action of leucine was observed in practical feeding with maize or sorghum as principal ingredients of the usual diets of monogastric animals. Maize and sorghum are particularly high in leucine, and growth impairment on a maize-rich diet can be overcome by addition of soybean meal which is rich in isoleucine.

Toxic Action of Amino Acids

This is seen when the adverse effect of an excess amino acid cannot be overcome by addition of another amino acid. Ingestion of methionine, tyrosine or tryptophan in large amounts, up to about two or three times the requirement, is followed by serious irregularities apart from growth depression. A dietary excess of tyrosine causes severe eye lesions. The toxic action of excess methionine has been attributed to inhibition of ATP synthesis, since methionine may be bound to adenosine, and to the formation of noxious breakdown products or methionine.

ELIMINATION OF NITROGEN FROM THE BODY

Synthesis of Urea and Uric Acid—the End-products of Protein Catabolism

Most of the ammonia arising in the mammalian body from the degradation of amino acids is excreted as urea. This is formed in the liver by a cyclic mechanism (discovered by Krebs), which is summarized in Figure 7.10 and described in biochemical texts. Ammonia and carbon dioxide first form carbamyl phosphate:

$$NH_4^+ + HCO_3^- + 2\,ATP \longrightarrow H_2N{-}CO{-}OPO_3H_2 + 2\,ADP + P_i$$

P_i = inorganic phosphate.

Carbamyl phosphate reacts in turn with ornithine to give citrulline, which acquires another nitrogen atom from aspartic acid to yield arginosuccinate. Aspartic acid is formed by transamination from oxaloacetate and glutamic acid, the latter in turn being regenerated by transamination of ketoglutaric acid with other amino acids (see above). Arginosuccinate splits into arginine and fumarate, the latter returning to the tricarboxylic acid cycle, while the arginine is finally split by hydrolysis to ornithine and urea, a reaction which completes the cycle.

Urea formation requires energy, four high-energy phosphate bonds being spent for the synthesis of each urea molecule formed from ammonia. Two mol of ATP are needed for carbamyl phosphate synthesis and 1 mol of ATP is required for the synthesis of arginosuccinic acid. Since, during the latter step ATP is hydrolysed as follows:

$$ATP \longrightarrow AMP + P\text{-}P$$

and the P-P (pyrophosphate) is then hydrolysed by pyrophosphatase action

$$P\text{-}P + H_2O \longrightarrow 2\,P_i$$

a total of four high-energy phosphate equivalents is required per 1 mol of urea formed. According to Blaxter and Martin (1962), a further 0.9 mol of ATP is consumed during the excretion of urea, so the total energy cost may be 4.9 mol of ATP (or 0.43 MJ) per 1 mol of urea. The high energy requirement for urea synthesis may be one of the main reasons for the great heat output observed after ingestion of protein excesses in animals. Feeding quantities of proteins above the needs of animals is wasteful because of the high energy requirement needed for deamination, urea formation and excretion, but it does not cause adverse effects in farm animals apart from decrease of food intake. On the other hand, ingestion of a surplus of certain essential amino acids is likely to cause serious disturbances (see above).

Birds, unlike mammals, excrete waste or excess nitrogen as uric acid rather than as urea. Whereas in humans urea accounts for 80–90% of the urinary nitrogen, in chickens uric acid is the excretory vehicle for 60–80% of the urinary nitrogen, of which ammonia accounts for 10–15%. Biosynthesis of uric acid in birds occurs in the liver and the kidney. The metabolic origin of the carbon and nitrogen atoms is shown in Figure 7.11. The N-atom in position 1 originates from aspartic acid, which in turn was formed by transamination in the same way as the aspartic acid acting in urea synthesis (see above). N-3 and N-9 come from glutamine, N-7 and C-4 and C-5 from glycine. Glycine

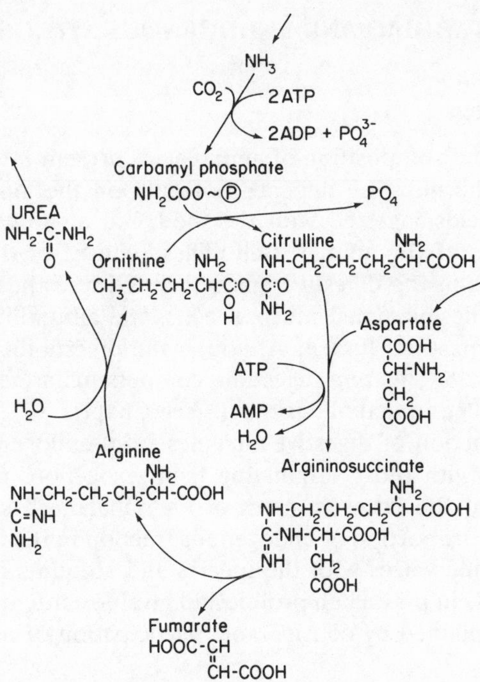

Figure 7.10. Biosynthesis of urea—urea cycle of Krebs (after Phillis, *Veterinary Physiology*, 1976, p. 626). *Reproduced by permission of W. B. Saunders Co.*

is an integral precursor of uric acid. For this reason chickens have a high glycine requirement. Although some is synthesized in the tissues, chickens also require dietary glycine to satisfy the needs for nitrogen excretion and tissue growth (see above). The final step of uric acid synthesis—oxidation of xanthine to uric acid—is controlled by the xanthine oxidase, a molybdenum-containing enzyme (see Chapter 10). The level of this enzyme in chick liver depends on the protein level of the diet.

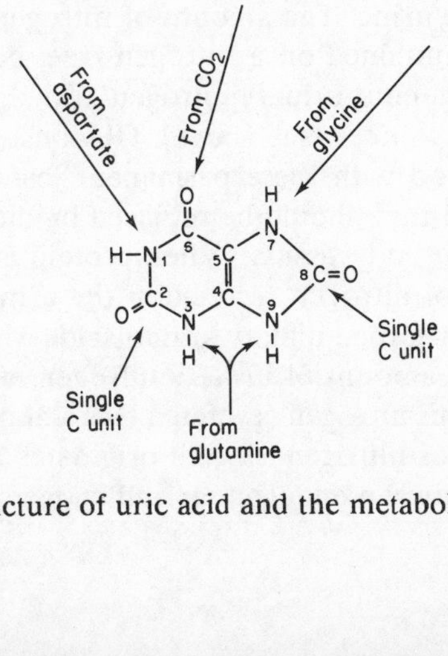

Figure 7.11. The structure of uric acid and the metabolic origin of each atom.

METABOLIC AND ENDOGENOUS NITROGEN

Metabolic Nitrogen

The amounts and composition of amino acids present in the chyme which passes through the intestine depends not only on the amounts of the respective amino acids ingested with the food, but also on the endogenous protein secretions of the animal itself. These consist of digestive enzymes and mucoproteins in the digestive juices secreted into the alimentary canal (see Chapter 3), desquamated mucosal cells, and albumin in the blood secreted into the intestinal lumen. Although the digestibility of endogenous proteins is rather low, its non-digestible components are excreted with the faeces and are called 'metabolic protein' (see Chapter 13). The amino acids liberated by the action of digestive enzymes from endogenous proteins are utilized together with those originating from exogenous food sources and might supplement the latter in cases of an unbalanced supply of dietary amino acids. The proportion of endogenous fraction in the total protein content of the intestine varies with the species and accounts for up to 50% in ducks and 20–80% in pigs. High proline and glycine contents of the intestinal chyme can be explained by considerable participation of mucoproteins and bile.

Endogenous Nitrogen

End-products of protein catabolism originating from body protein and food protein are excreted in the urine. As pointed out above, the proteins of body tissues are constantly being broken down and replaced; for example, a particular amino acid released by breakdown of liver protein will be incorporated into muscle protein, and vice versa. Thus body proteins exchange amino acids among themselves. The reutilization of amino acids released from tissues is not fully efficient and those amino acids not reutilized for biosynthesis are decomposed with release of amino nitrogen which is finally excreted as urea by the urine. The amount of nitrogen excreted daily in the urine of an animal maintained on a nitrogen-free, but otherwise adequate diet is defined as 'endogenous urinary nitrogen'. The 'metabolic faecal nitrogen' comes from body sources (see above). Obviously the organism cannot cope for a longer period with these permanent losses of endogenous and metabolic nitrogen and they should be replaced by dietary protein to maintain the integrity of the body tissuss. When protein is added to a nitrogen-free diet, the quantity of nitrogen secreted in the urine increases due to the partial breakdown of absorbed dietary amino acids which are not converted into body protein. The amount of urinary nitrogen excreted in the urine in excess of the endogenous nitrogen (excreted by an animal on a nitrogen-free diet), is called exogenous nitrogen, since it originates from ingested protein. An exact distinction between endogenous and exogenous nitrogen is not always possible since the ingestion of food protein causes an increase in the turnover of protein, resulting in increased excretion of the absolute amount of nitrogen originating from breakdown of body protein.

Nitrogen Balance

The overall metabolism of protein in the body can be summarized by nitrogen balance. This is the difference between nitrogen intake and nitrogen output

$$B = I - (U + F)$$

where B is the balance, I the nitrogen intake, U and F the nitrogen excretion in urine and faeces. For precise measurements small amounts of nitrogen lost in the shedding of hairs, feathers or hooves have to be added to excreted nitrogen. The nitrogen balance may be either positive (intake exceeds output) as in growing or pregnant animals or in those recuperating from illness, negative (output exceeds intake) as in protein malnutrition and in fasting and sick animals, or zero (nitrogen equilibrium where the nitrogen intake is equal to output).

BODY PROTEIN RESERVES

Not only growing and producing animals, but also adult, non-producing animals can show a positive nitrogen balance when they obtain amounts of protein exceeding requirements. Moderate amounts of excess protein can be taken up by animals independently of their function and age, while amounts exceeding a certain limit are oxidized and excreted. Small protein surpluses are deposited as protein reserves in various organs such as liver, muscles, etc. Protein reserves form up to 5–7% of the total body protein; they are labile, drawn upon in periods of starvation or reduced protein intake, and restored in times of plenty, thereby contributing to the free amino acid pools of the body during depleting processes. However, less metabolic importance should be attributed to the existence of protein reserves than to energy reserves (triglycerides, Chapter 6, or glycogen, Chapter 4) which are stored in larger amounts and in specific organs.

METHODS FOR MEASURING PROTEIN QUALITY OF DIETS FOR NON-RUMINANTS

Biological Assays

The primary function of dietary protein is to furnish a mixture of those amino acids required for synthesis of tissue proteins. Biological evaluation of proteins is often preferred over amino acid analyses since the results of biological assays reflect the ability of feed proteins to convert into body protein and their efficiency in maintenance (by replacement of body protein; see Chapter 17) and growth. The biological efficiency of dietary protein utilization depends not only on the balance of available amino acids, but also on the nitrogen and energy intake and upon the species of animal and its physiological state. Thus exact experimental conditions have been devised for the various biological assays. Biological methods for protein evaluation have been based on the following criteria: measurement of weight gain in growing animals; measurement of nitrogen retention by balance experiments; and measurement of nitrogen retention by body analysis.

Measurement of Weight Gain

Protein Efficiency Ratio (PER). This method normally uses growth of rats or chicks as a measure of the nutritive value of proteins. It is defined as the weight gain per unit weight of protein fed and may be calculated as follows:

$$\text{PER} = \frac{\text{Weight gain (g)}}{\text{Protein intake (g)}} \times 100$$

This simple method requires strict adherence to certain conditions: a level of 10% dietary protein, age of rats, duration of the assay (4 weeks). In Figure 7.12 the shape of the curve relating PER to the nitrogen content of the diet depends on the quality of the protein. With high-quality egg protein, the maximum PER is obtained for low nitrogen levels. Results are quite different when peanut or gluten protein are examined. The sharp decline in PER values resulting from increasing the content of high-quality protein in the diet has to be attributed to the stress caused by ingesting excess protein (see above).

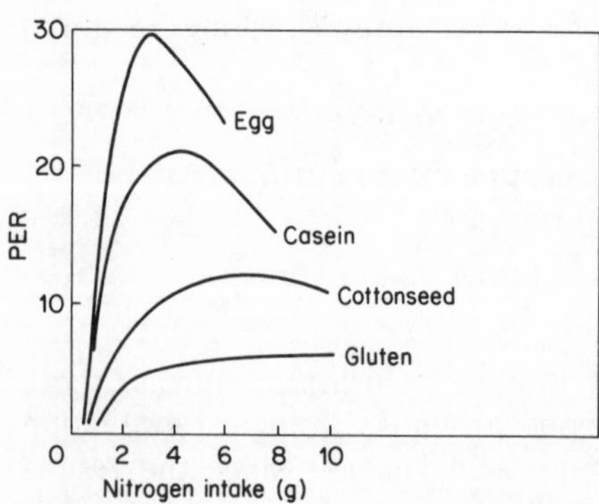

Figure 7.12. Relationship between protein efficiency ration (PER) and nitrogen intake. Growth of rats was measured from weaning over a period of four weeks while being fed different quantities of protein in a purified diet. (after Allison, in Munro and Allison, *Mammalian Protein Metabolism*, Vol. 1, 1964, p. 45). *Reproduced by permission of Academic Press.*

Net Protein Ratio (NPR). The PER assay makes no allowance for maintenance, but assumes that all protein consumed is for growth. In order to overcome this drawback the weight loss of a second group of rats fed a protein-free diet is included in the calculation of NPR. This permits evaluation of proteins for their ability to support maintenance as well as growth.

$$\text{NPR} = \frac{\text{Weight gain of test group (g)} - \text{weight loss of protein-free group (g)}}{\text{Weight of protein consumed by the test group (g)}} \times 100$$

The test group is fed a diet containing 10% protein for 10 days.

The Slope-ratio Assay of Hegsted. This test examines the correlation between live-weight gain and nitrogen intake. It involves feeding at various levels of nitrogen intake, including a protein-free diet. Weight gain or loss is plotted against nitrogen intake and the data are fitted to regression lines (Figure 7.13). The relationship between nitrogen intake and body weight is curvilinear, but the lower part of the curve is linear up to intakes that produce maximal growth. The value of the protein is proportional to the slope of the linear part of the regression line relating dose and response. The slopes are expressed in terms of percentages in relation to the slope obtained with a high-quality standard such as lactalbumin to which is assigned a potency of 100. The value thus determined is the Nitrogen Growth Index.

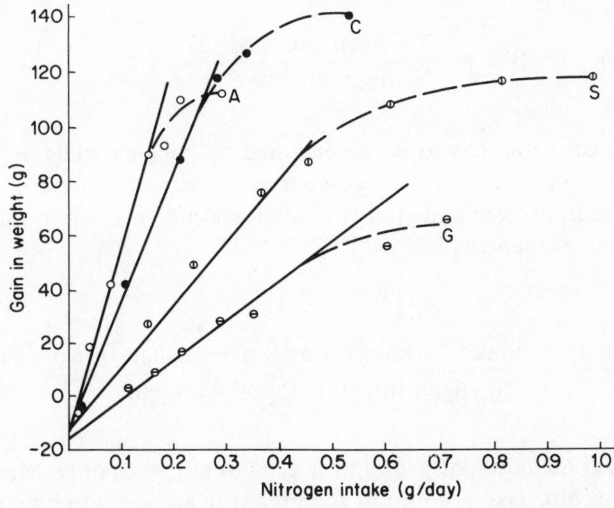

Figure 7.13. Slope–ratio assay of Hegsted—relationship between weight gain and nitrogen intake. Broken lines drawn according to experimental data. Solid lines are the regression lines calculated through the points that appear to be in the linear range. A = albumin; C = casein; S = soya protein; G = gluten. (From Hegsted and Chang, 1965, *J. Nutr.*, **85**, p. 159); adapted by Pyke and Brown, *Nutrition*, 2nd. ed., 1975, p. 867). Reproduced by permission of the publisher.

Gross Protein Value (GPV). This assay is used for the determination of the quality of protein feeds for chicks. The live-weight gain of chicks receiving a basal diet containing 8% protein is compared with the gain of chicks receiving the basal diet plus 3% of the protein to be examined (A) and with those receiving the basal diet plus 3% casein (A_0). The extra live-weight gain per unit of supplementary test protein expressed as a percentage of the extra live-weight gain per unit of supplementary casein is the gross protein value of the test protein:

$$\text{GPV} = \frac{A}{A_0} \times 100$$

The above four procedures, which depend upon measurement of live-weight gains, are not very exact. The gains are not uniform even in animals of the same species and age. Thus the live-weight gains may not be exactly related to the amounts of protein stored.

Measurement of Nitrogen Retention by Balance Methods

Biological Value (BV). The biological value of a dietary protein was defined by Thomas and Mitchell (1909) as the proportion of absorbed nitrogen (retained by the body for maintenance and growth) to the digested nitrogen:

$$\text{BV} = \frac{\text{Nitrogen retained (g)}}{\text{Nitrogen digested (g)}} \times 100$$

The biological value has to be determined by balance trials in which the nitrogen intake and the nitrogen excreted in faeces and urine are measured. Diets adequate in protein and energy content should be given to the animals. The calculation is made as follows:

$$\text{BV} = \frac{\text{Nitrogen intake} - (\text{faecal nitrogen} + \text{urinary nitrogen})}{\text{Nitrogen intake} - \text{faecal nitrogen}} \times 100$$

This formula gives values of protein for growth purposes only. More precise determinations also take account of maintenance and growth by considering the metabolic and endogenous losses of nitrogen (see above). Metabolic faecal nitrogen and endogenous urinary nitrogen represent nitrogen fractions that have been utilized by the animal, though they appear as excreta. They originate from body protein. The revised Thomas–Mitchell formula for the biological value of proteins is as follows:

$$BV = \frac{\text{N intake} - (\text{faecal N} - \text{metabolic N}) - (\text{urinary N} - \text{endogenous N})}{\text{N intake} - (\text{faecal N} - \text{metabolic N})} \times 100$$

The existence of metabolic nitrogen in faeces and of endogenous nitrogen in urine is independent of dietary nitrogen and can be determined with animals receiving a nitrogen-free diet. Except for rats, few animals will consume protein-free diets long enough for such a balance experiment to be carried out. Thus standard values for metabolic nitrogen (determined by extrapolation; see Chapter 13) and for endogenous nitrogen (calculated from metabolic weight of animals; see Chapter 17) are generally used. The biological values of some protein feeds for growing swine are given in Table 7.11.

Table 7.11 Biological values of the protein in various feeds for growing pigs (after Armstrong and Mitchell, 1955)

Feed	Biological value
Milk	95–97
Fish meal	74–89
Soybean meal	63–76
Cottonseed meal	63
Linseed meal	61
Maize	49–61
Barley	57–71
Beane	38

Certain unexpected influences of age and species of animal upon the biological value of certain proteins are seen in Table 7.12. The biological value of a particular protein for maintenance is generally higher than that for growth since larger amounts of essential amino acids are needed for growth than for maintenance. Divergent results have been obtained with respect to the biological values of casein and peanut flour. Both of these proteins are deficient in sulphur-containing amino acids and are thus of higher biological value for growing rats than for adult ones since the latter have a large requirement for these amino acids for intensive hair growth. The high value for wheat gluten in the adult rat compared to the growing rat and the adult human being is due to the lesser need for lysine by the adult rat.

Table 7.12 The biological value of some proteins for growing rats, mature rats and adult men (after Mitchell, 1950, quoted by Munro, 1964, *Mammalian Protein Metabolism*, vol. 2, p. 12. New York and London: Academic Press)

Protein	Growing rats	Mature rats	Adult humans
Egg albumin	97	94	91
Beef muscle	76	69	67
Casein	69	51	56
Peanut flour	54	46	56
Wheat gluten	40	65	42

Protein Replacement Value (PRV). This method avoids the use of nitrogen-free diets and measures the extent to which a test protein will give the same balance as (i.e. replace) an equal amount of egg protein. Two nitrogen balance determinations are carried out, one for egg protein as a standard of high quality and one for the protein under investigation. Great care must be taken to ensure that the experimental and control diets are isocaloric. The PRV is calculated as follows:

$$\text{PRV} = 100 - 100\left(\frac{A - B}{I}\right)$$

A and B denote the nitrogen balances of egg protein and test protein, respectively, and I represents the dietary protein intake which is the same for both groups. PRV measures the efficiency of utilization of the protein fed, and unlike the other methods not the utilization of the digested or absorbed protein.

Measurement of Nitrogen Retention by Body Analysis

Net Protein Utilization (NPU). The nutrition assay is the same as that for measuring the NPR, but in this case the gain in body weight is replaced by the measurement of nitrogen retention. The NPU method lasts 10 days and involves one group of animals (rats or chicks) on a protein-free diet, and another on a diet containing 10–13% of protein. At the close of the feeding period the animals are killed and carcasses are analysed for nitrogen or water. The nitrogen content of the animal may be deduced from its water content, since the ratio between body nitrogen and body water is constant in animals of the same species and age. The formula for calculating NPU is:

$$\text{NPU} = \frac{(\text{Body N of test group} - \text{body N of non-protein group} + \text{N consumed by non-protein group})}{\text{N consumed by test group}} \times 100$$

This method is similar to the biological value method since it is a measure of that proportion of food nitrogen which is retained by the test animals, and thus NPU = BV × digestibility. The BV can be estimated more easily by dividing NPU by digestibility than by balance experiments.

Net Protein Value (NPV). This is the product of NPU and the percentage of crude protein, and is a measure of the protein actually available for metabolism by the animal.

Biological assays have been found useful for the assessment of individual feedstuffs, for comparing the relative nutritive value of certain feedstuffs and for evaluating the effects of processing. However, the results of biological assays of individual feedstuffs cannot be applied to estimating the value of mixtures of feedstuffs, since the resultant value is simply a mean of the individual foods. When two or more proteins are given together, the mixture of amino acids will be better balanced and will have a higher value than either protein given alone.

Other Methods

Microbiological Methods. Certain microorganisms have amino acid requirements similar to those of animals. Rapid screening tests have been developed and involve measuring the growth of microorganisms in basal media which contain all nutrients required for growth (except amino acids) and the protein to be tested. As in rats, the response of the organism is dependent primarily on the amount of the limiting amino acid. Assays with the following microorganisms have been widely adopted: *Streptococcus zymogenes*, a lactic-acid-producing bacterium, which can be used after enzymatic predigestion of the sample; and the protozoa *Tetrahymena pyroformis*, which exerts a proteolytic activity and can be used without predigestion.

Biochemical Parameters as Indices for Protein Nutrition Value. Levels of amino acids in the blood are the result of the difference between uptake by tissues and the amounts coming from the diet plus those of endogenous origin (see above), and reflect the difference between need and dietary supplies. Low levels of a certain amino acid in blood plasma are indicative of its deficiency in the diet.

A negative relationship exists between the biological value of a protein and the blood urea level, since the nitrogen present in amino acids absorbed and not utilized for biosynthetic purposes is converted to urea. When the conditions prescribed for this test (nitrogen level of the diet and appropriate time interval between feeding and sampling of blood) are strictly kept, close agreement with the results of biological tests is obtained.

Chemical Measures of Protein Evaluation

Chemical Score

All amino acids must be provided simultaneously at the sites of protein synthesis and deficits of any one could limit the rate of protein synthesis (see above). Accordingly, the biological values of a dietary protein should be determined by the essential amino acid which is in greatest deficit when compared with a standard protein. The standard is egg protein or an amino acid mixture required for maximum growth rate. The content of each of the indispensable amino acids of a protein is expressed as a proportion of that in egg. The amino acid present in the smallest proportion was taken as the score. In wheat protein, for example, the most limiting amino acid is lysine. The lysine content of egg and wheat protein is 7.2% and 2.7%, respectively, and the chemical score is

$$2.7:7.2 = 0.37$$

There are some limitations to this method: (1) proteins totally lacking one amino acid can still have some biological value in supporting slow growth; and (2) there may be reduced availability of some amino acids, particularly in processed feedstuffs (see later).

Essential Amino Acids Index (EAA)

Oser proposed a similar method based on the contribution made by all ten essential amino acids rather than on one in the greatest deficit. Egg protein serves again as standard. a, b, and c are the percentages of the essential amino acids in the food protein and a_e, b_e, and c_e are the percentages of the respective amino acids in the whole egg protein. The essential amino acid index is the geometric mean of the ten egg ratios found by comparing the content of the essential amino acids in a feed protein with that found in whole egg protein. Algebraically the index is expressed as:

$$\text{EAA index} = \sqrt[n]{\frac{100\,a}{a_e} \times \frac{100\,b}{b_e} \times \frac{100\,c}{c_e} \ldots}$$

where n = number of amino acids involved in the calculation.

For computation it is convenient to express the equation in logarithmic form:

$$\log\,(\text{EAA index}) = \frac{1}{10}\left(\log\frac{100\,a}{a_e} + \log\frac{100\,b}{b_e} + \log\frac{100\,c}{c_e} + \ldots\right)$$

This method for expressing the nutritive value of food proteins enables an estimation to be made of the effect of fortifying proteins with amino acids, or of combining proteins of different value. These indices are more closely related to the biological values than the chemical scores.

Chemical Test for Available Lysine

The availability of lysine, which is the most limiting amino acid in many feedstuffs such as cereals, is affected by heat treatment since the free ε-amino group present in lysine (bound in peptide chains) reacts on heating with the aldehyde group of sugars or with the carboxyl group of amino acids. To assure successful use of soybean meal or cottonseed cake in the nutrition of monogastric animals, these feedstuffs have to undergo heat treatment. Overheating, however, would affect the availability of lysine (see above). According to the most useful method for determination of available lysine, devised by Carpenter (1973), fluoro-2,4-dinitrobenzene (FDNB) was allowed to react with free ε-NH_2 groups in intact protein and the ε-dinitrophenyllysine released after subsequent hydrolysis was measured colorimetrically. FDNB, however, does not react with amino groups attached to a sugar moiety and the concentration of the DNP-lysine corresponds to that of available lysine present in the sample.

FDNB + $H_2N-(CH_2)_4-$Protein $\xrightarrow{+OH^-}$ $O_2N-C_6H_3(NO_2)-NH-(CH_2)_4-$Protein $\xrightarrow[\text{hydrolysis}]{+HCl\ \text{Acid}}$ $O_2N-C_6H_3(NO_2)-NH-(CH_2)_4-CH(NH_2)COOH$ + Protein

(Dinitrophenyl) DNP-lysine

Lysine inside peptide chain

Dye-binding capacities of feedstuffs can be used for a quick determination of available lysine together with total basic amino acids and for estimation of the biological value of heat-processed soybean meals; blocking of the ε-amino group prevents the reaction with acid dyes.

BIBLIOGRAPHY

Books

Bender, D. A. (1975). *Amino Acid Metabolism.* New York: John Wiley.

Bigwood, E. J. (1972). *Protein and Amino Acid Function.* Oxford: Pergamon Press.

Buttery, P. J., and Lindsay, D. B. (1980). *Protein Deposition in Animals.* London: Butterworths.

Friedman, M. (1975). *Protein Nutritional Quality of Foods and Feeds*, 2 parts. New York: Marcel Dekker.

Munro, H. N., and Allison, J. B. (eds.) (1964). *Mammalian Protein Metabolism*, Vols. 1 and 2. New York: Academic Press.

Munro, H. N. (ed.) (1969–1970). *Mammalian Protein Metabolism*, Vols. 3 and 4. New York: Academic Press.

National Academy of Sciences (1974). *Improvement in Protein Nutriture.* Washington, DC: National Academy of Sciences.

Norton, G. (1978). *Plant Proteins.* London and Boston: Butterworths.

Porter, J. W. G., and Rolls, B. A. (1973). *Proteins in Human Nutrition.* London and New York: Academic Press.

Schepartz, B. (1973). *Regulation of Amino Acid Metabolism in Mammals.* Philadelphia: W. B. Saunders.

Waterlow, J. C., Garlic, P. J., and Milward, D. J. (1978). *Protein Turnover in Mammalian Tissue and in the Whole Body.* Amsterdam and New York: Elsevier.

Withaker, J. R., and Tannenbaum, St. R. (1977). *Food Proteins.* Westport: Avi Publishing Co.

European Association for Animal Production. International Symposia on 'Protein Metabolism and Nutrition':

First Symposium, Nottingham, 1976 (D. J. A. Cole *et al.*, eds), London, Butterworths. E.A.A.P. No. 16.

Second Symposium, The Netherlands, 1977 (S. Taminga, ed.). Wageningen: Centre for Agricultural Publishing and Documentation. E.A.A.P. No. 22.

Third Symposium, Germany, 1980 (H. J. Oslage and K. Rohr, eds). Braunschweig: Institute of Animal Nutrition. E.A.A.P. No. 27.

Fourth Symposium, France, 1983 (R. Pion *et al.*, eds). Paris: Institut National de la Recherche Agronomique. E.A.A.P. No. 31.

Articles

Protein and non-protein compounds in feedstuffs

Hegarty, M. P., and Peterson, P. J. (1973). Free amino acids, bound amino acids and amines. In *Chemistry and Biochemistry of Herbages* (G. W. Butler and R. W. Bailey, eds), Vol. 1, pp. 1–62. London and New York: Academic Press.

Lyttleton, J. W. (1973). Proteins and nucleic acids. In *Chemistry and Biochemistry of Herbages* (G. W. Butler and R. W. Bailey, eds), Vol. 1, pp. 63–105. London and New York: Academic Press.

Protein digestion and absorption in non-ruminants

Symposium on Protein digestion and amino acid absorption in *Fourth International Symposium on Protein Metabolism and Nutrition* (1983, loc cit.), Vol. 1, pp. 211–322.

Austic, R. E. (1985). Development and adaptation of protein digestion. *J. Nutr.*, **115**, 686–697.

Liener, I. E. (1979). The nutritional significance of plant protease inhibitors. *Proc. Nutr. Soc.*, **38**, 109–114.

Matthews, D. M. (1972). Intestinal absorption of amino acids and peptides. *Proc. Nutr. Soc.*, **31**, 171–177.

Matthews, D. M. (1975). Intestinal absorption of peptides. *Physiol. Rev.*, **55**, 538–608.

Silk, D. B. A., Grimble, G. K., and Rees, R. G. (1985). Protein digestion and amino acid and peptide absorption. *Proc. Nutr. Soc.*, **44**, 63–72.

Snook, J. T. (1973). Protein digestion. Nutritional and metabolic considerations. *World Rev. Nutr. Diet.*, **18**, 122–176.

Nutritive importance and metabolism of amino acids

Baker, D. W., and Boebel, K. P. (1981). Utilization of D- and L-isomers of arginine and histidine by chicks and rats. *J. Anim. Sci.*, **53**, 125–129.

Benevenga, N. J. (1984). Adverse effects of excessive consumption of amino acids. *Annu. Rev. Nutr.*, **4**, 157–181.

Evered, D. F. (1981). Advances in amino acid metabolism in mammals. *Biochem. Soc. Trans.*, **9**, 159–169.

Fehlig, P. (1975). Amino acid metabolism in man. *Annu. Rev. Biochem.*, **44**, 935–955.

Griminger, P. (1976). Protein metabolism in poultry. In *Avian Physiology* (P. D. Sturkie, ed.), pp. 233-251. New York: Springer-Verlag.

Harper, A. E. (1968). Diet and plasma amino acids. *Amer. J. Clin. Nutr.*, **21**, 358–366.

Harper, A. E. (1983). Some recent developments in the study of amino acid metabolism. *Proc. Nutr. Soc.*, **42**, 437–450.

Harper, A. E., Bonevenga, N. J. and Wohlhueter, R. (1970). Effect of ingestion of disproportionate amounts of amino acids. *Physiol. Rev.*, **50**, 428–558.

Harper, A. E., Miller, R. H., and Block, K. P. (1984). Branched-chain amino acid metabolism. *Annu. Rev. Nutr.*, **4**, 409–454.

Harper, A. E., and Rogers, Q. R. (1965). Amino acid imbalance. *Proc. Nutr. Soc.*, **24**, 173–190.

Lewis, D. (1972). Non-amino and amino nitrogen in nonruminant nutrition. *Fed. Proc.*, **31**, 1165–1171.

Lindsay, D. M. (1980). Amino acids as energy sources. *Proc. Nutr. Soc.*, **39**, 53–59.

Rogers, R. Q. (1976), Nutritional and metabolic effects of amino acids imbalance. In First Symposium on *Protein Metabolism and Nutrition (D. J. A. Cole et al., eds), pp. 279–302. London: Butterworths.*

Smith, R. H. (1980). Comparative amino acid requirements. *Proc. Nutr. Soc.*, **39**, 71–78.

Synthesis, degradation and turnover of proteins: Protein reserves

Fisher, H. *et al.* (1964). Protein reserves: evidence for their utilization under nutritional and disease stress conditions. *J. Nutr.*, **83**, 165–170.

Garlick, P. J. (1980). Assessment of protein metabolism in the intact animal. In *Protein Deposition in Animals* (P. J. Buttery and D. B. Lindsay, eds), pp. 51–68. London: Butterworths.

Goldberg, A. L. *et al.* (1980). Hormonal regulation of protein degradation and synthesis in skeletal muscle. *Fed. Proc.*, **39**, 31–36.

Lobley, E. G., Milne, V. *et al.* (1980). Whole body and tissue protein synthesis in cattle. *Br. J. Nutr.*, **43**, 491–502.

Milward, D. J. (1979). Protein deficiency, starvation and protein metabolism. *Proc. Nutr. Soc.*, **38**, 77–88.

Milward, D. J. *et al.* (1983). Protein turnover: the nature of the phenomena and its physiological regulation. Fourth Internat. Symposium on 'Protein metabolism and nutrition', loc cit., Vol. 1, pp. 69–96.

Reeds, P. J., and Lobley, G. E. (1980). Protein synthesis: are there real species differences. *Proc. Nutr. Soc.*, **39**, 43–52.

Reeds, P. J., and Fuller, M. F. (1983). Nutrient intake and protein turnover. *Proc. Nutr. Soc.*, **42**, 463–472.

Swick, R. W., and Benevenga, N. J. (1977). Labile protein reserves and protein turnover. *J. Dairy Sci.*, **60**, 505–515.

Elimination of nitrogen from the body

Allison, J. B., and Bird, J. W. G. (1964). Elimination of nitrogen from the body. In *Mammalian Protein Metabolism* (H. N. Munro and J. B. Allison, eds),Vol. 1, pp. 483–512. New York: Academic Press.

Blaxter, K. L., and Martin, A. K. (1962). The utilization of protein as a source of energy in fattening sheep. *Br. J. Nutr.*, **16**, 397–407.

Krebs, H. A., Hems, R., and Lund, P. (1973). Some regulatory mechanisms in the synthesis of urea in the mammalian liver. *Adv. Enzyme Regul.*, **11**, 361–377.

Methods for measuring protein quality in diets for non-ruminants

Allison, J. B. (1964). The nutritive value of dietary proteins. In *Mammalian Protein Metabolism (H. N. Munro and J. B. Allison, eds), Vol. 2, pp. 41–86. New York: Academic Press.*

Armstrong, D. G., and Mitchell, H. H. (1955). Protein nutrition and the utilization of dietary protein at different levels of intake by growing swine. *J. Anim. Sci.*, **15**, 49–68.

Bender, A. E. (1982). Evaluation of protein quality. Methodological considerations. *Proc. Nutr. Soc.*, **41**, 267–276.

Carpenter, K. I., and Booth, V. H. (1973). Damage to lysine in food processing: its measurement and significance. *Nutr. Abstr. Rev.*, **43**, 424–451.

Evans, E., and Witty, R. (1978). An assessment of methods used to determine protein quality. *World Rev. Nutr. Diet.*, **32**, 1–26.

Hegsted, D. M., and Chang, Y. (1965). Protein utilization in growing rats. 1. Relative growth rate as a bioassay procedure. *J. Nutr.*, **85**, 159–168.

Hegsted, D. M., and Chang, Y. (1969). Protein utilization in growing rats. 2. Evaluation of dietary protein quality in adult rats. *J. Nutr.*, **99**, 479–480.

McLaughan, I. M., and Campbell, I. A. (1969). Methodology of protein evaluation. In *Mammalian Protein Metabolism* (H. N. Munro, ed.), Vol. 3, pp. 391–422. New York: Academic Press.

Meade, R. J. (1972). Biological availability of amino acids. *J. Anim. Sci.*, **35**, 713–723.

CHAPTER 8

Metabolism of Protein in Ruminant Animals

As a result of the action of the rumen microorganisms, the mode of utilization of food proteins in ruminants differs significantly from that in monogastric species. As will be described in detail below, rumen microorganisms are characterized by their high potency for synthesizing all amino acids, including the essential ones required by the host animals. Thus, ruminant animals are almost completely independent of the quality of the ingested proteins. A part of the nitrogen in ruminant feeds may be supplied in the form of simple nitrogen compounds such as ammonium salts or urea.

Utilization of proteins ingested by ruminants takes place as follows:

(1) During passage of the food through the rumen, much of the protein is degraded to peptides by the action of proteases. The peptides are further catabolized to free amino acids and the latter to ammonia, fatty acids and carbon dioxide.

(2) The degradation products formed in the rumen, in particular ammonia, are used by microorganisms in the presence of adequate energy sources (carbohydrates) for the synthesis of proteins and other microbial cell constituents, such as nitrogen containing cell-wall components and nucleic acids.

(3) Part of the ammonia liberated in the rumen cannot be fixed by the microorganisms; it is absorbed into the animal's blood and transformed in the liver to urea (see Chapter 7), the greater part of which is not utilized by the animal and is excreted in the urine.

(4) Microbial cells (bacteria and protozoa) containing proteins as their main component pass, together with unaltered dietary proteins from the reticulorumen, through the omasum and abomasum into the small intestine. The proportion of the total dietary protein that is digested in the rumen varies from about 70–80% or more for most diets to 30–40% for less-soluble

proteins (see later). Between 30% and 80% of roughage protein is degraded in the rumen; the amount depends on residence time in the rumen and on the level of feeding. The digestion and absorption of microbial and dietary proteins occur in the small intestine of ruminant animals in a similar way to that in monogastric species (see Chapter 7). Microbial proteins and non-degraded food proteins are digested in the small intestine by proteases. Enzymes and other endogenous proteins (see Chapter 13) secreted into the gut are digested too. Amino acids originating from microbial protein, from feed and endogenous protein contribute to the flux of amino acids absorbed from the gut into the intestinal tissues and blood. The amino acids considered essential for ruminant animals appear to be the same as those essential for other mammals (see Chapter 7). The ruminant animal must depend on microbial protein plus dietary protein that escapes digestion in the rumen for its supply for essential amino acids.

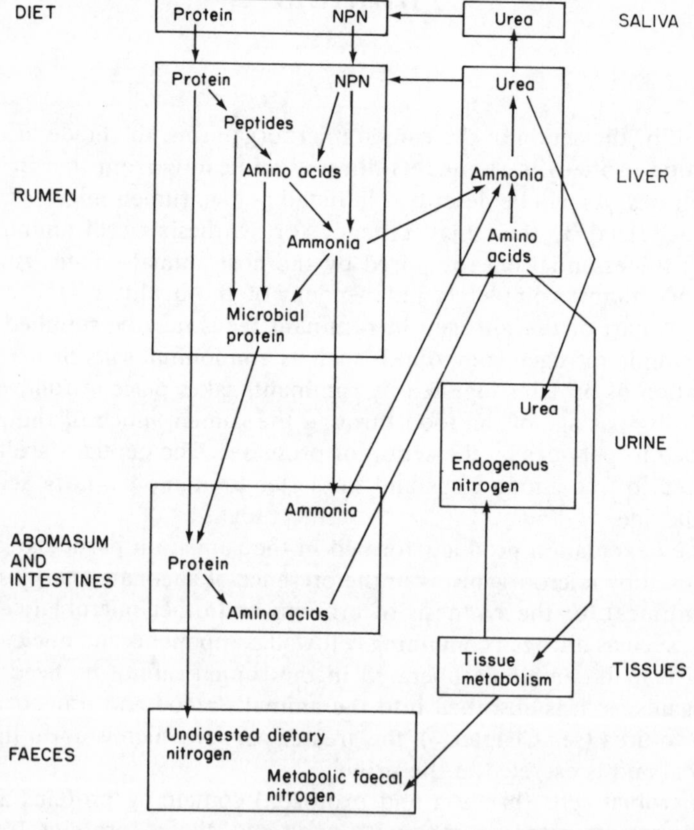

Figure 8.1. Pathways of digestion and metabolism of nitrogenous compounds in the ruminant (after Mercer and Amison, 1976, p. 411). NPN = non-protein nitrogen. *Reproduced by permission of Butterworths and the authors.*

DEGRADATION OF PROTEIN IN THE RUMEN

About 40% of rumen bacteria have proteolytic activity. The rumen bacterial proteases are cell-bound, but are located on the cell surface and thus are freely accessible to the substrate. The protozoa present in the rumen are equipped with powerful intracellular proteases (see later). These microbial protein-degrading enzymes operate at an optimal pH of 6–7. Proteins are degraded by rumen microbes according to the following steps leading finally to the production of ammonia, volatile fatty acids and carbon dioxide:

$$\text{Proteins} \xrightarrow{\text{Proteases}} \text{peptides} \xrightarrow{\text{Peptidases}} \text{amino acids} \xrightarrow{\text{Deamination}}$$

$$\text{carbon skeleton} + NH_3 \longrightarrow \text{volatile fatty acids} + CO_2 + (NH_3)$$

Deamination of the amino acids formed is considerable, as, for example, the formation of isobutyrate and ammonia from valine:

$$(CH_3)_2\text{CH.CH(NH}_2\text{).COOH} \longrightarrow (CH_3)_2\text{CH.CH.COOH} + NH_3 \longrightarrow (CH_3)_2\text{CH.COOH} + CO_2$$

Other branched-chain volatile fatty acids formed in the rumen by enzymatic breakdown of appropriate amino acids are

isovaleric acid $(CH_3.CH_2)(CH_3)\text{CH.COOH}$

from isoleucine $(CH_3.CH_2)(CH_3)\text{CH.CH(NH}_2\text{).COOH}$

or 2-methylbutyric acid $(CH_3)_2\text{CH.CH}_2\text{.COOH}$

from leucine

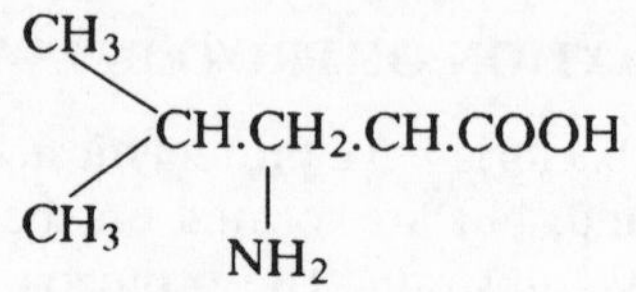

$$\begin{matrix} CH_3 \\ \end{matrix}\!\!\searrow CH.CH_2.\underset{\displaystyle | \atop NH_2}{CH}.COOH \qquad (CH_3)_2CH.CH_2.CH(NH_2).COOH$$

The presence of these branched fatty acids in the rumen content deserves comment as they serve as growth factors for microorganisms and as starting materials for the formation of long-chain branched fatty acids which are characteristic of microorganisms (see Chapter 6). It may be noted that a small amount of the volatile fatty acids, such as acetic, propionic and butyric acids, present in the rumen are formed by breakdown of amino acids; as shown in Chapter 5, the major part of these volatile fatty acids originates from carbohydrates. Moreover, the microbial breakdown of amino acids such as cysteine, glutamic acid and serine, leads to pyruvic acid, which is transformed into acetic, propionic and butyric acids via the usual mechanism as described in Chapter 4.

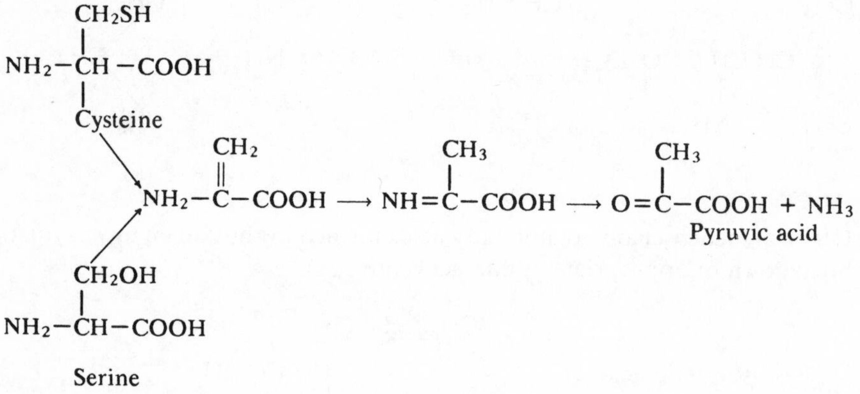

$$\underset{\text{Cysteine}}{NH_2-\overset{\displaystyle CH_2SH}{\overset{|}{CH}}-COOH} \searrow \qquad \underset{\text{Serine}}{NH_2-\overset{\displaystyle CH_2OH}{\overset{|}{CH}}-COOH} \nearrow$$

$$NH_2-\overset{\displaystyle CH_2}{\overset{\|}{C}}-COOH \longrightarrow NH=\overset{\displaystyle CH_3}{\overset{|}{C}}-COOH \longrightarrow \underset{\text{Pyruvic acid}}{O=\overset{\displaystyle CH_3}{\overset{|}{C}}-COOH} + NH_3$$

Composition and Size of the Nitrogen Pool in the Rumen Content

The organic matter, 500 g in a total volume of about 5 litres (in a sheep) contains 5–35 g of total nitrogen. About 90% of the total nitrogen present in the rumen content exists in an insoluble form; 60–90% of this is of microbial origin and the remainder represents smaller particles of undegraded dietary protein and also sloughed rumen epithelial cells. Nitrogen in the dissolved pool (about 10% of the total rumen nitrogen) consists mainly of ammonia nitrogen (average 70%) and the rest is a mixture of free amino acids and peptides. Ammonia nitrogen is present in the rumen in concentrations varying between 2 and 50 mg per 100 ml, depending on diet and time after feeding; maximal concentrations of ammonia are reached about 2 hours after ingestion of the protein-containing food. Amino acid nitrogen and peptide nitrogen are present in much lower amounts (usually less than 2 mg per 100 ml, about 6% of the nitrogen pool). Soluble protein nitrogen may show high values immediately after feeding, but very soon drops to values of about 4 mg per 100 ml (about 15% of the pool).

DESIRED AND NON-DESIRED ACTIONS OF RUMEN AMMONIA—BIOSYNTHESIS OF PROTEIN IN THE RUMEN

Ammonia is the main nitrogenous nutrient for rumen bacteria; they utilize ammonia in the presence of adequate dietary sources, mainly soluble carbohydrates, to synthesize the amino acids needed for their own protein requirements. Of the microbial nitrogen in the rumen, 50–80% has been estimated to be derived from ruminal ammonia. Recent studies using ^{15}N have suggested that bacteria can derive up to 30% of their protein from sources other than ammonia, such as peptides and amino acids; these amino acids are formed by bacterial proteolysis or they are intact dietary amino acids (Nolan *et al.*, 1976). Protozoa cannot utilize ammonia and derive their nitrogen by engulfing bacteria and digesting them with powerful intracellular proteases. Ammonia is also formed from urea of endogenous origin (see later) and from urea ingested with the food (see above). The breakdown of urea to ammonia and carbon dioxide by bacterial urease present in the rumen fluid is very rapid:

$$CO(NH_2)_2 + H_2O \longrightarrow CO_2 + 2\,NH_3$$

Rumen bacteria also liberate ammonia from non-protein nitrogen compounds present in foods, such as herbages or silages (see Chapter 7).

The ammonia produced by degradation of dietary protein is in turn incorporated into microbial cells. Its fixation by rumen microorganisms occurs according to processes opposite to the deamination of amino acids (see Chapter 7), i.e. the addition of ammonia to oxo acids. The most prevalent reaction is the formation of glutamic acid by reductive addition of ammonia to 2-oxoglutaric acid by the enzyme glutamate hydrogenase (see the description of the synthesis of non-essential amino acids occurring in the liver of monogastric animals; Chapter 7). Thus, the synthesis of glutamic acid occurs in ruminants, not only by their symbiotic microorganisms but also in tissues, such as liver, or in the epithelium of the rumen walls through which ammonia and 2-oxoglutaric acid diffuse simultaneously (see later). In the microorganisms, glutamic acid can be converted by transamination into other amino acids, both essential and non-essential. The oxo acids used by the microorganisms for synthesis of non-essential amino acids originate from the breakdown of carbohydrates (see Chapters 4 and 5). The carbon skeletons of essential amino acids are synthesized by reductive carboxylation reactions not likely to operate in aerobic organisms.

Another mechanism for the fixation of ammonia by microorganisms is by its addition to amino acids to form amides, the energy requirement of this reaction being provided by ATP (see Chapter 14):

$$\begin{array}{lcl} \mathrm{COOH} & & \mathrm{COOH} \\ | & & | \\ \mathrm{CHNH_2} & & \mathrm{CHNH_2} \\ | & & | \\ \mathrm{CH_2} & & \mathrm{CH_2} \\ | & & | \\ \mathrm{CH_2} & & \mathrm{CH_2} \\ | & & | \\ \mathrm{COOH} + \mathrm{NH_3} + \mathrm{ATP} & \xrightarrow[\text{synthetase}]{\text{Glutamine}} & \mathrm{CONH_2} + \mathrm{AMP} + \text{P-P} \end{array}$$

Glutamic acid glutamine

P-P = pyrophosphate

The amidic$-NH_2$ (in $-CONH_2$) is reactive and transforms oxo acids, produced in the rumen by degradation of carbohydrates, into amino acids.

In spite of the great importance of ammonia in the growth of rumen microorganisms, they can never completely utilize the ammonia present in the rumen as there is a limit to the amount of ammonia that can be fixed by these microorganisms. Rumen protein synthesis attains a maximum level when the ammonia concentration in the rumen content reaches 5 mM (9 mg per 100 ml). Results of some experiments are presented in Figure 8.2. With increasing nitrogen intake, there is a gradual increase in rumen ammonia concentration but total microbial protein production attains a maximum rate.

Rumen ammonia produced in excess of the ability of the microbes to assimilate it, is absorbed into the blood, carried into the liver and converted into urea. Part of the free ammonia present in the rumen is absorbed directly through the rumen epithelium into the blood; the remainder—in most situation the greater part of it—passes with the digesta into the lower digestive tract where it is finally absorbed through the epithelium into the blood and thence carried to the liver. Most of the urea formed in the liver is excreted through the kidney into the urine; a smaller part (up to about 20%) is recycled back into the rumen via the saliva or by direct diffusion from the blood through the rumen wall (see Figures 8.1 and 8.3). In the rumen, it is decomposed by bacterial urease to ammonia, which is utilized in microbial protein biosynthesis together with the ammonia produced from dietary proteins. The decomposition of urea is carried out to a small extent by ureolytic bacteria present in the rumen content but to a greater extent by the bacterial population adhering to the inner surface of the rumen wall. These adherent bacteria intercept the urea that passes across the rumen wall from the blood by simple diffusion. The extent to which the diffusion process is accelerated by bacterial urease will depend on the rumen ammonia concentration; the entry of urea into the reticulorumen is enhanced when the rumen ammonia concentration is low. Only under conditions of low rumen ammonia levels are relatively high levels of endogenous nitrogen recycled and able to serve as a secondary source of nitrogen for the ruminal organisms and to represent therefore a nitrogen-saving mechanism (Cheng and Costerton, 1979).

A considerable amount of ammonia is recycled via the breakdown and lysis of bacteria and by predation of bacteria by ruminal protozoa; approximately 30–40% of microbial cells are so degraded. According to recent measurements, the recycling of ammonia, which amounts to 10–50% of the ammonia flux, occurs chiefly in the rumen as a result of lysis and digestion of microorganisms.

As pointed out before, rumen protein synthesis attains a maximum level when rumen ammonia concentrations reach 9 mg per 100 ml (see Figure 8.2). This concentration is attained with diets containing 13% protein, and urea is well utilized when added together with sufficient amounts of carbohydrates to diets containing less than 13% protein. Feeding amounts of protein exceeding the requirements is wasteful and leads to excess rumen ammonia concentrations. Feeding of excess protein is not detrimental as ex-

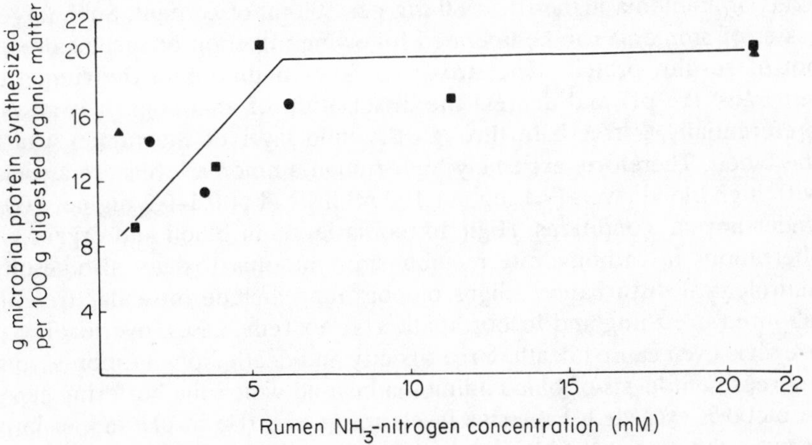

Figure 8.2. Relationship between rumen ammonia nitrogen concentration and protein production (from Buttery, 1977, p. 18). *Reproduced by permission of Butterworths and the authors.*

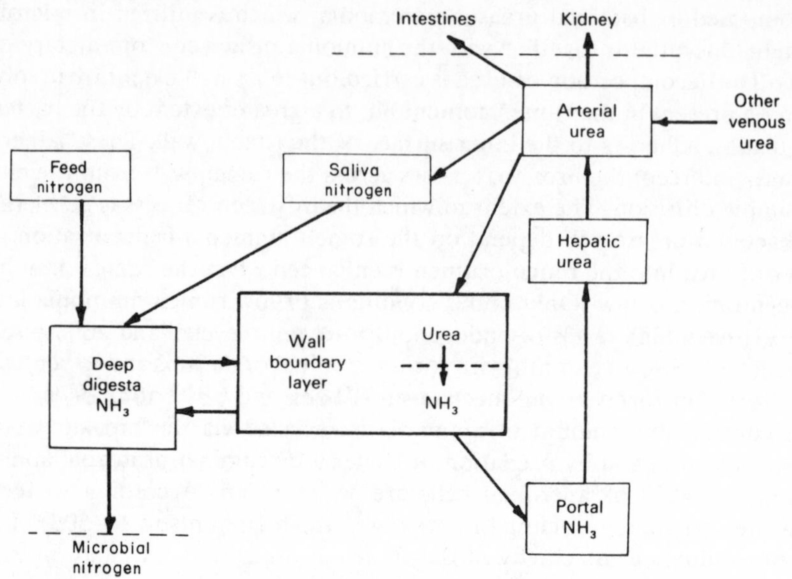

Figure 8.3. Nitrogen recycling to the rumen (from Egan, *Proc. Nutr.Soc.* **39**, 1980, p. 80). *Reproduced by permission of Cambridge University Press.*

cess ammonia is transformed by the liver into urea, the major part of which is excreted in the urine. However, feeding of exaggerated amounts of urea or ammonium salts to ruminant animals may be injurious, particularly when they do not consume sufficient amounts of carbohydrates. The mechanisms for detoxification of ammonia by conversion into urea cannot deal with large loads of ammonia higher than 80 mg per 100 ml of content. Such large excesses of ammonia can be liberated following ingestion of surplus urea, but not of surplus protein. The strong alkaline conditions in the rumen content raise the pH and depress the dissociation of ammonia. It is absorbed preferentially as free base through the lipid layer of the rumen wall into the blood. Therefore, extremely high rumen ammonia levels are associated with high blood levels (<1 mg per 100 ml instead of 0.1–0.2 mg per 100 ml) under normal conditions. High ammonia levels in blood and organs cause aberrations in carbohydrate metabolism; ammonia toxicity also results in neurological disturbances. Signs of poisoning include muscular trembling, laboured breathing and incoordination. In extreme cases, overloading with urea has even caused death. Urea toxicity and death from ammonia poisoning represent levels of blood ammonia beyond which the buffering capacity of blood is exceeded. Excessive levels result in a rise in pH and an impairment in the capacity of blood to expel carbon dioxide. On the other hand, as pointed out above, proper use of urea as a partial substitute for protein in ruminant nutrition has proved successful. It is advantageous to supply urea in liquid food supplements with higher concentrations of molasses and phosphoric acid since molasses serves as an energy source for microbial protein synthesis and phosphoric acid lowers the pH in the rumen content, which would be noxious to animals (see above).

Energy requirements for microbial protein synthesis are comparatively high. Relationships between the energy requirements, expressed in terms of high-energy phosphate bonds, and the amounts of microbial protein formed are the following: 3 mol of ATP are needed to bind ammonia to a nitrogen-free skeleton, i.e. to form 1 mol of amino acid; 5 mol of ATP are needed to bind 1 mol of amino acid to an existing protein. Considering 108 as the average molecular weight of an amino acid, 8 mol of ATP are needed for the synthesis of 108 g of protein or 1 mol of ATP for the synthesis of 13 g of protein; 2 mol of ATP and 1 mol of volatile fatty acids are produced by anaerobic breakdown of 100 g of glucose. The dry matter of microorganisms contains 75% protein. Thus, the energy needed for the synthesis of 15 g of microbial dry matter can be produced by the breakdown of 50 g of glucose. Experimental results show that the yield of microbial protein is variable, influenced by the growth rate of bacteria, and fluctuates between 90 and 230 g/kg of organic matter digested (Czerkawski, 1978).

Summary of Metabolic Pathways of Ammonia

The concentration of ammonia in the rumen content changes with the feeding regime and is regulated by the different metabolic pathways summarized in the following scheme:

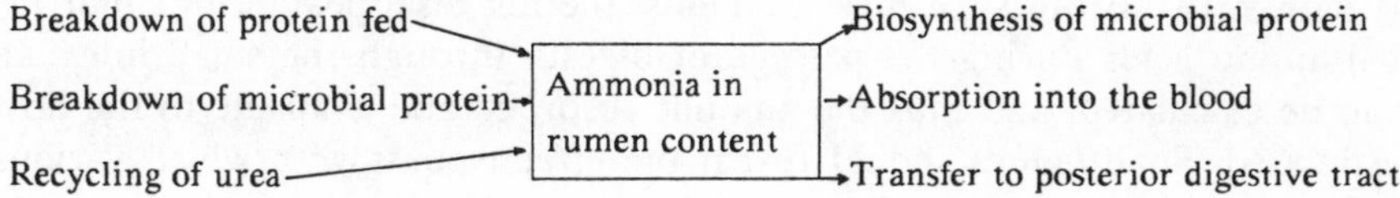

Quantitatively, the most important processes are the breakdown of dietary proteins to liberate ammonia and fixation of the ammonia by microorganisms. Fortunately, the degradative activities of the microbes are to a large extent balanced by their synthetic activities. Bacteria are very efficient at scavenging ammonia if it is in short supply, but will allow it to accumulate in the liquid phase of the rumen content if the supply exceeds their needs.

Requirements of Rumen Microorganisms in Addition to Proteins and Carbohydrates

In addition to proteins and carbohydrates, sources of phosphate and sulphur have to be supplied in the diet of ruminants in order to enable microorganisms to optimally synthesize proteins. Phosphate is needed for the synthesis of nucleic acids which contain 10–20% of the nitrogen present in microorganisms. Sulphur is needed for the synthesis of methionine and cysteine present in microbial protein. The diets for ruminants should contain nitrogen and sulphur at the ratio of 10:1; sulphur may be supplied as sulphate or as methionine and cysteine. The microorganisms form sulphide needed for the biosynthesis of cysteine and methionine by breakdown of sulphur-containing amino acids or by reduction of sulphate to sulphide:

$$SO_4^{2-} \xrightarrow{+H} S^{2-}.$$

The sulphur requirement of sheep for wool will be discussed in Chapter 17.

FATE OF NITROGENOUS COMPOUNDS ENTERING THE SMALL INTESTINE

It appears from Figures 8.1 and 8.3 that microbial protein, together with undegraded dietary protein, passes into the intestine. Smaller amounts of other nitrogenous materials entering the duodenum are proteins of metabolic origin (Chapter 7), microbial nucleic acids, and ammonia resulting from the microbial fermentation of nitrogenous materials in the rumen that was neither utilized for microbial cell synthesis nor absorbed before reaching the proximal duodenum. The processes of protein digestion and absorption occurring within the ruminant's small intestine are little different from those occurring in monogastric animals (Chapter 7). The amounts of amino acids disappearing from the intestine are determined by analysing digesta sampled through cannulae fitted into the proximal duodenum and terminal ileum (Chapter 13). From such measurements, the net disappearance of individual amino acids during the passage of digesta through the small intestine can be calculated, and thus the amount absorbed and available to the host estimated. Such values should reveal the effectiveness with which various dietary nitrogen sources contribute to amino acid supply for the host animal.

The microbial protein content of the mixtures or proteins in the duodenum can be measured by assays of components associated only with microorganisms, such as nucleic acids, 2,6-diaminopimelic acid

$$\begin{array}{c} HOOC-\underset{\displaystyle NH_2}{\underset{|}{CH}}-(CH_2)_3-\underset{\displaystyle NH_2}{\underset{|}{CH}}-COOH \end{array}$$

characteristic of bacteria, or aminoethylphosphonic acid

$$H_2N-CH_2-CH_2-\overset{\displaystyle O}{\overset{\|}{P}}(OH)_2,$$

characteristic of protozoa. An alternative approach would be to measure ^{15}N or ^{35}S incorporated into microbes.

Digestion processes occurring in the small intestine of ruminant animals differ from those in monogastric animals in the following respects. (1) The amount of metabolic nitrogen (i.e. that supplied in pancreatic juice and bile entering the duodenum, and as epithelial cells sloughed from the tract) in proportion to the amount of feed nitrogen is considerably greater in ruminant animals than in monogastric species. Thus, the metabolic nitrogen is a useful supplement to the dietary and microbial protein reaching the intestine, particularly when the diet is protein deficient (Chapter 13). (2) The rate of neutralization of the digesta flow entering the duodenum is slower in ruminants than in monogastric animals. (3) Activation and peak activity of the pancreatic proteases is delayed to the midjejunum (instead of the duodenum in monogastric animals; Chapter 7). Thus, in sheep, the section of the intestine 7–15 m distant from the pylorus, the area of peak pancreatic protease activity, is also the area of the most intensive absorption of amino acids (Ben-Ghedalia *et al.*, 1974). The supply of essential amino acids required for the tissue metabolism of the host animal is augmented by preferential absorption through the intestinal wall. As a result, 60.5% of the essential amino acids passing through the pylorus and 43% of the non-essential amino acids are absorbed.

The high activity of nucleases in pancreatic juice of ruminants seems to be adapted to the high content of nucleic acids in microbial cells and enables the release of nucleic phosphorus for recycling to the rumen via saliva (see above, p. 159). The other components of nucleic acids liberated by digestion, while almost useless, are by no means noxious to the animals.

The ratio between microbial and unchanged dietary protein in the flow entering the duodenum depends on the nature of the diet and particularly on the source of protein ingested. Nevertheless, 50–90% of the protein is likely to be microbial, the remainder being food protein that has escaped degradation in the rumen.

EVALUATION OF MICROBIAL PROTEIN

In order to investigate the composition and nutritive value of microbial protein, bacteria and protozoa have been isolated from the rumen content of animals fed a variety of diets. The amino acid composition of microbial protein shows a remarkable uniformity which is apparently uninfluenced by diet. With respect to the similar digestive processes occurring in the intestine of ruminants and monogastric animals, digestibility and biological value of dried bacteria and protozoa were determined by experiments with rats and mice; digestibilities of rumen bacterial and protozoal protein were 74% and 91%, and the biological values 71% and 80%. Similar values for the digestibility of microbial protein were found recently by infusion of ^{35}S-labelled microorganisms into the abomasum of sheep.

FACTORS CONTRIBUTING TO THE EFFICIENCY OF DIETARY PROTEINS FOR RUMINANTS

In spite of the quite high digestibility and biological value of microbial proteins, there is no advantage in giving easily degradable protein feeds to ruminants. On the contrary, it is, for the following reasons, preferable to use proteins relatively resistant to microbial degradation in the rumen and to have them digested mainly in the intestine as in monogastric animals. (1) The amino acid composition of many food proteins is more favourable than that of microbial proteins. (2) The transformation of food into microbial protein is associated with considerable loss of nitrogen due to the ruminal ammonia that is not utilized by bacteria and is excreted as urea (see above), and to the considerable energy input required for microbial protein synthesis and for excretion of non-utilized rumen ammonia as urea (Chapter 7). (3) Of the nitrogen contained in microorganisms 10–20% is present as nucleic acids and mucopeptides, components which are of very little value for the host animal (see above).

As pointed out above, there is considerable variability in the degradability of various sources of protein by rumen microorganisms. Ready susceptibility to ruminal attack is not desired in protein feeds. The speed and extent to which protein is degraded depend on many factors, such as solubility in ruminal liquid, protein structure, amount and source of carbohydrates in the diet, level of food intake and rate of passage through the rumen. Casein, the major milk protein, is very soluble and almost completely degraded when incubated in rumen contents. On the other hand, zein, a corn protein, is very insoluble and 40–60% escapes breakdown in the rumen. Eighty per cent of fairly soluble barley protein is converted to microbial protein. Mild heat treatment of many foodstuffs, such as toasting of soybean meal, reduces their solubility and improves the efficiency of this protein for ruminants (Tagari *et al.*, 1962). Reasons for improving the nutritive value of soybean meal for monogastric species, particularly poultry, by heat treatment which inactivates trypsin inhibitors, are not relevant for ruminants. However, excessive heating renders the protein less digestible even in the lower tract, thus making the product inferior as a protein source for ruminants. High solubility of proteins is not the only factor responsible for their susceptibility to rumen breakdown; readily soluble egg albumin, for example, is resistant to breakdown apparently because of its chemical structure.

As pointed out above, a major factor influencing the utilization of ruminal ammonia is the availability of carbohydrates. Nitrogen utilization may be improved as the amount of soluble carbohydrates, particularly starch, in the ration increases. Carbohydrates serve as energy sources for the microorganisms and also deliver the nitrogen-free skeletons needed for the microbial synthesis of amino acids. (Some undesirable side-effects of increased supply of starch, such as the reduction of milk fat content, are mentioned in Chapter 21.)

Proteins Protected from Microbial Degradation

One method proposed to reduce the breakdown of dietary protein in the rumen is to treat the proteins with formaldehyde. This reduces solubility by the formation of a complex between the aldehyde and a free amino group as in the Maillard reaction (Chapter 7). The formaldehyde–protein complex is fairly stable at pH 6.0 which prevails in the rumen, rendering it highly resistant to microbial attack. However, the complex decomposes readily at the acid pH of the abomasum, enabling the protein to be easily digested in the small intestine. The same method can also be used for protecting unsaturated fat from hydrogenation by ruminal action; for this purpose, fat particles are coated with a protein–formaldehyde emulsion (Chapter 6). Fresh green fodder and silages contain quite large amounts of soluble proteins and non-nitrogen protein compounds; too rapid liberation of ammonia from these feeds passing through the rumen can be prevented by treatment with formaldehyde before feeding.

Methionine, an amino acid likely to be limiting in rations for ruminant animals, can be used as a food additive when protected against decomposition in the rumen. This is accomplished by encapsulation in films of hydrogenated lipids which are insoluble in the rumen, but are readily broken down in the small intestine by the action of lipase and bile.

Early in the life of the calf, liquid foods, like milk, bypass the rumen through reflex closure of the oesophageal groove. This can be achieved in older animals by training the suckling habit when liquid suspensions of proteins feeds are offered to the animals (Orskov, 1977, 1982).

FUNCTIONS OF AMINO ACIDS ABSORBED BY RUMINANTS

The benefit which the ruminant derives from the intake of nitrogenous food components depends on the amino acid composition of the material reaching the small intestine and the amounts of amino acids absorbed from this organ.

The specific purpose of amino acids absorbed from the small intestine of ruminants, as in all other vertebrates, is protein synthesis, contributing to tissue growth and milk protein production. Because ingested carbohydrates are largely fermented in the rumen and absorption of glucose is very low relative to the host animals' metabolic requirements, the contribution of amino acids to the formation of glucose by gluconeogenesis is very important (Chapter 5). Amino acids, particularly glutamic acid and aspartic acid, serve in ruminants as an energy source for several purposes. Both, for example, are utilized extensively in the gut wall for the oxidative metabolism needed to ensure the metabolic activity in this tissue. The net uptake of both these amino acids into the portal blood plasma is almost zero, whereas the other amino acids, including all the essential ones, are absorbed into the portal plasma in highly significant quantities approximately at the same rate as the amino acids disappear from the gut lumen. The measurement of venoarterial amino acid concentration differences (by blood sampling through catheters implanted into blood vessels) has been used as an alternative procedure to that based on analysis of digesta sampled through intestinal cannulas (Chapter 13; Tagari and Bergman, 1978).

NITROGEN METABOLISM IN THE LARGE INTESTINE

The amount of nitrogen entering the caecum amounts to about 20% of total nitrogen intake; nitrogenous compounds enter the caecum from the upper digestive tract in feed residues, rumen microorganisms and endogenous materials. In addition, urea enters the large intestine directly from the blood.

Catabolic and anabolic microbial processes similar to those proceeding in the rumen occur in the large intestine. However, absorption of nitrogenous compounds other than ammonia from the large intestine is very limited. Ammonia absorbed from the caecum should not, however, be considered as a complete loss because it is converted to urea in the liver, and part of it is recycled to the rumen and incorporated into microbial protein. Variable portions of all three nitrogenous fractions originating from microbial, feed and endogenous proteins are excreted in the faeces (Chapter 13).

BIBLIOGRAPHY

Books

Orskov, O. R. (1982). *Protein Nutrition in Ruminants.* London and New York: Academic Press.

Owens, F. N. (1982). *Protein Requirements of Cattle.* Symposium, Oklahoma State University.

Articles

Symposium on nitrogen utilization by the ruminant (1973). *Proc. Nutr. Soc.*, **32**, 79–122.

Symposium on nitrogen metabolism (1975). In *Symposia on Ruminant Physiology, Digestion and Metabolism in the Ruminant.* 4th Symposium. Sydney, Australia 1974, p. 399–464.

Symposium on quantitative aspects of nitrogen metabolism in the rumen (1979). *J. Anim. Sci.*, **49**, 1604–1659.

Symposium on postruminal fate of nitrogenous compounds (1978). *Fed. Proc.*, **27**, 1222–1232.

Ben-Ghedalia, D., Tagari, H., and Bondi, A. (1976). The ileum as a site of protein digestion. *Br. J. Nutr.*, **36**, 211–217.

Ben-Ghedalia, D., Tagari, H., Bondi, A., and Tadmor, A. (1974). Protein digestion in the intestine of sheep. *Br. J. Nutr.*, **31**, 125–142.

Bergen, W. G. (1979). Free amino acids in blood of ruminants. Physiological and nutritional regulation. *J. Anim. Sci.*, **49**, 1577–1589.

Bergman, E. N., and Heitman, R. N. (1978). Metabolism of amino acids by the gut, liver, kidneys and peripheral tissues. *Fed. Proc.*, **37**, 1228–1232.

Buttery, P. I. (1977). Biochemical basis of rumen fermentation. In *Recent Advances in Animal Nutrition—1977* (W. Haresign and D. Lewis, eds), pp. 8–24. London and Boston: Butterworths.

Chalupa, W. (1975). Rumen bypass and protection of proteins and amino acids. *J. Dairy Sci.*, **58**, 1198–1218.

Cheng, K. J., and Costerton, J. W. (1979). Adherent rumen bacteria. Their role in the digestion of plant material, urea and epithelial cells. In *Digestive Physiology and Metabolism in Ruminants* (Y. Ruckbush and P. Thivend, eds). pp 227–250. Westport: Avi Publishing Co.

Czerkawski, J. W. (1978). Reassessment of efficiency of synthesis of microbial protein. *J. Dairy Sci.*, **61**, 1261–1273.

Demeyer, D. I., and Van Nevel, C. J. (1980). Nitrogen exchanges in the rumen. *Proc. Nutr. Soc.*, **39**, 89–96.

Egan, A. R. (1980). Host animal–rumen relationships. *Proc. Nutr. Soc.*, **39**, 79–88.

Fenderson, C. L., and Bergen, W. G. (1975). An assessment of essential amino acid requirements of growing steers. *J. Anim. Sci.*, **41**, 1759.

Hogan, J. P. (1975). Quantitative aspects of nitrogen utilization in ruminants. *J. Dairy Sci.*, **58**, 1154–1177.

Kennedy, P. M., and Milligan, L. D. (1980). The degradation and utilization of endogenous urea in the gastrointestinal tract of ruminants. A review. *Can. J. Anim. Sci.*, **60**, 205–221.

Lewis, D., and Mitchell, R. M. (1976). Amino acid requirements of ruminants. In *Protein Metabolism and Nutrition* (D. J. A. Cole *et al.*, eds), pp. 417–424. London: Butterworths.

Lindsay, D. B. (1979). Metabolism in the whole animal. *Proc. Nutr. Soc.*, **38**, 295–301.

McAllan, A. B. (1982). The fate of nucleic acids in ruminants. *Proc. Nutr. Soc.*, **41**, 309–318.

McMeniman, N. P., Ben-Ghedalia, D., and Armstrong, D. G. (1976). Nitrogen–energy interactions in rumen fermentation. In *Protein Metabolism and Nutrition* (D. J. A. Cole *et al.*, eds), pp. 217–230. London: Butterworths.

Mercer, I. R., and Annison, E. F. (1976). Utilization of nitrogen in ruminants. In *Protein Metabolism and Nutrition* (D. J. A. Cole *et al.*, eds), pp. 397–416. London: Butterworths.

Nolan, J. V., Norton, B. W., and Leng, R. A. (1976). Simulation of sheep nitrogen metabolism. *Br. J. Nutr.*, **35**, 127–147.

Owens, P. N., and Bergen, W. G. (1983). Nitrogen metabolism of ruminant animals. *J. Anim. Sci.*, **57**, Suppl. 2, 498–516.

Satter, L. D., and Roffler, R. E. (1977). Influence of nitrogen and carbohydrate inputs on rumen fermentation. In *Recent Advances in Animal Nutrition—1977* (W. Haresign and D. Lewis, eds), pp. 25–49. London and Boston: Butterworths.

Slyter, L. L., Satter, L. D., and Dinius, D. A. (1979). Effect of ruminal ammonia concentration on nitrogen utilization by steers. *J. Anim. Sci.*, **48**, *906–912*.

Stern, M. D., and Hoover, W. B. (1979). Methods for determining and factors affecting rumen microbial protein synthesis: a review. *J. Anim. Sci.*, **49**, 1590–1603.

Tagari, H., Ascarelli, I., and Bondi, A. (1962). The influence of heating on the nutrititive value of soya bean oil meal for ruminants. *Br. J. Nutr.*, **16**, 237–243.

Tagari, H., and Bergman, E. N. (1978). Intestinal disappearance and portal blood appearance of amino acids in sheep. *J. Nutr.*, **108**, 790–803.

CHAPTER 9

Water and its Importance in Nutrition

The primary importance of water as the major constituent of the animal body and the influence of various factors on the water content of the body such as species, age, dietary conditions, have been dealt with in Chapter 1.

There are three sources of water for the animal: (1) drinking water; (2) water contained in foods; (3) metabolic water. Green forages and silages contain 70–90% water and make a substantial contribution to the water requirement of animals. Dry food such as concentrates and hay contain between 7% and 15% water. The presence of more than 15% moisture in dry feedstuffs is not acceptable because of the subsequent diminution of the feeding value and the predisposition of moist feedstuffs to become mouldy or rotten.

Metabolic water is produced by metabolic processes in tissues, mainly by oxidation of nutrients. The three main kinds of nutrients produce different amounts of water. The oxidation of each gram of carbohydrate yields 0.6 g of water, each gram of fat yields 1.1 g of water, and each gram of protein 0.4 g of water. For most domestic animals metabolic water comprises only 5–10% of the total water intake. In certain conditions the metabolic water is the only source of water for animals. In such cases, and also in animals consuming less food than required, the production of metabolic water becomes more important, since depot fat and tissue protein are catabolized to supply energy.

Functions of Water

Many of the biological functions of water are dependent upon the property of water acting as solvent for numerous compounds. Water takes part in digestion (hydrolysis of proteins, fats and carbohydrates), in absorption of digested nutrients, in transport of metabolites in the body, and in excretion of waste products. Many catabolic and anabolic processes taking place inside the tissues involve the addition or release of water.

The regulation of body temperature is dependent partially on the high conductive property of water to distribute heat evenly within the body and eventually to remove by vaporization excess water released by metabolic reactions within the cells. Drastic changes in the body temperature are prevented by the high specific heat of water, i.e. by its high latent heat of vaporization, together with the high water content of the body.

Water Losses

Water is lost from the body constantly in the respired air by evaporation from the skin and periodically by excretion in urine and faeces. Water excreted in the urine acts as a solvent for excretory products excreted through the kidneys. The urine contains mainly breakdown products of protein (urea in mammals, uric acid in birds) and minerals (see Chapter 3). Urea in concentrated aqueous solutions would be toxic to the tissues. In urine the urea is diluted by water to harmless concentrations and finally excreted.

Protein consumption by birds involves a lower requirement for water than ingestion of protein by mammals (see Chapter 7) for two reasons: (1) the breakdown of protein to uric acid provides more metabolic water than its final catabolism to urea; and (2) uric acid, the final protein breakdown product of proteins in birds, is excreted in nearly solid form in the droppings.

Faecal water losses are considerably higher in ruminants than in other species, being about equal to urinary losses, whereas in man the faecal loss of water is only about 7–10% that of urinary water. Cattle that consume fibrous diets excrete faeces of 68–80% water. Sheep faeces, which form pellets, contain 50–70% water. The water losses in faeces of ruminants are small compared with the very large amounts of water secreted into the tract through saliva and the digestive juices. This may be explained by the fact that the greater part of the water secreted into the tract is reabsorbed. In diarrhoea large losses of water occur with the faeces.

Water Requirements

It is a well-known fact that animals are more sensitive to a lack of water than of food. The first noticeable effect of moderate restriction of water is a reduced intake of food. As a consequence of more severe restriction of water intake, weight loss is rapid as the body dehydrates. Dehydration involves the loss of water and electrolytes. Dehydration associated with the loss of 10% of water content of the body is considered severe and a 20% loss results in death, whereas animals are capable of living even after a loss of 40% of their dry body weight caused by starvation.

Water requirements are influenced by dietary and environmental factors. Water consumption is related to dry matter intake. The larger the proportion of undigested matter the greater is the loss of faeces and as a result a loss of the amount of water excreted in the faeces. Water requirements increase with the level of roughage. The water intake of adult cattle is 3–5 kg for 1 kg of dry matter intake; the water intake of suckling calves is much higher, being 6–7 kg per kg dry matter. Milking cows require additional amounts of water in order to provide adequate amounts for the secretion of large amounts of water in the milk; about 4–5 kg of water are required for each kg of milk produced. Egg laying, likewise, increases water requirement.

The presence of mineral salts, particularly sodium chloride, in the diet and ingestion of high-protein diets result in increased urinary excretion and as a result water consumption increases. Water containing 1.3–1.5% total dissolved salts is tolerated by cattle; water with higher loads of salts causes injury due mainly to the osmotic effect rather than to a toxic action of specific salts.

Water intake arises with increasing air temperature to counteract increased respiratory and sweat losses, while feed intake will be decreased. The water cycle of a large animal is exemplified in Figure 9.1 for a horse. Very large amounts of water are involved in the daily cycle, but water consumption may be comparatively limited due to reabsorption of water occurring to a great extent in the digestive tract.

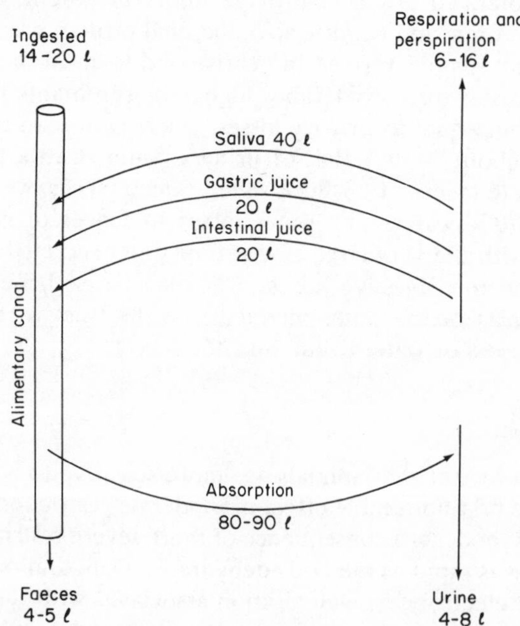

Figure 9.1. Metabolism of water in the body of a horse (after Nehring, 1959, *Animal Nutrition*, 7th edn (in German). Radebeul and Berlin: Neumann).

BIBLIOGRAPHY

Houpt, T. R. (1984). Water, electrolytes and excretion. In *Dukes' Physiology of Domestic Animals* (M. J. Swenson, ed.), 10th edn, pp. 486–506. Ithaca and London: Cornell University Press.

Leitch, I., and Thomson, J. S. (1944). The water economy of farm animals. *Nutr. Abst. Rev.*, **14**, 197–223.

Lepkovsky, S., Chari-Bitron, R., and Lyman, R. L. (1960). Food intake, water intake and body water regulation. *Poultry Sci.*, **39**, 390–394.

Macfarlane, W. V., and Howard, B. (1970). Water in the physiological ecology of the ruminant. In *Symposia on Physiology, Digestion and Metabolism in the Ruminant.* Third Symposium, pp. 362–374. Newcastle: Oriel Press.

Macfarlane, W. V. (1976). Water and electrolytes in domestic animals. In *Veterinary Physiology (I. W. Phillis, ed.), pp. 463–539. Philadelphia: W. B. Saunders.*

Robinson, I. R., and McCance, R. A. (1952). Water metabolism. *Ann. Rev. Physiol.*, **14**, 115–142.

Winchester, C. F., and Morris, M. I. (1956). Water intake rates of cattle. *J. Anim. Sci.*, **15**, 722–740.

CHAPTER 10

The Nutritional Importance of Minerals

Animal tissues and feeds contain about 45 mineral elements in widely varying amounts and concentrations. The following seven elements are found in the body in high concentrations (above 70 mg/kg live weight) and are termed macrominerals: calcium (Ca), phosphorus (P), magnesium (Mg), sodium (Na), potassium (K), chlorine (Cl) and sulphur (S). The body contains very small amounts of the approximately 40 other elements (less than 70 mg/kg live weight) known as microelements or trace elements. To date the following 15 of these elements have been shown to fulfil physiological functions in the body: iron (Fe), copper (Cu), cobalt (Co), manganese (Mn), zinc (Zn), iodine (I), selenium (Se), molybdenum (Mo), chromium (Cr), fluorine (F), tin (Sn), vanadium (V), silicon (Si), nickel (Ni), and arsenic (As). These elements are essential trace elements.

Proof that a certain element is essential is based upon animal experiments. It must be shown that purified diets (see introduction to Part 3) supplying all required nutrients and lacking only the specific element cause growth impairment in young animals and deficiency symptoms in both young and adult animals. These symptoms indicating biochemical changes in animal tissues can be reduced or prevented by adding the missing element to the experimental diets. In addition to the 22 essential macro- and microelements, plant and animal tissues contain an additional 23 mineral elements in low concentrations. No vital function has been assigned to them and they are thought to be present in the animal body because they are present in food. In the future, improved techniques may be able to demonstrate some definite function for some of these elements.

Minerals have three general functions:

(1) Calcium and phosphorus—the main structural components of bones and teeth—give rigidity and strength; magnesium, fluorine and silicon in bones and teeth also contribute to the mechanical stability of the body.

(2) Only small fractions of the calcium, magnesium and phosphorus and most of the sodium, potassium and chlorine are present as electrolytes in the body fluids and in the soft tissues. The electrolytes present in body fluids such as blood or cerebrospinal fluid serve important functions in maintaining acid–base balance and osmotic pressure; they regulate membrane permeability and exert characteristic effects on the excitability of muscles and nerves. The salts in the saliva, in the gastric and intestinal juices and in the rumen fluid produce in the digestive tract the medium appropriate for the action of enzymes and growth of microorganisms (see Chapter 3).

(3) Essential trace elements are integral components of certain enzymes and of other biologically important compounds, such as iron in haemoglobin, cobalt in vitamin B_{12} and iodine in the hormone thyroxine. Trace elements function also as activators of enzymes (see Chapter 7).

MINERAL CONTENT OF THE BODY AND ITS ORGANS

The amounts of each of the macrominerals present in the body of adult mammals and in their blood serum are given in Table 10.1. The proportions of each mineral, expressed as amount per 100 g of fat-free dry body substance, are very similar among species in adult mammals. Each organ—in accordance with its function—has a characteristic mineral composition, which again is very similar in all mammals. However, after a period of undernutrition or water deprivation there is quite a sharp rise in the mineral content (in the fat-free dry matter of the body). It should be noted that the sodium, potassium and chloride content of the body is constant during all the stages of development from embryo to full development, whereas the magnesium, calcium and phosphorus contents in the embryo (calculated on basis of the fat-free dry matter) are only one half of the respective content in the adult animal.

The levels of macrominerals in the blood serum, especially those of calcium, magnesium, potassium and chlorine, are maintained within rather narrow ranges by hormonal control mechanisms independently from the amounts provided by food. Such maintenance of steady-state concentrations of body constituents in the internal medium is called homeostasis; this phenomenon also exists with some organic compounds such as glucose (see Chapter 2). Control system regulating the rate of absorption of minerals from the digestive tract, the rate of their deposition and liberation from body reserves in the bones, and their rate of excretion via faeces and urine will be discussed below with the metabolism of individual minerals. On the

Table 10.1 Mineral contents of adult animal body and blood serum of mammals (according to Kirchgessner, *Animal Nutrition*, 3rd. ed. p. 145, 1978. Frankfurt: DLG)

	Calcium	Phosphorus	Magnesium	Sodium	Potassium	Chlorine
g/100 fat-free dry matter of body	1–2	0.7–1	0.05	0.15	0.30	0.1–0.15
Blood serum, mg/100 ml	10	4–7*	2–3	330	2	370

* Inorganic P.

other hand, the concentrations of essential trace minerals in blood tend to vary since there are no control mechanisms to cope with the greatly varying intake of minor elements with respect to their concentration in blood.

DIETARY SOURCES OF MINERALS

Farm animals derive most of their mineral nutrients from concentrate feeds and forages that they consume. Mineral supplements of animal origin (bone meal) or of geological origin (calcium phosphates, sodium chloride) are additional sources. Drinking water is only a minor source of minerals. Soil contamination of herbage can provide a further source for grazing animals.

Concentration of minerals in forage plants depends on the following factors: the species of plant, the composition of the soil in which the plant grows, the stage of maturity, climatic conditions, agricultural treatments, such as application of fertilizers or irrigation.

Environmental conditions of plant growth exert a much greater influence on the mineral composition of the vegetative parts of plants than on that of grain and seeds. There are much wider fluctuations in the content of minor elements in the same feedstuffs than in that of major elements. This is due to the wide variation in the trace element content of soils in different geographic areas and to the influence of varying soil conditions on the uptake of minerals by the plant.

Mode and Sites of Absorption of Minerals

Minerals are mainly absorbed as ions. The major sites of their absorption into the blood circulation are the small intestine and the anterior part of the large intestine. In ruminants some absorption also occurs through the walls of the rumen. Large amounts of minerals entering the digestive tract in digestive juices are reabsorbed together with those originating directly from the food.

The mode of excretion of minerals, whether faeces or urine, varies with the species of animal and the composition of the diet: ruminants tend to excrete calcium and phosphorus in the faeces whereas monogastric species excrete these elements mainly in the urine.

AVAILABILITY OF MINERALS TO ANIMALS

The evaluation of minerals contained in feeds and mineral supplements depends not only on their absolute mineral content but also on the extent of their absorption and utilization by the animal. Determination of the apparent digestibility of minerals (see Chapter 13) is not meaningful since the faecal output includes both unabsorbed and endogenous minerals. Minerals secreted in the saliva and digestive juices enter the intestinal lumen where they are handled with the same efficiency as minerals of dietary origin, i.e.

they are reabsorbed into the bloodstream. However, a part of the minerals secreted from the body is not absorbed and continue to the faeces; these are termed endogenous or metabolic minerals. The endogenous proportion of macrominerals present in the faeces of ruminants, particularly of calcium and phosphorus, may be considerable and even exceed the amount in the undigested portion. In monogastric animals endogenous minerals account for a small fraction of total faecal minerals. Only by determining the total balance (see Chapter 7) of the mineral in question, or by performing isotope experiments, is it possible to gain knowledge on the availability of minerals administered from dietary sources. If the balance is zero in mature animals or positive in growing animals, the feed source examined satisfies the needs. The opposite is true if the balance is negative. In order to eliminate the influence of changes in body stores on the result of balance experiments, use of comparative balance technique involves a comparison of mineral retention when balance are determined at two rates of intake. The availability of mineral can be estimated by dividing the difference in mineral balance at two intakes by the difference in its intake, according to the following equation:

$$\text{Percentage utilization} = \frac{\text{Improvement in balance}}{\text{Change in intake}} \times 100$$

Procedures with radioisotopes can differentiate between the unabsorbed mineral fractions of the faeces (originating directly from the ingested mineral) and the endogenous fraction of faeces, which has been absorbed, re-excreted by the gut and eliminated in the faeces (see Chapter 13).

The principles of the isotope method are as follows. The radioisotope of the element under study is first injected into the bloodstream while the test food is fed. The radioactive isotope introduced into the bloodstream is determined in the faeces soon after its injection and is found there in the same proportion to the stable mineral as in blood. After further ingestion of the test feed the radioisotope in the faeces becomes diluted in proportion to the exogenous fraction in the overall content of the element in the faeces. The calculation of the amount of endogenous mineral in the faeces and of the percentage utilization of the mineral can be calculated using the following equations:

Endogenous faecal phosphorus =

$$\frac{\text{Specific activity of faeces}}{\begin{array}{c}\text{Specific activity of endogenous phosphorus}\\ \text{(i.e. of urine or plasma)}\end{array}} \times 100$$

Percentage phosphorus utilization =

$$\frac{\text{Phosphorous intake} - \text{unabsorbed faecal phosphorus}}{\text{Phosphorus intake}} \times 100$$

Such measurements showed that 40–60% of phosphorus present in faeces of ruminants are of endogenous origin (see later).

EFFECTS OF DEFICIENCIES AND IMBALANCES OF MINERALS ON ANIMALS AND THEIR PREVENTION

It must be emphasized that the supply of some elements such as copper and cobalt to ruminants and of selenium and iron to all kinds of farm animals is essential, but that their ingestion above a certain level produces toxic symptoms.

The prolonged ingestion of diets that are deficient, unbalanced or excessively high in a certain mineral induces changes in its concentration in the animal's tissues from below or above the permissible limits. In such circumstances physiological functions can be adversely affected. Nutritional disorders may arise which, apart from external symptoms, result in retarded growth, decreased food utilization and productivity, and disturbances in fertility and general health. Such nutritional disorders range from severe mineral deficiency or toxicity with a high mortality to mild conditions which arise quite frequently from local mineral deficiencies. Fortunately serious cases of mineral deficiency or excess occur quite seldom under practical farm conditions. Minimum requirements and maximum tolerances have been determined for most essential elements (see later).

Accompanying inorganic and organic substances exert a greater influence on the utilization of certain minerals compared to the mutual interactions of organic nutrients (protein, fat and carbohydrates) when the latter are ingested in unusual proportions. Interrelationships between minerals or interactions between minerals and organic compounds may result in decreased or enhanced mineral utilization. A surplus of certain ions in the basic medium of the intestine may lead to precipitation of insoluble salts and to decreased availability of the respective minerals, for example, a surplus of phosphate can cause precipitation of calcium ions, and a surplus of molybdate can cause precipitation of copper ions. On the other hand, there are components of the food such as amino acids and peptides which improve the absorbability of certain minerals by formation of soluble chelates. Chelates are generally soluble compounds which are formed between an organic compound and a metallic ion. One of the most potent chelating agents is the synthetic compound EDTA (ethylenediaminetetraacetic acid). The addition of EDTA to poultry diets is used to increase the availability of zinc and manganese.

Prevention of Mineral Deficiencies and Imbalances

When the mineral requirements of animals are not met by a combination of the available feedstuffs, rations may be supplemented with concentrated sources of one or more mineral elements, or the use of commercial mineral mixtures. Supplementary minerals may also be provided by suitable licks containing appropriate proportions of the deficient elements, by treatment of drinking water with soluble salts or by injections of slowly absorbed organic compounds. Appropriate fertilizer treatment of the soil is also practised to improve the mineral composition of herbage.

MACROMINERALS

The Nutritive Importance of Calcium and Phosphorus

Distribution in the Body

These two elements are discussed together because they are closely associated with each other in their occurrence in feeds and in the animal body.

Approximately 99% of the calcium and 80% of the phosphorus of the body are present in bones and teeth; the remainder is distributed in soft tissues and body fluids. The approximate composition of normal adult bone, though somewhat variable according to species, age and state of nutrition, is water 45%, ash 25%, protein 20%, and fat 10%. The approximate composition of ash is calcium 36%, phosphorus 17%, and magnesium 0.8%. Calcium and phosphorus occur in bone ash in a ratio of 2 : 1. Bone has a crystallized phase composed largely of hydroxyapatite, $Ca_{10}(PO_4)_6(OH)_2$, and an amorphous phase containing $Ca_3(PO_4)_2$, $CaCO_3$, $Mg_3(PO_4)_2$ and small amounts of citrates, sodium, potassium, chlorine and fluorine. The organic matrix of bone in which the mineral salts are deposited consists of a mixture of proteins, mainly collagen.

The mineral composition of teeth is similar to that of bone. The enamel of the teeth, which is the hardest substance in the body, contains only 5% of water and 3.5% of organic matter. Small amounts of fluorine occurring in apatite, $Ca_{10}(PO_4)_6F_2$, are an integral part of the structure of teeth and bones.

Calcium and Phosphorus in Blood. Blood cells are devoid of calcium. Blood plasma of most animals contains 9–11 mg calcium per 100 ml, except the plasma of laying hens with 20–30 mg calcium per 100 ml (see Chapter 20). The calcium in the plasma of mammals exists in three forms: about 50% is present as free ions, 45% is bound to plasma proteins, and 5% is chelated with citrates and phosphates.

Whole blood contains 35–45 mg phosphorus per ml as orthophosphate (HPO_4^{2-} and $H_2PO_4^-$), most of which is in the cells. The levels of plasma inorganic phosphorus are between 4 and 9 mg per 100 ml; much of the plasma phosphate is ionized, but a small amount is complexed with proteins, lipids and carbohydrates.

Calcium and Phosphorus Metabolism in Bones. Bone is not a static depot of minerals serving only a structural function, but is in a dynamic state. Bones serve as a store of calcium and phosphorus which can be mobilized when the provision of these minerals is not sufficient to meet body requirements. Thus the mineral metabolism of bone involves not only the accretion of calcium and phosphorus during growth, but also their continuous exchange between the bones and the blood. The rate of exchange can be measured by the parenteral administration of radiocalcium (^{45}Ca) and its determination in the skeleton. The rate of exchange is most rapid in the spongy zones of the bones (trabeculae) and differs with that in compact bone (cortex). The rapidity with which the isotopes appear in bone is related to the enormous surface of bone crystal exposed to extracellular fluid.

Regulation of Calcium and Phosphorus Metabolism and Homeostasis

The overall metabolism of calcium, phosphorus and magnesium is schematically presented in Figure 10.1.

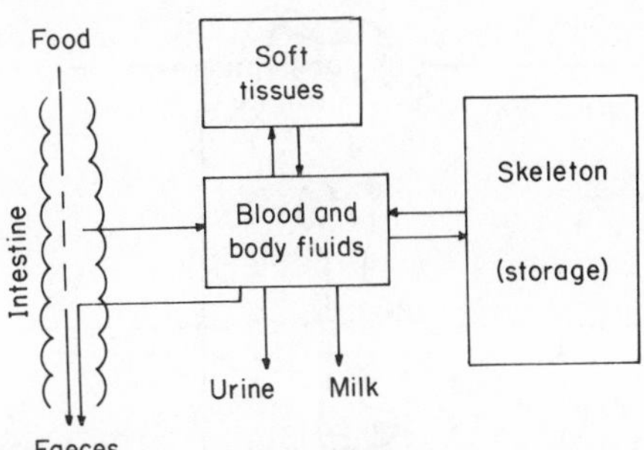

Figure 10.1. Schematic diagram of metabolism of calcium, phosphorus and magnesium.

Absorbed dietary macroelements (calcium, phosphorus, magnesium, potassium, sodium) reach the blood, which mediates also the exchange of calcium, magnesium and phosphorus between various organs. Calcium and phosphate concentrations in blood are maintained at constant levels by the regulatory actions of three hormones: parathyroid hormone (PTH), calcitonin and the active metabolite of vitamin D_3, 1,25-dihydroxycholecalciferol ($1,25(OH)_2D_3$, see Chapter 11). These hormones control the absorption of calcium and phosphorus from the gastrointestinal tract, influence their deposition and resorption in bone, and affect the extent of their excretion in faeces and urine.

When serum calcium is low because of an insufficient dietary supply or increased requirements of highly producing milk cows (see Chapter 21) or of laying hens (see Chapter 20), PTH is secreted. PTH is a single-chain polypeptide containing 84 amino acid residues and has a molecular weight of 8500. By means of a system of negative feedback, low concentrations of calcium in blood plasma stimulate the activity of the parathyroid gland and the secretion of PTH. This hormone leads to elevation of the blood calcium concentration by the following means: increased mobilization of calcium from the bones, increased tubular reabsorption of calcium and enhanced production of the hormonal form of vitamin D, $1,25(OH)_2D_3$. PTH plays an important role in the conversion of vitamin D into its active metabolite $1.25(OH)_2D$ which in turn stimulates calcium and phosphorus absorption from the intestine and which has a favourable effect on two reverse processes, bone resorption and bone formation (see Chapter 11). The various actions of PTH play a major role in regulating serum calcium concentrations (see Figure 10.2).

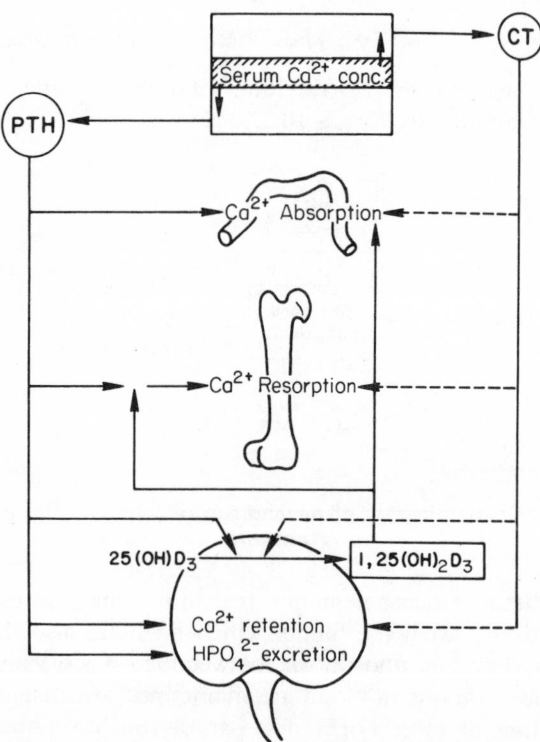

Figure 10.2. Calcium homeostatic mechanism involving parathormone (PTH), calcitonin (CT) and active vitamin D-$1,25(OH)_2D_3$. Solid line indicates stimulation and broken line shows inhibition (after Wasserman adapted by *Dukes' Physiology of Domestic Animals,* 7th edn, 1977, p. 748). *Reproduced by permission of Cornell University Press.*

The hormone calcitonin acts as a physiological antagonist of PTH and the hormonal form of vitamin D. Calcitonin, a polypeptide containing 32 amino acids, is secreted by the C-cells of the mammalian thyroid and by the ultimobranchial glands situated in the lower neck region of birds. When blood calcium levels are high, calcitonin is secreted and prevents bone calcium from being mobilized into the serum; in this way, and also by inhibiting the reabsorption of calcium ions in the kidney, a rapid decrease in blood calcium levels may be achieved. Furthermore, it appears that this hormone stimulates bone formation. The action of calcitonin, opposite to that of PTH, is similar to that of the insulin–glucagon system in controlling blood sugar (see Chapter 4). The concentration of calcitonin is directly proportional to plasma calcium concentration whereas that of PTH is inversely proportional. These relationships are shown in Figure 10.3.

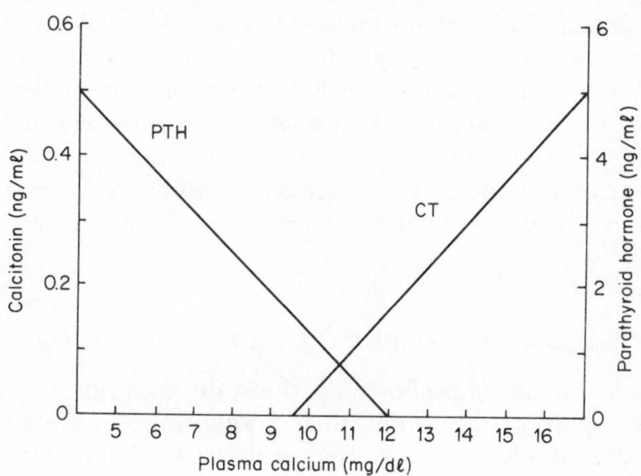

Figure 10.3. Effects of changes in serum calcium on the concentration of parathyroid hormone (PTH) and calcitonin (CT) in blood (after Potter, adapted by Pike, R. L., and Brown, M. L., *Nutrition: An Integrated Approach*, 2nd edn, 1975, p. 639. New York: John Wiley and Sons). *Reproduced by permission of the publisher.*

Plasma phosphate concentration is to a large extent regulated independently from that of calcium. Phosphate deprivation increases also the synthesis of $1,25(OH)_2D_3$, which in turn enhances phosphate absorption from the intestine. Another, even more important point of control of phosphate concentration is the kidney where the reabsorption of phosphate is increased by PTH and also by the active vitamin D metabolite. Furthermore, changes in plasma phosphate concentration may play a part in the regulation of plasma calcium, and vice versa, since inadequacy in dietary calcium or phosphate results in increased production of PTH and $1,25(OH)_2D_3$ and subsequently in an increased rate of absorption of both minerals.

Calcium and Phosphate in Soft Tissues and Body Fluids. The small proportions of body calcium (1%) and of phosphate (20%) present in soft tissues and body fluids have important functions. The importance of blood for supplying calcium and phosphorus needed for bone formation has been stressed above. As mentioned at the beginning of this chapter, calcium and phosphate ions participate in the regulation of the osmotic pressure and the acid–base balance of body and tissues fluids. However, calcium and phosphorus in body tissues and fluids have some specific functions: calcium controls the excitability of nerves and muscles, and it is needed for normal blood clotting as it must be present for the transformation of prothrombin to thrombin (see Chapter 11). The presence of calcium is necessary for the activation of certain enzymes such as trypsin and adenosinetriphosphatase.

Phosphorus has more known functions than any other element in the animal body. It functions in energy metabolism as a component of energy-rich compounds such as ADP, ATP and creatine phosphate (see Chapter 2). Metabolic reactions of carbohydrates, proteins and lipids occur through phosphorylated intermediate compounds. Phosphorus is a component of phospholipids which are important in lipid transport and metabolism and as constituent of cell membranes. Phosphate is a component of RNA and DNA, the vital cellular constituents essential for protein synthesis. Phosphate is a component of enzyme systems such as cocarboxylase and NAD (see Chapter 4). Phosphorus-containing proteins occur in milk (casein) and egg yolk (vitellin).

Absorption of Calcium and Phosphorus from the Digestive Tract

Adequate calcium and phosphorus nutrition depends not only on a sufficient dietary supply but also upon their mutual ratio in the diet and on the presence of other compounds or ions in the diet. Vitamin D is the most important of the compounds which affect the availability of calcium and phosphorus.

The major site of absorption of dietary calcium and phosphorus in most species is the duodenum, and considerable amounts of these minerals (of endogenous origin) are secreted into the lower part of the small intestine. The partition of calcium and phosphorus excretion between faeces and urine and the presence of large endogenous fractions of these minerals in the faeces of ruminants have been pointed out at the beginning of this chapter. The faecal losses of endogenous calcium and phosphorus as determined by balance trials with ^{45}Ca (see above) are independent of the amounts of these minerals ingested or absorbed and are directly proportional to body weight. The faecal loss of endogenous calcium for cattle is 16 mg/kg live weight, and that of phosphorus 10 mg/kg for weaned cattle and 4 mg/kg for milk-fed calves. The urinary loss is much smaller: 0.8 mg calcium and 2 mg phosphorus per kg live weight.

The absorption, i.e. the true digestibility of calcium in ruminants ranges from 22 to 55% with an average of 45%, whereas that of phosphorus is 55% (compare Chapter 21). The absorbability of calcium and phosphorus in cattle declines with age; calcium and phosphorus absorption from milk is considerably more efficient (about 90%) than from forage and concentrate mixtures.

Apart from the presence of vitamin D, the absorption of calcium and phosphorus depends on many factors which affect their solubility at the point of contact with the absorbing membranes. A large excess of either calcium or phosphorus interferes with the absorption of the other by decreasing the solubility product of calcium phosphate. In monogastric species a dietary calcium/phosphorus ratio of between 1 : 1 and 2 : 1 may well be optimal. The supply of vitamin D greatly reduces the significance of an adverse calcium/phosphorus ratio; the exposure of animals to sunlight is sufficient to produce the amount of vitamin D needed (see Chapter 11). Growing ruminants can tolerate a wide range of calcium/phosphorus even as large as 7 : 1. Pigs and growing poultry appear to be less tolerant of high calcium/phosphorus ratios. However, for laying hens the optimal ratio is considerably higher because of the great requirement of calcium for eggshell formation (see Chapter 20).

Large intakes of iron, aluminium and magnesium salts interfere with the absorption of phosphorus by forming insoluble phosphates. Likewise, in monogastric animals, high levels of fat increase faecal calcium losses through the formation of soaps. The availability of calcium in these animals is impaired by the presence of oxalic or phytic acid. These acids precipitate the calcium and prevent its absorption. Oxalates occur in sugarbeet and its byproducts, and phytates in cereal grains, particularly in bran and in oilcakes. Phosphorus present in phytates is poorly available to simple-stomached animals. In calculating the available phosphorus content of a feed for poultry, the phosphorus from inorganic supplements and in feedstuffs of animal origin is considered to be 100% available, while that from plant origin is assumed to be 30% available. Calcium in the presence of oxalates or phytates and phosphorus present in phytates are available to ruminants since oxalic acid is fully oxidized by microbial enzymes to carbon dioxide and water and phytates are hydrolysed by microbial phytases occurring in the rumen to inositol and phosphoric acid:

$$\begin{matrix} COOH \\ | \\ COOH \end{matrix} \longrightarrow CO_2 + CO + H_2O$$

Oxalic acid

$$\text{Phytic acid } (C_6H_6(O(P))_6) \xrightarrow{\text{Phytase}} \text{Inositol } (C_6H_6(OH)_6) + 6H_3PO_4$$

Phytic acid — Inositol

Manifestations of Deficiencies of Calcium and Phosphorus

A dietary deficiency of calcium and phosphorus, or lack of vitamin D which impairs their absorption and utilization (see above and Chapter 11), results in abnormalities of bones and teeth, subnormal growth and production, depressed appetite and efficiency of feed. The basic defect in these deficiencies is a reduction or failure in the mineralization process of bone while the synthesis of the matrix continues. The bones of affected animals have a characteristically low ash content, and are soft and unable to maintain their normal shape. Bone abnormalities caused by mineral disorders may occur at any age, but more common in young animals. *Rickets* is the term used to denote the defective calcification of bones in young animals of all species (see Chapter 11). Rickets is characterized by malformed bones, enlarged joints, lameness, fractures and stiffness of gait. In mature animals the disease is termed *osteomalacia* and may be caused by the excessive mobilization of minerals from bone. *Osteoporosis* is another failure of bone metabolism in mature animals caused by a deficiency of calcium. In osteoporosis the mineral content of bone is normal—in contrast to osteomalacia—but the absolute mass of bone is reduced. Osteoporosis is prevalent in old age in humans, particularly in women. It facilitates the fracture of limbs and is followed by long periods of convalescence. A deficiency of calcium is not the only cause of this disease; in osteoporosis the resorption of bone exceeds its formation and the synthesis of bone matrix is defective.

Milk fever, a metabolic disease of cows and ewes which is associated with severe hypocalcaemia, is discussed in Chapter 24.

Animals are more sensitive to a deficiency of phosphorus than to one of calcium because bone mineral is less readily mobilized to maintain the level of serum phosphorus than that of calcium. Thus a low level of serum inorganic phosphate may indicate a deficiency of phosphorus. Such deficiencies have been frequently observed in livestock grazing on phosphorus-deficient soil; the phosphorus level of herbage may be lower than that in normal pasture by an order of magnitude, i.e. about 0.04% rather than 0.40%. The first symptom of phosphorus deficiency is probably *anorexia* (reduced appetite). Naturally occurring deficiencies of phosphorus are seldom uncomplicated, and affected animals may develop a craving to chew and ingest miscellaneous non-nutritive objects like soil, wood and bones. This type of behaviour has been termed *pica*. Pica is not a specific sign of a phosphorus deficiency; it may also be brought about by a deficiency of salt, potassium, fibre, etc., in the diet.

The calf crop of beef cows and the lamb crop of ewes grazing on phosphorus-deficient pasture are very low. Reports from South Africa on the dramatic increase of the calf crop of beef cows supplemented with bone meal have led to the belief that phosphorus is more important than other minerals in promoting fertility (see Chapter 19). This belief has now been tempered by recognition of the fact that a severe phosphorus deficiency reduces feed intake and feed utilization and may produce very complex secondary deficiencies.

Effect of Excess Calcium and Phosphorus

An excess of dietary phosphorus in relation to calcium may result in a bone disorder called nutritional *secondary hyperparathyroidism.* An excess of phosphorus depresses calcium absorption and thus the blood level of calcium drops. This effect stimulates the release of the hormone PTH which causes calcium to be removed from the bone to replenish the blood level (see above). The demineralized skeleton is replaced by fibrous connective tissue. Nutritional secondary hyperparathyroidism occurs in horses fed large amounts of grain without calcium supplementation.

Another disorder caused by ingestion of excess phosphorus is urolithiasis. This is the formation of stones in the kidney or bladder with the resultant obstruction of urine excretion.

Excess calcium in the diet reduces the absorption and utilization of minerals, particularly of phosphate and trace minerals (see later). Parakeratosis, a zinc-deficiency disease occurring in pigs fed marginal amounts of zinc and an excess of calcium is dealt with later in this chapter.

Feedstuffs as Sources of Calcium and Phosphorus

Different feeds vary widely in their content of calcium and phosphorus. Animal by-products such as fish meal and meat meal are comparatively rich sources of both minerals, particularly products of lower quality containing much bone. All the cereal grains and their by-products, oilseeds and oilcakes are low in calcium and rich in phosphorus. Forages, particularly legumes, have higher levels of calcium than of phosphorus. The calcium and phosphorus content of forages is low during the vegetative period; thus the mineral content of green and preserved forages depends on the vegetative stage of the plants when harvested. Grazing animals or those fed high-roughage diets are more likely to be deficient in phosphorus than in calcium because of the low phosphorus content of forages. Animals on diets high in concentrates are very likely to be deficient in calcium, because of the high phosphorus/calcium ratio in grain and oil meals.

In practice, rations may be deficient in phosphorus or calcium and should be supplemented with minerals. The composition of the common supplements of both minerals is given in Table 10.2. The phosphorus and calcium from all the sources presented in the table are readily available for poultry and all other species of farm animals. Untreated rock phosphate and superphosphate may be harmful because of the fluorine present. Technical methods for the defluorination of rock phosphates have been developed. Phosphorus and calcium in most defluorinated phosphates are fully available. Some overheated products are less well utilized by animals because of the formation of pyro- and metaphosphate. Limestone, calcite or oyster shells containing almost pure calcium carbonate are used for supplementing calcium-deficient diets.

Table 10.2 Composition of phosphorus and calcium supplements

Supplement	Calcium (%)	Phosphorus (%)
Bone meal	24–28	12–14
Tricalcium phosphate	34	14
Difluorinated dicalcium phosphate	20–24	16.5–18.5
Monocalcium phosphate	20–21	14–21
Monosodium phosphate		22
Diammonium phosphate		20
Calcite	34–38	

Magnesium

Magnesium is largely associated with calcium and phosphorus in the body. About 70% of the magnesium present in the body is located in the skeleton. It represents 0.5–0.7% of the bone ash in all animals; the calcium/magnesium ratio in the bones is about 55 : 1. About one-third of the magnesium in the bones is bound to phosphate; the remainder is absorbed on the surface of the mineral structure. About 30% of the magnesium present in the body is distributed in the soft tissues and fluids. Like potassium, in soft tissues it is present mainly within the cells. About 75% of the magnesium in blood is in the red blood cells; blood serum contains 2–4 mg of ionized magnesium per 100 ml, as well as smaller amounts of protein-bound magnesium. The magnesium ions present in blood serum are continuously being exchanged with magnesium absorbed on the bone surface.

In addition to being an essential constituent of bones and teeth, magnesium is required for oxidative phosphorylation leading to ATP formation. Thus it is involved in the metabolism of carbohydrates and lipids and in the biosynthesis of proteins.

Most of the commonly fed diets for farm animals contain amounts of magnesium sufficient to meet body needs. Ruminants require diets containing 0.20% magnesium in the dry matter and monogastric species only 0.05%. Magnesium is absorbed in monogastric species through the small and large intestine and in ruminants mainly through the reticulorumen. Availability values for magnesium in milk as high as 70% have been found in suckling ruminants, but decline in mature ruminants to 35% for magnesium present in concentrate feeds and to about 20% for that in forages. The coefficient of absorption of magnesium by monogastric species is about 50. A part of the magnesium present in green plants is found in bound form as chlorophyll, from which it is liberated in the acid medium prevailing in the true stomach or the rumen.

Diets too low in magnesium are supplemented with magnesium oxide or carbonate. The absorption in ruminants is about 70% for magnesium from both these mineral sources. Absorption in the rumen may be depressed by high levels of potassium, ammonia and phosphates and by certain organic acids occurring in plants. High levels of ammonia in rumen fluid are caused by excessive amounts of highly soluble proteins or non-protein nitrogen compounds ingested (see Chapter 8).

Magnesium deficiencies are characterized by a decline in the magnesium concentration of cerebrospinal fluid and serum, and acute neuromuscular disorders such as hyperirritability, muscular incoordination and convulsions. Under practical conditions, two types of magnesium deficiency occur in cattle. The first is that in calves fed over a prolonged period with an all-milk diet that is deficient in magnesium. The second type, called 'grass tetany', which occurs in lactating cows, will be described in Chapter 24.

A deficiency of magnesium in the diet of laying hens results in a rapid decline in egg production and a withdrawal of magnesium from bones. Under practical conditions, an excess of magnesium in poultry diets is more likely than a deficiency. Such an excess may be caused by the use of dolomitic (magnesium containing) limestone as a source of calcium in poultry diets (see Chapter 20). The presence of more than 1% of magnesium depresses the growth rate of chickens and reduces egg production and eggshell thickness.

Sodium, Potassium and Chloride

Distribution in the Body and Functions

These three minerals are considered together because of the similarity of their functions and distribution in the body. In contrast to calcium, phosphorus and magnesium, which are stored in the skeleton, sodium, potassium and chloride occur largely in body fluids and soft tissues. These three minerals maintain osmotic pressure, regulate acid–base equilibria and control water metabolism in the tissues. They are essential for the operation of enzyme systems. Neural and muscular conduction and transmission are highly dependent on proper levels of sodium and potassium and of magnesium. Nutritionally, sodium, potassium and chloride are often considered to be of minor importance because meeting body needs presents no difficulty, and the dangers of excessive intake exist only under special conditions.

The body content of potassium, sodium and chloride in mammals is given in Table 10.1. Sodium occurs largely in the body fluids and bones and potassium mainly in muscles and nervous tissue. Sodium is present largely in extracellular fluid with less than 10% within the cells. Most of the potassium is found inside the cells. An appreciable proportion of the extracellular sodium is adsorbed on the inorganic crystals of the bones. In the blood, sodium makes up over 90% of the bases of the serum and thus it is the major cation in acid–base regulation. While in blood the level of sodium is higher than that of potassium by an order of magnitude, in milk the reverse is true. Potassium occurs in the erythrocytes at a concentration about 25 times that of the plasma. Chloride is found within the cells and in the body fluids, including the gastric secretion where it occurs as hydrochloric acid.

Absorption. Sodium, potassium and chloride ions are quantitatively absorbed from the gastrointestinal tract of all farm animals. Large proportions of these absorbed ions are of endogenous origin, i.e. they enter the gastrointestinal tract in the saliva and digestive juices. The endogenous secretion of these ions may exceed several times their intake in food. Sodium, potassium and chloride are absorbed mainly in the duodenum, and to a lesser extent in the stomach, lower intestine and colon.

Supply in the Food and Metabolic Regulation

The requirements of all farm animals for sodium and chloride are of the order of 0.1–0.2% of the dry matter (for each). The requirement for potassium is about 0.2–0.5% in monogastric species and slightly higher, 0.6–0.8%, in ruminants. As a rule, forages are higher in potassium than concentrates; even so, the common rations of all the farm animal species usually meet the requirements for potassium. However, lactating cows fed high-concentrate diets may be marginally deficient in potassium. Sodium, on the other hand, and to a lesser extent chloride, are not always present in common diets in sufficient amounts. Therefore, it is common practice to supplement rations with common salt.

An excess of chloride in the ration may cause acidosis and an excess of sodium may cause alkalosis. Therefore it is desirable to feed rations supplying approximately equivalent amounts of sodium and chloride. This is more important in monogastric species and in young animals and particularly so in poultry, because of their limited capacity to regulate acid–base balance.

The ratio of potassium to sodium in common rations is not critical under most circumstances. High concentrations of potassium and subnormal levels of sodium in the ration release the hormone aldosterone which enhances the tubular reabsorption of sodium in the kidney and the excretion of potassium. Thus the levels of sodium and potassium in serum are maintained within normal limits. This mechanism successfully regulates the levels of these two ions, even in ruminants grazing on very lush pasture which may supply a potassium/sodium ratio as high as 20. However, excessively high levels of potassium in the diet, about 3% of the dry matter, in very lush grasses, may severely depress the absorption of magnesium and may increase the incidence of hypomagnesaemia in ruminants (see Chapter 24).

Deficiency of Sodium Chloride. The salt requirement of mature non-producing animals is low. The ability of animals to conserve both sodium and chloride by reducing endogenous losses in urine and faeces is rather high and the symptoms of a salt deficiency develop rather slowly. However, when salt-deficient diets are fed to growing animals or to lactating cows and to laying hens, the symptoms develop more rapidly. The sodium and chloride content of milk, eggs and added body tissue tend to be fixed. Therefore, a salt deficiency in growing, lactating or laying farm animals curtails production even if all the other nutrients are supplied in sufficient amounts. Sweat contains a considerable amount of salt. Therefore, sweating, whether caused by hot weather or by heavy physical work, or both, greatly increases the salt requirement of animals. This is particularly so in horses, which sweat profusely and may produce sweat containing as much as 4% sodium chloride.

Excess of Sodium Chloride. Excessive ingestion of sodium chloride may be harmful when the supply of drinking water is limited. It increases the osmotic pressure of body fluids and enhances the retention of water. The results of salt poisoning are thirst, weakness and oedema. Young animals are less tolerant than old ones; 5% salt in the mash may produce limited mortality in poultry at 3 months of age but a heavy death rate in baby chicks. Salt poisoning is also common in pigs. Mature cattle and sheep can tolerate a comparatively high amount of salt (even 5% of the dry matter intake) provided water is readily available.

Sulphur

Most of the sulphur in the animal body and in its food occurs in proteins that contain the sulphur-containing amino acids cystine, cysteine and methionine (see Chapter 7); only a small amount of sulphur is present in inorganic form, mainly sulphates. The body contains approximately 0.15% of sulphur. The blood contains small amounts of sulphate. Wool is rich in cystine and contains about 4% of sulphur. Furthermore, sulphur is a constituent of several biologically important compounds such as the two vitamins thiamine and biotin (see Chapter 12), glutathione which is present in all cells and plays a role in the oxidation–reduction process, the hormone insulin, coenzyme A (see Chapter 6) and chondroitin sulphate which serves a structural function in connective tissues. Sulphur needed in the body for synthesis of these compounds appears to be available from sulphur-containing amino acids. The end products of catabolism of organic sulphur compounds are sulphuric acid and taurine. Sulphates and smaller amounts of organic derivatives of sulphuric acid mentioned in Chapter 3 are excreted in urine and taurine in the bile (see Chapter 6).

Most diets for all kinds of farm animals are adequate in sulphur-containing proteins. Sulphates can be used by rumen microorganisms for protein biosynthesis. Ruminants depending largely on non-protein nitrogen sources such as urea to replace food proteins may need supplemental inorganic sulphates to correct this deficiency (see Chapter 8). For efficient utilization of urea in lactating cows a nitrogen/sulphur ratio of 12 : 1 in the diet is adequate and a ratio of 10 : 1 in all other kinds of ruminants.

In cases of scarcity of sulphur-containing amino acids in the diet of monogastric species, part of the sulphur requirement may be met by inorganic sulphates, but the cystine-sparing effect of sulphates in poultry appears to be small.

Excessive amounts of sulphates will decrease feed intake and adversely affect the animals by reducing the availability of other minerals such as zinc and manganese. The estimated sulphur requirement for cattle is 0.20%; the maximum should be limited to 0.35% of the diet with no more than 0.20% from added sulphates

MICROMINERALS

Iron

Iron fulfils most important functions in the body, but in general there seems to be no need to supply additional iron to farm animals, particularly not to adult ones, since in practice diets contain the required amounts of this mineral.

Distribution in the Body and Functions of Iron

The bodies of the different species contain 60–70 mg iron per kg live weight, of which 60–70% is in the form of haemoglobin, 3% in myoglobin, 26% in storage and less than 1% in iron transport compounds.

Haemoglobin, myoglobin and several respiratory enzymes contain iron chelated in the form of a porphyrin complex, haem, which is linked to a protein component which varies with each of these biologically active compounds (see biochemical texts). *Haemoglobin* functions as an oxygen carrier in the respiration process because the bonds between iron and globin stabilize the iron in the ferrous state and allow it to be reversibly bound to oxygen. Haemoglobin transports oxygen between the lungs and the tissues. Blood haemoglobin concentration varies with species, age and sex, values ranging between 10 and 18 g per 100 ml in the different species.

Myoglobin, a combination of haem and muscle globin, has a greater affinity for oxygen and serves as a oxygen store within muscle. Myoglobin is necessary for the functioning of muscle (see Chapter 22). Iron occurs in a number of enzymes including iron-containing flavoprotein enzymes (xanthine oxidase, succinic dehydrogenase; see Chapter 12) and in haemoprotein enzymes such as the cytochrome enzymes, catalase and peroxidase. The latter enzymes are essential for oxygen utilization at the cellular level.

The reserve of storage iron in the body occurs in two non-haem compounds *ferritin* and *homosiderin*, which are located predominantly in the liver, spleen and kidneys. Ferritin is a non-haem protein compound (apoferritin) containing up to 20% iron; homosiderin consists mainly of ferric hydroxide in a protein-free aggregate and may contain up to 35% iron. Iron is present in blood serum attached to a colourless protein called *transferrin*, a glycoprotein which appears to be a carrier for iron in the same way that haemoglobin acts as a carrier for oxygen.

Red blood cells and haemoglobin are constantly being destroyed and replaced. Thus iron undergoes a very active metabolism in the body. Iron released by normal destruction of red cells is made use of in the resynthesis of haemoglobin which occurs in bone marrow to replace catabolized haemoglobin. The rate of haemoglobin synthesis is controlled by a glycoprotein hormone, erythroprotein, which is produced in the kidney and is present in the circulating blood. The average life span of haemoglobin is about 2 months in most species, with the exception of 4 months in man. The iron liberated from haemoglobin can be used 9–10 times for the resynthesis of haemoglobin. Only small amounts of iron escaping from this cycle are excreted by faeces (in bile liquor) and urine. For this reason, and because of the efficient recycling of iron, the requirements of this mineral for farm animals are comparatively low (25–100 mg per kg dietary dry matter) compared with its content in the body (200–300 mg per kg dry matter).

Absorption of Iron

Iron is poorly absorbed from the digestive tract. In monogastrics it is mainly absorbed in the duodenum. Iron is absorbed from the intestinal lumen into the mucosal cells. Within the cells it is partly bound to transferrin and partly to ferritin. At the serosal surface, it becomes attached to transferrin for transport in the plasma. Any iron not transported to plasma is retained in the cell until it is sloughed off, thus returning to the lumen of the gastrointestinal tract. The coefficient of absorption of iron is very low, about 5–10% in adult animals, and is affected by various factors: the age and iron status of the animal, the amount and chemical form of the iron ingested, and the amounts in the diet of other compounds with which iron interacts.

(1) The iron absorption is related to the body needs and is more efficient in young than in mature animals.

(2) Haem compounds present in feedstuffs of animal origin such as fishmeal are better absorbed than iron from plant food sources containing mainly inorganic iron salts.

(3) The extent of iron absorption is influenced by chelates, some of which (ascorbic acid or cysteine) enhance the absorption and others inhibit absorption. Iron absorption is decreased by other divalent ions (zinc, manganese, cobalt) which are believed to compete for binding sites in the intestinal mucosa. Phosphates and phytates interfere with the absorption of iron by the

formation of insoluble iron salts (see above). Copper plays a key role in the utilization of iron since copper is found in the enzyme ferroxidase which facilitates the release of iron from ferritin in the intestinal mucosal cells (see later).

A scheme for iron metabolism occurring in the body is presented in Figure 10.4, and includes processes such as biosynthesis and catabolism of haemoglobin, absorption, transport and storage of iron and also recycling of iron.

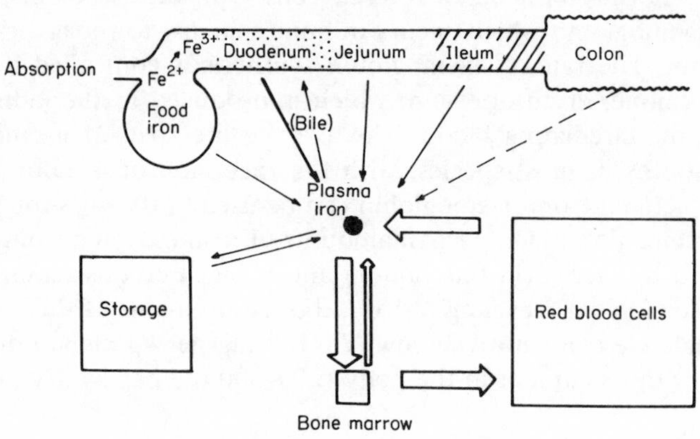

Figure 10.4. Metabolism of iron in the body.

Requirements of Iron and Iron Deficiency

Iron requirements are low in adult animals, as mentioned above (25–40 p.p.m. on a dry basis in diets for ruminants, and 80 p.p.m. in diets for pigs), but are higher in the following physiological states.

(1) The factors responsible for the higher iron requirements of very young milk-fed animals are the relative low stores of iron at birth, the low level of iron in milk of cows and sows (0.5–1.0 p.p.m.) and the high growth rate early in life. Iron reserves in the neonatal calf (different from those in the piglet) are sufficient to prevent serious iron deficiency. In fact, the pale colour characteristic of good veal is caused by iron deficiency. Milk feeding induces a moderate deficiency in calves and a critical one in piglets, which are more susceptible to iron deficiency. The iron requirements are about 125 p.p.m. in piglets' diet (on a dry basis) and 40 p.p.m. in that for calves. Extraneous sources of iron should be provided to piglets only, orally or intramuscularly, in order to prevent the particular features of iron deficiency (poor appetite, subnormal growth, and diarrhoea).

(2) Dosing of pregnant females with iron compounds may be used as a way of increasing haemoglobin levels in blood and stores of iron in the body of newborn animals, but the iron content of milk is not increased by feeding.

(3) The demands for iron during egg-laying are very high because an egg contains 1 mg of iron. However, the usual laying rations cover the iron requirements without the need for specific iron supplements.

Anaemia is the main symptom of iron deficiency with a depletion of iron stores in the body, i.e. a reduction in the number of red blood cells and a reduced haemoglobin content of the blood. Anaemia may result from interference with the production of haemoglobin, from increased destruction of haemoglobin or from blood loss. For example, parasitic infestation of sheep involving severe blood loss produces anaemia. Thus anaemia may result for reasons of diet or for pathological or hereditary reasons. Blood haemoglobin concentrations 25% below the normal range clearly indicate anaemia. The decreased resistance to infection that characterizes iron deficiency is well established.

Sources of Iron

Concentrates and roughage generally contain iron in sufficient quantities to meet dietary needs. Leguminous plant species are richer in iron (200–400 p.p.m. on a dry basis) than grasses (about 40 p.p.m.). Cereal grains contain between 30 and 60 p.p.m. and oilseed meals from 100 to 200 p.p.m. Feeds of animal origin, other than milk, are rich sources of iron.

Copper

Copper is essential for growth and for the prevention of a wide range of clinical and pathological disorders in all types of farm animals. Copper deficiency in grazing cattle and sheep is recognized as a major practical problem in many parts of the world. It results either from too little copper or from influences of interfering substances present in pastures plants such as molybdenum or sulphates.

Functions of Copper in the Animal Body

Copper is an integral part of several enzymes with an oxidase function such as ferroxidase, monoamine oxidase, cytochrome oxidase and tyrosinase. The functions of these enzymes in the animal body will be described below and their activities will be related to the disorders that develop in the copper-deficient animal. The copper-containing enzyme ferroxidase (or caeruloplasmin) is involved in iron utilization since it is necessary for the formation of transferrin, the transport vehicle for iron (see above), and for the incorporation of iron into haemoglobin. Thus copper is necessary for the utilization of iron in haemoglobin synthesis, and anaemia becomes evident in copper deficiency as in iron deficiency (see above).

Monoamine oxidase, an enzyme containing 4 atoms of copper per molecule, catalyses the oxidative deamination of several monoamines to aldehydes and is involved in maintaining the structural integrity of bone tissues and blood vessels. Under normal conditions lysine is transformed into desmosine by the copper-containing lysyl oxidase (amine oxidase) and the adequately cross-linked protein structure of collagen and elastin is achieved by incorporation of desmosine. In copper-deficient animals the activity of the copper enzyme amine oxidase is reduced and cross-linking of collagen and elastin is impaired. This leads to diminished stability and strength of bone collagen. Thus gross bone disorders with spontaneous fractures and severe deformations occur in copper-deficient ruminants, pigs and chicks. Reduction in the desmosine content of elastin in the aorta of copper-deficient chicks and pigs leads to a derangement in the elastic tissues of the aorta and dangerous ruptures of the major blood vessels.

Formation of the myelin sheath that surrounds nerves is impaired by copper deficiency since some copper-containing enzymes such as cytochrome oxidase and amine oxidase are involved in the synthesis of myelin lipid compounds, mainly cholesterol. Nervous disorders of lambs born from copper-deficient ewes are characterized by lack of coordinated movement and a stiff, staggering gait. Some lambs are paralysed when born and soon die. Such cases of *neonatal ataxia* occur in different parts of the world and are given local names such as 'swayback' in England and 'lechtsucht' in The Netherlands.

Pigmentation failure and marked changes in growth and physical appearance of hair of cattle and wool of sheep commonly occur in copper deficiency. The mechanism is related to the conversion of tyrosine to the dark pigment melanin which is catalysed by copper-containing tyrosinase.

Copper deficiency leads to less crimp in the wool until the fibres emerge as almost straight, hair-like growths. The characteristic physical properties of the wool, including crimp, are dependent on disulphide bridges in two adjacent cysteine residues. The precise biochemical involvement of copper in the oxidation of $-SH$ to $-S-S-$ groups is not yet known.

Severe diarrhoea occurring in copper-deficient ruminants has been associated with mucosal atrophy in the small intestine which may have produced a malabsorption syndrome.

Copper deficiency may also result in impaired reproduction (see Chapter 19).

Utilization of Copper in Animals

In general copper is poorly absorbed, with only 5–10% of the ingested copper being absorbed and retained. Most of the faecal copper is unabsorbed dietary copper, but some of it comes from the bile which is the major pathway for the secretion of endogenous copper. The absorbability of copper and therefore the requirements for this mineral are powerfully influenced by other dietary components.

Copper absorption is markedly depressed by contamination of herbage with calcium carbonate. Molybdenum and sulphate exert also an adverse effect on copper utilization by ruminants. The limiting effect of molybdenum and sulphate on copper retention may be explained by the following interaction of copper, molybdenum and sulphate. Sulphide is formed by microorganisms from sulphate (see Chapter 8); the sulphide then reacts with molybdate to form thiomolybdate which in turn combines with copper ions to form insoluble copper thiomolybdate, thereby limiting the absorption of dietary copper. The absorption of copper is reduced by a factor of 2–4 as molybdenum and sulphur concentrations are increased within the normal range for these elements in herbage (see Figure 10.5). Such increases in the concentration of molybdenum and sulphur in diets may be responsible for

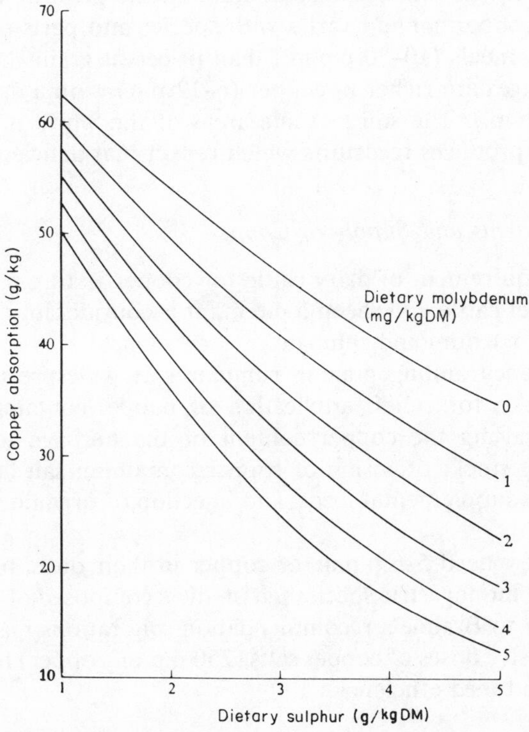

Figure 10.5. Effects of changes in the dietary contents of molybdenum and sulphate on the absorption of copper (after Suttle, N. F. *Micronutrients as regulators of metabolism*, in: Rook, J. A. F., and Thomas, P. C., *Nutritional Physiology of Farm Animals*, 1983, p. 440). *Reproduced by permission of Longman Group Ltd.*

the induction of copper deficiency in lambs. On the other hand, the toxic level of copper is increased by very low levels of molybdenum and sulphate (see later).

Distribution of Copper in the Animal Body

This varies with species and the copper status of the animal. The highest copper concentrations are found in liver, brain, kidneys, heart and hair. The liver concentration is higher in sheep, cattle and ducks (100–400 p.p.m. in fresh matter) than in horses, pigs, chicks and turkeys (10–50 p.p.m.). Since the liver is the main storage organ for copper it provides a useful index of the copper status, more reliable than the copper content of blood plasma.

Distribution of Copper in Feedstuffs

Copper is absorbed from the soil into plants and its content in plants depends on that in the soil, the type of soil and the growing conditions of the plants. The copper content varies with species and parts of plants; it is higher in oilseed meals (10–30 p.p.m.) than in cereal grains (4–10 p.p.m.). Leguminous forages are richer in copper (6–12 p.p.m. on a dry basis) than grasses (4–10 p.p.m.). The soil in some areas of the world is quite low in copper and thus produces feedstuffs which reflect that deficiency.

Dietary Requirements and Supply of Copper

The dietary requirement of dairy cattle for copper is 10 p.p.m. (on a dry basis) and for beef cattle and sheep 5 p.p.m.; it is considerably higher in the presence of molybdenum and sulphate.

Copper deficiency among grazing ruminants is widespread. There are several possibilities for relief: application of copper-containing fertilizer as a means of raising the copper content of the herbage to levels adequate for grazing stock; provision of copper-containing salt licks; addition of copper salts in supplemental feeds; and injection of organic complexes of copper.

Pigs and poultry need 5–6 p.p.m. of copper in their diets, but deficiency does not occur in monogastric species fed on diets composed of natural feedstuffs. According to Braude's recommendation, pig rations may be supplemented with massive doses of copper salts (250 p.p.m. copper) for producing increased gain and feed efficiency.

Copper Poisoning

Pigs can tolerate the massive doses of copper (250 p.p.m. in the diet) recommended by Braude for stimulating growth. The tolerance to copper varies considerably among species. Pigs are highly tolerant and cattle are less so, whereas sheep are particularly susceptible to copper toxicity. When excessive copper is consumed, ruminants accumulate high amounts in the liver. The toxic symptoms are due to the liberation of large amounts of copper from the liver to the blood causing considerable haemolysis. Chronic copper poisoning occurs in grazing sheep in parts of Australia as consequence of excessive copper intake of the herbage or of normal copper intakes (up to 20 p.p.m.) together with very low levels of molybdenum (0.1–0.2 p.p.m.), or in association with subclinical poisoning with alkaloids contained in the heliotrope plant. Consumption of this plant may damage the liver and make it more susceptible to accumulation of copper.

Cobalt

The first evidence that cobalt is a dietary essential for ruminants was obtained 50 years ago in Australia (see Underwood, 1977). This research established that supplemental dietary cobalt was effective in curing or preventing deficiency diseases of grazing ruminants.

Metabolism of Cobalt in Ruminants

The only established physiological function of cobalt is its role as an integral part of the vitamin B_{12} molecule (see Chapter 12). Cobalt is required by the microorganisms of the rumen for biosynthesis of this vitamin, which in turn is needed by the tissues of the host animal. There is evidence for this mode of action of cobalt since injections of preformed vitamin B_{12} into the blood of cobalt-deficient ruminants alleviate the deficiency symptoms of these animals. Oral administration of preformed vitamin B_{12} is less effective than injections because of the poor absorption of this vitamin by the ruminant. On the other hand, parenteral injections of cobalt salts are not effective since insufficient amounts of cobalt entering by this route reach the rumen where microbial synthesis takes place.

Cobalt deficiency trouble in ruminants is actually a vitamin B_{12} deficiency. When dietary cobalt is inadequate, rumen microorganisms are not able to synthesize sufficient amounts of vitamin B_{12} and thus the animals exhibit symptoms of cobalt deficiency. The ruminants make inefficient use of the dietary cobalt. As was shown by experiments with isotopic cobalt, 80% of the ingested cobalt is excreted in the faeces. The rumen microorganisms convert a high proportion of the absorbed cobalt into physiologically inactive vitamin-B_{12}-like compounds that can neither be used nor absorbed by the host animal. Only 3% of the cobalt intake is converted into vitamin B_{12}, which is absorbed into the blood and transferred to tissues of the host animal. The low efficiency with which cobalt is converted into the vitamin is inversely proportional to the cobalt intake and in deficient animals it is raised to 13.5%.

Vitamin B_{12} is involved in the breakdown of propionic acid which is produced by fermentation in the rumen and serves as a main source of energy in these animals (see Chapter 5). The metabolism of propionic acid occurs mainly in the liver and is depressed in cobalt-deficient ruminants since a coenzyme form of vitamin B_{12} is required in the conversion of methylmalonate to succinate, an intermediate step in the metabolism of propionate. The decreased feed intake responsible for the symptoms of cobalt deficiency results from the inability of vitamin-B_{12}-deficient animals to metabolize propionate. The voluntary food intake may be in an inverse relationship to the propionate concentration in blood, which is elevated in vitamin-B_{12}-deficient ruminants (see Chapter 23).

Cobalt Deficiency in Ruminants

Cobalt deficiency occurs in grazing ruminants with varying degrees of severity. Serious cobalt deficiency arises in areas of insufficient plant-available soil cobalt such as occurs in Australia, Great Britain and The Netherlands; these deficiencies have been given local names (wasting disease or coast disease) or the scientific designation 'enzootic marasmus'. Even acute cobalt deficiencies are characterized by non-specific disorders such as decrease of feed intake, loss of body weight, reduced growth, wasting of the skeletal muscles, emaciation, fatty degeneration of the liver, etc. Some areas in the world are affected by a deficiency of both cobalt and copper.

The clinical manifestations of cobalt deficiency are similar to those of general malnutrition and are not sufficiently specific to permit diagnosis. Liver and blood vitamin B_{12} and responses of animals to cobalt have been used as biochemical means of diagnosing cobalt deficiency. The livers of normal healthy sheep and cattle usually contain 0.2–0.3 p.p.m. Cobalt on a dry basis and values of less than 0.08 p.p.m. indicate cobalt deficiency.

The Requirements for Cobalt and its Supply

The critical level below which cobalt deficiency signs appear in dairy cattle and sheep is 0.07 p.p.m. in the dry matter. Beef cattle and lambs are more susceptible to cobalt deficiency and exhibit signs of deficiency with food containing 0.10 p.p.m. It may be noted that the requirements for cobalt account for only 1% of those for copper. Even in cobalt-deficient areas adequate intake can be ensured by including cobalt salts in fertilizers or by supplementation of the diet with cobalt salts. Furthermore, cobalt deficiency in ruminants has been successfully alleviated by placing small pellets containing cobalt oxide in the rumen to yield a relatively steady supply of cobalt to rumen microorganisms.

Microbial synthesis of vitamin B_{12} occurs to a small extent also in the lower intestine of non-ruminants. Thus, it is not possible to meet the requirements for vitamin B_{12} of non-ruminants by dietary cobalt. The latter animals require preformed vitamin B_{12} to meet their metabolic needs (see Chapter 12).

Manganese

Distribution in the Body

Manganese is distributed throughout the body, but the total quantity is much lower than for other elements, amounting to only 20% of the copper content and to 1% that of zinc. It may be noted that the concentration of manganese in soils exceeds 15-fold that of zinc and the concentrations of each of these elements in plants are almost equal. This situation is the result of the low absorption rates of manganese by plants and animals (see later). Manganese is not concentrated in any specific organ of the animal body, but is found in higher concentrations in bone, liver, kidney and pancreas (1–3 p.p.m. of fresh tissue) than in skeletal muscle (0.1–0.2 p.p.m.).

Functions of Manganese

As a component of various enzymes manganese fulfils specific biochemical functions in the body. It is an integral part of enzymes that act in the synthesis of chondroitin sulphate which is a component of mucopolysaccharides present in the organic matrix of bone. Manganese is required in two steps of this biosynthesis. (1) The action of galactotransferase which incorporates galactose (from UDP-galactose) into the galactose–galactose–xylose trisaccharide; this trisaccharide serves as the linkage between the polysaccharide and the protein associated with it through the hydroxyl group of serine. (2) A polymerization enzyme which is responsible for the formation of the polymeric polysaccharide present in chondroitin sulphate. Since mucopolysaccharides are vital structural components of the organic matter in bone, these findings provide a likely biochemical explanation for the skeletal defects associated with manganese deficiency (see later).

Manganese is a component of the enzyme pyruvate carboxylase and as such it is involved in the metabolism of carbohydrates and fats. The following reaction (see Chapter 12) has been proposed for this enzyme:

$$\text{Enzyme–biotin} + CO_2 + \text{pyruvate} \longrightarrow \text{enzyme–biotin} + \text{oxaloacetate}$$

Manganese is necessary as a cofactor in the enzyme that catalyses the conversion of mevalonic acid to squalene and is needed for biosynthesis of cholesterol (see Chapter 6). It stimulates the activity of arginase and may play a regulatory role in urea production (see Chapter 7). The manganese-containing superoxide dismutase which catalyses the dismutation of the free radical O_2^- to hydrogen peroxide and oxygen, functions as a defence against the deleterious reactions of this free radical and protects the integrity of cell membranes.

Absorption of Manganese

Manganese is poorly absorbed (5–10%), and excessive calcium and phosphorus depress the absorption of manganese even more since calcium phosphate precipitated in the gut may be capable of absorbing manganese. Manganese is excreted mainly in the faeces; this also includes manganese of endogenous origin secreted via bile and pancreatic juices.

Manganese Deficiency Symptoms

Manganese deficiency has been found in poultry, pigs and ruminants. The defects of this deficiency include skeletal deformations, retarded growth, disturbance of reproductive function and abnormalities of the new-born.

Manganese deficiency in the diet of growing chicks is manifested as the disease *perosis*, a malformation of the leg (see PLATE IIA). Perosis in chicks is the most commonly observed manganese deficiency. Perosis is characterized by enlarged and malformed hock joints, twisting and bending of the tibia and thickening and shortening of the long bones. If the manganese-deficient diet is fed from the first day of life, these signs appear at the second week. At a later stage the Achilles tendon slips sideways from its groove at the back of the hock joint, and the continued use of this tendon by the chick pulls the leg below the hock joint sideways and backwards to give the typical syndrome of perosis. Manganese is not the only factor involved in the production of perosis; other factors such as choline or biotin may cause a similar condition (see Chapter 11). Choline, in addition to manganese, is necessary to prevent perosis in birds. Excess of calcium and phosphorus may aggravate perosis, probably by interfering with manganese absorption (see above).

Manganese deficiency in the diet of breeding hens causes a serious disorder in embryonic chicks, called nutritional 'chondrodystrophy'. The latter condition is characterized—similarly to perosis—by shortened and thickened wings and legs of the embryo and parrot beak resulting from a disproportional shortening of the lower mandible and high mortality.

Poor eggshell formation, particularly reduced eggshell strength, may result from a reduction in the mucopolysaccharide content of the shell matrix due to a deficient supply of manganese.

Skeletal abnormalities occurring in manganese-deficient pigs are also characterized by lameness and enlarged hock joints with crooked and shortened legs.

Diets low in manganese for cows and goats cause delayed oestrus and conception and increased abortion.

Dietary Requirements and Supply of Manganese

The manganese requirement of poultry is higher than that of other farm animals. It has become common practice to increase the manganese content of poultry diets to 50 p.p.m. by addition of manganese salts, and in diets for growing pigs to 40 p.p.m. Most constituents of these diets contain inadequate amounts of manganese. Rich sources are wheat bran (115 p.p.m.) and rice bran (200 p.p.m.) since manganese is concentrated in the exterior layers of cereal grains. Manganese deficiency in poultry is prevented by incorporation of manganese salts into the mineral supplements or into the whole ground rations. The supplementary manganese is readily transmitted by hens to the egg for use by the developing chick embryo. The oxide, chloride, sulphate, oxide and carbonate of manganese are highly available as sources of manganese in poultry rations; the manganese in some carbonate or silicate ores is essentially unavailable.

Manganese contents of forages vary considerably; however, the manganese contents of practical ruminant diets neither produce deficiencies nor surpass the maximum tolerable levels.

Zinc

Distribution of Zinc in the Body and its Functions

Zinc distribution in the body is rather homogeneous. The average zinc content of the animal body is approximately 30 p.p.m.; higher concentrations are present in bones, liver, kidney, skin, hair and wool, and particularly in some eye tissues and in male sex organs. Zinc is an integral constituent of various enzymes including lactate, malate and glutamate dehydrogenases, alkaline phosphatase (see above), carboxypeptidases A and B (see Chapter 7) and carbonic anhydrase. Zinc as a constituent of RNA and DNA polymerases is involved in protein biosynthesis. Carbonic anhydrase present in erythrocytes catalyses the synthesis and breakdown of carbonic acid:

$$CO_2 + H_2O \rightleftharpoons H_2CO_3 \rightleftharpoons H^+ + HCO_3^-$$

This enzyme plays an important role in maintaining the desired acid–base equilibrium of the body, in calcification of bone and in eggshell formation (see Chapter 20).

Zinc functions in several enzyme systems as a cofactor. The tissue distribution of zinc is roughly associated with the tissue distribution of the zinc-containing enzymes. The high concentration of zinc in the pancreas is related to the presence of zinc-containing digestive enzymes and of zinc bound to insulin.

Zinc Requirements and Digestive Absorption

Most practical diets supply sufficient zinc for farm animals (30–40 p.p.m. on the basis of dry matter). Of the zinc contained in feedstuffs 15–30% is absorbed by all kinds of farm animals; it is absorbed mainly in the small intestine.

The absorption of zinc is depressed by various accompanying substances such as calcium and copper ions which compete for the absorption sites (see Chapter 3) and aggravate signs of zinc deficiency. When 250 p.p.m. of copper are added as a growth stimulant to diets of growing pigs (see above), an additional 150 p.p.m. of zinc may be necessary. The absorption of zinc by monogastric species is also depressed by phytic acid which is present in some concentrate feedstuffs such as soybean meal or maize grains. The phytates bind the zinc in a form that is not readily released and absorbed. The availability of zinc for absorption from combination with phytic acid is increased by autoclaving or by treatment with the synthetic compound ethylendiamintetracetic acid (EDTA) which forms highly soluble and absorbable chelates with zinc.

Symptoms of Zinc Deficiency

Zinc-containing enzymes are involved in primary processes of protein metabolism and cell division and the following manifestations of zinc deficiency were observed in farm animals: subnormal growth, depressed food consumption, poor food conversion, impaired reproductive performance and abnormalities of the skin and its outgrowths (hair, wool or feathers).

Skin lesions are the most conspicuous sign of zinc deficiency. In swine a dermatitis called parakeratosis is an important zinc deficiency disease found under practical conditions. Thickening or hyperkeratinization of the epithelial cells of the skin mostly occur around the eyes and mouth and on the lower parts of the legs of pigs. In former times pigs fed maize–soybean diets were frequently affected by parakeratosis, but nowadays this sickness is prevented by the addition of zinc to such diets. Ruminants grazing in zinc-poor areas show parakeratosis, i.e. damage of epidermal tissues and open lesions of the skin, and this renders the animals more accessible to infections. In young chicks the symptoms of zinc deficiency include scaling of the skin, very poor feathering and frizzled feathers.

Decrease of egg production, decreased hatchability and impaired development of chick embryos occur when hens are subjected to zinc deficiency. In chick embryos affected by zinc deficiency, skeletal development is grossly impaired, the long bones becoming shortened and thicker. The zinc concentration in bones of all kinds of zinc-deficient animals is below normal. Zinc clearly plays a part in the events leading to calcification, but at what point is not yet known.

Wound healing is impaired in zinc-deficient animals. Zinc is preferentially concentrated in healing tissues, in skin and muscle wounds and in bone fractures as well. This suggests a heightened metabolic demand for zinc in tissue synthesis during the healing process. The decrease in collagen synthesis observed in zinc deficiency explains the slower wound healing in the absence of sufficient zinc.

The adverse effect of zinc deficiency on vitamin A metabolism noted in Chapter 11 may be explained by the decrease in the synthesis of the retinol-binding protein caused by lack of zinc.

Iodine

The mode of action of iodine differs principally from that of all other trace elements. The only known function of iodine is as a constituent of the hormone thyroxine. This hormone is the only one that contains an inorganic constituent. Thyroxine is synthesized and stored in the thyroid gland. Eighty per cent of the iodine in the body is present in the thyroid gland, which forms only 0.2% of the whole body weight; this is the most spectacular example of the concentration of a trace element in a tissue. From the thyroid gland the hormone is transferred by the circulation to the various tissues where it regulates the primary processes of energy metabolism occurring in the cells. The normal development of young animals, good fertility and efficient production of animal products depend on the correct secretion of thyroxine. When the diet contains insufficient amounts of iodine or constituents which interfere with the use of iodine by the thyroid gland, animal productivity is adversely affected. The main indication of such a deficiency is an enlargement of the thyroid gland, called goitre.

Metabolism of Iodine

Iodides ingested with the food are absorbed from the gastrointestinal tract into the circulation. The thyroid gland exhibits a remarkable capacity to concentrate iodides, which are very actively trapped by the vesicles. After trapping, the iodide is oxidized by peroxidase to elemental iodine as a preliminary step to its incorporation into organic combination:

$$2\,I^- + H_2O_2 \longrightarrow H_2O + \tfrac{1}{2}O_2 + I_2$$

The elemental iodine combines with the tyrosine residues within thyroglobulin—a colloid protein inside the thyroid vessels—to form 3-monoiodotyrosine and 3,5-diiodotyrosine (see Figure 10.6). Two diiodotyrosine molecules undergo oxidative condensation with the release of an alanine residue and the formation of thyroxine (T_4), or one mono- and one diiodotyrosine are coupled to form one molecule of triiodothyronine (T_3). Thyronines, the active hormones of the thyroid (T_3 and T_4) are compounds of two molecules of tyrosine. The mono- and diiodotyrosines are without biological activity. The active hormones, T_3 and T_4 (stored in the colloid bound to thyroglobulin), are released through proteolysis from the thyroglobulin and then pass into the circulation where they are transported to the sites of action. T_3, which has four times the potency of T_4, circulates in a much lower concentration in the blood than T_4. The active thyronines are present in plasma bound to carrier proteins and to a very small extent in the free state. About 80% of the thyroid hormones entering the tissues are broken down by deiodinating enzymes and the iodine so liberated is recycled within the body for reutilization.

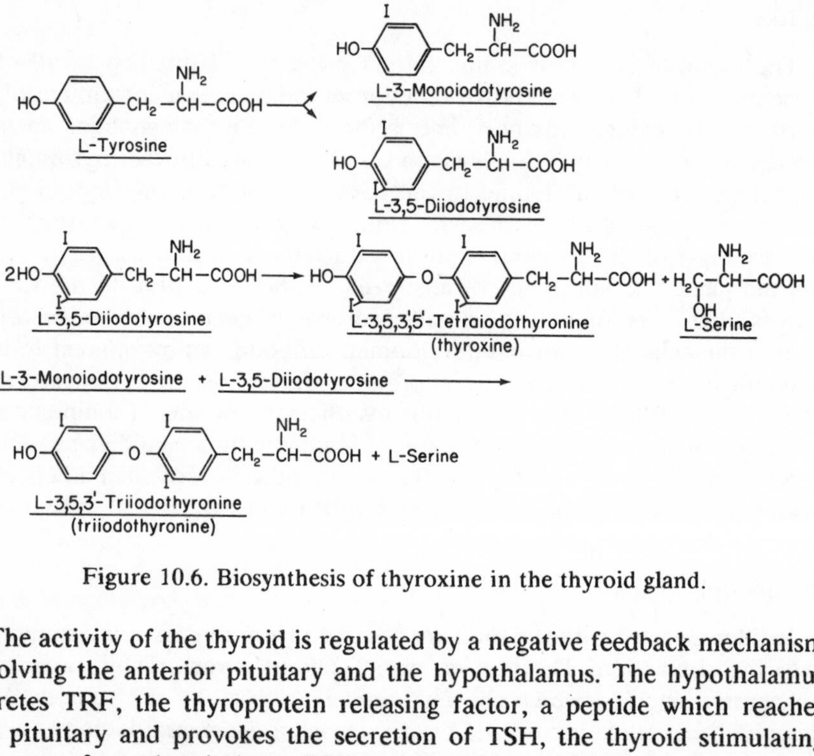

L-Tyrosine → L-3-Monoiodotyrosine + L-3,5-Diiodotyrosine

2 L-3,5-Diiodotyrosine → L-3,5,3',5'-Tetraiodothyronine (thyroxine) + L-Serine

L-3-Monoiodotyrosine + L-3,5-Diiodotyrosine → L-3,5,3'-Triiodothyronine (triiodothyronine) + L-Serine

Figure 10.6. Biosynthesis of thyroxine in the thyroid gland.

The activity of the thyroid is regulated by a negative feedback mechanism involving the anterior pituitary and the hypothalamus. The hypothalamus secretes TRF, the thyroprotein releasing factor, a peptide which reaches the pituitary and provokes the secretion of TSH, the thyroid stimulating hormone, from the pituitary. TSH is the primary regulator of the synthesis and of the secretion of the thyroid hormones. The thyroid hormones in turn inhibit the release of both, TRF by the hypothalamus and TSH by the pituitary, in this way controlling the levels of the thyroid hormones. In the absence of adequate thyroxine for inhibiting TSH release, the thyroid gland becomes hyperactive and enlarged. This condition is known as goitre (see later).

Iodine Deficiency and Functions

The primary function of iodine, because of its presence in the thyroid hormones, is the control of the rate of cellular oxidation. The thyroid hormones accelerate cellular reactions in nearly all cells of the body, resulting in increased oxygen consumption and an increase in basal metabolic rate (see Chapter 17). Therefore thyroxine dysfunction has a profound effect on the animal. When the diet contains an insufficient amount of iodine, the production of thyroxine is decreased. The enlargement of the thyroid, goitre—the most characteristic manifestation of iodine deficiency—is regarded as a compensatory effect of this tissue to offset deficient production of thyroxine. The appearance of goitre may result as an advanced stage of deficiency. Decreases in the concentrations of bound and free thyroxine in blood are indicative of lower iodine deficiency (see above).

Impaired reproductive performance and and stunted growth are the most marked consequences of reduced thyroid function. Changes in the skin of mammals or in the plumage of birds are among the most constant features of iodine deficiency. Pigs and calves born to iodine-deficient mothers are frequently weak and hairless and have thick, pulpy skins. Reduced quantity and quality of wool may be associated with goitre of sheep.

Goitrogenic Substances

Thyroid insufficiency is not necessarily a reflection of inadequate supplies of dietary iodine. It may be due to the presence of dietary constituents called goitrogens which interfere with thyroid hormone synthesis. Goitrogens may act at two levels. (1) Thiocyanate and perchlorate ions act by inhibiting the selective concentration of iodide by the thyroid; their actions are reversible by iodide. (2) Thiocarbamides (thiourea and thiouracil) contain —SH groups which prevent the oxidation of iodide to free iodine and therefore its incorporation into tyrosine as precursor of the thyroactive substances. Unlike goitre produced by thiocyanates, the goitrogenic action of thiocarbamides can only be partially controlled by supplemental iodine. Goitrogenic activity has been found in many plants including cruciferous plants (rape seed, kale or cabbage), soybeans, linseed and groundnuts. The mildly goitrogenic principle present in soybeans or groundnuts is eliminated in processing the beans to produce oilseed meal.

Supply of Iodine

Iodine deficiency is a regional problem. The existence of iodine-low soils can be attributed to the small accumulation of iodine of marine origin due to the long distance of such areas from the sea. Supplementation of iodine-deficient diets containing less than about 0.04 p.p.m. in the dry matter, is accomplished either by use of salt licks containing potassium iodide or by incorporating iodized salt (containing 0.01% potassium iodide) into mineral mixtures or concentrates. Salt licks containing potassium iodide are subject to considerable losses by volatilization or by leaching from the surface layer of block salt, particularly in hot, humid climates. It is preferable to use the more stable potassium iodate instead of potassium iodide. Care must be exercised with such salt mixtures to avoid excess iodine. The acute effects of very large doses of iodide include saturation of the transport mechanism for iodide ions, subsequent inhibition of thyroid hormone synthesis and finally goitre as in iodine-deficient animals.

Among the trace elements iodine is unique in the ease of transfer of ingested iodine to milk and eggs; their iodine content is responsive to changes in dietary iodine intakes and can be increased considerably by feeding with sufficiently large amounts of iodine.

With regard to the role of thyroxine in controlling metabolism, the ingestion of thyroactive iodinated proteins (iodinated casein) has been proposed for stimulating growth of meat animals, and wool, egg and milk production by creating a mild hyperthyroidic state. But because of some disadvantages, little use has been made of these feed additives in practice (see Chapter 21).

Selenium

The nutritional significance of selenium as an essential element has arisen only relatively recently—in the 1950s—although its toxic effects were already observed in the middle of the nineteenth century. The beneficial level for animals is less than 0.1 p.p.m. on a dry matter basis, but levels of 3–5 p.p.m. are already toxic. Levels of soil selenium vary widely in different areas of the world, some suffering from deficiency and others having an excess. Selenium as an essential nutrient element bears a functional relationship to vitamin E in that both selenium and vitamin E are involved in the defence of the cell against oxidative damage through reactive metabolites of lipids.

Biochemical Functions of Selenium

The following two biochemical functions of selenium have become apparent from recent research.

(1) Selenium is an integral part of the blood enzyme gluthathione peroxidase (GSH-P_x) which contains four atoms of selenium per molecule. This enzyme catalyses the reduction of hydrogen peroxide and of peroxides formed from fatty acids (see Chapter 7) according to the general reaction

$$\text{ROOH} + 2\,\text{GSH} \longrightarrow \text{R}-\text{OH} + \text{HOH} + \text{GSSG}$$

where ROOH are lipid peroxides and R– OH are non-damaging hydroxy compounds. The selenium-containing enzyme GSH-P_x acts by destroying peroxides before they can attack the cellular membranes, whereas vitamin E functions by preventing the formation of these peroxides (see Chapter 11).

GSH-P_x occurs in blood and many organs, and its activity is particularly high in liver and erythrocytes. The activity of this enzyme in blood plasma can be used for assessing body selenium status and correlates with selenium intake.

(2) Selenium is also required for normal pancreatic integrity and function. Therefore through its effect on pancreatic lipase production selenium is involved in the normal absorption of lipids and vitamin E. In the absence of selenium, the pancreas degenerates and fails to produce lipase, resulting in a lack of monoglycerides which are needed for the formation of lipid–bile salt micelles required for vitamin E absorption (see Chapter 6).

Selenium Deficiency Diseases

The following metabolic diseases of animals and birds are caused by deficiencies of selenium or vitamin E or of both these nutrients: nutritional muscular dystrophy, mulberry heart disease, exudative diathesis and pancreatic fibrosis. These diseases are preventable by selenium or vitamin E. With diets low in vitamin E, the requirements for selenium are increased and vice versa. Since, as shown above, selenium fulfils some specific function not related to GSH-P_x, there is a limit to which vitamin E can substitute for selenium and then only selenium will be effective. On the other hand, encephalomalacia in chicks and liver necrosis in rats are examples of cases that are responsive to vitamin E only (see Chapter 11).

Nutritional muscular dystrophy (NMD) or white muscle disease is a disease of the striated muscles that occurs in a wide range of animals: lambs, calves, pigs and chickens. The disease is most common in lambs and calves reared in selenium-deficient areas. The animals with white muscle disease have chalky white striations and degeneration of muscles (see PLATE IIB). The result of the degeneration of muscles is stiffness and difficulty in locomotion (see PLATE IIIA). Animals with NMD show a drastic rise in blood plasma concentrations of several enzymes that are normally intracellular, but are released into the plasma because of cell damage occurring in animals affected by NMD. These enzymes include glutamic–oxaloacetic transaminase (GOT, see Chapter 7) and lactate dehydrogenase (LDH). This is an index of damage to muscle or heart tissue resulting from leakage of these enzymes into the blood, although is not specific for NMD.

NMD is characterized by abnormally low selenium and GSH-P_x concentrations in blood and tissues. If the heart muscle is affected, sudden death may occur. This is an increasingly common problem in growing pigs fed on rations low in selenium and is often referred to as 'mulberry heart disease'.

The disease exudative diathesis of chicks fed diets deficient in selenium and vitamin E is characterized by oedema, i.e. accumulation of subcutaneous fluids on the breast (see PLATE IIIB). The fluids arise from abnormal permeability of the capillary walls. In the absence of adequate amounts of selenium and subsequently of glutathione peroxidase, the capillary walls suffer oxidative damage. Exudative diathesis can be largely cured or prevented by vitamin E or by selenium; vitamin E can have a sparing action, though it cannot fully replace selenium (see Chapter 11). Thus, in case of total lack of selenium, addition of this element is necessary for the formation of glutathione peroxidase, so as to assure the integrity of cell membranes and prevent exudative diathesis.

Severe selenium deficiency, even in the presence of plenty of vitamin E in the diet, results in atrophy of the pancreas of chicks, in addition to poor growth and feathering. The pancreas is the target organ for selenium deficiency in chicks as the other organs are unaffected if vitamin E is adequate. The pancreatic atrophy is associated with an impairment of lipid and vitamin E absorption, as outlined above.

In all species of farm animals selenium deficiency results in impaired reproductive performance in males and females. In poultry particularly, there is reduced egg production and hatchability of fertile eggs. A selenium-deficient diet may also cause infertility in ewes due to early embryonal deaths, and in dairy cows because of retained placentae (see Chapter 19).

'Ill thrift' occurs in sheep and cattle raised on selenium-deficient pastures and can be prevented by selenium treatment with considerable improvement in growth and wool yield.

Requirements and Sources of Selenium

The dietary requirements of selenium vary with the form of selenium ingested, the criteria of adequacy employed and the other constituents of the diet, particularly the vitamin E level. The recommended selenium requirement for most species is 0.1 p.p.m.; in turkeys it is 0.2 p.p.m.

The selenium concentrations in concentrate feeds and particularly in herbages vary widely, depending on the plant species and selenium status of the soils on which they grow. Variations exist among different chemical forms of selenium in their availability. The most common inorganic forms of selenium are selenates (SeO_4^{2-}) and selenites (SeO_3^{2-}) which are the selenium analogues of sulphates and sulphites. The dominant forms of selenium in feeds and forages are selenomethionine, $CH_3.Se\ CH_2.CH_2.CHNH_2.COOH$, and selenocysteine, $HSe\ CH_2.CHNH_2.COOH$, the seleno analogues of methionine and cysteine. In ruminants a large percentage of the ingested selenium appears to be incorporated into these seleno analogues. These appear to be absorbed by the animals and deposited in the tissues in the form of selenoamino acids. These inorganic forms are less available than the organic forms (25% versus 60–90% in the chicken).

The procedures available for providing selenium supplements to selenium-deficient animals include injections or oral dosing with selenium salts, provision of salt mixtures or licks containing small amounts of selenium salts or use of feeds obtained from selenium-rich areas.

Selenium Poisoning

Selenium toxity is now recognized in localized areas in many countries. The range between the required amounts (about 0.1 p.p.m.) and toxic levels (3–5 p.p.m.) is very narrow, the toxic range being of the order of 20 times the requirement. Harmful selenium intake by animals arises from the ingestion of forage plants with relatively high selenium concentrations due to the presence of excessive selenium levels in the soils. Soils containing more than 0.5 p.p.m. are potentially dangerous to livestock as the plants grown on such soils may contain more than 4 p.p.m. of selenium. Plant species such as the weeds astragalus and asters, which accumulate 400–800 p.p.m. selenium are termed 'indicator plants' because they can be used in identifying selenium-bearing soils. Indicator plants absorb selenium from soils in which it is present in a form relatively unavailable to most of the usual forage plants, and they play an important role in producing chronic poisoning of grazing animals, since such plants when they die return selenium to the soil in organic forms that are available to other species.

'Alkali disease' and 'blind staggers' are chronic and acute forms of selenium poisoning which affect horses, cattle, sheep and pigs. The symptoms of the chronic condition (occurring from prolonged consumption of feeds containing 5–40 p.p.m.) are dullness and lack of vitality, emaciation, roughness of coat, soreness and sloughing of the hooves (see PLATE IV) and erosion of the joints of the long bones, leading to lameness. Reduction in reproductive performance and retardation of growth are the main economic effects of chronic selenium poisoning. In acute poisoning caused by consumption of considerable amounts of indicator plants, the animals suffer blindness, abdominal pain, some degree of paralysis and finally death due to respiratory failure.

In chickens, excess levels of selenium reduce growth rate, egg production and hatchability, and cause embryonic abnormalities.

Since selenium is deposited in all the tissues of the body, except adipose tissue, high concentrations of selenium in hair, blood, urine, milk and eggs provide evidence of excessive intake.

Some procedures have been proposed to counteract selenium toxicity: (1) addition of sulphates to the soil to reduce selenium uptake by plants, (2) addition of certain substances to the feeds (sulphates, excess protein, particularly as linseed meal, arsenite or organic arsenic compounds) to reduce selenium absorption or increase excretion and thus limit its accumulation in the tissues. The applicability of these procedures is limited. It may be preferable to restrict seleniferous areas to the production of industrial crops rather than forage.

Molybdenum

Molybdenum is another element that has been found to be essential in traces but where higher levels are toxic. Molybdenum has been identified as a constituent of the enzymes xanthine oxidase, aldehyde oxidase and sulphite oxidase. The chick has a particular need for xanthine oxidase for uric acid formation (see Chapter 7) in contrast to the rat in which, like in other mammalians, the principal end-product of nitrogen metabolism is urea. Nevertheless low-molybdenum diets were found to be well tolerated by chicks, but when tungsten (a molybdenum antagonist) was added to the diet the growth was retarded and tissue xanthine oxidase levels lowered. A nutritional role of molybdenum in the growth of lambs has been reported and this has been explained by stimulation of cellulose breakdown. The absolute molybdenum requirement is very low and the needs by farm animals are met by their usual rations.

The toxicity of molybdenum is related to copper deficiency as molybdenum antagonizes the availability of copper (see above). Excessive levels of molybdenum and of sulphate as well, induce symptoms of copper deficiency in ruminants, even with an adequate intake of copper. On the other hand, chronic copper poisoning has been observed under conditions of moderate copper intake and very low dietary levels of molybdenum and sulphate.

The major symptoms of molybdenum toxicity are those of copper deficiency. The tolerance of farm animals to high molybdenum intake varies among different species and the copper and sulphur content of the diet. Cattle are the least tolerant species, followed by sheep, whereas horses and pigs are the most tolerant farm animals. Severe molybdenosis occurs in cattle grazing pastures containing 20–100 p.p.m. molybdenum compared with 3–5 p.p.m. or less in healthy herbage. The prominent manifestations of molybdenum toxicity in cattle—similar to those of copper deficiency—are diarrhoea, scouring, harsh, staring coats and weight loss. This condition, first recognized in England and termed there ‘teart’ or ‘peat scours’, may be counteracted by orally or intravenously administered copper.

Fluoride

Fluoride has been recognized as an essential nutrient for human beings and laboratory animals. The incidence of dental caries in the human population was found to be significantly higher in areas where the water supply was virtually free from fluoride than in those where the water contained 1.0–1.5 p.p.m. fluoride. Fluoride is a constant minor constituent of bones and teeth; also soft tissues and body fluids contain minute amounts. The small quantities of fluoride present in every practical diet with the drinking water meet the requirements. Thus, fluorine deficiency should never be a problem with farm animals.

Fluoride is important in animal nutrition because of its toxic effects. Toxic amounts are ingested by farm animals (1) from the use of non-defluorinated rock phosphates (containing up to 3–4% fluoride) as mineral supplements (see above), (2) from the consumption of forages contaminated with fluoride fumes released by smelting plants, (3) from drinking water that in some areas contains too much fluoride.

When excessive amounts of fluorides are ingested bones and teeth accumulate abnormal amounts that induce changes. Teeth changes may include mottling, staining, excessive wear and defects in the enamel such as erosion or pitting. The affected bones appear rough, porous and chalky white when compared with the smooth lustre of normal bones. Dental lesions and joint abnormalities bring about reduced food consumption due to the inability of the animals to gather and masticate fodder.

In dairy cattle fluorine toxicosis is associated with levels of 5000–10 000 p.p.m. fluoride in bones, whereas the fluorine concentration in bones of all kinds of normal adult farm animals lies within the range 300–600 p.p.m. The calcium/phosphorus ratio of fluorotic bones is normal, the carbonate content decreases but that of magnesium increases. This indicates that the fluoride ion replaces carbonate, but not phosphate, in the apatite crystals (see above) and that there may be precipitation of some fluoride as calcium fluoride.

Tolerance to dietary fluoride depends on its chemical form, on the species and age of the animal and the duration and continuity of intake. The highly soluble sodium fluoride is more toxic than calcium fluoride and other insoluble products such as fluoride-containing rock phosphates. Data on the safe levels of soluble and insoluble fluorides for different species are given in Table 10.3. Poultry will tolerate a considerably higher level than other farm animals. This higher tolerance of poultry is probably due to a low absorption and increased excretion of fluoride.

Table 10.3 Safe levels of fluoride in the total diet (after Hays and Swenson, Minerals. In *Dukes' Physiology of Domestic Animals,* Ninth Ed., 1977, p. 401)

Species	Sodium fluoride or other soluble fluoride (p.p.m.)	Rock phosphate (p.p.m.)
Dairy cow	30–50	60–100
Beef cow	40–50	65–100
Sheep	70–100	100–200
Pig	70–100	100–200
Chicken	150–300	300–400
Turkeys	300–400	

Other Minerals

The following elements have been shown to be essential nutrients for rats and chicks: silicon, chromium, tin and nickel. Rats or chicks fed on purified diets showed an increased growth rate from the addition of each of these elements. The required level is low and there is no evidence that the usual laboratory- or farm-animals fed diets have a deficiency of any of these minerals.

Silicon

Silica, the oxide of silicon, is found in the ash of most plant and animal tissues, in small quantities. The element is believed to function as a biological linking agent, possibly as an ether derivative of silicic acid of the type $R_1-O-Si-O-R_2$ and contributes to the structural integrity of connective tissues and bones and of plants as well. Silicon is an integral component of the mucopolysaccharides (see Chapter 4), of cartilage and collagen. In silicon-deficient rats and chicks, bone abnormalities occur because of an impairment of mucopolysaccharide synthesis and formation of cartilage.

In plants silica serves as a structural component complementing lignin to strengthen and rigidify plant cell walls. Silicon dioxide is not digestible and exerts a negative effect upon cell wall digestibility (see Chapter 13). In some feedstuffs silicon dioxide is found in large amounts as, for example, in rice hulls, rice straw and wheat straw. High levels of silicon dioxide in the diet of farm animals may be detrimental. Ruminants grazing in grasses high in silicon dioxide suffer from kidney stones, the main route of excretion of absorbed silicon dioxide being in the urine.

Chromium

Chromium has been shown to be essential for glucose metabolism in rats and probably acts as a cofactor with insulin. Nothing is known of the chromium requirements of farm animals.

Vanadium

This is an essential trace element for chicks and rats. If chicks consume a diet low in vanadium, the rate of body and feather growth decreases. High intakes of vanadium inhibit cholesterol synthesis; vanadium counteracts the stimulation of cholesterol synthesis by manganese.

Heavy Metals

The following heavy metals are of interest in animal agriculture because of their toxicity: lead, mercury, arsenic and cadmium.

Lead

Lead toxicity is considered the most common cause of death by accidental poisoning in cattle. The main source of excess lead causing acute poisoning of farm animals was paint until the use of lead-based pigments in paints was restricted. Chronic lead poisoning may occur when cattle, sheep or horses eat forage contaminated by lead from fumes and dust emitted from agricultural spray residues or from industrial lead operations. Horses appear to be more susceptible to lead poisoning than cattle: lead concentrations of 80 p.p.m. in forage are toxic to horses and 200 p.p.m. to cattle. Lead toxicity affects organs and tissues, particularly the skeleton in which the greatest part of the absorbed lead is deposited. The absorption and retention of ingested lead is greatly reduced by high dietary intakes of calcium, phosphorus, iron, copper and zinc, whereas low intakes of calcium and phosphorus increase absorption of lead. Anaemia, a characteristic symptom of lead poisoning, arises from its interference in biosynthesis of haem.

Arsenic

The well-known toxicity of this element may be explained by its attachment to the sulphydryl groups of protein thereby blocking enzyme actions. Arsenic can be a major source of livestock poisoning in areas surrounding smelters and where arsenicals are used to control weeds and insects. Generally, inorganic arsenicals are more toxic than organic compounds. Several organic arsenicals are recognized as growth stimulants for pigs and poultry and their action resembles that of antibiotics (see Chapter 18).

Mercury

Cases of mercury poisoning of farm animals occurred through feeding seed grains treated with mercury fungicides. Mercury, like arsenic, has a high affinity for − SH groups and inhibits the activity of SH-containing enzymes. Organic mercury compounds are more toxic than inorganic mercury salts. All these compounds result in liver and kidney accumulation of mercury, kidney necrosis and death.

Cadmium

Pollution deriving from industrial uses of cadmium has been shown to increase the cadmium content of forages more than 40-fold. Cadmium is known as a metabolic antagonist to zinc, copper and calcium. The toxic effects of cadmium are similar to those occurring in zinc deficiency.

Nitrates

Nitrates occurring occasionally in weeds and some forages, particularly after application of great quantities of N fertilizer are, *per se*, almost harmless. The toxic effect of nitrate in ruminants is caused by its reduction to nitrate in the rumen (see Chapter 5). Nitrate converts the haemoglobin in the blood into met-haemoglobin which is unable to act as an oxygen carrier.

BIBLIOGRAPHY

Books

Bronner, F., and Coburn, J. W. (1982). *Disorders of Mineral Metabolism, Vol. 1, Trace Minerals.* New York and London: Academic Press.

Comar, C. L., and Bronner, F. (1962–1969). *Mineral Metabolism. An Advanced Treatise,* 3 vols. New York: Academic Press.

Hoekstra, W. G., Suttie, J. W., Ganther, H. E., and Mertz, W. (eds) (1974). *Trace Element Metabolism in Animals,* 2nd edn. Baltimore: University Park Press.

National Academy of Sciences–National Research Council (1980). *Mineral Tolerance of Domestic Animals.* Washington, DC.

Underwood E. J. (1977). *Trace Elements in Human and Animal Nutrition,* 4th edn. New York and London: Academic Press.

Underwood, E. J. (1981). *The Mineral Nutrition of Livestock.* London: Commonwealth Agricultural Bureau.

Major mineral elements

Articles

Adams, R. S. (1975). Variability in mineral and trace-element content of dairy cattle feeds. *J. Dairy Sci.,* **58**, 1538–1548.

Ammerman, C. B., and Goodrich, R. D. (1983). Advances in mineral nutrition in ruminants. *J. Anim. Sci.,* **57**, Suppl. 2, 519–533.

Miller, W. I. (1975). New Concepts and developments in metabolism and homeostasis of inorganic elements in dairy cattle. *J. Dairy Sci.,* **58**, 1549–1560.

Pamp, D. E., Goodrich, R. D., and Meiske, J. C. (1976). A review of the practice of feeding minerals free choice. *World Rev. Anim. Prod.,* **12**, 13–31.

Peeler, H. T. (1972). Biological availability of nutrients in feeds: availability of major mineral ions. *J. Anim. Sci.,* **35**, 695–712.

Scott, D., and McLean, A. F. (1981). Control of mineral absorption in ruminants. *Proc. Nutr. Soc.,* **40**, 257–266.

Calcium and phosphorus

Book

Irving, J. T. (1973). *Calcium and Phosphorus Metabolism.* London and New York: Academic Press.

Articles

Borle, A. B. (1974). Calcium and phosphate metabolism. *Annu. Rev. Physiol.,* **36**, 361–390.

Braithwaite, G. D. (1976). Calcium and phosphorus metabolism in ruminants with special reference to parturient paresis. *J. Dairy Res.,* **43**, 501–520.

Hemingway, R. G., and Fishwick, G. (1976). Sources, availability and use of supplementary phosphorus for ruminants. In *Feed Energy Sources for Livestock* (H. Swan and D. Lewis, eds), pp. 95–116. London: Butterworths.

Hurwitz, S. (1976). Absorption of calcium and other minerals. In *Digestion in the Fowl* (K. N. Boorman and C. W. Freeman, eds), pp. 157–178. Edinburgh: British Poultry Science Ltd.

Hurwitz, S. (1978). Calcium metabolism in birds. In *Chemical Zoology* (M. Florkin and B. T. Scheer, eds), Vol. 10, pp. 273–306. New York: Academic Press.

Preston, R. L., and Pfander, W. H. (1964). Phosphorus metabolism in lambs fed varying phosphorus intakes. *J. Nutr.*, **83**, 369–378.

Taylor, T. G. (1979). Availability of phosphorus in animal feeds. In *Recent Advances in Animal Nutrition—1979* (W. Haresign and D. Lewis, eds), pp. 23–34. London and Boston: Butterworths.

Wasserman, R. H. (1984). Bones. In *Duke's Physiology of Domestic Animals* (M. J. Swenson, ed.), 10th edn, pp. 467–485. Ithaca and London: Cornell University Press.

Magnesium

Fontenot, J. P. (1983). *Role of Magnesium in Animal Nutrition.* Blacksbury: Virginia Polytechnic Institute and State University.

Tomas, F. M., and Potter, B. J. (1976). The site of magnesium absorption from the ruminant stomach. *Br. J. Nutr.*, **36**, 37–45.

Sodium, potassium and chloride

Hurwitz, S., Cohen, J., Bar, A., and Bornstein, S. (1973). Sodium and chloride requirement of the chick. Relationship to acid–base balance. *Poult. Sci.*, **52**, 903–909.

Meneely, G. R., and Aines, P. D. (1976). Sodium and potassium. In *Present Knowledge in Nutrition*, 4th edn. New York: Nutrition Foundation.

Mongin, P. (1981). Recent advances in dietary anion–cation balance in poultry. In *Recent Advances in Animal Nutrition–1981* (W. Haresign, ed.), pp. 109–120. London: Butterworths.

Trace elements

Bremner, J., and Davies, N. T. (1980). Dietary composition and the absorption of trace elements by ruminants. In *Symposia on Physiology, Digestion, and Metabolism in Ruminants.* 4th Symposium, pp 409–427.

Forbes, R. M., and Erdman, J. W., Jr. (1983). Bioavailability of trace mineral elements. *Annu. Rev. Nutr.*, **3**, 289–307.

Pond, W. G. (1975). Mineral interrelationships in nutrition: practical implications. *Cornelian Vet.*, **65**, 441–456.

Suttle, N. F. (1983). Micronutrients as regulators of metabolism. In *Nutritional Physiology of Farm Animals* (I. A. F. Rook and P. C. Thomas, eds), pp. 415–453. London and New York: Longman.

Iron

Linder, M. C., and Munro, H. N. (1976). The mechanism of iron absorption and its regulation. *Fed. Proc.*, **36**, 2015–2023.

Thomas, J. W. (1970). Metabolism of iron and manganese. *J. Dairy Sci.*, **53**, 1107–1123.

Copper

Cooke, B. C. (1983). Copper in animal feeds. In *Recent Advances in Animal Nutrition—1983*, pp. 209–226. London and Boston: Butterworths.

Braude, R. (1976). Growth-promoting substances. *Proc. Nutr. Soc.*, **35**, 377–382.
Ward, G. M. (1978). Molybdenum toxity and hypocuprosis in ruminants. A review. *J. Anim. Sci.*, **49**, 268–276.

Cobalt

Mills, C. F. (1981). Cobalt deficiency and cobalt requirements of ruminants. In *Recent Advances in Animal Nutrition—1981* (W. Haresign, ed.), pp. 129–142. London: Butterworths.

Manganese

Leach, R. M., and Liburn, M. S. (1978)..Manganese metabolism and its function. *World Rev. Nutr. Diet.*, **32**, 123–134.

Zinc

Chesters, J. K. (1978). Biochemical functions of zinc in animals. *World Rev. Nutr. Diet.*, **32**, 135–164.
Lönndearl, B., Keen, C. L., and Hurley, L. S. (1984). Zinc binding ligands and complexes in zinc metabolism. In *Advances in Nutritional Research* (H. H. Draper, ed.), Vol. 6, pp. 135–168. New York and London: Plenum.

Selenium

Combs, F. G., Jr., and Combs, S. B. (1984). The nutritional biochemistry of selenium. *Annu. Rev. Nutr.*, **4**, 257–280.
Hoekstra, W. G. (1975). Biochemical function of selenium and its relation to vitamin E. *Fed. Proc.*, **34**, 2083–2089.
Julien, W. E., Jones, J. E., and Moxon, A. L. (1976). Selenium and vitamin E and incidence of retained placenta in parturient dairy cows. *J. Dairy Sci.*, **59**, 1954–1960.
Rickaby, C. D. (1981). The selenium requirements of ruminants. In *Recent Advances in Animal Nutrition—1981*, pp. 121–128. London: Butterworths.
Scott, M. L. (1976). Selenium and vitamin E in poultry rations. In *Feed Energy Sources for Livestock* (H. Swan and D. Lewis, eds), pp. 117–129. London: Butterworths.

Fluorine

National Research Council (1974). *Effects of Fluorides in Animals.* Washington, DC: National Academy of Sciences.

Subba Rao, G. (1984). Dietary intake and bioavailability of fluoride. *Annu. Rev. Nutr.*, **4**, 115–136.

Chromium

Girson, C. T. (1977). The metabolic significance of dietary chromium. In *Advances in Nutrition Research* (H. H. Draper, ed.), Vo. 1, pp. 23–54. New York and London: Plenum.

CHAPTER 11

Vitamins

GENERAL INTRODUCTION

In addition to proteins, fats, carbohydrates, minerals and water, specific organic compounds, termed vitamins, are required in relatively minute amounts for the normal functioning of the animal body, for maintenance, growth, health and production. The vitamins are not feedstuffs in the ordinary sense, but rather function catalytically. Vitamins must be provided from exogenous sources, mainly with the diet or from biosynthesis by microorganisms harboured in the gastrointestinal tract. Many plants and microorganisms are capable of synthesizing vitamins, a feature not shared by animals whose biosynthetic capacity is in general much more limited. The definition that vitamins are organic compounds which must be supplied to animals in small amounts from outside is not completely satisfactory, since in some instances the required vitamins are synthesized within the body tissues (see later).

Each vitamin is required for specific metabolic reactions within the cells. When a particular vitamin is omitted from the diet, the respective biochemical reaction in which the vitamin participates cannot take place and specific symptoms of an avitaminosis are produced. Such absolute vitamin deficiencies may not occur under practical farming conditions, but rather marginal deficiencies leading to non-specific symptoms, such as loss of appetite, unthrifty appearance, reduced growth and feed utilization. The course of a deficiency varies widely among animal species.

The various vitamins differ greatly in chemical structure and metabolic function. The list in Table 11.1 gives the presently accepted designations for the vitamins. Before the discovery of their chemical nature the vitamins were called simply by letters of the alphabet. Although the designation by letters is generally still accepted, for some vitamins the present tendency is to use chemical or trivial names.

Table 11.1 Vitamins important in animal nutrition

Vitamin	Chemical name
Fat-Soluble vitamins	
A	Retinol
D_2*	Ergocalciferol
D_3*	Cholecalciferol
E*	Tocopherol
K*	Phylloquinone
Water-soluble vitamins	
B_1	Thiamine
B_2	Riboflavin
B_3	Nicotinamide (niacin)
B_6	Pyridoxine
	Pantothenic acid
	Biotin
	Folacin (folic acid)
	Choline
B_{12}	Cyanocobalamin
	Inositol
C	Ascorbic acid

* Some compounds of similar structure as vitamins D, E and K possess vitamin activity.

The vitamins are divided on the basis of their solubility properties into fat-soluble and water-soluble vitamins. The fat-soluble vitamins include vitamins A, D, E and K, while the members of the B complex and vitamin C are classified as water-soluble vitamins. In the early years of vitamin research only one B vitamin was known. Later research revealed that B was not a single vitamin, but a mixture of several vitamins which were designated with number subscripts to the letters (B_1, B_2, etc.).

The division of vitamins on the basis of their solubility in fat or water is a useful classification considering their functions and metabolic pathways in the body. All the vitamins of the water-soluble B group function as coenzymes. A coenzyme may be defined as a small molecule loosely bound to a protein carrier, the apoenzyme, to provide an enzyme (called holoenzyme):

$$\text{Coenzyme} + \text{apoenzyme} \longrightarrow \text{enzyme}$$

The coenzyme can actually be the vitamin as such, or the vitamin after undergoing some chemical change (within the animal's tissues), or the vitamin bound to another small molecular-weight compound. The enzymes containing B vitamins catalyse the oxidation of carbohydrates, fatty acid and amino acids, reactions vital for energy production. They function also in biosyntheses of important cellular components. The apoenzyme confers substrate specificity to the system (see Chapter 12).

The fat-soluble vitamins fulfil specific and independent functions: vitamin A is necessary for the maintenance of vision, vitamin D for the utilization of calcium in the body and vitamin K is part of the system involved in blood clotting. Moreover, these fat-soluble vitamins are concerned in some stage of biosynthesis of certain proteins which have important biological functions. Vitamins C and E act in a more general fashion in preventing the oxidation of sensitive biological substances found in the cells.

Properties of Fat-soluble Vitamins Compared with Water-soluble Vitamins

Fat-soluble vitamins are absorbed from the intestinal tract along with fats after incorporation into micelles. Conditions favourable to fat absorption such as adequate bile flow and small particle size will also improve the absorption of fat-soluble vitamins. B vitamins are readily absorbed mostly by passive diffusion. The unique mode of absorption of vitamin B_{12} will be dealt with in Chapter 12.

Fat-soluble and water-soluble vitamins also differ in their capacity to be stored in the body. Fat-containing tissues (liver, adipose) are able to store fat-soluble vitamins, which may serve the body as a source of supply during depletion periods. The water-soluble vitamins, however, are not stored in the body in any significant amounts and thus frequent dietary intake is dictated in order to avoid deficiencies.

The difference in the excretion pathway also reflects the solubility difference of each group of vitamins. The fat-soluble vitamins are primarily excreted in the faeces via the bile, and the water-soluble vitamins chiefly in the urine. Some water-soluble B vitamins originating from bacterial synthesis may also be present in the faeces.

Excess amounts of fat-soluble vitamins—more than 500 times the daily recommended allowances—produce severe toxic effects, but not a surplus of B vitamins, since the organism is capable of excreting large amounts of these in the urine.

Provitamins

There are certain compounds which are not themselves vitamins and which function as vitamins only after undergoing a chemical change. After their ingestion with the diet these compounds are converted in the cells of the body to vitamins. They are precursors of vitamins and are called provitamins. The best example is the carotenes—plant pigments which are converted in the intestinal wall to vitamin A (retinol; see later). Other examples are cutaneous provitamin D which is converted by irradiation to vitamin D, or the amino acid tryptophan which can be converted to niacin (see Chapter 12).

Antivitamins

Certain compounds similar in structure to the vitamins, mainly B vitamins, can replace the vitamin in the holoenzyme (see above) by attaching themselves to the apoenzyme, thereby preventing the true vitamin from combining with the apoenzyme that normally would combine with the vitamin. The result is inhibition or complete prevention of the vitamin functioning properly. Compounds of this type which resemble particular vitamins—the majority are synthetic substances—compete with the vitamin within the animal body and are termed antivitamins. Since many microorganisms are dependent upon B vitamins as essential growth factors, the antivitamins inhibit their development. Thus antivitamins have been used to produce vitamin deficiencies in experimental animals or medically as drugs against pathogenic bacteria.

Determination of Vitamins

Biological, microbiological and chemical assays and assays using radioisotopes have been developed for the determination of vitamins in foods and biological samples. Biological assays measure the quantity of vitamins available to animals. Chemical determination may accurately measure the total quantity of vitamin present in a sample. Biological methods are time-consuming and expensive and the results are valid only for the animal species tested. Biological tests are still used to check the other methods.

Biological tests are carried out according to the principles of feeding tests outlined in Chapter 13. Groups of animals are first depleted of the vitamin by a diet lacking it. Then known amounts of the vitamin are administered in a series of levels to groups of the depleted animals. From a response in growth or other appropriate criterion, a standard response curve is prepared. The bioassay may also be based on the prevention or cure of some deficiency symptoms known to develop in the absence of the vitamin.

Microbiological assays are adequate for the determination of B vitamins. Some microorganisms having a specific requirement for these vitamins can be used as test organisms, preferentially the lactic acid bacteria. The latter produce a metabolic product, lactic acid, which is readily determined by titration. Using a basal medium complete in all respects except for the vitamin under test, growth response of the bacteria or the amount of lactic acid formed is compared quantitatively in standard and unknown solutions. However, the microbiological tests have the disadvantage that the vitamin must first be extracted from the sample in water-soluble form before being added to the growth medium used for the microorganism. The principles and procedures of microbiological tests for vitamins simulate those recommended for amino acids (see Chapter 7).

Microbial Synthesis of Vitamins in Rumen and Intestine

In ruminants, microbial synthesis of the B vitamins and vitamin K occurs in the rumen. These vitamins are highly available to the host animals as they pass down the tract through areas of efficient digestion and absorption. Therefore, in ruminants, microbial production of B vitamins and of vitamin K is sufficient to cover their needs. Microbial synthesis of these vitamins occurs also in the caecum and colon of both ruminants and monogastric animals, but the extent of absorption from the lower tract is very limited. Single-stomached animals, including young ruminants, prior to the development of their rumen require dietary sources of the B vitamins (see Chapter 18).

Although some species of microorganisms can synthesize vitamins, other species compete with the host for vitamins. Also the latter needs must be met by dietary vitamins. Intestinally produced vitamins become more available to animals (rabbit, rat and chick) that eat their own faeces (coprophagy) and thus recycle the products of the lower gut. The requirements for B vitamins, like those of other nutrients, are elevated in animals affected with infectious diseases due to decreased efficiency in the absorption through the intestinal wall, and the metabolic rate which is enhanced because of increased body temperature. Oral administration of antibiotics or sulphonamides to sick animals may depress the microbial synthesis of vitamins in the gut and further increase the need for dietary supplementation of B vitamins.

Requirements for Vitamins

The term requirement is not used in a uniform way. It covers requirements from minimum to optimum requirements. The minimum requirement of a vitamin is that amount needed daily to prevent deficiency symptoms. The requirement is highly dependent on the criterion used as a reference factor, such as measurement of a plasma profile, body reserves (in the case of fat-soluble vitamins), or visible deficiency symptoms.

The term allowance or simply requirement is used to express the daily intakes of a vitamin that should cover both the minimum requirement and a safety margin. The optimum requirement is defined by the intake necessary for maximum production. Requirements are expressed in mg or international units (UI; see later) per kilo live weight or per animal or on the basis of the amount of feed (mg per kg feed).

Distribution of Vitamins in Feedstuffs

Most feedstuffs contain vitamins; concentrate feeds particularly contain B vitamins and forages are mainly sources of carotene. The vitamin content of feedstuffs varies widely according to the type of feedstuff, parts of plants (seeds, leaves, stems), growing conditions, season of harvesting and means of preparation of feeds. Many vitamins are destroyed by oxidation, a process enhanced by heat, air, light and traces of heavy metals. Therefore conditions of storage and processing of feeds will influence their value as vitamin sources.

In order to assure adequate supply of vitamins, one must make up the difference between the requirements of animals (according to species and performance) and the vitamin content of the diet by supplementation with pure synthetic vitamins or commercial vitamin concentrates.

Fat-soluble Vitamins

VITAMIN A AND PROVITAMIN A*

Supply of vitamin A to farm animals appears to be more critical than that of any other vitamin, and it is the only vitamin which may be deficient to ruminants under farm conditions.

Vitamin A is necessary for vision, reproduction, development of epithelia, growth and health. The exact role of vitamin A in these functions has not been exactly elucidated except for its well-documented role in vision (see later).

Chemical Structure of Vitamin A and Derivatives

Vitamin A can be provided by the vitamin itself, called retinol, by some less active analogues and by its precursors, carotenes, in particular β-carotene. Retinol and its derivatives are only found in sources of animal origin. Carotenes are principally found in association with chlorophyll in all photosynthetic tissues (leaves) and in lower concentrations in roots, seeds, fruits and flowers.

Retinol is an unsaturated 20-carbon cyclic alcohol which consists of a β-ionone (trimethylcyclohexene) nucleus and an unsaturated side-chain. Five conjugated double bonds are present in the retinol molecule, including the double bond in the β-ionone ring which is in conjugation with those in the side-chain (Figure 11.1). Vitamin A can occur as free vitamin A alcohol (retinol) or esterified with a fatty acid, usually palmitic acid. Vitamin A is deposited in animal tissues mainly as vitamin A ester, and it accumulates particularly in the liver; vitamin A alcohol is the principal form in blood.

---

* In collaboration with D. Sklan.

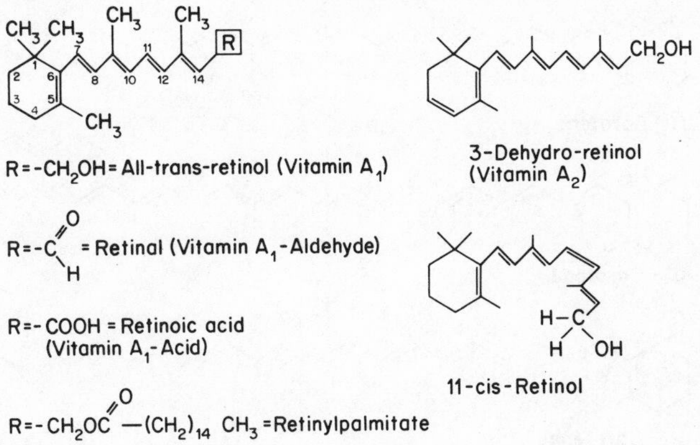

R = -CH_2OH = All-trans-retinol (Vitamin A_1)

R = -CHO = Retinal (Vitamin A_1-Aldehyde)

R = -COOH = Retinoic acid (Vitamin A_1-Acid)

R = -$CH_2OC(O)-(CH_2)_{14}CH_3$ = Retinylpalmitate

3-Dehydro-retinol (Vitamin A_2)

11-cis-Retinol

Figure 11.1. Chemical structure of retinol and some derivatives.

Because of the many double bonds present in retinol, different isomeric forms exist with different biological activities. All-*trans*-retinol is the most active and abundant form. Vitamin A_2 has an additional double bond between C-3 and C-4 of the ring and is found in fresh-water fish, but has less activity in mammals and birds than retinol. Vitamin A derivatives with terminal aldehyde (retinal) or carboxyl groups (retinoic acid) are active biological intermediates (see later).

Structure of Carotenes

Several carotenoids, yellow C_{40} compounds with structures based on a tetraterpenoid skeleton, show vitamin A activity. The formulae of some carotenoids present in different plant tissues are given in Figure 11.2. The main carotenoid hydrocarbons present in leaves are α-, β- and γ-carotenes and some oxygenated carotenoids (xanthophylls or luteins) are also found in leaves and green vegetables. The ability of carotenes to act as provitamin A is contingent upon conversion to retinol by the animal and depends on the presence of β-ionone. Carotenes containing at least one β-ionone ring can be converted to retinol by animals. The most effective and important carotenoid is β-carotene, which contains two β-ionone rings. α-carotene and γ-carotene, however, cannot be converted by animals to retinol with the same efficiency as β-carotene, and the α-ionone ring of α-carotene cannot be converted in the body to β-ionone, and the open chain structure of γ-carotene cannot be cyclized by animals. Thus α-carotene and γ-carotene are transformed into retinol with half the efficiency of β-carotene.

The biological activity of the β-ionone ring present in carotenes is suspended by the introduction of an hydroxyl group. Cryptoxanthine, a pigment

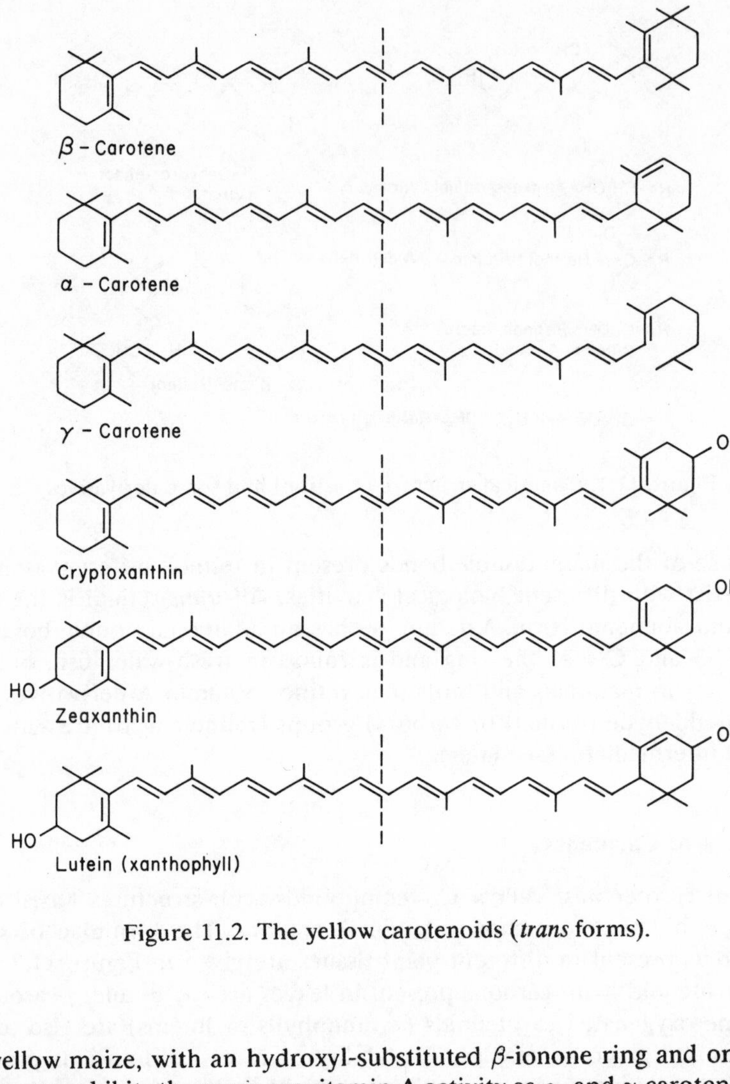

β - Carotene

α - Carotene

γ - Carotene

OH

Cryptoxanthin

OH

HO

Zeaxanthin

OH

HO

Lutein (xanthophyll)

Figure 11.2. The yellow carotenoids (*trans* forms).

of yellow maize, with an hydroxyl-substituted β-ionone ring and one intact β-ionone exhibits the same provitamin A activity as α- and γ-carotene. Zeaxanthin, another yellow carotenoid present in yellow maize, is composed of two hydroxyl-substituted β-ionone rings, and is devoid of any activity as a pro-vitamin. Xanthophyll, also called lutein, which occurs in all green plant tissues, contains one β-ionone ring and one α-ionone ring, though both are hydroxyl substituted. This compound also is devoid of pro-vitamin activity. The mixture of α-, β- and γ-carotene accompanying chlorophyll in green plants obtains more than 90% β-carotene. Carotenoids are the major pigments of carrots which contain some 60% of β-carotene. Carotenoids are also found in animals in fat deposits, milk fat, egg yolk, blood, liver, corpus luteum of cows and in testes of bulls.

Conversion of β-carotene to Retinol

The conversion of β-carotene to retinol in the animal body is effected by two enzymes: a copper-containing carotenoid dioxygenase which oxidatively cleaves the β-carotene molecule at the central 15,15′ double bond to form mainly two molecules of retinaldehyde, but also other fragments, and retinaldehyde reductase, a zinc-containing enzyme which reduces retinaldehyde, in the presence of NADH, to retinol. Both these processes occur in calves only in the intestinal mucosa where the carotene is absorbed from the intestinal lumen. In other animals some conversion also occurs after absorption in liver and kidney. These reactions are summarized in Figure 11.3.

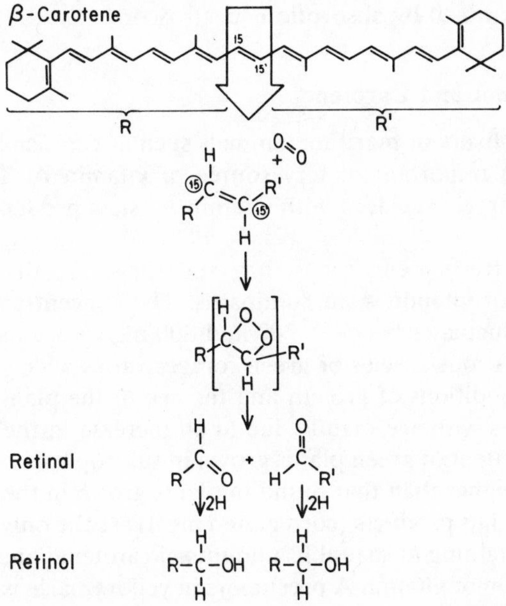

Figure 11.3. Cleavage of carotene to retinal and its reduction to retinol.

Activity Standards for Vitamin A

Vitamin A activity was originally expressed in international units (IU). This unit was defined as the daily dose required to produce a weight gain of 3 g/week in young rats between the fourth and eighth weeks of life. However, with the advent of accurate analytical procedures, the unit of vitamin A has been based on exact amounts equivalent to 0.3 μg of crystalline retinol (or 0.344 μg of retinyl acetate or 0.55 μg of retinyl palmitate). The β-carotene molecule is essentially a double retinol structure, and, theoretically, its oxidative cleavage should yield two retinol molecules. Biologically, however, the activity of β-carotene is only about half that of retinol for rats, and even less for other species (see later). This disparity between chemical structure and biological activity of β-carotene is due in part to incomplete absorption and in part to non-stoichiometry of the reaction due to the formation of oxidized metabolites of retinol. The quantity of β-carotene equivalent to one unit of vitamin A is thus 0.6 μg.

Chemical assays of retinol were based on the blue colour produced by dehydration reactions with antimony trichloride or trifluoroacetic or trichloroacetic acid; more modern methods rely on spectrofluorimetric measurements, or high-performance liquid chromatographic separation followed by retinol quantitation using absorption or fluorimetric detection. Carotenes are usually determined by absorption spectroscopy.

Sources of Retinol and Carotene

Oils from the livers of maritime animals such as cod-fish and shark, have been used as an important dietary source of vitamin A. These oils have, however, been largely replaced with vitamin A esters produced by industrial synthesis.

Carotenes in fresh green crops, hay and silage are the most common dietary sources of vitamin A in ruminants. The concentration of carotene in green crops fluctuates between 200 and 800 mg/kg dry matter. However, the content of various species of green forages varies widely and is strongly influenced by conditions of growth and the age of the plants. The carotene content decreases with age mainly due to an increase in the stalk/leaf ratio. The carotene content of green plants grown in subtropical regions during the rainy season is higher than that found in plants grown in the hot dry season. Yellow maize and its products (corn gluten meal) are the only cereal or cereal by-products containing measurable amounts of carotenes and cryptoxanthin. The concentration of vitamin A precursors in yellow maize is lower than that found in the dry matter of green crops.

Instability of Carotene and Vitamin A

Both retinol and carotene are readily oxidizable owing to the presence of the conjugated double bond system. Retinol is particularly unstable on exposure to light or heat, especially in the presence of heavy metal ions and water. Under these conditions rapid oxidation takes place and the biological activity is lost. Esterified vitamin A is more stable than retinol.

Trace metals present in feed mixtures and hydroperoxides of fatty acids present in the lipid fraction of animal feeds have a deleterious effect on vitamin A stability, the unsaturated, easily oxidized fats being the most destructive (Chapter 6). The stability of vitamin A added to feeds is low if the vitamin presents a large surface area for oxidation. This oxidation can be somewhat alleviated by the addition of a synthetic antioxidant such as ethoxyquin to the vitamin A.

A stable preparation of vitamin A intensively used in animal feeding involves coating. In this process the vitamin A is embedded, together with an antioxidant such as ethoxyquin or BHT, in a gelatin–carbohydrate matrix to form a beadlet. In the gastrointestinal tract the beadlet softens and disintegrates as the matrix is digested and releases the vitamin A which is thus fully biologically available. This form of vitamin A preparation, when added to feed mixtures, results in less than 3–4% loss of vitamin A per month on storage of animal feeds under low moisture conditions. Storage of feed under high moisture conditions can double or even treble this value. Pelleting feeds destabilizes the gelatin beadlet somewhat and results in greater loss of vitamin A on storage.

Carotenoids in plant material are subject to enzymatic oxidation through the lipoxygenase system, and to non-enzymic processes accelerated by direct sunlight. Slow drying of forages in the sun—for hay preparation—may cause an up to 80% destruction of carotene. High-temperature short-term heat treatment (artificial dehydration of green crops such as lucerne) deactivates the lipoxygenase and enables conservation of the carotene. Carotene is usually well preserved in silage, with total losses of about 10% typical of ensiling processes. Wilting prior to ensiling causes greater carotene losses of up to 30%.

Utilization of Carotene in Different Species of Animals

Absorption of β-carotene depends on adequate fat being present in the ration. However, utilization of β-carotene varies considerably with species, as is evident from Table 11.2. It should be noted that the conversion factors given in Table 11.2 are only applicable in the low range of β-carotene intake (<30 μg/kg live weight) as conversion efficiency decreases with increasing intake. In sheep, conversion efficiency decreases from 25% at a carotene intake of 20 μg/kg, to 7% at an intake of 160 μg/kg.

In rat, pig, goat, sheep, rabbit and dog, almost all of the carotene absorbed into the intestinal mucosa is cleaved to retinol and very little β-carotene is absorbed into the circulation. In man, cattle, horses and poultry some carotenoids escape conversion and are taken up intact into the circulation. Cattle preferentially absorb carotenes whereas birds absorb mainly xanthophylls (oxycarotenoids) which do not posses vitamin A activity. The absorbed carotenoids are transported in the blood plasma, carried to the tissues and contribute to the pigmentation of meat, eggs and milk. The animal species capable of absorbing carotene have yellow body and milk fat, whereas species that do not absorb carotene, such as sheep and goats, have white fat and no detectable carotene in blood or milk. In cattle, differences

Table 11.2 Efficiency of conversion of carotene to vitamin A in different species

Species	Conversion efficiency (%)	IU vitamin A equivalent to 1 mg β-carotene
Rat	100*	1667
Chicken	100	1667
Pig	30	500
Cattle	24	400
Sheep	30	500
Horse	33	555
Man	33	555
Dog	67	1111

* Efficiency of conversion has been set at 100% for the rat (based mainly on Beeson, 1965).

between breeds are noted with respect to carotene absorption. Holsteins absorb less carotenes than Jerseys and Guernseys as is clearly manifested in the yellow tint of the milk fat of the latter breeds.

Pigmentation of egg yolk and poultry carcasses are subjects of marked consumer preference and, as such, sources of xanthophylls such as lucerne meal or synthetic carotenoids are often added to poultry diets.

The efficiency of utilization of added synthetic carotene is dependent on the size of the particle formed in the gastrointestinal tract; the smaller the particle the higher the uptake. Aqueous colloidal dispersion, containing surface-active agents, may give better results than oily suspensions.

Efficiency of conversion of carotene to vitamin A decreases with increasing intake of carotene or of vitamin A. The carotene content of cattle tissues, blood and milk and of egg yolk is lowered with increased dietary vitamin A. It has been suggested that the carotene and vitamin A are competitive inhibitors of absorption (Chapter 3).

Utilization of Carotene by Ruminants

The poor utilization of carotene by ruminants has been attributed by some authors to degradation during passage through the rumen. This, however, has not been confirmed. In experiments where labelled carotene was supplied orally to sheep and cattle, faecal recovery of over 90% of the intake was observed. This limited uptake is probably due to limited solubilization of carotene in the small intestine because the composition of the lipids present in the intestine of ruminants differs from the respective lipids in monogastric species (see Chapter 6). Therefore the limited micellar solubilization and absorption of carotene in ruminants may be attributed to the presence of high levels of free fatty acids and little glycerides in the ruminant's small intestine which would reduce the micellar solubilization of β-carotene. (Fernandez-Cohen *et al.*, 1976).

The amount of carotene in green crops is generally sufficient to meet the vitamin A requirements of ruminants, despite the low utilization of carotene. However, when pastures fail during prolonged drought or when ruminants are stall-fed poor quality roughages and/or concentrate feeds, their diets are supplemented with stabilized vitamin A concentrates or retinol is administered to high-producing cattle by massive intramuscular injections of vitamin A emulsified in aqueous dispersions. Dry stabilized retinol is also used as a feed additive to milk substitutes for calves and lambs (Chapter 18).

Metabolism of Vitamin A

Intestinal Absorption and Liver Storage of Vitamin A

Dietary vitamin A from animal or synthetic sources is usually in the form of retinyl esters, the palmitate being the most abundant ester. These esters are hydrolysed in the intestinal lumen by pancreatic hydrolases and by that in the brush border, and retinol is taken up from the micellar phase into the intestinal mucosal cells (see Chapter 6). Bile acids are essential for this process, as well as for solubilization and uptake of carotene. In the mucosal cell β-carotene is converted to retinol by soluble enzymes. Retinol

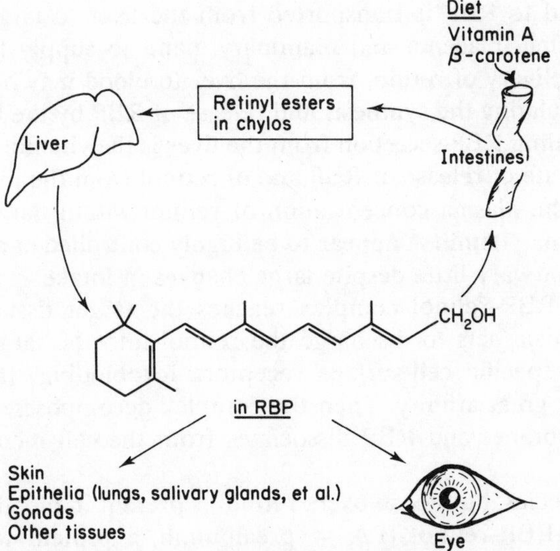

Figure 11.4. Processes involved in absorption, storage and transport of retinol in the body (according to Goodman, 1979). *Fed. Proc.*, **38**, 2502. *Reproduced by permission of the Federation of American Societies for Experimental Biology.*

originating from preformed retinol or from carotene is esterified (inside the mucosal cell) mainly with palmitate. The retinyl esters are then incorporated into chylomicrons which in birds enter the blood circulation directly and in mammals via the lymphatic system. Retinyl esters remain associated with the chylomicrons and are taken up by the liver. In the liver hydrolysis of the retinyl esters occurs during uptake, and vitamin A is stored following re-esterification within fat-storing cells in the liver. The liver contains the major vitamin A stores of the animal; adult animals generally have some 50–300 μg/g liver. The amount of vitamin A stored in the liver is influenced, in addition to intake, by sex, age, rate of production and health. A high rate of productivity increases the requirement for vitamin A and leaves less for storage, as, for instance, in lactating cows. The liver can store enough vitamin A to protect the animal during long periods of vitamin A scarcity. Ruminants on good pasture can store reserves to cover their needs during seasons when their diets may be deficient. The depletion rate is less in older animals than in younger ones.

Plasma Transport of Vitamin A

On its release from the liver, retinyl palmitate is hydrolysed to retinol which is transported in the plasma bound to a specific transport protein called retinol-binding protein (RBP). This binding of retinol to RBP renders the retinol stable in aqueous solution towards chemical and enzymatic attack. Retinol bound to RBP is transported from the liver to target tissues such as eye, intestine, placenta and mammary gland to supply their metabolic needs. The delivery of retinol from the liver to blood may be controlled by processes regulating the synthesis and release of RBP by the liver. One such factor controlling RBP secretion from the liver is the vitamin A status of the animal. The rate of release of RBP and of retinol from the liver are adapted to maintain the plasma concentration of retinol within narrow limits. The levels of plasma vitamin A appear to be highly controlled at about 30–40 μg per 100 ml and vary little despite large changes in intake.

When the RBP–retinol complex reaches the target tissues (see above) some mechanism acts to discharge the retinol into the target cells. These cells possess specific cell-surface receptors for binding the RBP–retinol complex with great affinity. Then the complex decomposes, retinol crosses the cell membranes and RBP dissociates from the cell membranes and is degraded.

In many species, but not in oxen, retinol is present in blood as the ternary complex PA–RBP–retinol (PA = prealbumin, a protein with a molecular weight 40 000–50 000; RBP has a molecular weight of 21 000). Retinol is even further stabilized by the formation of this large complex. Retinol when released from this ternary complex is well adapted to uptake by target tissues.

Intracellular Metabolism of Vitamin A

Within the cells vitamin A is always associated with specific cellular proteins, enzymes and lipids. Vitamin A in such a complex is the substrate for enzymatic transformation. As mentioned, retinol is readily esterified and retinyl esters are hydrolysed in most cells.

Enzymatic Oxidation of the Alcohol Group in Retinol and Fate of the Oxidation Products. Retinol may also be reversibly oxidized to retinaldehyde (retinal) and then irreversibly to retinoic acid (see Figure 11.1). Retinoic acid supports the biological activities of vitamin A with the exception of vision and reproduction. Both retinol and retinoic acid are conjugated in the liver to water-soluble β-glucuronides which are excreted through the bile and hence via the faeces. Some oxidized chain-shortened derivatives of retinoic acid are excreted in the urine. Of the absorbed vitamin A the part not stored in the liver is rapidly metabolized and excreted in faeces and urine.

Function of Retinol in Glycoprotein Synthesis. A further series of reactions in which retinol appears to be involved is the attachment of sugars to membrane glycoproteins, thereby the biosynthesis of these compounds will be completed. The first step of the biochemical pathway established by G. Wolf (see Figure 11.5) is the phosphorylation of retinol. Then the retinyl phosphate (R-P) reacts with a nucleotide to form glycosylretinyl phosphate. The last step is the transfer of a monosugar to the growing oligosaccharide chain of glycoproteins. Glycoproteins of cell surfaces were found to be defective in vitamin-A-deficient animals as a result of underglycosylation. Alterations in the normal function of epithelium caused by vitamin A deficiency will be described later.

Other intracellular processes in which retinol is involved are RNA—and consequently protein—biosynthesis; this may explain some of the cellular changes observed in vitamin A deficiency.

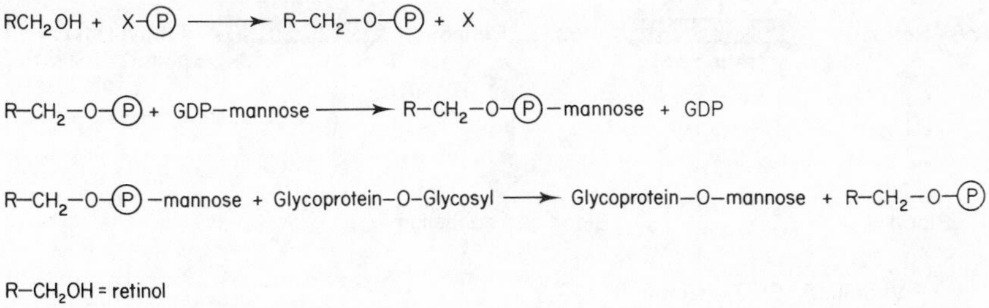

$$RCH_2OH + X{-}Ⓟ \longrightarrow R{-}CH_2{-}O{-}Ⓟ + X$$

$$R{-}CH_2{-}O{-}Ⓟ + GDP{-}mannose \longrightarrow R{-}CH_2{-}O{-}Ⓟ{-}mannose + GDP$$

$$R{-}CH_2{-}O{-}Ⓟ{-}mannose + Glycoprotein{-}O{-}Glycosyl \longrightarrow Glycoprotein{-}O{-}mannose + R{-}CH_2{-}O{-}Ⓟ$$

$R{-}CH_2OH$ = retinol

Figure 11.5. Biosythesis of membrane glycoproteins (according to G. Wolf, Vitamin A. In: *Human Nutrition, (R. B. Alfin-Slater and D. Kritchevsky, eds.)* 1980. Vol. 3A, p. 160. *Reproduced by permission of Plenum-Press.* 1980). Formation of retinyl phosphate and mannosylretinyl phosphate, and transfer of mannose to glycoprotein.

Physiological Functions of Vitamin A and Deficiency Effects

Vitamin A and Vision

Retinol functions in the eye in the transmission of light stimuli to the brain. The reactions involved is visual cycles were first elegantly described by G. Wald, and are shown in Figure 11.6. 11-*cis*-Retinal is combined with the protein opsin in both rods and cones in the light-sensitive pigments (rhodopsin or iodopsin) of the retina. This pigment is then bleached by light isomerizing the *cis*-retinal to *trans*-retinal which is bound less strongly to the visual pigment, and *trans*-retinal is accordingly released. The isomerization of *cis*-retinal triggers a nerve impulse, and the energy derived from this reaction is transferred to the brain via the optic nerve and the seeing of colour is registered. In the dark, the pigment is regenerated; this requires that *trans*-retinal is isomerized to the *cis* form, which occurs following reduction to retinol. Some of the vitamin A is degraded in the overall process to retinoic acid. Vitamin A is mainly stored in the pigment epithelium of the retina and inadequate intake of vitamin A is usually first manifested in defective dark adaptation and night-blindness due to insufficient vitamin A present for rhodopsin regeneration. Prolonged lack of rhodopsin may result

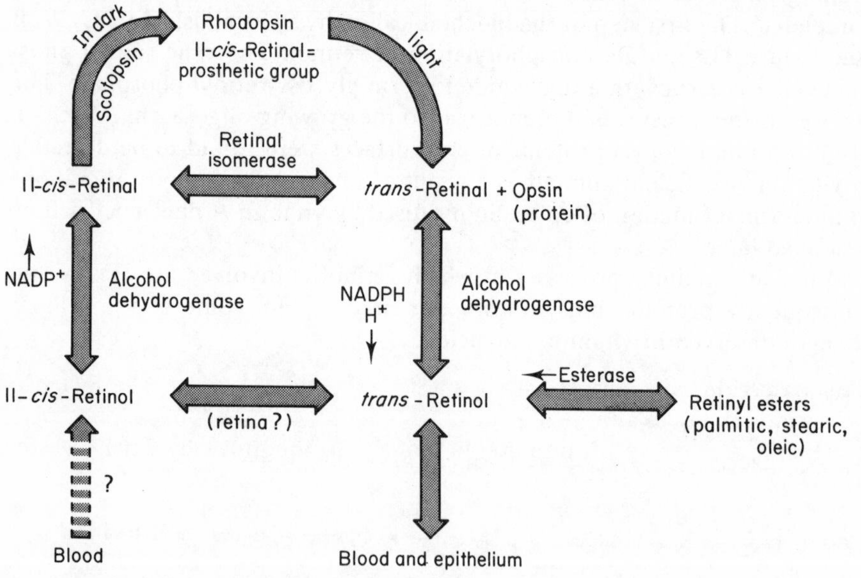

Figure 11.6. Vitamin A and vision—the visual cycle in the eye (adapted from Wald, 1968); after Scott *et al. Nutrition of the Chicken*, 2nd. ed. 1976 p. 132) *Reproduced by permission of Scott & Associates.*

in degeneration of the retina and ultimately in irreversible blindness. In addition, vitamin A deficiency also results in other defects in vision which will be discussed later.

Maintenance of the Integrity of Normal Epithelial Tissue

The involvement of retinal in glycoprotein biosynthesis has been pointed out previously. Glycoproteins of cell surfaces were found to be defective in vitamin-A-deficient animals as a result of underglycosylation. Thus, in vitamin A deficiency epithelia cells form an extensive keratinized layer, the normal mucus-secreting epithelium becomes replaced by a squamous keratinizing tissue. Vitamin A is required for the maintenance of the epithelium lining of all those canals and cavities of the body which communicate with the external air, such as the alimentary, respiratory and the genito-urinary tract, as well as the corneal epithelium and the soft tissues around the eyes. The stratified squamous epithelium formed in vitamin-A-deficient animals blocks the ducts of the salivary, tear and intestinal mucus glands causing them to be distended with secretions and the glands may cease to function.

Loss of appetite in vitamin A deficiency may be connected with changes in the epithelial cells. Infiltration of keratin into the pores of the taste buds and creamy white pustules were observed in the roof of the mouth and in the oesophagus in vitamin-A-deficient hens (see PLATE VA). A watery discharge from the nostrils and eyes results from vitamin A deficiency in chicks and calves, and the eyelids are often stuck together. As the deficiency continues, the eyes become so filled with this white exudate that it is impossible for the animal to see. Keratinization of corneal epithelium occurring in vitamin A deficiency is an important cause of vision failure which is termed xerophthalmia.

The breakdown of the mucous membranes and their keratinization lower the resistance of animals to the entrance of infective microorganisms into the tissues. Therefore, when affected by lack of vitamin A, cattle become more susceptible to cold and pneumonia and birds to coccidosis. Vitamin A supplement improves the performance of infected chicks. The increased requirement of vitamin A by birds with parasite infections is probably a reflection of poor absorption due to damage to the mucosa of the digestive tract.

Bone Growth and Cerebrospinal Fluid Pressure

Vitamin A deficiency causes retardation of bone growth. In vitamin A deficiency, the bones are altered in shape, becoming cancellous, shorter and thickened. The failure in bone development is probably associated with the influence of vitamin A on the biosynthesis of mucopolysaccharides (glycoproteins) which are major constituents of organic matter in bones. Thus, vitamin A enables the proper growth of the cartilage matrix upon which bone is deposited. A failure of the spine and some other bones to develop—due to vitamin A deficiency—results in mechanical pressures on the nerves. Bone changes constricting the optic nerve are responsible for permanent blindness in cattle. This is another possible involvement of vitamin A in the visual process.

Closely related to bone changes is the rise of cerebrospinal fluid (CSF) pressure which occurs in vitamin A deficiency. The rise in CSF pressure is caused by impaired CSF absorption, possibly due to thickening of the connective tissue matrix of the cerebral dura mater. The increase in CSF pressure can be used as a measure of vitamin A deficiency in calves. It is assumed that high CSF pressure is responsible for the staggering gait and for convulsions in advanced vitamin A deficiency.

Vitamin A Deficiency and Growth

As has been previously described, vitamin A deficiency affects growth of bones and epithelium as well. Thus it is not surprising that in vitamin A deficiency growth is retarded. Weanling rats on a vitamin-deficient diet grow until the initial liver stores of vitamin A are exhausted, attaining at this point a 'weight plateau'. After a short time at constant weight, a decline commences ending in death. This has been in part attributed to depression of appetite, previously described. Depression of food consumption has been considered to be an early sign of vitamin A deficiency and the adjustment of a 'weight plateau' in rats has been used as a confirmation of deficiency. Supplements of vitamin A result in renewal of growth. Bieri *et al.* (1968) noted that germ-free rats reached a weight plateau later, and survived longer than conventional ones (272 instead of 54 days). These experiments clearly show that the early death of conventional rats is a consequence of bacterial infection and that the growth of rats is directly dependent on the supply of vitamin A. Influence of growth rate of farm animals on vitamin A requirement will be dealt in Chapter 18.

Vitamin A and Reproduction, Fetal and Neonatal Vitamin A Metabolism

The reproductive apparatus in both males and females is sensitive to vitamin A deficiency, and this can be explained partly by changes in the epithelia involved. In males, vitamin A deficiency leads to degeneration of germinal epithelium and cessation of spermatogenesis. In the female vitamin A is required for maintenance of vaginal epithelium, and deficiency in pregnancy results in resorption of the placenta or spontaneous abortions. Under vitamin-A-deficient conditions malformed fetuses may result. However, in general, vitamin A transfer from the mother to the fetus through the placenta is highly regulated, with minimal changes in fetal vitamin A status occurring in response to large variations in maternal intake. Generally the neonate has low stores of vitamin A before suckling, which increase following ingestion of colostrum and milk. The relatively high vitamin A requirements of the mammal neonate is generally met by the high concentration of vitamin A in colostrum (see Chapter 18).

In contrast, in birds the reserves of vitamin A available to the newly hatched chick from the yolk are sufficient to prevent the onset of deficiency for several weeks when a vitamin-A-deficient diet is fed to the young chick. The levels of vitamin A in the yolk can be increased by feeding higher dietary levels of vitamin A to the hens (see Chapters 18 and 20).

It may be noted in the female plasma RBP levels rise with oestrogen levels during the breeding season of sheep and birds. This may reflect an increased demand for vitamin A for reproduction, or in birds for transfer to the egg (Glover *et al.*, 1980; 1983).

Dietary Factors Affecting Vitamin A

Protein. Adequate dietary protein is required in order to synthesize the proteins connected with both the absorption and transport of carotenes and retinol (such as pancreatic retinyl hydrolase, RBP or carotene 15,15′-dioxygenase). On the other hand, the vitamin A requirement increases with the supply of diets high in protein or energy which accelerate the growth rate. There are clear indications that maintenance of liver stores of vitamin A in calves given diets consisting mainly of barley requires about 50% more vitamin A than calves receiving mainly roughage.

Fat. Small amounts of fat as usually contained in animal diets are necessary for micellar absorption of vitamin A or carotene from the intestinal lumen.

Zinc. A linkage between vitamin A and zinc metabolism has become apparent from recent research. Zinc may limit the synthesis of RBP because of its general role in protein synthesis (see Chapter 10). Zinc-deficient animals have low circulating RBP and retinol levels. This defect is due to limited RBP–retinol release from the liver. An additional area of zinc–vitamin A interaction is the oxidation–reduction of retinol in eye and other peripheral tissues. The conversion of retinol to retinal involves catalysis by a zinc metalloenzyme. This explains the abnormal dark adaptation and night blindness found in zinc deficiency.

Vitamin E. Vitamin E and other antioxidants present in the diet may protect the sensitive conjugated double bond system of vitamin A from oxidation during storage of feed mixtures, and also in the intestinal lumen and inside tissues. Thus, vitamin E exerts an apparent sparing effect on vitamin A. Converse action of surplus vitamin A enhancing the appearance of vitamin E deficiency will be dealt with in the section on vitamin E.

Mineral Oils. The presence of poorly absorbed or non-absorbed hydrocarbons such as mineral oil depresses vitamin A and carotene absorption. This is due to the partitioning of vitamin A into the hydrophobic phase, thus reducing its availability for absorption. This is relevant to the use of paraffin or mineral oils in the controlling of bloat in grazing cattle.

Chlorinated Naphthalenes. A disease of obscure aetiology termed X-disease was common among cattle in the period 1940–1962. The symptoms closely paralleled vitamin A deficiency, including loss of appetite, thickening of the skin, profuse lacrimation and salivation and sterility. Outbreaks of the disease were traced to contamination of the feed with mineral oil from lubrication in machinery. Chlorinated naphthalenes added to lubricating oils to enhance the mechanical properties appeared to reduce the conversion of carotene to vitamin A in the intestine. The addition of chlorinated naphthalenes for agricultural purposes is now illegal in many countries.

Effect of Excess Vitamin A (Hypervitaminosis A)

The organism has a good mechanism for protecting itself against the membranolytic action of such highly surface-active compounds as retinol or its esters; it binds the vitamin by specific retinol-binding proteins in plasma, on the cell surface and intracellularly. The toxic range is reached when daily intakes of vitamin A approach 50–500 times the requirement. Then liver storage and serum transport capacities are exceeded and symptoms of hypervitaminosis appear. These include hepatic disorders culminating in fatty liver. Bone fragility also increases due to resorption of bone matrix and cartilage, and bone length and thickness decreases. As in bone formation, hypervitaminosis A causes the inverse of what happens to CSF pressure in vitamin A deficiency. This declines sharply as a result of increases in membrane permeability.

VITAMIN E *

Vitamin E was discovered in the early twenties in vegetable oils such as wheat germ oil as a factor required for reproduction in rats. It was given the name tocopherol (*tokos* = childbirth, *phero* = to bear, *ol* = an alcohol or phenol). The functions of vitamin E are not restricted to those of reproduction. The vitamin is present in minute amounts in every cell. The most widely accepted biochemical function for this vitamin is its role as an antioxidant. It is nature's best fat-soluble 'biological' antioxidant and functions in protecting cell membranes as well as other nutrients, such as polyunsaturated fatty acids (see Chapter 6) or vitamin A, from destructive oxidation. Thus vitamin E deficiency results in certain pathological conditions to be described below.

* In collaboration with P. Budowski.

Structure and Properties of Vitamin E

Vitamin E includes two groups, tocopherols and tocotrienols, which are all produced in green plants. The term 'vitamin E' refers to any mixture of physiologically active tocopherols. The structure of α-tocopherol, the most widely distributed and biologically the most active compound, is presented in Figure 11.7. The four forms of tocopherols, designated α-, β-, γ- and δ-tocopherols, differ only in the number of methyl groups and their position around the benzene ring (see Table 11.3). The difference between tocopherols and tocotrienols is due to the unsaturation of the side-chain in the tocotrienols. Only the tocopherols are of major physiological importance. α-tocopherol is the only one present in the animal body and in food of animal origin.

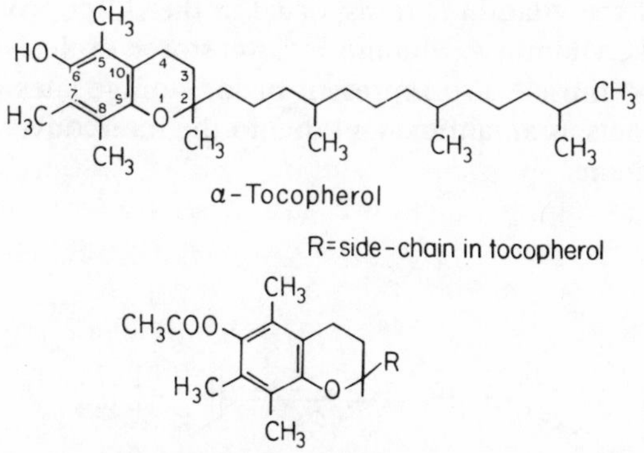

α-Tocopherol

R=side-chain in tocopherol

α-Tocopheryl-acetate

Figure 11.7. Formulae of α-tocopherol and α-tocopheryl acetate.

Table 11.3 Chemical structure and biological activity of various tocopherols

	Structure	Biological activity*
α-Tocopherol	5,7,8-Trimethyltocol†	100
β-Tocopherol	5,8-Dimethyltocol	15–40
γ-Tocopherol	7,8-Dimethyltocol	8–20
δ-Tocopherol	8-Methyltocol	0.3–0.7

* Relative values based on antioxidant and antisterility activities.

† Tocopherol with an unsubstituted phenol ring is designated tocol.

The tocopherols are very resistant to heat, but are readily oxidized, for example by rancid fats or in presence of minerals supplied with the diet. Since esterification improves the stability of tocopherol, it is usually provided commercially as the acetate ester (see Figure 11.7). One IU of vitamin E is equivalent to 1 mg of synthetic *d,l*-α-tocopherol acetate. The *d*-isomer is more active than the *l*-form; *d*-α-tocopheryl acetate possess an activity of 1.36 IU per mg, while that of *d*-α-tocopherol is 1.49 IU per mg and that of *d,l*-α-tocopherol 1.1 IU, The figures also show that the free form is more active than the acetate.

Metabolism of Vitamin E

Absorption of this vitamin occurs, like that of vitamin A and D, through the micelles formed in the intestine from the dietary fat only to the extent of 20–40%. Most of the vitamin is transported in the β-lipoprotein fraction of the plasma. Unlike vitamin A, vitamin E is not stored exclusively in the liver; adipose tissue and muscle also represent major storage sites (see above).

α-Tocopherol acts as an antioxidant due to the interconversions shown in the following scheme:

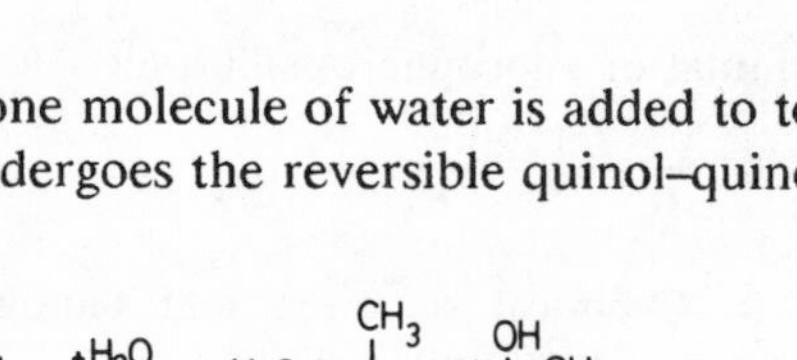

$$+ 2H^+ \underset{-\,2e}{\overset{+\,2e}{\rightleftharpoons}}$$

In the first stage one molecule of water is added to tocopherol and then the quinol formed undergoes the reversible quinol–quinone conversion:

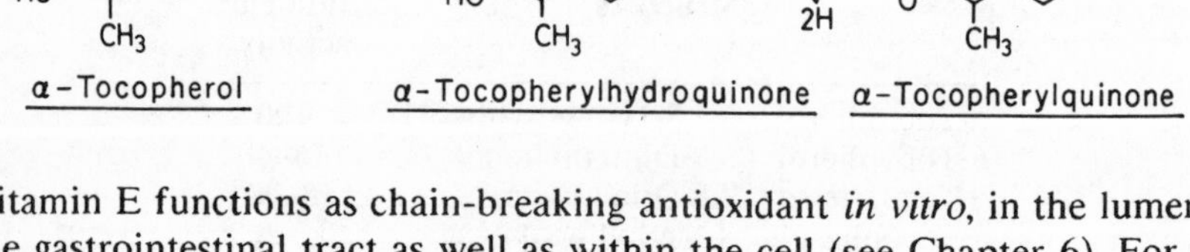

α-Tocopherol α-Tocopherylhydroquinone α-Tocopherylquinone

Vitamin E functions as chain-breaking antioxidant *in vitro*, in the lumen of the gastrointestinal tract as well as within the cell (see Chapter 6). For this purpose vitamin E (α-tocopherol) donates a phenolic hydrogen atom to a free radical formed from unsaturated fatty acid:

$$ROO^{\bullet} + \alpha\text{-tocopherol} \longrightarrow ROOH + \text{oxidized } \alpha\text{-tocopherol}$$

thus inactivating the unpaired electron of the radical, and at the same time undergoing conversion to the quinone form. Consequently vitamin A, carotene, vitamin C and polyunsaturated fatty acids, particularly those present in tissue lipids (mainly phospholipids), are protected from free-radical attack and oxidations. Since free-radical-catalysed lipid peroxidation damages cellular and ultracellular structures and changes the permeability properties of the cell membranes, this process is thought to occur in vitamin E deficiency and account for the resulting diverse pathological effects of vitamin E deficiency.

In the event that insufficient vitamin E is available to trap radicals and prevent peroxide formation, the selenium-containing enzyme glutathione peroxidase plays an additional protective role. As outlined in Chapter 10, glutathione peroxidase metabolizes polyunsaturated fatty acid peroxides, as well as hydrogen peroxide; vitamin E and selenium cooperate in the protection of subcellular membranes against damage caused by peroxide. The vitamin can be regarded as the first line of defence in preventing oxidation of polyunsaturated fatty acids, the selenium-containing enzyme acting as the second line of defence in destroying peroxides which are formed (because of insufficient action of vitamin E) before they can damage membranes.

Vitamin E appears to reduce the selenium requirement of animals by preventing the chain-reactive autoxidation of lipids, particularly of those within the membrane, thereby inhibiting the production of peroxides; this reduces the amount of selenium-containing glutathione peroxidase needed to destroy the peroxides formed in the cell.

Vitamin E Deficiency Diseases of Poultry

Chickens show three distinct vitamin E deficiency diseases: encephalomalacia, exudative diathesis, and nutritional muscular dystrophy. It is of great interest that these diseases are prevented, or cured, not only by vitamin E but also by other specific compounds—certain synthetic antioxidants, selenium and sulphur-containing amino acids—different for each disease. Therefore, some conclusions may be drawn concerning the mechanism responsible for each of these diseases.

Encephalomalacia. This is also called 'crazy chick disease', and is a nervous syndrome characterized by ataxia, i.e. lack of coordination, involuntary movements of the legs and head twisting (see PLATE VB). The cerebellum is softened and swollen because of haemorrhages and oedema (PLATE VIA). The disease occurs in young chickens fed a vitamin-E-deficient diet containing certain polyunsaturated fatty acids such as linoleic acid. The incidence and severity of this disease are markedly increased with the degree of peroxidation of the fats in the diet. This disease responds to vitamin E or certain fat-soluble antioxidants, but not to dietary selenium or sulphur-containing amino acids. In the prevention of encephalomalacia, there is no doubt that vitamin E functions as an antioxidant, and the disease can be prevented by synthetic antioxidants, provided these are absorbed and reach the cell membranes, where they prevent oxidation of polyunsaturated fatty acids.

Exudative Diathesis. Exudative diathesis of chicks fed diets deficient in selenium and vitamin E has already been discussed in the section on selenium (Chapter 10). This disease can be largely prevented by selenium; vitamin E can exert a sparing action, though it cannot fully replace selenium as mentioned previously (Chapter 10). Exudative diathesis is not effected by the levels of antioxidants or cystine in the diet.

Nutritional Muscular Dystrophy. On diets low in vitamin E, sulphur-containing amino acids and sometimes selenium, and high in polyunsaturated fatty acids, some species of animals develop muscular dystrophy as outlined in Chapter 10. This condition in chicks is characterized by degeneration of muscle fibres (myopathy), particularly those of the breast and legs. The amino acid cysteine has a primary role in the prevention of muscular dystrophy in chicks, as the incidence of this disease is increased by dietary factors that deplete body cysteine and decreased by factors that spare it. Similar signs of muscular dystrophy occur in vitamin-E-deficient ducks and turkeys. In turkeys it is characterized by lesions in the muscular wall of the gizzard. The disease is prevented in chicks by supplement of either vitamin E or cysteine, in ducks only by vitamin E, while in turkeys selenium is the primary factor responsible for the prevention of muscular dystrophy.

Other Symptoms of Vitamin E Deficiency in Poultry and Other Farm Animals

Combined deficiency diseases of vitamin E and selenium are observed on farm animals and are discussed in the section on selenium (Chapter 10).

Ceroid Pigmentation. The yellow–brown coloration in adipose tissue or liver that originates from lipid oxidation *in vivo* is another characteristic symptom of vitamin E deficiency.

Erythrocyte Haemolysis. The structural integrity of the membranes is deranged by attack of radicals produced in vitamin-E-deficient animals and thus the erythrocytes from vitamin-E-depleted animals may be more susceptible to haemolysis by hydrogen peroxide. This increased tendency of erythrocytes to haemolyse serves as a sensitive measure of vitamin E deficiency in several species.

Vitamin E and Reproduction. As mentioned above, the first recognized symptom of vitamin E deficiency was reproductive failure in female rats by death and subsequent resorption of the fetus. Testicular degeneration occurs with vitamin-E-deficient diets in male rats and cocks. However, in humans and ruminants, sterility has not been shown to be linked with a deficiency of vitamin E, possibly because of practical difficulties in carrying out long-term depletion experiments.

While practical rations for chickens contain enough vitamin E for normal egg laying and hatchability, breeding turkeys fed practical diets require additional vitamin E for maximum hatchability (see Chapter 20).

In ruminants, serum and tissue levels are low in newborn calves because of inefficient placental transfer, but colostrum and milk provide considerable amounts of vitamin E (see Chapter 21). Since the quantities of vitamin E present in the tissues of the newborn calf and in the mother's milk are influenced by the mother's diet, pregnant cows should be given diets containing adequate amounts of vitamin E.

Sources of Vitamin E and its Supply to Farm Animals

The amounts of total tocopherols and of α-tocopherol, the most active tocopherol, in various feedstuffs are given in Table 11.4. α-Tocopherol is practically the only tocopherol present in green plants and in animal products. Its content is quite high in green plants, but low in feedstuffs of animal origin. The germs of cereals, particularly the germ oil as well as other plant oils are very rich sources of tocopherols, but in some of these oils the ratio of α-tocopherol to other less active tocopherols is quite low. Under most conditions the common feedstuffs appear to supply adequate amounts of tocopherols to most species of farm animals. In order to ensure the required supply of this vitamin to growing and breeding birds, α-tocopheryl acetate is regularly added with vitamin concentrates designed for these animals. The addition of tocopherol to the diets of broilers and poults also improves the

Table 11.4 Approximate contents of total tocopherols and α-tocopherol in various feedstuffs (mg tocopherol in 1 kg dry matter)

Feedstuff	Total tocopherol	α-Tocopherol
Various green forages	200–400	220–400
Lucerne meal	190–250	180–240
Wheat grain	30–35	15–18
Maize grain	4–10	0.5–3.0
Soybean meal	3–6	1
Fish meal	21	21
Wheat germ oil	1700–5000	800–1200
Safflower oil	500	350
Soybean oil	1000	100
Soybean soapstock	1000	100

palatability of the meat, as the tocopherol deposited in the tissues inhibits the oxidation of unsaturated fatty acids (see Chapter 18). Likewise, vitamin E reduces the incidence of oxidized flavour in milk (see Chapter 21).

Since vitamin E is rapidly destroyed by oxidized fat, fats high in polyunsaturated fatty acids increase the requirement for this vitamin. The dependence of vitamin E requirement on the polyunsaturated fatty acid supply has been shown in experimental animals and in humans, but seems to be present in farm animals as well. For example, the daily requirement for tocopherol, which is 5 mg/kg dietary DM for calves raised on milk, may be increased to 10–50 mg by inclusion of plant oils in milk substitutes (see Chapter 18).

VITAMIN D

Vitamin D is required for the regulation of calcium and phosphorus metabolism and consequently for calcification of growing bones. There is no need for a dietary source where animals are exposed to sunlight for at least a short time during the day since ultraviolet light converts the provitamin present in the skin to vitamin D. Nevertheless, this vitamin is inactive per se and its biological activity is primarily due to one of its metabolites. This metabolite is considered as the hormonal form of vitamin D since it is produced in one organ, the kidney, and elicits effects in other tissues (intestine and bones) and its production is feedback regulated. Vitamin D is unique among the vitamins in as much it is the only vitamin known to be a precursor of a hormone and its action is truly hormonal.

Chemistry of Vitamin D

About 25 biological forms of this vitamin are known which all are steroid derivatives, but only two—vitamin D_2 and vitamin D_3—are important in animal nutrition. (The term vitamin D_1 was originally suggested by earlier workers for an active compound, but this was later found to consist mainly of vitamin D_2 and some impurities. Therefore the name vitamin D_1 has been abolished.) The structures of vitamins D_2 and D_3, which are shown in Figure 11.8, have the same steroid nucleus and differ only by the side-chain attached to C-17. Vitamins D_2 and D_3 are formed by ultraviolet irradiation of the steroids ergosterol and 7-dehydrocholesterol, respectively. Ergosterol is produced in plants, also in moulds and yeasts, and 7-dehydrocholesterol in animal tissues. The precursors have no biological activity until the B-ring is opened between the 9- and 10-positions of the steroids. Exposure of animals to sunlight for a short time during the day is sufficient to convert provitamin D_3 present in the skin to vitamin D_3 which eliminates the need for a dietary source. Vitamin D_2 is almost as effective as vitamin D_3 for mammals, but vitamin D_2 is virtually inactive in birds. Thus poultry feeds must be fortified with vitamin D of animal origin or synthetic vitamin D_3.

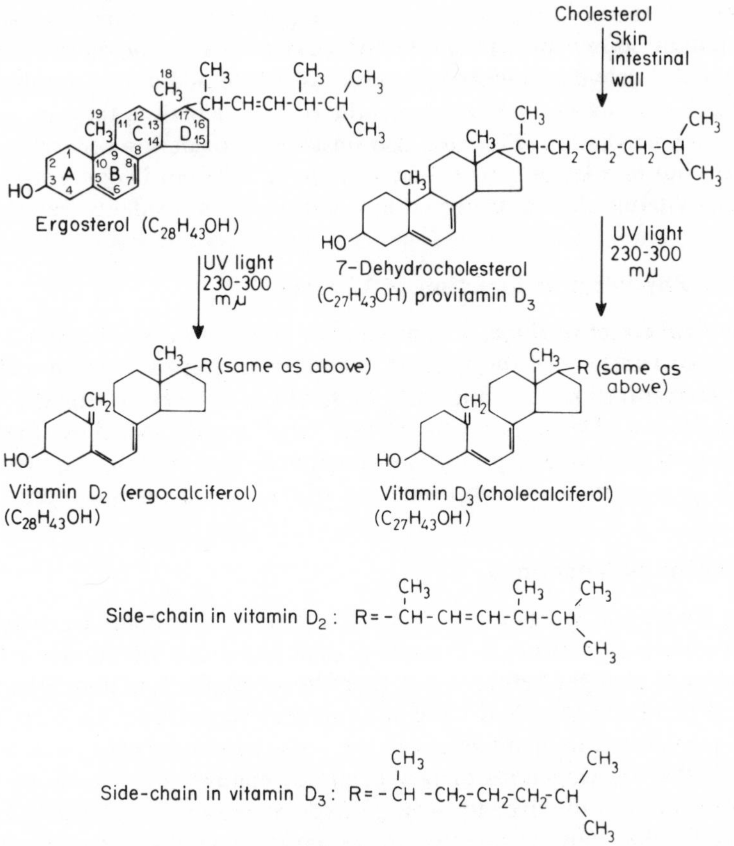

Side-chain in vitamin D_2: $R = -CH(CH_3)-CH=CH-CH(CH_3)-CH(CH_3)_2$

Side-chain in vitamin D_3: $R = -CH(CH_3)-CH_2-CH_2-CH_2-CH(CH_3)_2$

Figure 11.8. Transformation of provitamin D to vitamin D by ultraviolet irradiation.

Sources of Vitamin D

Vitamin D requirements are often expressed in terms of units. One IU of vitamin D is defined as the activity of 0.025 μg of vitamin D_3 for all kinds of animals and of vitamin D_2 for mammals only. Provitamins D_2 and D_3 are both widely distributed in nature, but the distribution of vitamin D_2 and D_3 in foods is very limited. The best sources of vitamin D_3 are fish liver and fish liver oils (10 000–4 000 000 IU in 100 g of fish oil according to the species of fish). Among animal products, milk and egg yolks contain the vitamin, but the amount present depends on the season in which they are produced and on the amount of vitamin supplied in the diet of cows or hens (average of 260 IU in 100 g of egg and 1–4 IU in 100 g of milk). It is possible to increase the vitamin content of milk by fortification.

Ergosterol, provitamin D_2, occurs commonly in plants and is transformed into vitamin D_2 during sun-drying of harvested forages; 1 kg of hay contains between 800 and 1700 IU. Yeast is rich in ergosterol and its irradiation results in a potent source of vitamin D for human beings and most mammals. Irradiated 7-dehydrocholesterol is used as a feed additive to poultry diets. Although vitamin D_2 and D_3 are produced by irradiation, they are unstable upon over-irradiation. They are also unstable to oxidation occurring during storage, but to a lesser extent than vitamin A. Vitamin D concentrates, like those of vitamin A, are protected by addition of antioxidants (see above).

Overall Physiological Function of Vitamin D

The final effect of the action of vitamin D is the mineralization of bone. The major function of the vitamin is to elevate plasma calcium and phosphate concentrations by supersaturation which is necessary for the normal mineralization of bone (see Chapter 8). This is accomplished by stimulation of intestinal calcium and phosphate absorption, their mobilization from previously formed bone and improvement of their renal reabsorption.

Metabolism of Vitamin D

The absorption of vitamin D occurs like that of other lipids, by micellar solution from the intestine. It is now well established that vitamin D undergoes biochemical changes before it can exert its metabolic functions (see Figure 11.9). Vitamin D absorbed from the intestine passes into the bloodstream and is transferred to many tissues of the body, particularly the liver. Vitamin D_3 (whether of exogenous dietary origin or endogenous cutaneous origin) is metabolized in the liver to form 25-hydroxyvitamin D_3 [25(OH)D_3]. The 25(OH)D_3 does not act directly on the target tissues, but must be further modified; it is transported to the kidney where it is hydroxylated, enzymatically, on C-1 to produce 1,25-dehydroxyvitamin D_3 [1,25$(OH)_2D_3$]. The latter compound is the physiologically active form of vitamin D.

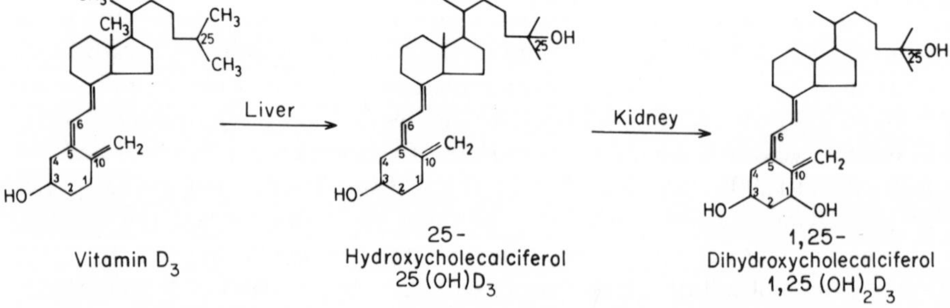

Figure 11.9. Metabolic transformation of vitamin D.

$1,25(OH)_2D_3$ is considered to be the hormonal form of vitamin D and meets the usual criteria of a hormone; i.e. it is produced on one site (in the kidney) and causes a response elsewhere (in intestine and bones as target tissues) and its production is feedback regulated (see Figure 10.2).

In the liver $25(OH)D_3$ limits its own formation by inhibition of the 25-hydroxylase. This control restricts the excessive formation of active vitamin D derivatives when the vitamin is present in slight surplus over the required amount. Apart from this, the controlled conversion of vitamin D_3 conserves the vitamin D for further use because vitamin D and not its active derivatives can be stored in the body for as long as several months. (Vitamin D may be stored in all lipid-rich tissues of the body, and contrary to vitamin A not primarily in the liver.)

Whereas the 25-hydroxylation is not regulated by serum calcium and phosphorus levels, the rate of formation of the active metabolite $1,25(OH)_2D_3$ is related to the calcium and phosphorus needs of the animal and is feedback regulated by serum concentrations of calcium and phosphate. In order to increase the calcium level of plasma more $1,25(OH)_2D_3$ is needed. High calcium levels decrease the production of $1,25(OH)_2D_3$ and low calcium levels stimulate it. Moreover, the lack of calcium causes the secretion of parathyroid hormone which in turn stimulates the enzymatic conversion of $25(OH)D_3$ to $1,25(OH)_2D_3$. In addition, parathyroid hormone enhances directly the absorption of calcium from the intestines and its utilization for bone formation.

The function of the active vitamin D metabolite in the regulation of phosphate metabolism has been pointed out in Chapter 10.

High serum $1,25(OH)_2D_3$ levels are found in egg-laying birds as calcium is being mobilized from medullary bone for the formation of eggshell (see Chapter 20).

As calcium levels in blood serum become normal or excessive, the formation of the active metabolite $1,25(OH)_2D_3$ is shut down. Under these circumstances a further regulatory action takes place in the kidney, the induction of 24-hydroxylation which leads to the formation of $24,25(OH)_2D_3$ from $25(OH)D_3$ and of $1,24,25(OH)_2D_3$ from $1,25(OH)_2D_3$. The 24-hydroxylation leads to inactivation of potent vitamin D metabolites.

The vitamin D metabolites, $25(OH)D_3$ and $1,25(OH)_2D_3$ are transported in the plasma bound to transport proteins which have a higher affinity for these compounds than for the respective vitamin D_2 derivatives, and this may explain the lower biological activity of D_2 even in mammals. Vitamin D_2 undergoes in mammals analogue transformations such as those of vitamin D_3 described above which have been observed in chicks and rats.

The major site of action of $1,25(OH)_2D_3$ is the intestine where the absorption of calcium is increased. The exact mechanism of this action is not entirely clear. It is believed that $1,25(OH)_2D_3$ binds to a specific cytoplasmic binding protein which translocates the active vitamin into the duodenal epithelial cell nucleus. $1,25(OH)_2D_3$ induces the formation of messenger-

RNA for the synthesis of another specific protein such as calcium-binding protein. This molecular mode of action of $1,25(OH)_2D_3$ resembles that of steroid hormones. Calcium-binding protein is concentrated in the brush border of the intestinal mucosa; the entry of calcium into the intestinal cell occurs across the brush border. Calcium-binding protein helps to maintain calcium in solution and there exists a high correlation between calcium-binding protein and $1,25(OH)_2D_3$; however, the actual participation of this protein in the calcium-transport process has not been established. Calcium-binding protein also occurs in other tissues associated with calcium movement, including kidney, brain and the shell gland of the laying hen (see Chapter 20).

Vitamin D Requirements and Symptoms of Deficiency

Different species require different forms and amounts of vitamin D as already indicated. Chicks have a much higher requirement per unit of body weight (100 IU/kg live weight) than pigs (25 IU/kg) and pigs a higher requirement than cattle (5 IU/kg). Turkeys have an even higher requirement than chicks. Part of the difference in species requirement may be related to growth rate, since animals with a very rapid growth rate, and consequently with a very rapid rate of bone formation, are more susceptible to symptoms of deficiency.

The need for supplementing the diet of ruminants with vitamin D is not so great as for pigs and poultry. Ruminants can receive adequate amounts of vitamin D from irradiation (see above) while grazing or from sun-cured hays. In animals that do not have direct access to sunlight, particularly confined farm birds, the endogenous vitamin D production does not meet the requirement and their diets must be supplemented with vitamin D concentrates. Supplementation of calf rations is indicated when there is minimum exposure to sunlight. The use of vitamin D and of its metabolites for milk fever control will be dealt with in Chapter 24. This disease and metabolic bone disorders respond to very small doses of synthetic $1,\alpha(OH)D_3$ and $1,25(OH)_2D_3$.

Vitamin D requirements of animals depend upon the calcium and phosphorus content of the diets. If the amounts of these minerals or the ratio between them in the diet (see Chapter 10) is suboptimal, the requirement for vitamin D increases. However, vitamin D will not compensate for severe deficiencies of calcium or phosphorus. The vitamin D requirement of monogastric species is also increased when the diet contains sources of poorly available phosphorus such as phytate (see Chapter 10).

The typical signs of vitamin D deficiency—rickets in young animals and osteomalacia in adults—caused by inadequate calcification of bone are described in Chapter 10. There is excess formation of demineralized bone matrix, due in large part to insufficient absorption of calcium and phosphate. Clinical symptoms differ somewhat in different species. In calves a swelling occurs in the metacarpal and metatarsal bones, bending of the forelegs and arching of the back (PLATE VIB). In pigs the symptoms are enlarged joints, broken bones and lameness. In chicks the joints become enlarged, and the beak becomes soft and rubbery and can be easily bent (PLATE VIIA). The most characteristic internal signs of vitamin D deficiency in chicks are a beading of the ribs at their junction with the spinal column, and crooked backbone (PLATE VIIB). In laying hens the same signs are observed and the eggs become thin-shelled, hatchability is reduced and finally egg production decreased.

Overdosage with Vitamin D

Administration of vitamin D for a few months in amounts exceeding at least ten times its requirements, leads to a toxicity termed hypervitaminosis D. Surplus of vitamin D induces symptoms the reverse of those observed in rickets. The vitamin ingested in large doses overcomes the control limiting the transformation of the vitamin into its active metabolites by hydroxylation (see above). As result bone resorption is increased, thus the bones become brittle and deformed. Calcium mobilized from the bones is deposited in soft tissues such as joints, lung arterioles and kidneys. The kidney damage is particularly serious.

Grazing animals (cattle and horses) develop symptoms of vitamin D toxicity from consuming some wild plants growing in several parts of the world. Such toxic pasture plants are *Cestrum diurnum* (wild jasmine) and *Solanum malacoxylon.* The ingestion of these plants induces harmful symptoms similar to those of hypervitaminosis D, such as deposition of calcium salts in soft tissues and bone troubles. These plants contain the active derivatives of vitamin D, $1,25(OH)_2D_3$, or very similar compounds as glycosides. The glycoside is cleaved within the gastrointestinal tract to sugar and the active compound $1,25(OH)_2D_3$ by the action of hydrolytic enzymes probably from bacterial origin. Comparatively small amounts of $1,25(OH)_2D_3$ absorbed from the gastrointestinal tract cause toxicosis since the feedback regulation of conversion of $25(OH)D_3$ to $1,25(OH)_2D_3$ by the kidney enzyme (see above) is by-passed when preformed $1,25(OH)_2D_3$ was ingested. Poisoning of animals by these plants has caused great economic losses.

VITAMIN K

In 1930, Henryk Dam of Denmark noted that chicks kept on an ether-extracted diet became anaemic and developed a syndrome characterized by subcutaneous and intramuscular haemorrhage and prolonged blood-clotting time. These symptoms were relieved by a factor present in the ether extract of the feedstuffs and the new substance was called vitamin K from the Danish word for coagulation.

Structure of Vitamin K

There are several different compounds, all derivatives of naphthoquinone, that have vitamin K activity (see Figure 11.10). Vitamin K_1 (phylloquinone) occurs in green plants and in seeds; vitamin K_2 (menaquinone) is a product of bacterial synthesis taking place in the rumen and lower intestine. The carbon atom at the 3-position in vitamin K_1 is substituted with a phythyl side-chain, which also occurs in vitamin E (see above) and in chlorophyll. Some K_2 compounds of microbial origin exist with side-chains containing 30–45 carbon atoms. Synthetic naphthoquinone derivatives without a side-chain at C_3 such as menadione and menadione-sodium bisulphite, water-soluble compounds, are even more active biologically than vitamins K_1 or K_2.

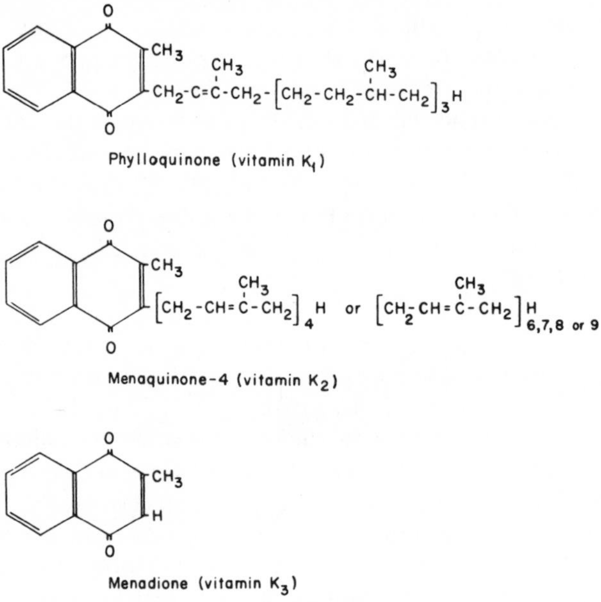

O, O, CH_3, $CH_2-C{=}C(CH_3)-CH_2-[CH_2-CH_2-CH(CH_3)-CH_2]_3H$

Phylloquinone (vitamin K_1)

O, O, CH_3, $[CH_2-CH{=}C(CH_3)-CH_2]_4H$ or $[CH_2-CH{=}C(CH_3)-CH_2]_{6,7,8 \text{ or } 9}H$

Menaquinone-4 (vitamin K_2)

O, O, CH_3, H

Menadione (vitamin K_3)

Figure 11.10. Structure of vitamin K.

Metabolic Functions of Vitamin K

Vitamin K is required for normal blood clotting. The clotting time is a good index of the vitamin K status. A normal clotting time of a few seconds may be extended to several minutes for blood from vitamin-K-deficient animals.

The major chemical defence against blood loss is the formation of the blood clot. The colourless protein mainly responsible for coagulation is fibrin which is formed form its soluble precursor, fibrinogen. This transformation is catalysed by an enzyme called thrombin, which cleaves off two small peptide chains from the fibrinogen molecule:

$$\text{Fibrinogen} \xrightarrow{\text{Thrombin}} \text{fibrin} + \text{peptide}$$

Finally, the liberated fibrin is transformed enzymatically in the presence of Ca^{2+} into the insoluble fibrin clot.

Thrombin itself is formed from prothrombin, an enzymatically inactive precursor of thrombin. Vitamin K is required for the synthesis of prothrombin in the liver and also for the synthesis of the following three additional factors involved in the conversion of prothrombin to thrombin: Stuart factor, plasma thromboplastin and tissue thromboplastin. The mode of action of vitamin K is to convert the precursors, including that of prothrombin, or preprothrombin to the active compounds by carboxylation of peptide-bound glutamate residues to *p*-carboxyglutamate residues:

$$HOOC.CH_2.CH_2.CHNH_2.COOH \xrightarrow{+CO_2} \begin{matrix} HOOC \\ HOOC \end{matrix}\!\!>\!CH.CH_2.CHNH_2.COOH$$

This reaction converts the vitamin-K-dependent zymogens into Ca^{2+}-binding proteins—an essential step in the activation of these zymogens. The completed prothrombin, which contains glycosyl groups in addition to those of *p*-carboxyglutamic acid, is secreted from the liver into the blood where it fulfils its essential function in the clotting mechanism. The involvement of vitamin K in various steps of the clotting mechanism is shown in the scheme given in Figure 11.11.

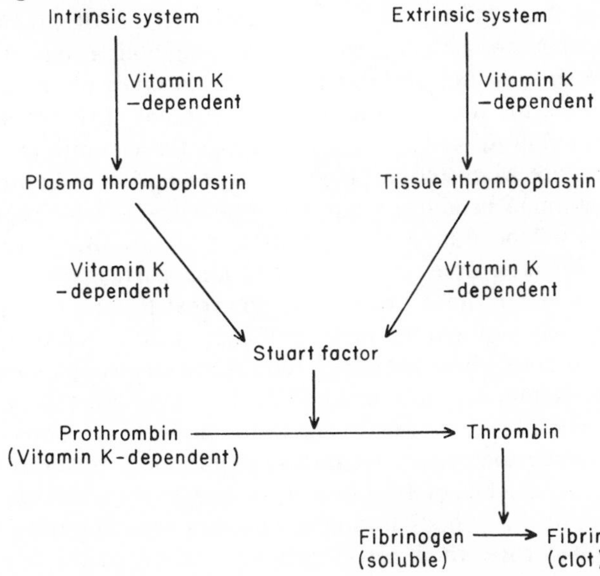

Figure 11.11. Involvement of vitamin K in blood clotting mechanism.

The Stuart factor is needed for the transformation of prothrombin into thrombin. The activity of the Stuart factor depends on two kinds of thromboplastin: plasma thromboplastin, which is activated by factors present in blood (intrinsic system), and tissue thromboplastin, which is activated by injury (extrinsic system). In vitamin-K-deficient animals the activity of the coagulation factors is depressed and the administration of vitamin K brings about a prompt response and return of coagulation factors to normal activity. The clotting mechanism is inhibited by oxalate and citrate which precipitate Ca^{2+}, and by heparin which blocks the formation of the Stuart factor.

Vitamin K Deficiencies

Symptoms of vitamin K deficiency occur in practice only in intensively reared young chicks, since other animals are likely to obtain their vitamin K requirements from bacterial synthesis in the gut. Since deficiency results in the prolongation of the clotting time, as mentioned above, deficient chicks exhibit subcutaneous haemorrhages on the breast, legs, wings or on the surface of the intestine (see PLATE VIIIA). Some deficient chicks may bleed to death from a slight bruise or other injury. Adult hens deficient in vitamin K produce eggs low in this vitamin and chicks hatched from such eggs are deficient and may bleed to death from slight injury.

Supply of Vitamin K to Farm Animals

Like most of the B vitamins (see Chapter 12), vitamin K is synthesized by microorganisms in the gastrointestinal tract. In ruminants the rumen is the principal site of this synthetic activity and the amounts of vitamin K produced and absorbed are adequate to meet the metabolic needs of the host. In monogastric species the synthesis of vitamin K takes place in the large intestine, i.e. below the zone of intensive absorption of nutrients, so that the vitamin is excreted in the faeces. In rats and pigs the vitamins are absorbed when the faeces are eaten (coprophagy), but prevention of coprophagy in monogastric mammals produces vitamin K deficiency. Otherwise, mammals do not necessarily depend on a dietary source of vitamin K. However, addition of vitamin K has become routine in pig rations, particularly for newborn pigs housed in wire-bottomed cages in which access to faeces is prevented.

Birds, particularly young ones, have such a short intestinal tract and harbour so few microorganisms that they require dietary supplementation of vitamin K. The vitamin K requirement of chicks is very low (50 μg/kg food). Antimicrobial agents that suppress bacterial vitamin K synthesis in the intestine may increase the dietary vitamin K requirement to up to ten times that needed in the absence of such drugs. An attack of coccidiosis produces a triple stress upon the blood-clotting mechanism resulting from (1) treatment with sulpha drugs which arrest bacterial activity in the intestine, (2) reduced feed intake, hence a reduced supply of vitamin K, and (3) coccidiosis, which injures the intestinal wall and reduces absorption of all nutrients including vitamin K. When additional vitamin K is required, poultry diets are supplemented with vitamin K-rich feedstuffs such as lucerne meal, or with synthetic vitamin K concentrates such as menadione.

Antagonists of Vitamin K

A well-known natural antagonist of vitamin K is dicoumarol (see formula) which is produced by moulds that attack sweet clover. Consumption of weather-damaged hay or silage made from sweet clover and containing dicoumarol causes massive internal haemorrhages, and even death, in calves. Outbreak of this disease, called sweet-clover poisoning, may be prevented by administering vitamin K. The synthetic antagonist warfarin (see formula) inhibits the vitamin-K-dependent carboxylation of prothrombin precursor in the liver (see above) and is widely used as a rat poison. Sulphonamides, administered at sufficient levels, inhibit the synthesis of vitamin K in the intestine of monogastric species (see above).

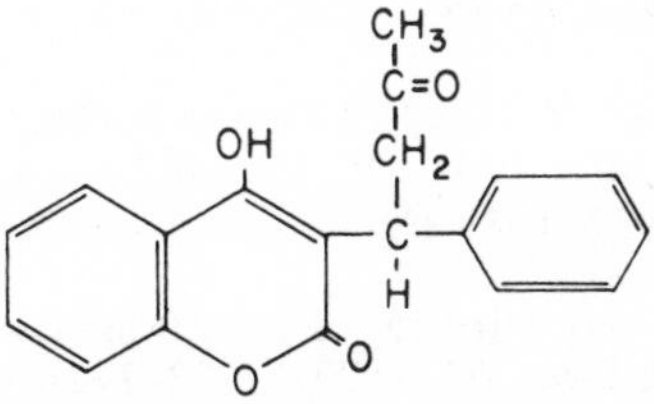

Warfarin

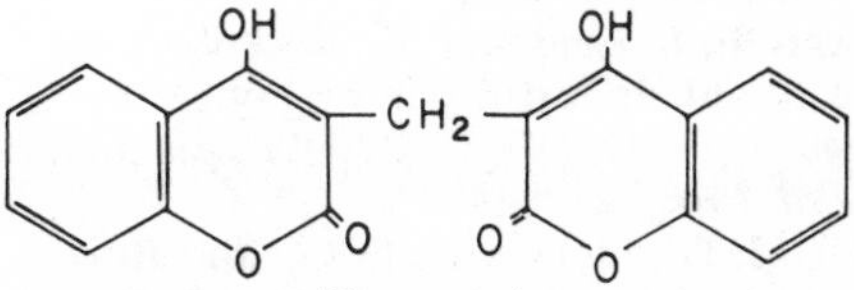

Dicoumarol
3,3'-methylenebis (4-hydroxycoumarin)

Use of Vitamin K in Human Medicine

Since antagonists of vitamin K (dicoumarol, warfarin) decreases the ability of blood to coagulate, they are used in the treatment of thrombosis. In some human diseases the need for vitamin K is increased and supplemental doses are given (1) to decrease the risk of bleeding during operations, (2) in cases where absorption is impaired, as in obstructive jaundice due to impaired bile production, and (3) in haemorrhagic diseases of the newborn. The blood of infants at birth contains less than the usual amount of prothrombin and this deficiency is increased further by the lack of microbes in the intestine of the newborn.

BIBLIOGRAPHY

Books

DeLuca, H. F. (1978). *The Fat-soluble Vitamins,* 2nd edn. New York: Plenum.
F. Hoffman-La Roche (1976). *Vitamin Compendium.* Basle: F. Hoffman-La Roche.

Morton, R. A. (1970). *Fat-soluble Vitamins.* Oxford: Pergamon Press.
Sebrell, W. H., Jr., and Harris, R. S. (1967–1979). *The Vitamins (Chemistry, Physiology, Pathology, Methods)*, 7 vols. New York and London: Academic Press.

Articles

Calman, K. C., *et al.* (1981). Symposium on vitamin status in man and animals (vitamins C, A, D, E and biotin). *Proc. Nutr. Soc.*, **40**, 147–194.

Vitamin A (retinol) and carotenes

Books

Bauernfeind, J. C. (1981). *Carotenes as Colorants and Vitamin A Precursors.* New York: Academic Press.
Isler, O. (1971). *Carotenoids.* Basle: Birkhauser.
Moore, Th. (1957). *Vitamin A.* Amsterdam: Elsevier.

Articles

Symposium on vitamin A in nutrition and disease. (1983). *Proc. Nutr. Soc.*, **42**, 1–101.
Ascarelli, I. (1969). Absorption and transport of vitamin A in chicks. *Am. J. Clin. Nutr.*, **22**, 913–922.
Ascarelli, I., Edelman, Z., Rosenberg, M., and Folman, Y. (1985). Effect of dietary carotene on fertility in high-yielding dairy cows. *Anim. Prod.*, **40**, 195–207.
Beeson, W. H. (1965). Relative potencies of vitamin A and carotene for animals. *Fed. Proc.*, **24**, 924–926.
Bieri, J. G., McDaniel, E. G., and Rogers, W. E. (1968). Survival of germfree rats without vitamin A. *Science*, **163**, 574–575; see *Fed. Proc.*, **30** (1970), 1773–1778.
Bondi, A., and Sklan, D. (1984). Vitamin A and carotene in animal nutrition. *Prog. Food Nutr. Sci.*, **8**, 165–191.
Donoghue, S. (1982). Vitamin A metabolism in ruminants and horses. *Proc. Cornell Nutr. Conf., 1982*, pp. 7–12.
Eaton, H. D. (1969). Chromic bovine hypo- and hypervitaminosis and cerebrospinal fluid pressure. *Am. J. Clin. Nutr.*, **22**, 1070–1080.
Fernandez-Cohen, S., Budowski, P., Ascarelli, I., Neumark, H., and Bondi, A. (1976). Low utilization of carotene by sheep. *Int. J. Vit. Nutr. Res.*, **46**, 439–454.
Glover, J. (1983). Factors affecting vitamin A transport in animals and man. *Proc. Nutr. Soc.*, **42**, 19–30.
Glover, J., Heaf, D. J., and Large, S. (1980). Seasonal changes in retinol-binding holoprotein concentration in Japanese quail. *Br. J. Nutr.*, **43**, 357–366.
Goodman, S. De Witt (1980). Vitamin A metabolism. *Fed. Proc.*, **39**, 2716–2723.
Olson, J. A. (1972). The biological role of vitamin A in maintaining epithelial tissues. *Israel J. Med. Sci.*, **8**, 1170–1178.
Simpson, K. L. (1983). Relative value of carotenoids as precursors of vitamin A. *Proc. Nutr. Soc.*, **42**, 7–18.
Thompson, S. Y. (1975). Role of carotene and vitamin A in animal feeding. *World Rev. Nutr. Diet.*, **21**, 225–280.
Ullrey, D. E. (1972). Biological availability of fat-soluble vitamins: vitamin A and carotene. *J. Anim. Sci.*, **35**, 648–657.
Wald, G. (1968). The molecular basis of visual excitation. *Nature*, **219**, 800–807.
Wald, G., and Hubbard, R. (1970). The chemistry of vision. In *Fat-soluble Vitamins* (R. A. Morton, ed.), pp. 267–278. Oxford: Pergamon Press.

Wolf, G. (1977). Retinol-linked sugars in glycoprotein synthesis. *Nutr. Rev.*, **35**, 97–100.

Wolf, G. (1980). Vitamin A. In *Human Nutrition* (R. B. Alfin-Slater and D. Kritchevsky, eds), Vol. 3A, pp. 97–204. New York: Plenum.

Vitamin E

Book

Machlin, L. J. (1980). *Vitamin E, A Comprehensive Treatise.* New York: Marcel Dekker.

Articles

Bartov, I., Budowski, P. and Bornstein, S. (1966). Relation between α-tocopherol content of the breeder diet and that of the newly hatched chick. *Poult. Sci.*, **45**, 512–519.

Bauernfeind, J. C. (1977). The tocopherol content of food and influencing factors. CRC. *Crit. Rev. Food Sci. Nutr.*, **8**, 337–383.

Bieri, J. G., and Farrell, P. M. (1976). Vitamin E. *Vit. Horm.*, **34**, 31–76.

Lannek, N., and Lindberg, P. (1975). Vitamin E and selenium deficiencies of domestic animals. *Adv. Vet. Sci. Comp. Med.*, **19**, 127–164.

Scott, M. L. (1980). Advances in our understanding of vitamin E. *Fed. Proc.*, **39**, 2736–2739.

Ullrey, D. E. (1981). Vitamin E for swine. *J. Anim. Sci.*, **53**, 1039.

Vitamin D

DeLuca, H. F. (1979). *Vitamin D Metabolism and Function.* Heidelberg: Springer-Verlag.

Lawson, E. D. (1978). *Vitamin D.* London and New York: Academic Press.

Norman, A. W. (1979). *Vitamin D. The Calcium Homeostatic Steroid Hormone.* New York: Academic Press.

Articles

Bronner, F. (1982). Symposium on vitamin D and membrane structure. *Fed. Proc.*, **41**, 60–87.

DeLuca, H. F. (1979). The vitamin D system in the regulation of calcium and phosphorus metabolism. *Nutr. Rev.*, **37**, 161–193.

Haddad, J. G. (1982). Vitamin D binding proteins. In *Advances in Nutrition Research* (H. H. Draper, ed.), Vol. 4, pp. 35–58. New York and London: Plenum.

Henry, H. L., and Norman, A. W. (1984). Vitamin D metabolism and biological action. *Annu. Rev. Nutr.*, **4**, 493–520.

Marx, S. J., Liberman, U., and Eil, C. (1983). Calciferols: actions and deficiencies in action. *Vit. Horm.*, **40**, 235–308.

Wasserman, R. H. (1975). Metabolism, function and clinical aspects of vitamin D. *Cornell Vet.*, **65**, 3–25.

Wasserman, R. H. (1975). Active vitamin D-like substances in *Solanium malacoxylon* and other calcinogenic plants. *Nutr. Rev.*, **33**, 1–5.

Vitamin K

Griminger, P., and Brubacher, G. (1966). The transfer of vitamin K and menadione from the hen to the egg. *Poult. Sci.*, **45**, 511–519.

Olson, R. E. (1984). Function and metabolism of vitamin K. *Annu. Rev. Nutr.*, **4**, 281–337.

Suttie, J. W. (1977). Role of vitamin K in the synthesis of clotting factors. In *Advances in Nutrition Research* (H. H. Draper, ed.), Vol. 1, pp. 1-22. New York and London: Plenum.

Suttie, J. W., *et al.* (1978). Symposium on vitamin K. *Fed. Proc.*, **37**, 2598–2626.

Suttie, J. W. (1980). The metabolic role of vitamin K. *Fed. Proc.*, **39**, 2730–2736.

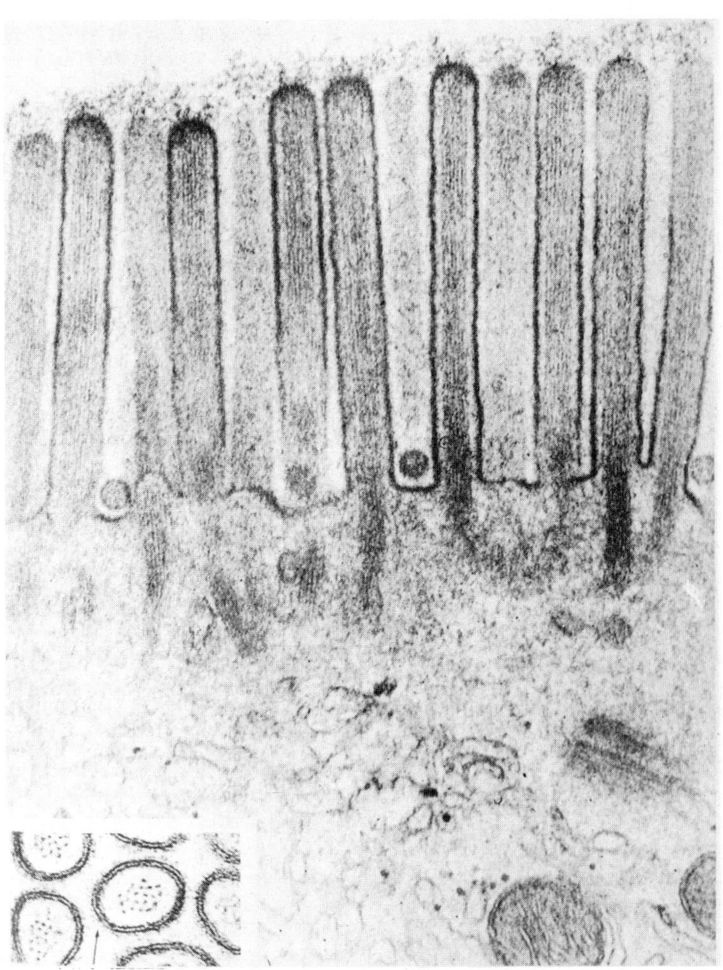

Plate I Electron micrograph of a typical brush border from an absorptive cell (from Madge, *The Mammalian Alimentary System*. 1975, p. 74). *Reproduced by permission of the author and Edward Arnold*

Plate II

Plate II A Perosis in chicken fed a manganese-deficient diet (after Scott *et al.*, *Nutrition of the Chicken*, 2nd ed., 1979, p. 318). *Reproduced by permission of Scott & Associates*

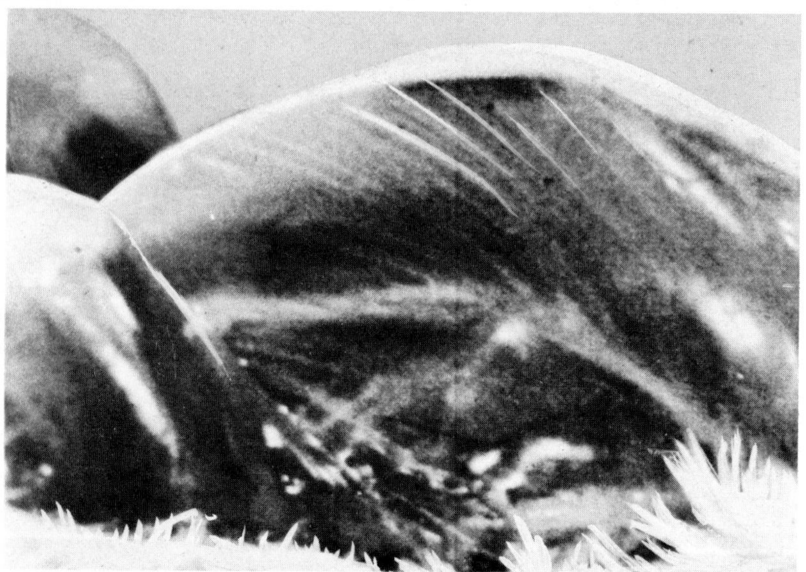

Plate II B Striations in the breast muscle of a chicken affected by a diet deficient in selenium and vitamin E (nutritional muscular dystrophy) (after Roche, *Vitamin E Deficiency in Poultry*, Basle, 1984). *© British Crown Copyright 1985, reproduced with the kind permission of Her Majesty's Stationery Office*

Plate III

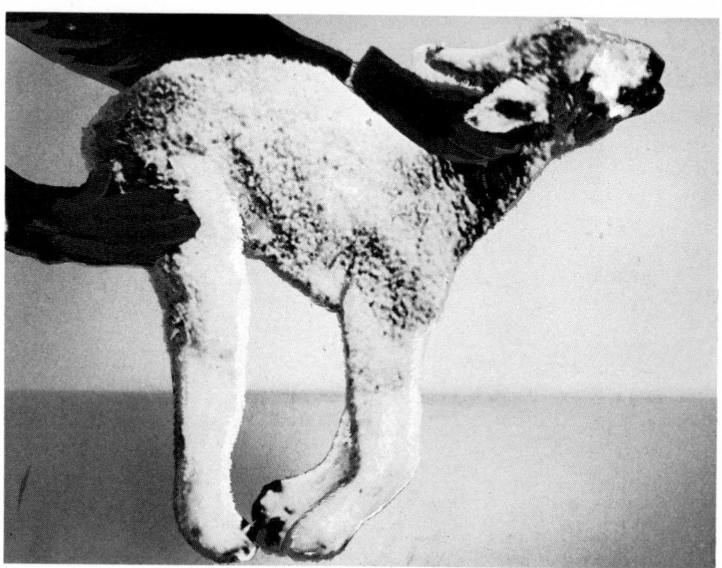

Plate III A Poor muscles and lack of viability in a new born lamb due to selenium and vitamin E deficiency (after Roche, *Vitamin E Deficiency in Ruminants*, 1983)

Plate III B Exudative diathesis in chicks fed a selenium- and vitamin-E-deficient diet (after Scott *et al.*, *Nutrition of the Chicken*, 2nd edn. 1979, p. 164). *Reproduced by permission of Scott & Associates*

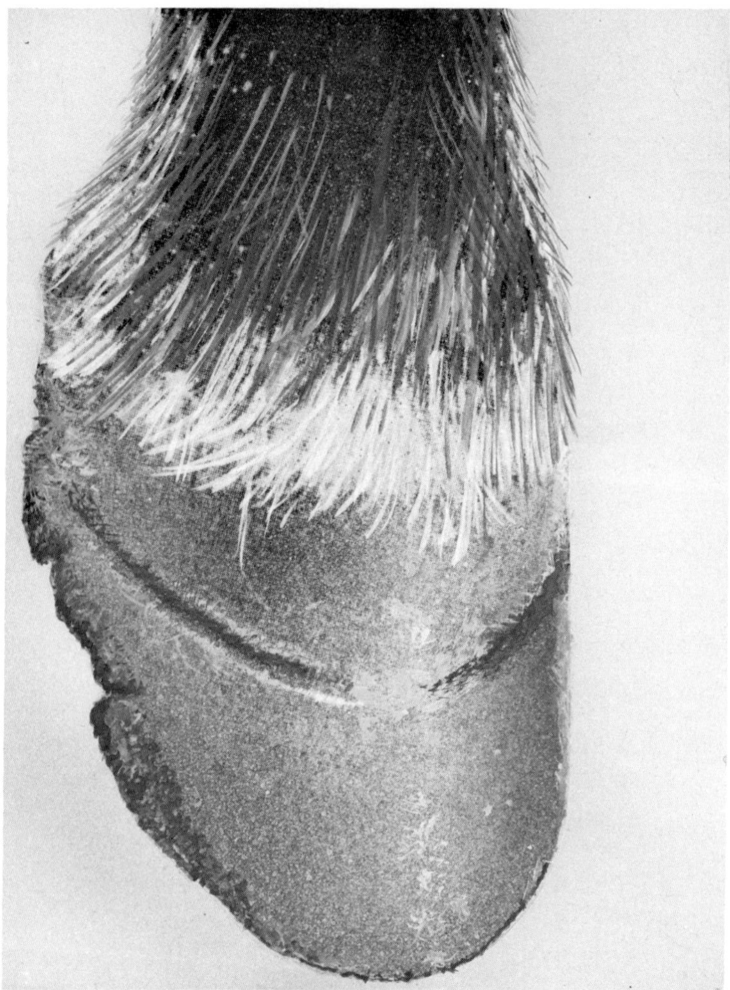

Plate IV Sloughing of a hoof from a horse with selenium toxicity. *By permission from* Animal Nutrition, *7th edition, by L. A. Maynard* et al. *Copyright 1979 MacGraw-Hill Book Co.*

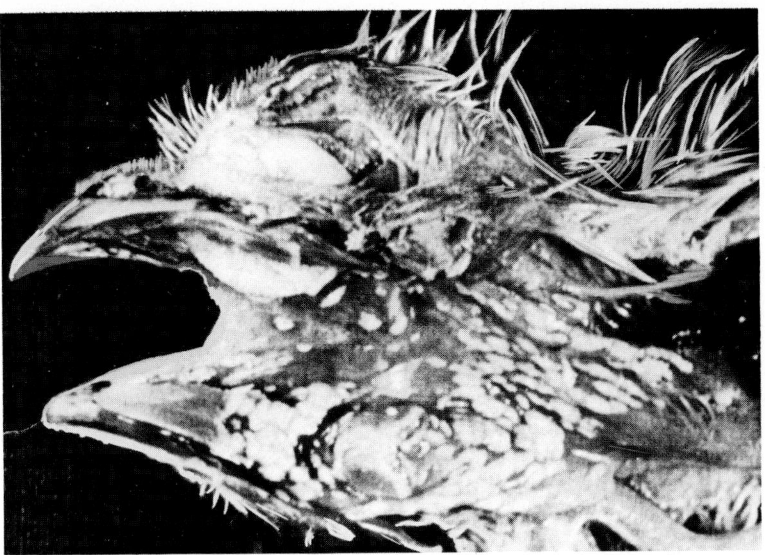

Plate V A Pustules in the roof of the mouth and in the oesophagus resulting from vitamin A deficiency in the chicken (from Scott *et al.*, *Nutrition of the Chicken*, 2nd edn, 1976, p. 140. Ithaca: Scott & Associates. *Reproduced by permission of Scott & Associates*

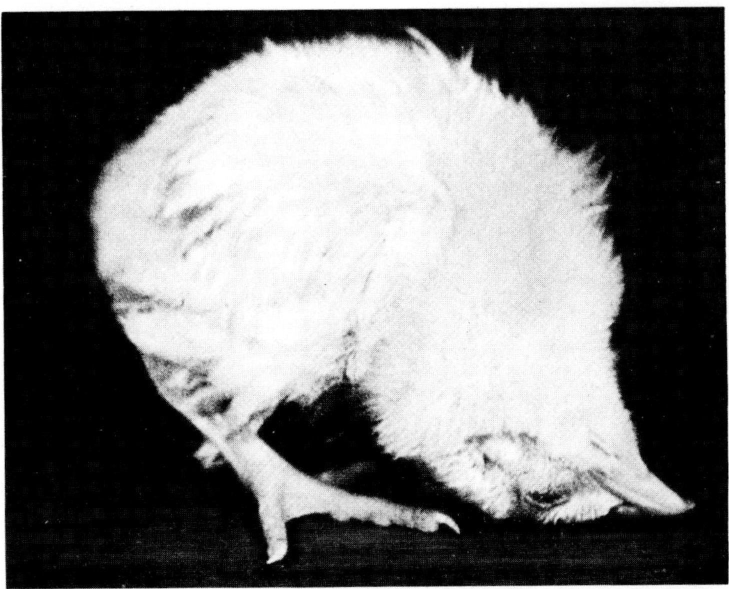

Plate V B Encephalomalacia, the crazy chick disease (after Hoffman-La Roche, *Vitamin Compendium*, 1976, p. 75)

Plate VI

Plate VI A Cerebellum of a chicken with encephalomalacia, softened swollen because of haemorrhages and oedema (courtesy of Prof. P. Budowski, Rehovot, Israel)

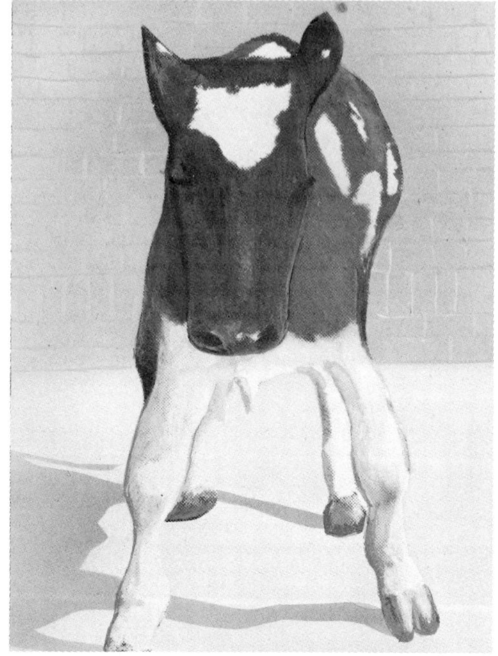

Plate VI B Calf suffering from rickets (from Church, *Digestive Physiology and Nutrition of Ruminants*, Vol. 2, 1979, p. 242). *Reproduced by permission of Prof. D. C. Church*

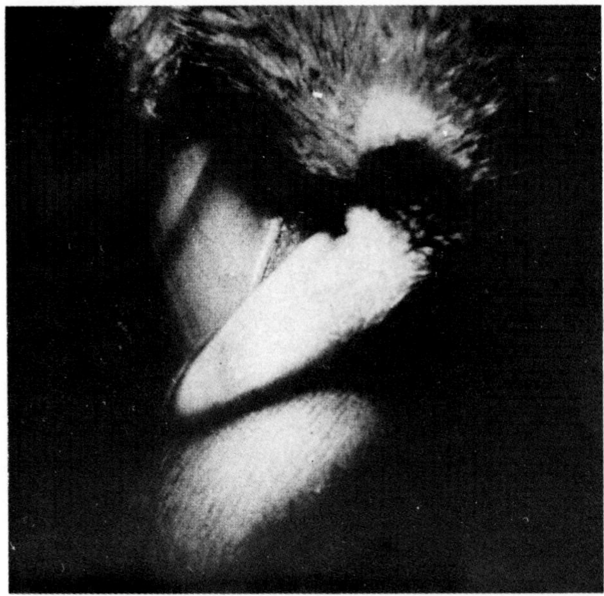

Plate VII A Vitamin D deficiency in chicks—soft and rubbery beak (from NRC, *Nutrient Requirements of Poultry*, 7th ed., 1977, p. 13). *Reproduced by permission of National Academy Press*

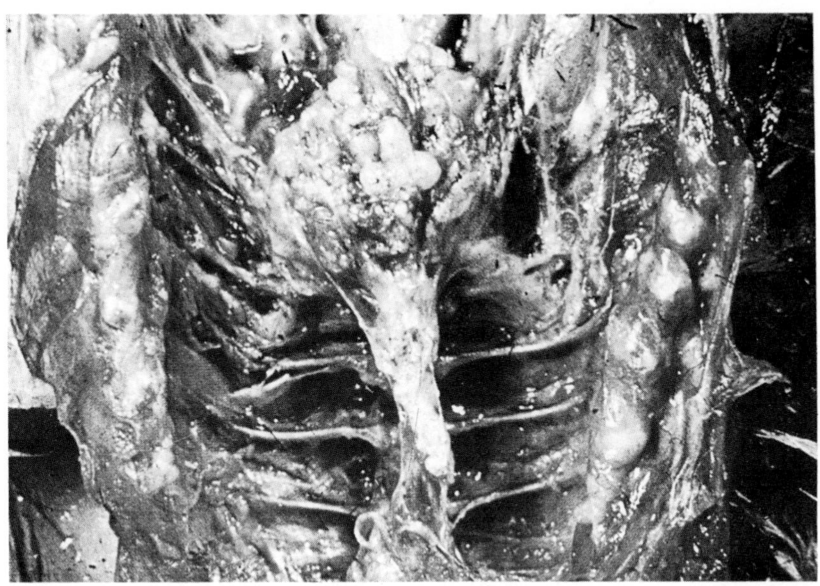

Plate VII B 'Rickety rosary' swelling on the ribs, a characteristic symptom of rickets in chickens (after Hoffman-La Roche, *Vitamin Compendium*, 1976, p. 66). © *British Crown Copyright 1985, reproduced with the kind permission of Her Majesty's Stationery Office*

Plate VIII

Plate VIII A Vitamin K deficiency—spontaneous haemorrhages (*Vitamin Manual*, 1957, p. 79. Michigan: Upjohn)

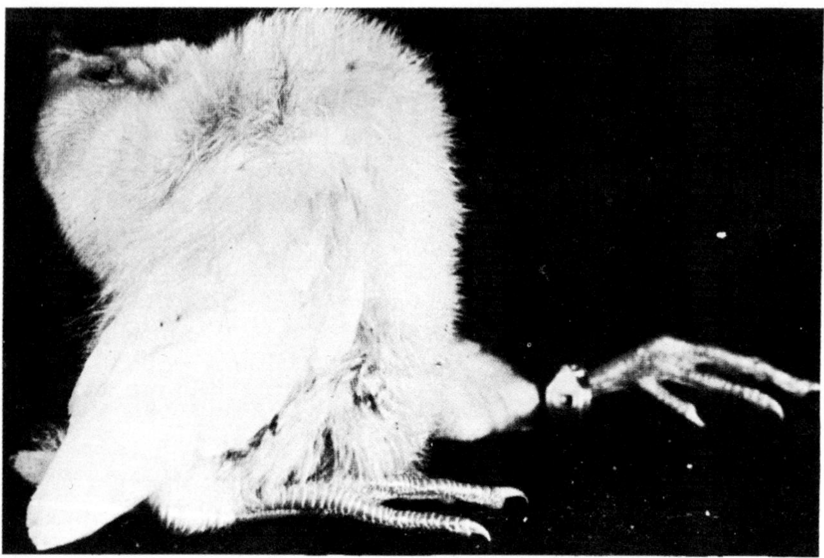

Plate VIII B Polyneuritis in a thiamine-deficient chick (after Scott *et al.*, *Nutrition of the Chicken*, 2nd ed., 1976, p. 192). Ithaca, N.Y. *Reproduced by permission of Scott & Associates*

Plate IX

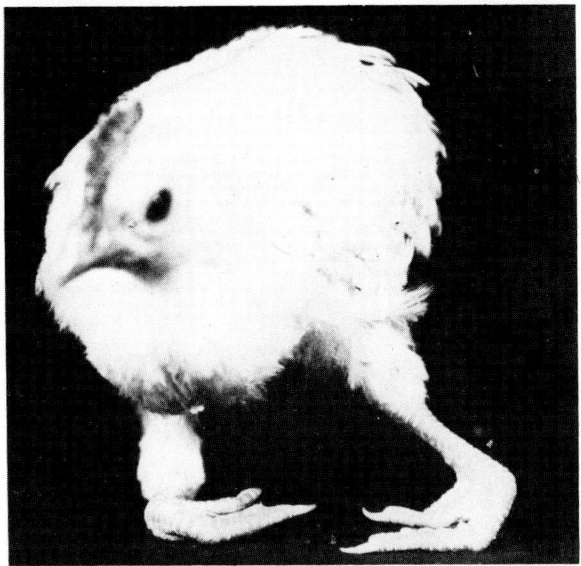

Plate IX A Riboflavin deficiency in a young chick (from Scott *et al.*, *Nutrition of the Chicken*, 1976, p. 200). *Reproduced by permission of Scott & Associates*

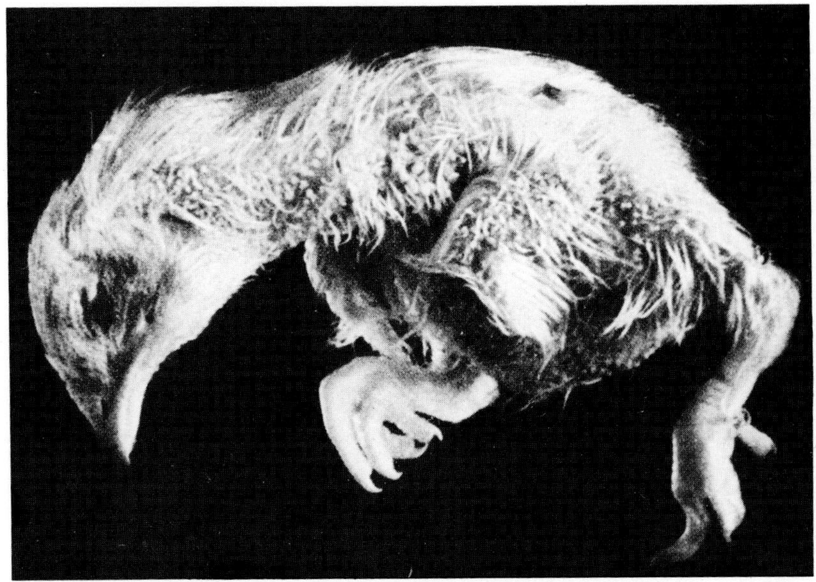

Plate IX B Riboflavin deficiency in 19-day-old chick embryo, resulting from such a deficiency in the diet of the breeding hen (after McDonald *et al.*, *Animal Nutrition*, 2nd ed., 1973, p. 77). *Reproduced by permission of Longman Group Ltd*

Plate X

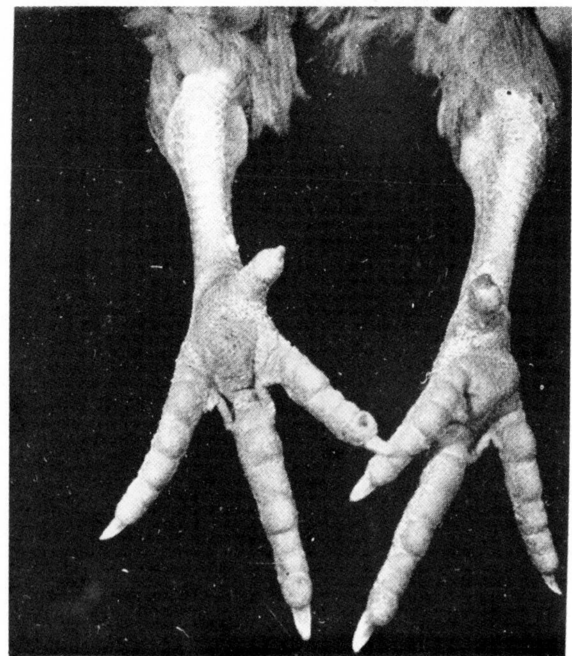

Plate X A Enlargement of the hock joint of a chick fed a niacin-deficient diet (after Scott *et al.*, *Nutrition of the Chicken* 1976, p. 210). *Reproduced by permission of Scott & Associates*

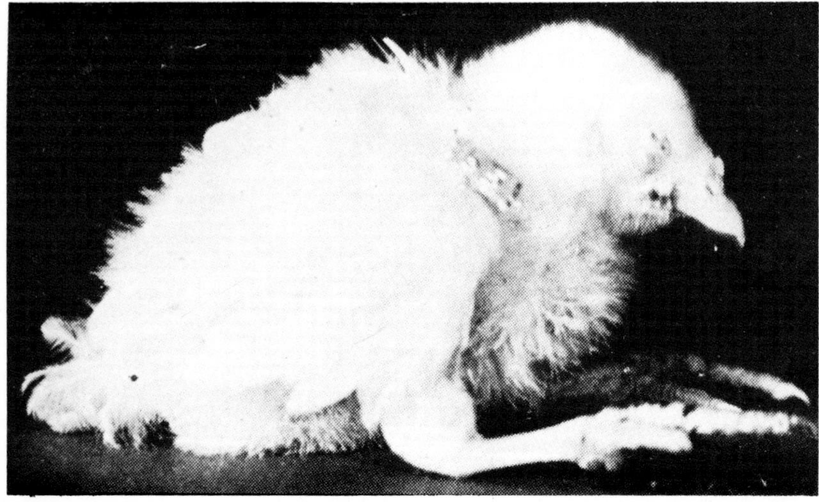

Plate X B An advanced stage of pantothenic acid deficiency in a chick (after Scott *et al.*, *Nutrition of the Chicken*, 1976, p. 22). *Reproduced by permission of Scott & Associates*

Plate XI

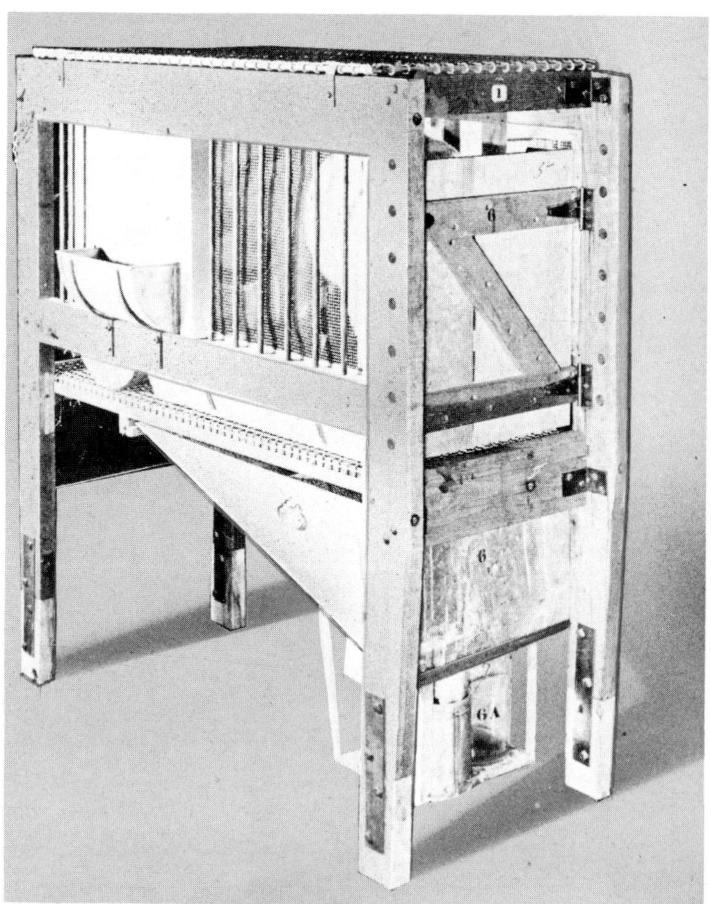

Plate XI Metabolism cage for the collection of faeces in digestibility trials with sheep (from Schneider and Flatt, *The Evaluation of Feeds through Digestibility Experiments*, 1975, p. 21). *Reproduced by permission of the University of Georgia Press.*

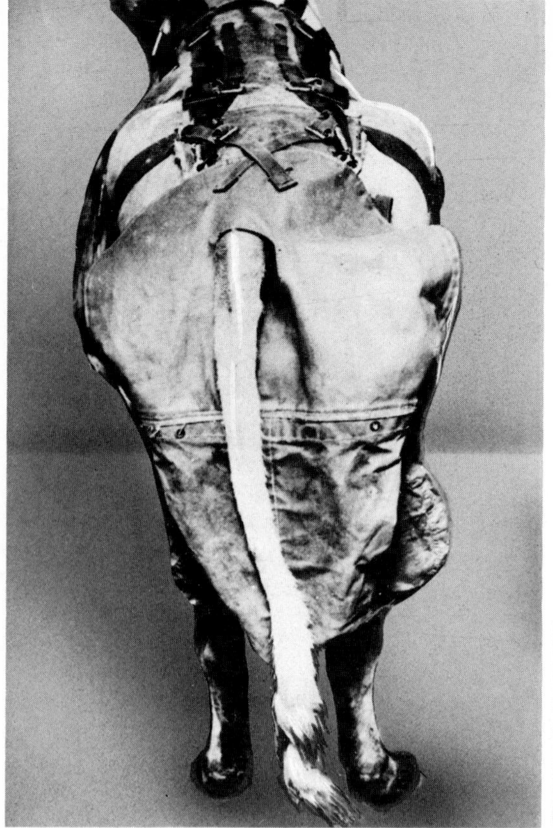

Plate XII A Apparatus for collection of faeces in digestibilit trials with pasturing animals (fron Schneider and Flatt, 1975, p. 37) *Reproduced by permission of th University of Georgia Press*

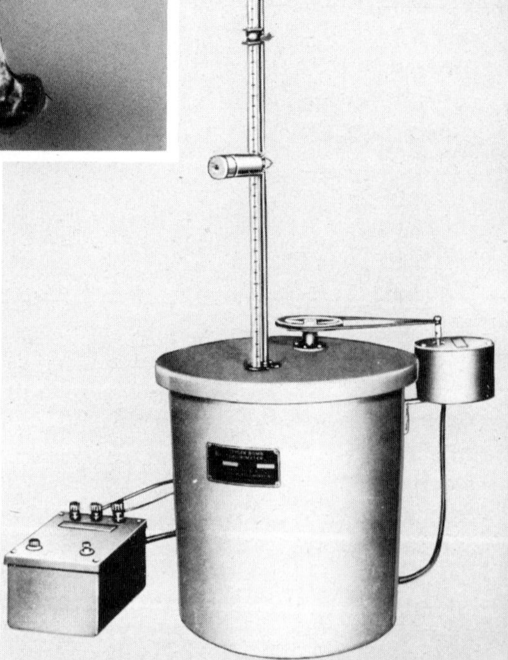

Plate XII B A bomb calorimeter

CHAPTER 12

*Water-soluble Vitamins**

Classification and general properties of B vitamins are given at the beginning of Chapter 11. The B vitamins function primarily as parts of the enzyme systems that catalyse the metabolism of carbohydrates, fats and proteins (see Chapters 4–7). The participation of B vitamins in enzymatic reactions is described in biochemical textbooks and the details will not be dealt with here.

B vitamins are present in feeds mostly in bound form as coenzymes, and they have to be liberated in the digestive tract by appropriate enzymes before they can be utilized. The transport of vitamins from the digestive tract into the bloodstream and finally into target tissues may require specific transport proteins for each particular vitamin.

B vitamins are present in all plant and animal cells and in significant quantities in a wide variety of feeds. Thus, rations of pigs and poultry have to be supplemented only with certain vitamins, as it will be pointed out in this chapter. In animals with a functional rumen and in horses, the requirement for B vitamins is met entirely from the microbial synthesis occurring in the rumen and lower gastrointestinal tract, respectively (see Chapters 11 and 22). As explained at the beginning of Chapter 11 and in Chapter 18, the requirements of preruminants are usually satisfied by milk, but single vitamin B deficiencies have been identified in calves and lambs fed purified diets.

---

* In collaboration with I. Ascarelli.

THIAMINE

Thiamine

Vitamin B_1 (thiamine) consists of a pyrimidine ring joined by a methylene bridge to a thiazole ring. Thiamine is absorbed from the small intestine and carried to the liver where it is phosphorylated by means of ATP to form a coenzyme, thiamine pyrophosphate (TPP) or cocarboxylase:

$$\text{Thiamine} + \text{ATP} \longrightarrow \text{thiamine pyrophosphate} + \text{AMP}$$

TPP is the coenzyme for the enzymic decarboxylation of α-keto acids, the most important of which is the oxidative decarboxylation of pyruvate to acetate.

Pyruvate, after undergoing oxidative decarboxylation, forms acetyl-CoA which enters the tricarboxylic acid cycle. Cocarboxylase acts in conjunction with nicotinamide adenine dinucleotide, coenzyme A (which contains pantothenic acid) and lipoic acid in this reaction:

$$\underset{\text{Pyruvic acid}}{CH_3.CO.COOH} + \underset{\text{coenzyme A}}{\text{CoA}} + NAD^+ \xrightarrow[\text{LTPP}]{\text{TPP}} \underset{\text{acetyl-CoA}}{CH_3.CO.CoA} + CO_2 + NADH + H^+$$

LTPP = lipothiamide (containing lipoic acid and TPP)

Another reaction in which TPP plays an essential role is the oxidation of glucose via the pentose shunt; this is the only pathway in which the body can produce ribose for RNA and results also in the formation of NADPH (reduced nicotinamide adenine dinucleotide phosphate) which is needed in the biosynthesis of long-chain fatty acids (see Chapters 6 and 21) and other reductive functions.

The decarboxylation of pyruvic acid to acetic acid is essential for the utilization of carbohydrates in order to provide energy for body processes and for the conversion of carbohydrates into fat. Thus, thiamine requirements of monogastric species are related to the amount of carbohydrates ingested. High-fat diets decrease the requirements because fatty acids can enter the tricarboxylic acid cycle without going through pyruvic acid.

Symptoms of Deficiency

Because relatively high levels of thiamine are present in feedstuffs, particularly in cereal grains, practical rations for pigs and poultry easily meet the requirements; in practice pigs and poultry are unlikely to suffer from thiamine deficiency.

Since thiamine is poorly stored in the body, monogastric species need a regular supply, except pigs which can store fairly large quantities of thiamine in their tissues.

Because carbohydrate metabolism in thiamine-deficient animals is impaired, pyruvic and lactic acids accumulate in blood and brain tissues and cause neurological disorders. Thus, neurological symptoms may be causally related to thiamine deficiency. The classic diseases—beriberi in man and polyneuritis in birds—represent a late stage of the deficiency resulting from a peripheral neuritis. The neurological signs only appear in the final stages. The first symptoms to appear in thiamine-deficient chicks and pigs are anorexia (lack of appetite), bradycardia (slowing of the heart beat), enlargement of the heart and gastrointestinal troubles, and result from general weakness. As the deficiency progresses in chicks, apparent paralysis of the muscles occurs, the chicken sits on its flexed legs and draws back its head in a star-gazing position (see PLATE VIIIB). Retraction of the head is due to paralysis of the anterior neck muscle.

Not only can thiamine deficiency be produced in preruminants (see Chapter 18), but it has also been observed in beef cattle and lambs fed diets high in soluble carbohydrate and low in fibre. This disease, called polioencephalomalacia, is characterized by lesions of the nervous system, such as disturbances of vision and tetanus and can be cured by intravenous administration of thiamine.

A thiamine deficiency in foxes, called Chastek paralysis, is caused by feeding raw fish. This condition is due to an enzyme, thiaminase, present in raw fish which splits the thiamine molecule into two compounds and thus renders it inactive.

Some synthetic compounds, similar in structure to thiamine, when added to the diet act as antivitamins, i.e. as antagonists of thiamine (see above). One of these compounds, the coccidiostat amprolium, acts by interfering with the thiamine metabolism of the coccidia, while at recommended levels it does not interfere with the thiamine metabolism of the chicken. Overdosage, however, could cause thiamine deficiency.

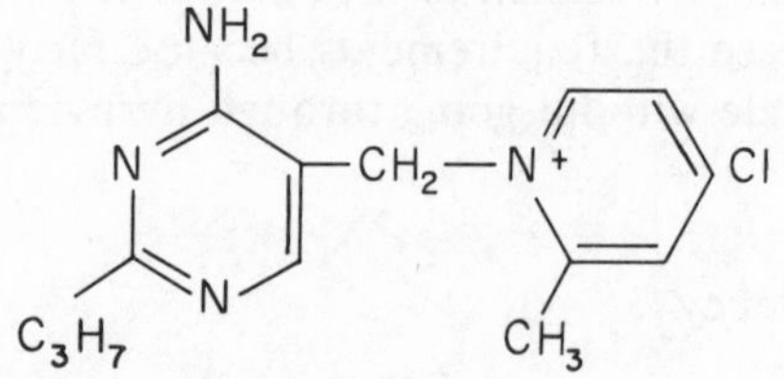

Amprolium

Although in ruminants and horses the thiamine requirement is met from microbial synthesis in the rumen and lower gastrointestinal tract, respectively, thiamine deficiency occurs in these animals by feeding bracken ferns. The causative agent in fern poisoning is apparently a natural antithiamine and not a thiaminase as assumed previously.

RIBOFLAVIN

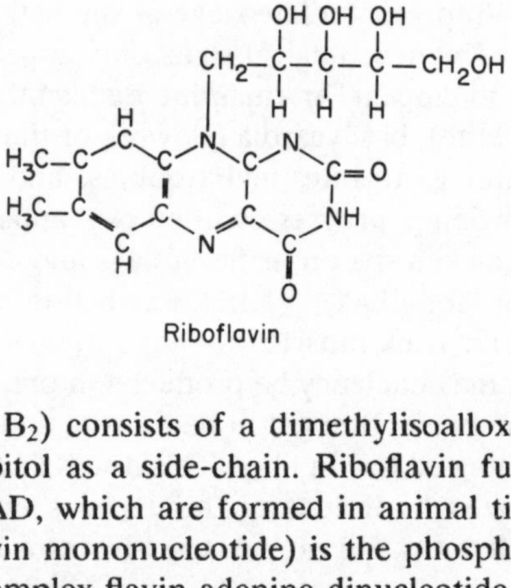

Riboflavin

Riboflavin (vitamin B_2) consists of a dimethylisoalloxazine ring combined with the alcohol ribitol as a side-chain. Riboflavin functions in the coenzymes FMN and FAD, which are formed in animal tissues, particularly in the liver; FMN (flavin mononucleotide) is the phosphate of riboflavin and FAD is the more complex flavin-adenine dinucleotide. FMN and FAD occur as coenzymes in the flavoprotein enzyme systems. The flavoproteins are concerned with chemical reactions involving the transport of hydrogen; the main function of the flavoproteins is to transfer hydrogen in the respiratory chain between the nicotinic acid-containing coenzymes, NAD and NADP (see later) and the cytochromes. Thus, the riboflavin-containing coenzymes are needed in every living cell for the breakdown of energy-yielding nutrients. In addition to this role, riboflavin is a component of the following enzymes: L-amino acid oxidase (see Chapter 7), xanthine oxidase, succinic dehydrogenase, the enzymes involved in the saturation and desaturation of fatty acids (see Chapter 6), and others.

Deficiency Symptoms

Deficiency of riboflavin will lead to an impairment of energy-producing processes. Thus, deficient animals will consume 15–20% more energy than control animals because of incomplete oxidation. The primary general effect, as with most of the water-soluble vitamin deficiencies, is reduced growth rate in young animals.

Various pathological lesions accompany a riboflavin deficiency. The two most affected tissues in riboflavin deficiency in the chick are the epithelium and the nerve sheaths of some of the main nerve trunks. The sciatic nerve may be enlarged up to six times its normal diameter; damage to the myelin sheath pinches the nerve, causing a permanent stimulus which results in the 'curled-toe paralysis', i.e. the chicks walk on their hocks with the toes curled inwards (see PLATE IXA).

In the adult hen, egg production is not affected by riboflavin deficiency; however, hatchability becomes poor within two weeks of feeding a deficient diet, but can be returned to normal within only one week after addition of riboflavin to the diet. In breeding hens fed a riboflavin-deficient diet, a peak in embryonic mortality develops about the twelfth day, while in many of those that survive the down follicle fails to rupture. The down feather continues to grow inside the follicle giving the characteristic clubbed down condition as shown in PLATE IXB.

A riboflavin deficiency in swine causes thickening of the skin, crooked and stiff legs, accompanied by skin eruptions, lens opacities and cataracts; furthermore, anorexia, vomiting and birth of weak piglets have been reported in deficient swine.

Antiriboflavin compounds have been obtained by chemical changes in either the alloxazine nucleus or the ribityl side-chain of the riboflavin molecule. These riboflavin antagonists inhibit growth of pathogenic bacteria and are of great medical importance, as, for example, atabrine which is an antimalarial.

Riboflavin Requirements

Diets for chicks and breeding hens must be supplemented with riboflavin concentrates. However, the riboflavin content of the usual rations is always sufficient for laying hens. The riboflavin requirement for poultry decreases with age, and chicks affected by riboflavin deficiency may recover from the respective symptoms during the proceeding growth period, even though the riboflavin content of the diet is not altered.

Supplemental riboflavin is generally included in the diet of pigs, since the usual cereal grain–soybean meal diets are marginally riboflavin deficient, particularly for young pigs and for reproductive females. The suggestion that the riboflavin requirement of pigs increases with elevated protein and fat levels in the diet, is in accordance with the above-mentioned functions of riboflavin-containing coenzymes in protein and fat metabolism.

NIACIN

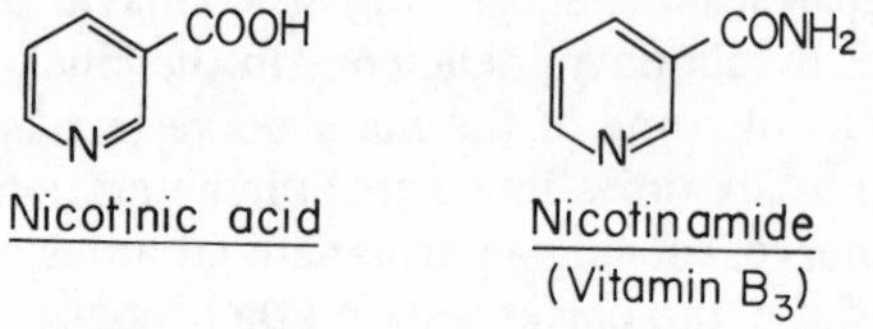

COOH CONH$_2$

N N

Nicotinic acid Nicotinamide (Vitamin B_3)

Nicotinic acid and nicotinamide, which are biologically equivalent vitamins, are both referred to as niacin. Niacin is readily absorbed from the intestine and incorporated into two important coenzymes which act as codehydrogenases: NAD or nicotinamide-adenine dinucleotide, and NADP or nicotinamide-adenine dinucleotide phosphate. These niacin-containing enzymes together with the flavoproteins are important links in the stepwise transfer of hydrogen (or electrons) from substrates to molecular oxygen, resulting in water formation. The coenzymes NAD and NADP are involved in carbohydrate, fat and protein metabolism and are especially important in metabolic reactions that provide energy to the animal, particularly the processes by which high-energy phosphate bonds are formed. The influence of these coenzymes is so widespread in cellular metabolic processes that a lack of the vitamin results in major damage to cellular respiration.

Metabolism

Niacin requirements are influenced by the following two phenomena:

(1) Many feedstuffs, particularly cereal grains and to lesser extent oil cakes, contain niacin in a bound form which is unavailable to monogastric species; only 10% of the niacin present in maize and 40% of that in oil cakes is available to rats, pigs or chickens. Treatment of these feeds with alkaline solutions releases niacin from its bound form and makes it fully available.

(2) The presence of tryptophan, an essential amino acid, in the diet can lower an animal's requirement for niacin since it is synthesized in the animal body from tryptophan. If tryptophan is provided in the diet in excess of the amount needed for tissue protein synthesis, the excess can be used as a precursor for nicotinic acid. The extent to which tryptophan is converted

to nicotinic acid is low and varies with the species of animal and with dietary conditions. The synthesis of 1 mg of niacin requires about 60 mg of tryptophan for man, pigs and chicks, and 35 mg for rats. For the chemical pathways by which the transformation of tryptophan into nicotinic acid occurs, the reader is referred to biochemical texts. It may be noted that for several steps of the biosynthesis of nicotinic acid, the vitamins pyridoxine and riboflavin are required; in their absence side-reactions will prevail and no nicotinic acid will be formed. Thus, prevention of nicotinic acid deficiency through its biosynthesis depends not only on adequate dietary supply of tryptophan, but also on adequate levels of pyridoxine and riboflavin.

The ability of some species of animals to synthesize niacin from tryptophan is also due to the inherent levels of picolinic acid carboxylase, which diverts one of the intermediates to a side-reaction instead of allowing this compound to condense to quinolinic acid, the immediate precursor of nicotinic acid. For example, the activity of picolinic acid carboxylase in the liver of ducks is three times that of chicks, and the niacin requirement of ducks is correspondingly higher than in chicks.

Another interrelationship that can affect the biosynthesis of niacin is amino acid imbalance, which may be caused by the presence of excess leucine; a considerable excess of leucine is present in jowar, a type of millet, and some excess in maize. There are thus three reasons why the use of maize for supplying niacin is critical: (1) the presence of niacin in a bound form, (2) a low content of tryptophan, and (3) the presence of excess leucine. Indians in Mexico subsisting on diets composed predominantly of maize are not affected by niacin deficiency since treatment of maize with lime before incorporation into the diet is common practice in these areas.

The amount of niacin stored in the body, mainly in the liver, is not great (like other B vitamins); the urine is the primary pathway of excretion. In ruminants niacin is excreted unchanged. In monogastric species (pigs and rats) the principal excretory products are the methylated metabolites N'-methylnicotinamide and N'-methyl-2-pyridone-5-carboxylamide. In the chicken nicotinic acid is conjugated with ornithine as dinicotinyl ornithine. Measurement of the urinary excretion of niacin metabolites is useful in establishing a diagnosis of niacin deficiency.

Symptoms of Deficiency

Niacin deficiency leads to pellagra—a metabolic disease of humans and all species of farm animals. In pellagra, the lack of NAD and NADP seems to be responsible in some way for the epithelial lesions that are extended to internal surfaces such as the mouth and the gastrointestinal tract. Pellagra in humans is characterized by a fiery red tongue; in dogs and pigs the most typical symptom is the 'black tongue'. Other symptoms of deficiency in pigs such as dermatitis, diarrhoea, anorexia and decreased rate of gain (occasionally vomiting) are commonly associated with the lack of this vitamin, but partly they may be caused by concomitant deficiencies of other B vitamins.

In chicks a deficient diet results in inflammation of the mouth, diarrhoea and poor feathering, but the main symptom is an enlargement of the hock joint and bowing of the legs similar to perosis (see PLATE XA and Chapter 10). The difference between this condition and perosis is that in nicotinic acid deficiency the tendon of Achilles does not slip. Turkeys and ducks show similar, but much more severe symptoms.

Compared to the chick, the turkey poult, duckling and gosling have high requirements for niacin. The higher needs of these species for this vitamin and the frequent use of a chick growing diet for them may result in the development of leg weakness.

PYRIDOXINE

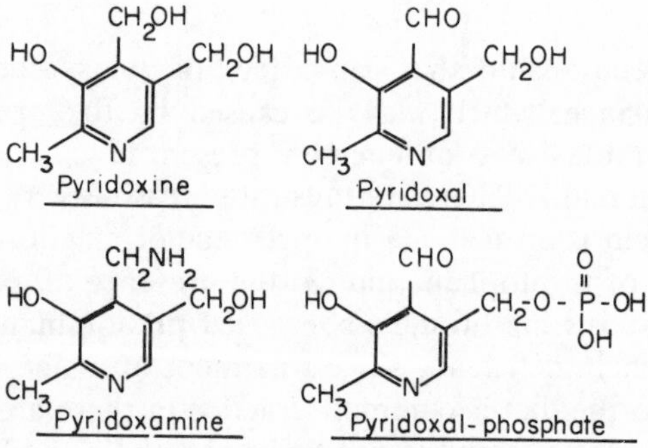

Pyridoxine

Pyridoxal

Pyridoxamine

Pyridoxal-phosphate

The name pyridoxine (or pyridoxol) given to vitamin B_6 reflects its structure as a pyridine derivative. Three biologically active forms of vitamin B_6 are now recognized: pyridoxine or pyridoxol, the primary alcohol form; pyridoxal, the aldehyde form; and pyridoxamine, the amine form. The active form of vitamin B_6 is the coenzyme pyridoxal phosphate also referred to as codecarboxylase. All three forms of the vitamin are converted in the animal body to the active coenzyme.

Pyridoxal phosphate functions as a coenzyme in more than 50 different enzyme systems. These enzymes are concerned primarily in protein metabolism, many of them are involved in the non-oxidative degradation of amino acids (see Chapter 7) and include transamination, deamination, decarboxylation and desulphhydration. The production of neuroactive amines like serotonin from tryptophan and *p*-aminobutyric acid from glutamic acid are carried out by pyridoxine-dependent enzymes. These reactions are the basis of the neurological disorders that accompany the deficiency of this vitamin (see later). Pyridoxal phosphate is also needed for the synthesis of δ-aminolaevulinic acid which is a precursor in haem biosynthesis. Thus, vitamin B_6 is somehow involved in blood cell formation. As mentioned above, in the degradation of tryptophan to niacin one of the steps is B_6-dependent, and under conditions of B_6 deficiency abnormal metabolites will be excreted in the urine, especially xanthurenic acid; the level of the latter acid can be used in the detection of B_6 deficiency in pigs.

Furthermore pyridoxal phosphate acts as a coenzyme for glycogen phosphorylase which catalyses the initial step in the breakdown of glucose (see Chapter 4). Low levels of glucose in the blood occur as a consequence of B_6 deficiency. This coenzyme is also involved in the metabolism of essential fatty acids (see Chapter 6).

Symptoms of Pyridoxine Deficiency

Although pyridoxine deficiency is unlikely in farm animals, deficiency symptoms may be provoked by feeding purified diets lacking in vitamin or by addition of structural analogues with antivitamin activity.

The symptoms of vitamin B_6 deficiency follow a reasonably uniform pattern among animal species: alterations in the skin and nervous tissues, anaemia and retardation of growth. Acrodynia, a characteristic form of dermatitis (lesions in the skin) in the ears and paws of the pyridoxine-deficient rat, is not observed in other species. In the chick there is anorexia and poor growth; the nervous symptoms include jerky movements of the legs when walking and spasmodic convulsions more intense than those caused by tocopherol deficiency, followed by complete exhaustion. By feeding 75% of the requirement, a borderline deficiency is obtained that produces perosis (see Chapter 10) without any nervous symptoms. In adult chickens reproduction is affected, both in egg production and hatchability in the female and involution of testes and comb in the male.

Pyridoxine-deficient pigs are affected by poor growth, convulsions and microcytic, hypochromic anaemia with high deposits of iron in the spleen.

Like some other B vitamins, a pyridoxine deficiency reduces the immune response of the pig.

A high amount of oxalic acid is excreted in the urine of pyridoxine-deficient animals; the acid is derived from disturbed metabolism of alanine, serine and other compounds which under normal conditions are metabolized to compounds other than oxalic acid.

Requirements for Pyridoxine

Because of the wide distribution of this vitamin in feedstuffs, the usual rations provide sufficient amounts of it for all types of monogastric species, even for heavy breeds of chickens which show a much higher requirement than White Leghorn chicks. This requirement in chicks increases also with the protein content of the diet or with that of certain amino acids such as methionine or tryptophan since the vitamin is strongly involved in amino acid metabolism (see above). Its requirement is also increased in the presence of linseed oil meal; this adverse effect can be eliminated by autoclaving the meal or by increasing the levels of pyridoxine in the diet.

PANTOTHENIC ACID

$$HO{-}CH_2{-}C(CH_3)_2{-}CH(OH){-}C({=}O){-}NH{-}CH_2{-}CH_2{-}COOH$$

Pantothenic acid

Pantothenic acid is a peptide of the amino acid β-alanine and a butyric acid derivative. This vitamin is widely distributed; its name is derived from the Greek *pantothen* meaning from everywhere, indicating its ubiquitous distribution. Free pantothenic acid is an unstable, highly hygroscopic, viscous oil. The solid calcium salt is the most common form in which the vitamin is added to diets. The synthetic product is a racemate of the *d* and *l* optically active isomers. Only the *d*-form occurs in nature and is biologically active; the racemic form has one half the biological activity of the natural product. Pantothenate is fairly stable in feedstuffs during prolonged storage; there are considerable losses due to heating during processing, but the usual pelleting causes only small losses.

The general and important role of pantothenic acid is as a constituent of coenzyme A (CoA). The active group of coenzyme A is the sulphydryl group and to point out this property, its name is sometimes shortened to

CoASH. The combination of CoA with two-carbon fragments is an essential step in activating these fragments for acetate transfer. As pointed out in preceding chapters, carbohydrates, fats and some amino acids are converted to acetyl-CoA (CH_3.CO.CoA) before they are oxidized in the tricarboxyl acid cycle, thus the name coenzyme A means coenzyme for acetylation. The formation of acetyl-CoA is an essential step in the synthesis of fats and steroids. Furthermore, CoA can activate carbon fragments larger than acetate so that it is involved in acyl transfers.

All living cells contain CoA due to its essential functions. Pantothenic acid is widely distributed in feedstuffs, mostly in bound form as CoA. Rich sources are liver, egg yolk, groundnuts and yeast.

Deficiency Symptoms

The chief lesions of pantothenic acid deficiency in all species involve the nervous system, the adrenal cortex and the skin; thus, a correct supply of pantothenic acid is important in coping with stress and disease.

In chickens the first symptoms of pantothenic acid deficiency are retardation of growth and impaired feather development. The further signs of deficiency appearing in young chicks are very similar to those of biotin deficiency: severe dermatitis with crusty, scab-like lesions in the corner of the mouth and around the eyes with the eyelids frequently stuck together; dermatitis may also appear at the bottom of the feet, feathers break and there is myelin degeneration of the spinal chord and degenerating fibres are seen in all segments of the chord (see PLATE XB).

Hatching is strongly affected by deficiency of this vitamin. When its supply is borderline, the hatched chicks are very weak and fail to survive. In severe deficiency the hatchability declines and the chicks cannot emerge from the egg.

The utilization of metabolic energy has been shown to be impaired in growing chickens; because of the deficiency of pantothenic acids, more heat was liberated, leaving less energy for productive purposes, thus causing significant decreases in the amount of protein and fat stored in the body (Beagle and Begin, 1976).

Pigs deficient in this vitamin show a scurfy skin, thin hair, secretion around the eyes, gastrointestinal troubles and abnormal gait with stiffened hind legs called 'goose stepping', which results from degeneration and demyelination of dorsal nerves.

Very active synthetic antagonists of this vitamin have been developed that can induce deficiencies.

Requirements for Pantothenic Acid

Requirements for egg production are low and practical diets contain sufficient amounts. However, the requirements are considerable in order to assure normal growth of young chicks and good hatchability. Thus, calcium pantothenate is usually added to rations for chicks and breeding hens (see above). The requirements of turkeys are almost twice those of chicks. Supplemental pantothenate is generally required in cereal- and soybean-meal-based diets for pigs.

BIOTIN

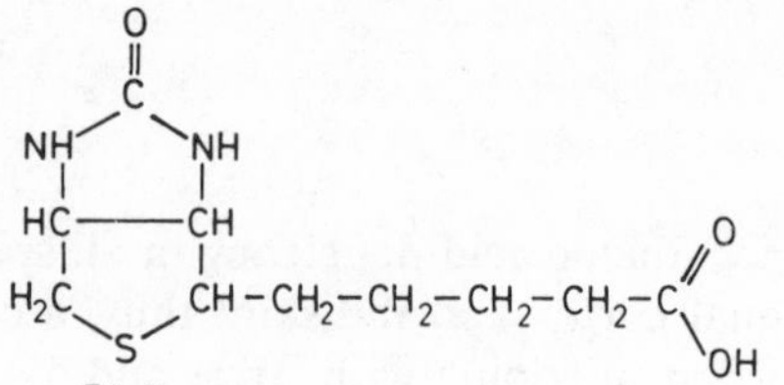

Biotin

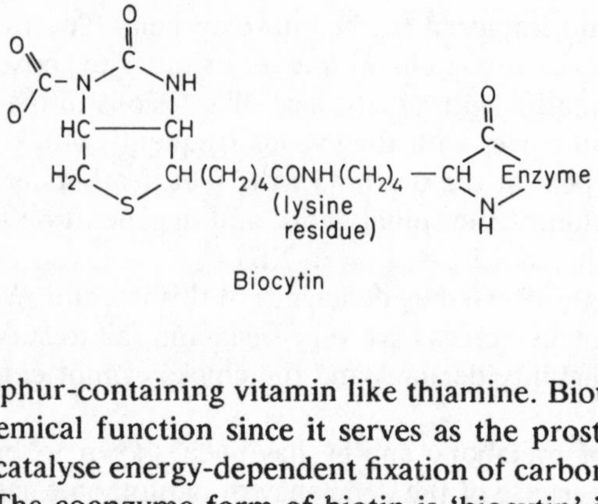

Biocytin

Biotin is a sulphur-containing vitamin like thiamine. Biotin performs a very important chemical function since it serves as the prosthetic group of enzymes which catalyse energy-dependent fixation of carbon dioxide to various compounds. The coenzyme form of biotin is 'biocytin' in which the biotin molecule is bound to the ϵ-amino group of a lysine residue of the apoenzyme. The structure of the CO_2–biotin compound is shown above.

Very important processes for carbon dioxide fixation are the following: (1) Conversion of acetic acid to malonic acid; this is an essential step in the synthesis of long-chain fatty acids (see Chapters 6 and 24). (2) Conversion of propionic acid to methylmalonic acid plays a key role in gluconeogenesis, an important process in all animals, especially in ruminants (see Chapters 4 and 5). (3) Conversion of pyruvate to oxaloacetate, a very important step in the production of energy through the tricarboxylic acid cycle (see Chapters 4 and 6). (4) Furthermore, biotin appears to be active in the deamination of amino acids, urea synthesis, etc. (see Chapter 7).

Biotin Sources and Symptoms of Deficiency

In spite of the need for biotin in important metabolic processes, cases of biotin deficiency are rare. Also in monogastric species the small amount required is usually supplied in the diet and is augmented by microbial synthesis in the digestive tract, particularly in those animals that practise coprophagy. The biotin present in the usual feedstuffs is sufficient, though cereals are poor sources and only 40–50% of the biotin present in wheat, barley and also fish meal are available to animals; however, the biotin in maize, oil cakes and green crops is fully available.

Biotin deficiency can be produced in monogastric species by feeding diets containing high levels of raw egg white, since egg white contains avidin, a protein which combines with biotin and prevents its absorption from the intestine. Avidin is denatured by moist heat, thereby preventing its binding with biotin, thus the feeding of cooked egg white does not cause biotin deficiency (see Chapter 20). It may be noted that egg yolk is very rich in biotin, therefore the ingestion of whole egg, even uncooked, does not cause biotin deficiency in man.

Other dietary factors inducing biotin deficiency are the addition of sulfa drugs, which eliminate microbial synthesis, and the presence of rancid fat, which inactivates biotin.

Deficiency of biotin in the diet of the chick results in retarded growth and brittle feathering. The lesions in biotin deficiency in chicks are similar to those observed in pantothenic acid deficiency (see above), but dermatitis appears first on the feet. The bottoms of the feet become rough; the toes become necrotic and slough off. Later the dermal lesions spread to the eyes and to the area around the beak.

Biotin is also one of the nutritional factors, along with manganese, whose supply is needed for preventing perosis (see Chapter 10). A low supply of biotin in a breeder ration causes reduced egg hatchability and embryonic malformations.

Turkeys are very susceptible to biotin deficiency. The biotin requirements of turkey poults are 2–3 times those of the chicken. A generous supply of biotin to poults helps to prevent leg disorders in growing turkeys.

It has been found recently that fatty liver and kidney syndrome, a fatal condition in chicks and poults, is a metabolic disorder associated with biotin deficiency. It was suggested that due to loss of activity of pyruvate carboxylase (see above) the gluconeogenic activity in the liver is depressed; in response to low blood glucose levels, mobilization of free fatty acids occurs and infiltrate into tissues.

Biotin deficiency symptoms in pigs are similar to those induced by scarcity of other B vitamins: dermatitis, which can spread over the entire body, and cracks in the feet making walking difficult.

FOLIC ACID

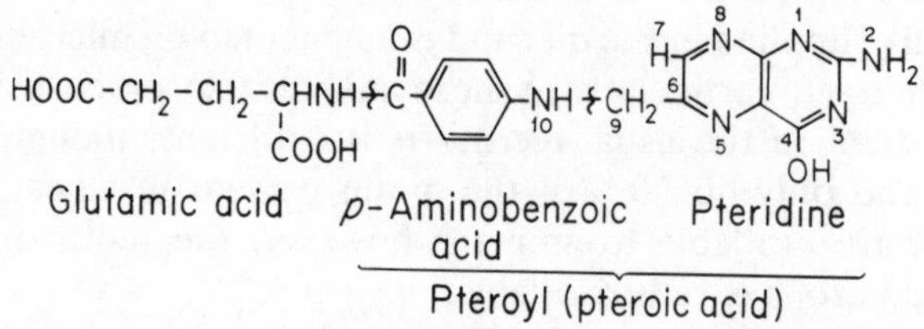

Glutamic acid *p*-Aminobenzoic acid Pteridine

Pteroyl (pteroic acid)

Folic acid (folacin)

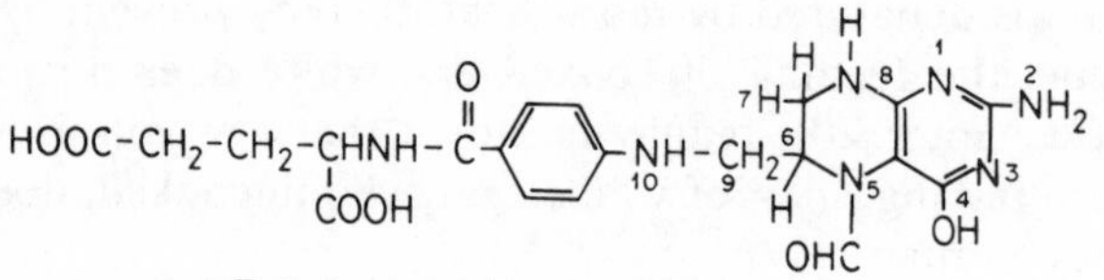

Folinic acid (leucovorin)

As shown by its formula, folic acid, also known as pteroylglutamic acid, consists of three basic components: a pteridine nucleus, a *p*-aminobenzoic acid residue, and a molecule of glutamic acid. Additional molecules of glutamic acid (3–7) may be linked to folic acid by peptide bonds. These folic acid glutamates appear to be a biologically inactive storage form. To be of use in the body, folate must undergo enzymatic breakdown by which all but one of the glutamic acid molecules split off, thereby giving pteroylglutamic acid. In the organism folic acid is reduced to its coenzyme form, tetrahydrofolic acid or folinic acid. The coenzyme functions in various biological reactions in the transfer of single-carbon units, a role analogous to that of CoA in the transfer of two-carbon units (see above). The transfer of C_1 units (formyl or methyl groups) is effected metabolically by folinic acid or by biotin or with the participation of vitamin B_{12} (see later). In this capacity folate coenzymes are required in the following processes: (1) interconversion of the amino acids serine and glycine (see Chapter 7), (2) synthesis of purines, (3) degradation of histidine, and (4) synthesis of methyl groups for compounds such as choline and methionine.

Symptoms of Deficiency

As mentioned above, folic acid is involved in the biosynthesis of purines. This synthesis is depressed in folic acid deficiency and deficient purine synthesis results in a deficiency of nucleoprotein for blood-cell maturation; as a consequence, anaemia and pale muscle will occur. The effects of folic acid deficiency on the metabolism, together with the changes in the blood picture, lead to weakness, loss of appetite and diarrhoea. There is, furthermore, an increased susceptibility to infections as the synthesis of immunoglobulins is impaired.

In addition to an anaemic appearance, signs of folic acid deficiency in chicks are a retardation of growth, poor feathering and deficient pigmentation of the feathers. A deficiency in the breeding diet not only reduces egg hatchability, it also causes embryo mortality and malformation of the surviving embryos. Nervous disorders have been observed in deficient turkeys.

Sources and Requirements of Folic Acid

Folic acid is distributed in feedstuffs of plant and animal origin. The folic acid content of ingredients commonly used plus that produced by bacterial synthesis in the intestine are sufficient to meet the requirements of monogastric farm animals. Supplementation of the diet with this vitamin has been recommended for chicks and particularly for turkeys.

The folic acid requirement of animals may be increased when sulfa drugs are included in their diets. The microbial synthesis of folic acid is inhibited in the presence of sulfa drugs because of a similarity in the chemical structure of *p*-aminobenzoic acid and sulfa drugs.

VITAMIN B_{12}

Research carried out in the 1940s and 1950s revealed that the same compound present in foods of animal origin and in microorganisms is responsible for the following three biological functions—the compound was identified as vitamin B_{12}: (1) The growth rate of single-stomached animals, particularly poultry, is considerably improved by addition of foods of animal origin (fish meal, meat meal or milk powder) to diets entirely of plant origin. This growth-stimulating effect of animal protein feeds may be attributed to the presence of the animal protein factor later recognized as vitamin B_{12} (see Chapter 7). (2) Human beings suffering from the dangerous condition called pernicious anaemia can be cured if they include liver in their diets. The principal active ingredient in liver has been identified as vitamin B_{12}. (3) Optimal growth rate of certain types of lactobacilli (*Lactobacillus lacti* Dorner) will be attained only in the presence of a particular factor in their basal nutrient medium. This growth factor, first isolated from yeasts, was recognized as vitamin B_{12}.

Sources

The origin of vitamin B_{12} in nature is almost exclusively microbial synthesis. It is synthesized by a wide range of bacteria and in small amounts by yeasts and fungi. It is widely distributed in foods and feeds of animal origin, liver being a particularly rich source. Its presence in the tissues of animals is due to the ingestion of animal products or from intestinal or rumen synthesis. Faeces are a quite rich source since colonic microorganisms make large amounts of the vitamin. Plant products are practically devoid of it.

The Chemical Structure of Vitamin B_{12} and its Metabolic Functions

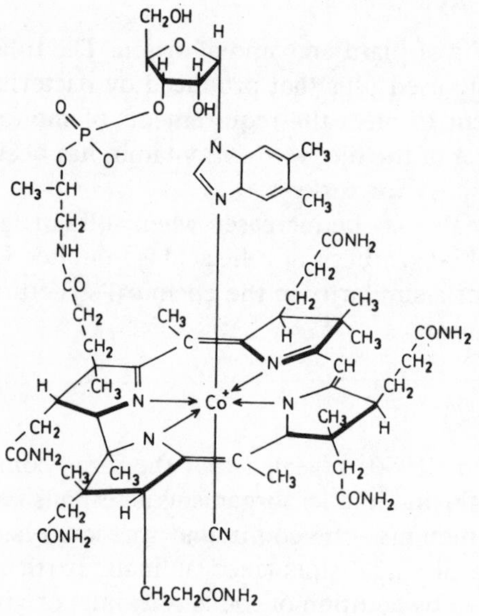

Vitamin B_{12}

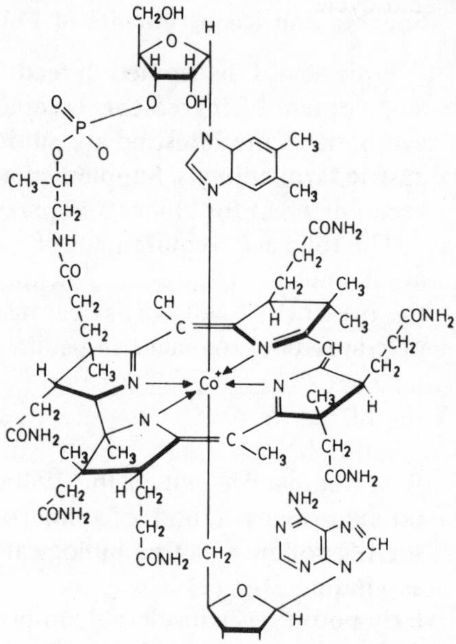

Coenzyme B_{12}

Vitamin B_{12} is the only vitamin that contains a metal ion, cobalt. Its structure comprises four reduced pyrrole rings linked together and designated 'corrin' because they are the core of the molecule. Another condition for vitamin B_{12} activity is the presence of a cobalt atom in the centre of the corrin nucleus. The name cobalamin has been used for such compounds. Vitamin B_{12} is always isolated from its natural sources as cyanocobalamin in which the cyano group is attached to the cobalt atom. It is now widely suspected that this is an artifact and must be converted in the body to one of the naturally occurring forms such as hydroxycobalamin in which the cyano group is replaced by a hydroxy group. The vitamin fulfils its metabolic functions after being transformed to coenzyme B_{12}, in which a molecule of 5′-deoxyadenosine is attached to the cobalt in place of the cyanide. The particular bond between the cobalt atom and a carbon atom of the pentose is considered to be responsible for the ability of the coenzyme to perform several important metabolic functions. The following biochemical processes which depend on the presence of coenzyme B_{12} occur in bacteria and in animal tissues as well.

(1) Methylmalonyl Mutase. Coenzyme B_{12} is the active principle in methylmalonyl mutase, the enzyme needed for the transformation of methylmalonyl-CoA to succinyl-CoA (see Figure 5.4). The latter reaction is an essential step in the utilization of propionic acid (produced in the rumen) for gluconeogenesis or for energy production through the tricarboxylic acid cycle:

$$\text{Propionyl-CoA} + CO_2 + \text{ATP} \xrightarrow{\text{Biotin}} \text{methylmalonyl-CoA} \xrightarrow{B_{12}\text{ coenzyme}} \text{succinyl-CoA} \begin{cases} \to \text{glucose} \\ \to CO_2 + H_2O \end{cases}$$

Thus, a B_{12} deficiency impairs the utilization of propionate and therefore methyl malonate has been found in increased amounts in the urine of cobalt-deficient sheep. The depression in the utilization of propionic acid may be one of the reasons why sheep are so seriously affected by B_{12}, respectively cobalt deficiency (see Chapter 10).

(2) Vitamin B_{12} Transmethylase. Vitamin B_{12} and folinic acid are involved in transmethylation processes. Transmethylation is a condition for the synthesis of a number of amino acids; for example, the formation of methionine from homocysteine (see Chapter 7). In this biosynthesis of methionine a methyl group is transferred from a donor, 5-methyltetrahydrofolic acid, to an acceptor, homocysteine, giving rise to methionine according to the following scheme:

$$\text{5-Methyltetrahydrofolic acid} + \text{homocysteine} \xrightarrow[\text{Methyltransferase}]{B_{12}\text{ coenzyme}} \text{tetrahydrofolic acid} + \text{methionine}$$

In the absence of vitamin B_{12} the rate of reaction falls, accompanied by diminished regeneration of tetrahydrofolic acid and consequently deficiencies of folic acid and methionine are produced.

(3) DNA Biosynthesis. Vitamin B_{12} and folinic acid are required for the biosynthesis of deoxyribonucleic acid (DNA). Therefore, inadequate ingestion of both these vitamins leads to a decrease in cell division and consequently to nutritional anaemias.

Absorption and Transport of Vitamin B_{12}

The absorption of vitamin B_{12} from the intestine requires the presence of a carrier substance IF, a glycoprotein which is secreted into the stomach. This substance has been designated the intrinsic factor (IF) since it is synthesized in the body itself and functions together with ingested vitamin B_{12}; in the absence of IF, vitamin B_{12} deficiency occurs. The B_{12}.IF complex formed in the gastrointestinal tract passes down to the ileum and is subsequently absorbed through the mucosal cell membranes by pynocytosis (see Chapter 3). Then the complex is broken down and the vitamin B_{12} released is transported in the serum bound to other carrier proteins. The amount of B_{12} not needed for immediate use is stored in the liver. Some of the vitamin, particularly when supplied in large quantities by the diet, may be absorbed in its free form by a diffusion mechanism. It should be noted that the nature of the intrinsic factor varies according to the animal species; for example, IF isolated from pig stomach inhibits vitamin B_{12} absorption in poultry.

The conversion of vitamin B_{12} to its metabolically active coenzyme form takes place mainly in the liver. Unlike other B vitamins, vitamin B_{12} that is not required for immediate utilization is stored in the liver and other organs.

Supply and Deficiency Symptoms

In Monogastric Species

The supply of vitamin B_{12} in the diet of humans and monogastric species (poultry and pigs) is of greatest importance. However, symptoms of frank deficiency are rarely seen under practical farm conditions. The requirement for the small amounts needed is easily covered by addition of feedstuffs of animal origin to the diet and/or by incorporation of vitamin B_{12} concentrates; pure vitamin B_{12} is obtained as a by-product of the preparation of certain antibiotics. Vitamin B_{12} synthesized by intestinal bacteria is of limited value since this synthesis occurs in the lower intestine beyond the sites of active absorption. Pigs and poultry housed with access to litter ingest considerable amounts of B_{12} from the litter.

The absolute amount of vitamin B_{12} required is very small compared to other vitamins; for example, 1 kg of feed for chicks should contain about 9 μg versus 4 mg of riboflavin. The absolute requirement may vary somewhat according to the composition of the diet. Excess fat and protein in the diet increase the need for the vitamin through its functions in the metabolism of protein and fat. Because of the involvement of vitamin B_{12} in the biosynthesis of methyl groups (see above), the need for it is less when the diet contains an abundance of compounds that can supply methyl groups, e.g. methionine or choline.

In growing chicks and broilers a deficiency of vitamin B_{12} reduces the gain of body weight and may be followed by secondary symptoms, not particularly specific for this vitamin, such as nervous disorders or defective feathering.

In laying birds a deficiency of vitamin B_{12} reduces egg production and particularly hatchability; impaired viability of the embryos goes along with decreasing concentrations of the vitamin in egg yolk.

Baby pigs show a comparatively high requirement for vitamin B_{12} (20 μg per kg feed). Growth failure is the most evident effect of B_{12} deficiency in growing pigs apart from visible symptoms (rough skin and hair coat, nervous disorders, incoordination of hind legs). The reproductive performance of sows (litter size, weight and viability of the piglets) depends amongst others on the supply of the required amount of vitamin B_{12}.

In Ruminants

As pointed out in Chapter 10, cobalt is an essential trace element for ruminants and its only known physiological function is its role in the microbial synthesis of vitamin B_{12} in the rumen. The symptoms of cobalt deficiency in ruminants appear to be entirely due to the lack of vitamin B_{12}. Various types of rumen bacteria produce vitamin B_{12} as an essential growth factor, but the minimum requirements for fermentation are much lower than those required for the animal's tissue level. At the tissue level the vitamin is needed as part of a coenzyme enabling the utilization of propionic acid (see above). The vitamin is synthesized adequately in the rumen, provided that sufficient cobalt is available. Several dietary factors may influence the ruminal synthesis of vitamin B_{12}; restriction of roughage may decrease the synthesis of vitamin B_{12}.

Whereas inorganic cobalt salts satisfy the B_{12} requirements of ruminating animals, in preruminant animals preformed vitamin B_{12} mostly provided by milk, is required.

CHOLINE

$$HO \cdot CH_2 \cdot CH_2 - N(OH)(CH_3)_3$$

Choline

Choline does not readily qualify as a vitamin because it is actually a structural component of fat and nerve tissue and is required by a number of animals species (chick and pig) in the diet at levels far greater than those of the other vitamins. Moreover, no evidence exists for the role of choline as a cofactor in enzymatic reactions. Choline is widely distributed in feedstuffs of plant and animal origin, mainly in a bound form as a component of phospholipids (see Chapter 6). Relatively rich sources of choline are soybean meal, fish meal and dry yeasts.

Functions

Various physiological functions of choline are known: (1) It is essential for the formation of acetylcholine, a compound that is physiologically important in the transmission of nerve impulses. (2) Choline is needed together with other nutrients, particularly manganese salts, for preventing the bone disease of chickens called perosis (see Chapter 10); choline seems to be required for this purpose since lecithins are constituents of the organic matter present in bones and cartilage. (3) Choline, methionine and betaine are lipotropic substances, i.e. they prevent abnormal accumulation of fat in the liver by promoting the removal of triglycerides from the liver through their transformation to lecithins. (4) Choline furnishes labile methyl groups for a number of metabolic processes such as the formation of methionine from homocysteine (see above) and of creatine present in muscles (see Chapters 7 and 22). In addition to its interrelationship with methionine, vitamin B_{12} and folic acid are involved in the transfer of methyl groups (see above).

Supply and Requirements

Most animals species can produce by biosynthesis the required amount of choline—even laying hens, although eggs contain considerable amounts of choline found almost entirely in bound form as phospholipids (see Chapter 20). In growing animals, chicks and pigs, the rate of biosynthesis of choline is not sufficient to satisfy their needs, and if there is a deficiency of choline the growth rate may be reduced. As mentioned above, the requirements for choline are much higher than those for other B vitamins; for example, chicks require 200 times the amount of riboflavin. Nevertheless, the usual feedstuffs generally supply the choline requirements for growing animals. However, it is an accepted practice to add choline chloride to diets of growing chicks.

It may be noted that the requirements for choline decrease when the diets contain a surplus of other substances involved in the transfer of methyl groups (methionine, folic acid or vitamin B_{12}).

ADDITIONAL METABOLICALLY IMPORTANT COMPOUNDS

There are some additional compounds which fulfil important functions as coenzymes in the metabolic processes occurring in animal tissues; carnitine, lipoic acid and myoinositol are some examples. However, they cannot strictly be defined as vitamins because these compounds are provided through biosynthesis in the body and dietary deficiency is never encountered.

Carnitine has been discovered as an essential growth factor for certain insects and is involved in the fatty acid metabolism of higher animals. *Lipoic acid*, together with thiamine, constitutes part of an enzymatic system which catalyses the decarboxylation of pyruvic acid (see above). *Myoinositol* acts in a similar way to choline as a lipotropic substance and stimulates the intestinal and tissue synthesis of biotin.

Unidentified Growth Factors

Some additional factors are known which influence favourably the growth rate of chicks and pigs and the hatchability of eggs. Such a response has been obtained by the addition of certain natural feedstuffs or fractions prepared from them to fully adequate diets containing purified nutrients and all the recognized vitamins and essential minerals. The responsible, but as yet unidentified factors are present in fresh or dried grass or grass juice (grass factor), in fish meal and fish solubles (fish factor) and in milk products (whey factor). Further evidence is required of their activity as vitamins.

VITAMIN C (ASCORBIC ACID)

Vitamin C or ascorbic acid is a hexuronic acid, a hexose derivative; it is easily oxidized to form dehydroascorbic acid, which is just as easily reduced back to its original form. Further oxidation of dehydroascorbic acid results in the formation of diketogulonic acid, an inactive compound; its formation is the result of an irreversible reaction. This reaction proceeds easily under conditions of light or heat or in presence of traces of heavy metals and explains the instability of vitamin C in foods.

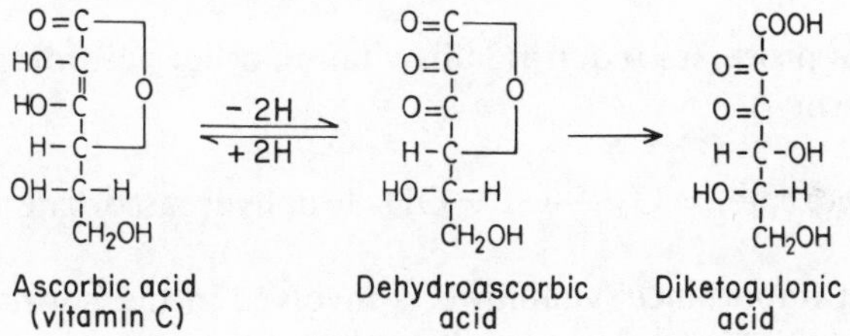

Ascorbic acid (vitamin C) $\underset{+2H}{\overset{-2H}{\rightleftharpoons}}$ Dehydroascorbic acid $\longrightarrow$ Diketogulonic acid

Most mammalian and avian species, including farm animals, are able to synthesize ascorbic acid from glucose in adequate amounts by the following reactions:

D-Glucose ⟶ D-glucuronic acid ⟶ L-gulonic acid ⟶ L-gulonolactone ⟶ L-ascorbic acid

Ascorbic acid was found to be a dietary requirement of only man, monkeys, guinea pigs and some species of birds, fishes and insects. The inability to synthesize ascorbic acid is apparently due to the absence of gulonolactone oxidase, the enzyme responsible for the final step in the biosynthesis of ascorbic acid; thus, the mentioned species develop vitamin C deficiency when dietary ascorbic acid is withheld. In mammals this biosynthesis occurs in the liver, in birds in the kidney. Generally it is unnecessary to supply vitamin C in the diet of farm animals (see later). Ruminants utilize only that vitamin C that is produced in their tissues because the vitamin C ingested with the diet is totally destroyed as it passes through the rumen; for this reason it is impossible to increase the vitamin C content of milk by increasing the quantity of the vitamin in the cow's diet. Among the foods of animal origin, milk is the only one which supplies man with vitamin C. The best sources of this vitamin for man are fresh fruit, particularly citrus fruits, leafy vegetables, tomatoes, green peppers and potatoes.

Functions of Vitamin C and Symptoms of Deficiency

The typical deficiency disease in humans is scurvy characterized by capillar fragility and resulting in haemorrhages in the body, loosening of the teeth, weak bones and anaemia. The defects can be explained by a failure in the formation of collagen, the interfibrillar ground substance of connective tissue. The formation of structurally correct collagen is related to the function of vitamin C in regulating the hydroxylation of proline and lysine bound in polypeptide chains. Collagen in connective tissue of healthy persons contains about 15% hydroxyproline and 1% hydroxylysine. Collagen produced in scorbutic persons contains very little hydroxyproline. This sharp reduction in the hydroxyproline content is accompanied by structural changes of the collagen that are associated with the described symptoms of scurvy.

Hydroxylation processes requiring this vitamin occur according to the following mechanism:

$$\text{R—H} + \text{ascorbate} + O_2 \longrightarrow \text{R—OH} + \text{dehydroascorbate} + H_2O$$

and are of great importance; vitamin C is involved in the normal oxidation and hence metabolism of tyrosine and in the synthesis of steroid hormones in the adrenal cortex.

Furthermore, vitamin C functions as an antioxidant and may act in the same processes as vitamin E; it participates in the conversion of folic acid to folinic acid (see above) and enhances the absorption of iron from the gut, presumably by maintaining iron in the reduced, and hence more absorbable, form.

Although the requirement of farm animals is met by tissue synthesis (see above), it is often suggested that stress caused by adverse environmental conditions (elevated temperature and air humidity, or subclinical diseases) may increase the need for vitamin C and under these circumstances animals may respond to dietary vitamin. It has been observed that retarded growth rate of chicks and decreased egg production of hens kept in very hot climates could be overcome by the addition of ascorbic acid to the diets. It may be mentioned that in the first few weeks of life, calves have only a limited capacity to synthesize ascorbic acid; their supply of the vitamin usually comes from milk (see Chapter 18).

BIBLIOGRAPHY

Vitamin B complex: General

Cormier, M. (1977). Regulatory mechanisms of energy needs. Vitamins in energy utilization. *Prog. Food Nutr. Sci.*, **2], 347–356.**

Goldsmith, G. A. (1975). Vitamin B complex. *Prog. Food Nutr. Sci.*, **1**, 559–610.

Gries, C. L., and Scott, M. L. (1972). The pathologies of thiamin, riboflavin, pantothenic acid and niacin deficiencies in the chick. *J. Nutr.*, **102**, 1269–1285.

Rose, R. C. (1980). Water soluble vitamin absorption in intestine. *Annu. Rev. Physiol.*, **42**, 157–171.

Roth-Maier, A., and Kirchgessner, M. (1977). B-vitamins in ruminant nutrition. *Livestock Prod.*, **4**, 177–189.

Thiamine

Brin, M. (1967). Functional evaluation of nutritional status. Thiamin. In *Newer Methods of Nutritional Biochemistry* (A. A. Albanese, ed.), Vol. 3, pp. 407–446. New York and London: Academic Press.

Evans, W. C. (1975). Thiaminases and their effects on animals. *Vit. Horm.*, **33**, 467–504.

Riboflavin

Foy, H., and Mbaya, V. (1977). Riboflavin. *Prog. Food Nutr. Sci.*, **2**, 357–394.

Wyatt, R. D., *et al.* (1973). A new description of riboflavin deficiency syndrome in chicks. *Poult. Sci.*, **52**, 237–244.

Niacin

Gopalon, C., and Rao, K. S. J. (1975). Pellagra and amino acid imbalance. *Vit. Horm.*, **33**, 505–528.

Henderson, LaVell, M. (1983). Niacin. *Annu. Rev. Nutr.*, **3**, 289–307.

Kodicek, E. (1976). Some problems connected with availability of niacin in cereals. *Bibl. Nutr. Dieta*, **23**, 86–87.

Pyridoxine

Coursin, D. B. (1975). Vitamin B_6 and pantothenic acid. *Prog. Food Nutr. Sci.*, **1**, 183–190.

Dakshinamuri, K. (1982). Neurobiology of pyridoxine. *Adv. Nutr. Res.*, **4**, 143–180.

Fuller, H. L. (1964). Vitamin B_6 in farm animal nutrition. *Vit. Horm.*, **22**, 659–677.

Ink, S. L., and Henderson, M. L. (1984). Vitamin B_6 metabolism. *Annu. Rev. Nutr.*, **4**, 455–470.

Pantothenic acid

Beagle, W. S., and Begin, J. J. (1976). The effects of pantothenic acid in the diet of growing chicks on energy utilization and body composition. *Poult. Sci.*, **55**, 950–957.

Biotin

Murthy, P. N., and Mistry, S. P. (1977). Biotin. *Prog. Food Nutr. Sci.*, **2**, 405–455.

Scheiner, J., and De Ritter, E. (1976). Biotin content of feedstuffs. *J. Agr. Food Chem.*, **23**, 1157–1162.

Whitehead, C. C. (1977). The use of biotin in poultry nutrition. *World Poult. Sci. J.*, **33**, 140–154.

Vitamin B_{12}

Elliot, J. M. (1980). Propionate metabolism and vitamin B_{12}. In *Digestive Physiology and Metabolism in Ruminants* (Y. Ruckebush and P. Thivend, eds), pp. 485–504. Lancaster: MTP Press.

Farguharson, J., and Adams, J. F.(1976). The forms of vitamin B_{12} in food. *Br. J. Nutr.*, **36**, 127–141.

Grasbeck, R., and Salonen, E. M. (1976). Vitamin B_{12}. *Prog. Food Nutr. Sci.*, **2**, 193–232.

Seetharam, B., and Alpers, D. H. (1982). Absorption and transport of cobalamin (vitamin B_{12}). *Annu. Rev. Nutr.*, **2**, 243–269.

Shane, B., and Stokstad, E. L. R. (1984). The interrelationships among folate, vitamin B_{12} and methionine metabolism. *Adv. Nutr. Res.*, **6**, 133–170.

Choline

Kuksio, A., and Mookerjea (1978). Choline. *Nutr. Rev.*, **36**, 201–207.

Part III

MEASUREMENT OF THE UTILIZATION OF NUTRIENTS AND ENERGY OF FOOD

Introduction: The Design of Feeding Experiments

By D. Drori

As pointed out in Chapter 1, the results of chemical analyses of feedstuffs require supplemental information about the utilization of the feed before its nutritional value can be determined. Quantitative relationships between the input of a feed and animal production can be inferred from the results of feeding experiments.

Usually, in feeding experiments the value of a feed or a defined diet is compared to that of another of known efficiency. The value of the tested feed may be expressed in terms of its digestibility, metabolizability (see Chapter 14) or its effect on production. Production may be expressed in terms of energy, or more simply as body weight gain, amount of milk, eggs or wool produced. Since the composition of body weight gain in large animals is difficult to measure, nitrogen balance trials are often performed (see Chapter 7). Such trials produce an indirect measure of nitrogen retained in the body. From the nitrogen retention the amount of retained protein may be estimated. Often more than one dependent variable is measured in a feeding experiment, in order to establish the relationship between the digestibility, metabolizability and productive value of a feed or a diet.

Although for the purpose of feed evaluation animal feeding experiments are more reliable and applicable than in vitro feed analysis, their execution is fraught with difficulties. Within any population of living organisms not entirely inbred, there is variation due to both genetic and environmental factors. This means that, even under presumably similar conditions, individual organisms do not perform in an identical way. A commonly used term for the variation among organisms that the biologist is unable to control is *biological variation.*

The sources of uncontrolled variation in feeding experiments are also of hereditary and environmental origin. It is well known that farm animals have widely different quantitative and qualitative genetic characteristics. Potential growth rate, the production of milk or eggs, the partition of feed energy between body tissue and milk or eggs, and the composition of milk, are only a few examples. As a rule, farm animals are not inbred and even individuals within one herd or flock of the same breed show great genetic variation. Even in well-planned experiments environmental factors are seldom perfectly controlled and introduce further variation. Age and long reproductive cycles further increase the variation at any given time, especially in lactating females. Environmental factors may introduce variation which may persist long after the factors have ceased to act. Such *after-effects* may be of short or long duration. Prepartum feeding or twinning may affect performance in the subsequent lactation, and neonatal or early effects may affect performance for a lifetime. Complex long-term experiments are performed to study after-effects, usually in a factorial design (see below).

In order to design efficient experiments, the uncontrolled variation must be minimized. This is done in experimental biology by using appropriate methods of statistical design and analysis and by using techniques that reduce errors of measurement.

Many textbooks on statistical methods in biology and agriculture are available, but most of these have been written by plant scientists rather than animal scientists (see references). Unfortunately, both the statistical and the biological problems are more complex in animals. Furthermore, the difficulties increase with the size of the animal. In large farm animals of both sexes the number of experimental observations is necessarily limited by their availability, and in females by the fact that they have long reproductive cycles. Ideally, among lactating females only those of the same breed, age and parity are comparable for experimental purposes and, even then, only if they are at about the same stage of lactation and pregnancy. Undoubtedly, the design of feeding experiments in lactating dairy cows is more problematic than in any other class of stock.

A large number of experimental designs and techniques are used in nutrition research. In the following section only general guidelines for the design and execution of feeding experiments are presented. In principle, the purpose of proper experimental design and technique is to minimize the experimental error that is caused by uncontrolled variation and inaccuracies of measurement.

GUIDELINES FOR THE DESIGN OF FEEDING EXPERIMENTS

Diets

The most important factor in a feeding experiment is the composition of the experimental diets. The guiding principle for the formulation of two or more experimental diets in one experiment is that they must be similar and adequately balanced in every respect except for one or more ingredients tested in the experiment. In order to measure the relative energy value of a diet, it must supply adequate levels of protein, minerals and vitamins; to measure the utilization of protein, the diet must supply adequate levels of energy, minerals and vitamins. However, if the sole purpose of the experiment is to determine the required concentration of a vitamin, mineral or amino acid, diets may be fed which contain different levels of the specific nutrient ranging from the deficient to the excessive. Unless an experiment specifically deals with the *interaction* between two or more nutrients, it is desirable to vary only one ingredient at a time.

Animals

The choice of animals for a feeding experiment is of major importance. Certainly, the animals must belong to the population to which the results must be applied. It is also common knowledge that the control group and the treatment group(s) must be as similar as possible before the experiment is begun. In order to increase the power of the experiment, the animals chosen must be members of a homogeneous group. This is so because the experimental error, which is a yardstick which determines whether the differences between any two treatments are meaningful (statistically significant), is greatly affected by the variability of the animals. A very heterogeneous group of animals is unfit for experimental work because it produces a large experimental error and inconclusive results unless the number of animals is very great.

Statistical Design

Proper statistical design also increases the power of an experiment. In simple growth trials with two or more treatments, it is best to assign the same number of animals to each treatment. If the age and body weight of all the available animals are uniform, it is best to distribute them at random. Randomization tends to balance latent differences between the experimental groups. In healthy farm animals these differences are mainly of a genetic source. If variation in body weight is large, very light and very heavy animals may be discarded. Alternatively, the animals may be stratified into weight groups (e.g. light, medium and heavy) and these in turn may be distributed at random, one weight-group at a time, to the treatments.

Since increasing the number of observations reduces the error of mean estimates, it is desirable to measure both input and output on an individual basis. In practice, the observation of animal output on an individual basis is relatively easy, but that of input, i.e. food intake, is more difficult. Although statistical theory favours the individual control of food intake, other considerations militate against it. The isolation of an animal may reduce its food intake and thereby its general performance; this often occurs in lambs (see Chapter 23). In addition, especially in small animals, the labour and expense involved in the control of individual food intake would be prohibitive. From the practical point of view, it would appear that group feeding as done on the farm produces the most applicable results. Therefore, group feeding is often used in experimental work. The statistical compromise between individual and group feeding is *replication* in lots. In a typical growth trial using 300 chicks and three dietary treatments, the chicks may be allotted into 15 groups of 20 birds and five replicates, i.e. five lots of 20 birds may be assigned at random to each treatment. In such an experiment, assuming no mortality, 100 observations per treatment may be obtained for weight gain but only five for food intake and food efficiency, i.e. the ratio of body weight gain to food consumed. Assuming that the total number of animals for an experiment is fixed, replication increases the number of certain observations by breaking up a large group into several small ones. Replication also serves to reduce chance environmental variation which may affect the treatment groups unequally and bias the experimental results.

The length of a feeding experiment is determined by practical considerations. A growth trial must be long enough to permit any potential differences between treatments to be expressed. In poultry, pigs, lambs and young calves, growth trials may be continued for weeks and in fattening cattle for months.

The significance of the differences among the performance means of treatments is usually evaluated by analysis of variance. However, in some experiments, the effects on performance of graded levels of nutrients (vitamins, minerals, amino acids, etc.) are measured. Regression analysis may be used to evaluate the results by calculating a dose–response relationship. It is important to realize that such relationships may approximate a straight line only up to a point. Above that point the relationship may approach a plateau or reach a peak and decline. In the latter case, performance is optimal within a limited range; this often occurs in experiments with amino acids (see Chapter 7). Therefore, linear regression analysis must be used with care, and curvilinear relationships must be considered.

Simple designs, as described above, are usually suitable for growth trials and also for production trials with layers. However, in experiments dealing with milk production in ruminants, conditions are different (see Chapter 21). At any given time, a herd of lactating cows represents a very heterogeneous group and a simple design may be rather inefficient when the effects of diets on production are studied rather than the nature of the lactation curve. Even if a number of cows sufficient for an experiment calf at about the same time, variation among cows and the experimental error would be very large. To reduce the experimental error *changeover designs* are used. In a changeover design each animal is submitted sequentially to more than one treatment. Therefore, there is no fixed control group although one treatment may be so designated. In a changeover design each animal serves as its own control; with appropriate statistical analysis, the experimental error, including the effect of animal variation, is reduced and efficient experiments may be performed with a relatively small number of animals.

In the simplest changeover design, the reversal trial, two treatments designated A and B are used with two groups of cows, in an experiment extending over two periods of several weeks each, as indicated below.

	Period 1	Period 2
Group 1	A	B
Group 2	B	A

In this type of design, the dependent variable submitted to statistical analysis is the change in milk production within cows which is coincident with the change in treatments, i.e. the difference in mean production between periods 1 and 2.

The efficiency of a reversal trial may be greatly increased by reversing the treatments a second time. This design is referred to as a *double reversal* or *switchback trial.* It extends over three consecutive periods as indicated below.

	Period 1	Period 2	Period 3
Group 1	A	B	A
Group 2	B	A	B

In a double reversal trial the variable submitted to statistical analysis is the deviation in milk production in period 2 from the mean of periods 1 and 3. Double reversal trials can be easily extended to more than two treatments. Their efficiency is very high; small differences between treatments can be detected with accuracy.

Another common changeover design, adapted from agronomic experiments, is the *latin square.* In this design the number of treatments, of groups and of periods is identical, as indicated below.

		Periods			
		1	2	3	4
Groups	1	C	D	A	B
	2	B	A	C	D
	3	A	B	D	C
	4	D	C	B	A

The major disadvantage of changeover trials with lactating cows is the risk of nutritional after-effects, also referred to as *carryover effects.* These may be minimized by extending the planned experimental periods and discarding the results of the first week or fortnight of each period.

Other, more complicated changeover designs, have been suggested especially for work with dairy cows.

If the specific purpose of an experiment is to study short- or long-term after-effects, or the simultaneous effects of more than one variable including their possible interactions, the *factorial design* may be used. This design permits the combination of several comparisons in one experiment, preserving the separate examination of each factor and the simultaneous determination of interactions which may occur between the factors. An example of a factorial design would be a lactation experiment using N levels of feeding in the prepartum period and M levels of feeding during the lactation. The total number of sequences (subtreatments) would be $N \times M$. In a growth trial, N levels of variable A may be combined with M levels of variable B to produce $N \times M$ experimental diets fed in one experiment. Provided the number of animals available for an experiment is sufficiently large, any number of variables and levels may be combined into a single experiment, although the subtreatments must be replicated. Factorial experiments often produce a large amount of information and present the ideal design for studying interactions between nutrients.

Experimental Techniques

Miscellaneous experimental techniques may be used to increase the efficiency of feeding experiments. These usually decrease the environmental variation. However, a technique referred to as split-litter reduces both the genetic and environmental variation.

Split-litter

In species that produce litters, i.e. more than one offspring at a time, the young from any one litter may be placed on treatments in such a fashion that each litter is represented in each treatment group. This procedure tends to reduce the mean genetic differences among the various treatment groups and the maternal and pre-experimental environmental effects acting on them, because littermates are genetically related and are usually raised by the same dam in the same environment. This technique is practical in pigs, which produce large litters (and in many laboratory rodents). Only offspring of the same sex are comparable for experimental purposes. In beef and dairy cattle, monozygous twins have been used in experiments with two treatments; this is a particular case of the split-litter technique. Monozygous twins are of the same sex and genetically identical. Thus, they practically eliminate the genetic variation between two treatments. A feeding experiment with monozygous twins may be 10–20 times more efficient than one using animals chosen at random from the same herd and breed. Experiments with monozygous twins have shown the pervasive influence of genetic make-up on production traits. Unfortunately, monozygous twins are rare and obtaining them for experimental purposes requires an organized effort (as done in New Zealand). Furthermore, they may not perform as well as animals born singly, probably because twins develop more slowly in utero. Therefore, although the use of monozygous twins in cattle is statistically very efficient, it has been criticized for poorly representing the population.

Pair-feeding

In quantitative feeding experiments voluntary feed intake is a dependent variable and must be measured, because, for most practical purposes, increasing voluntary feed intake improves performance. However, in experiments specially planned to investigate deficiencies of specific nutrients (vitamins, minerals, amino acids or protein), diets deficient in the respective nutrients may be used. These deficient diets may greatly reduce the voluntary feed intake and thereby exaggerate the effects of the deficiency. In order to distinguish the effects of the deficiency proper from those of reduced feed intake, it is useful to limit the intake of the control, i.e. the non-deficient diet, to equal that of the deficient diet. This technique is known as *pair-feeding.* As a rule, control animals pair-fed to a deficient group, perform at a level lower than controls fed free choice, but do not suffer from the symptoms shown by the deficient group. Pair-feeding has been consistently practised by the nutritionists who discovered the known vitamins and essential minerals.

The Measurement and Evaluation of Body Weight Gain

Measuring body weight gain is a simple procedure but evaluating it is difficult because gain consists of tissue water, protein, fat and minerals in variable proportions and may include gastrointestinal contents (fill), urine in the bladder and, in lactating females, milk in the udder. Consequently, body weight gain is an inherently variable and inaccurate parameter, especially in mature ruminants which carry a large amount of fill. Many simple to complex techniques are used to decrease the experimental error of changes in body weight and to facilitate their interpretation.

The simplest technique to reduce the experimental error of body weight is repeated weighing. Animals in a growth trial may be weighed several times at intervals of several hours or a day, in order to obtain a more reliable initial and final weight for any period. Feed and water may be withheld from the experimental animals for 24 hours in order to reduce fill of gut and bladder to a minimum. This practice may have some after-effects but it efficiently reduces the experimental error of body weight observations.

The most accurate way of evaluating body weight gain is the slaughter trial. Representative samples of animals may be slaughtered at the beginning and at the end of a trial and chemical analyses may be performed on the ground carcasses for all the constituents (water, protein, fat and minerals). Slaughter trials are expensive and require much labour, especially when using large animals (see Chapters 1 and 18). Therefore, in the latter, compromise methods are most often used. Only one half of the carcass may be analysed. The composition of the carcass may be estimated from the characteristics of certain cuts or organs using known mathematical relationships between the chosen characteristics and the whole animal. Many such useful relationships have been established, and in currently published research often only the characteristics of certain cuts or organs are reported. A few examples are empty carcass weight, the thickness of subcutaneous fat, the area of a cut through a muscle at a standardized location and the weight of certain muscles and well-defined separable masses of adipose tissue.

Since the composition of the young animals used in a feeding experiment may already be known, slaughter data may be obtained only at the termination of the experiment.

A variety of sophisticated chemical methods have been developed to estimate the composition of live animals but these are not commonly used in feeding trials. Indirect calorimetry is essentially such a method (see Chapters 15 and 18).

SPECIAL PROBLEMS AND TECHNIQUES

Purified Diets

Diets using purified ingredients are used to study the deficiencies and the requirements of specific nutrients. The purified ingredients may be pure chemicals, e.g. amino acids, vitamins, minerals, glucose, fats, washed vitamin-free casein, etc., or semi-purified ingredients like plant proteins, raw casein or fibres. Purified or semi-purified diets are less well accepted by animals than diets consisting of natural ingredients and are most problematic in ruminants, which require coarse fibrous food to maintain normal rumen function.

Germ-free Animals in Nutrition Research

The survival of mature ruminants probably depends on the presence of microoganisms in the gut, especially in the rumen and in the caecum. However, the nutritional requirements of many monogastric species are also modified by the presence of bacteria in the gut which produce vitamins of the B complex group and other by-products. Chicks, several species of rodents and a few immature ruminants have been used in the germ-free state, i.e. in the total absence of microorganisms and in isolation, to study their absolute nutritional requirements and interactions between the microflora and the host. Germ-free animals are more useful for basic nutrition research than for evaluating feeds for farm animals.

BIBLIOGRAPHY

Books

Dunn, O. J., and Clark, V. A. (1974). *Applied Statistics: Analysis of Variance and Regression.* New York: John Wiley.

Little, T. M., and Hills, F. J. (1972). *Statistical Methods in Agricultural Research.* Berkeley: Agriculture Extension, University of California.

Senedecor, G. W., and Cochran, W. G. (1980). *Statistical Methods.* Iowa: Iowa State Press.

Steel, R. G. D., and Torrie, J. H. (1980). *Principles and Procedures of Statistics.* New York: McGraw-Hill.

Articles

Gill, J. L. (1981). Evolution of statistical design and analysis of experiments. *J. Dairy Sci.*, **64**, 1494–1519.

Lucas, H. L. (1957). Extra period, latin square, changeover design. *J. Dairy Sci.*, **40**, 225–239.

Morton, J. R., and Ridgman, W. J. (1977). Problems in the statistical design and analysis of feeding trials. *Proc. Nutr. Soc.*, **36**, 173–177.

Roberts, P. (1983). The number of replicates and other considerations in the design of field trials. In *Recent Advances in Animal Nutrition—1983* (W. Haresign, ed.), pp. 3–12. London: Butterworths.

Waldo, D. R. (1976). An evaluation of multiple comparison procedures. *J. Anim. Sci.*, **42**, 539–544.

Wostman, B. S. (1981). The germfree animal in nutrition studies. *Annu. Rev. Nutr.*, **1**, 257–279.

CHAPTER 13

The Digestibility of Feedstuffs and its Importance for their Evaluation

The actual value of a feed for an animal cannot be determined by chemical analysis alone, but only after allowances are made for the inevitable losses that occur during digestion, absorption and metabolism. Undigested residues of feedstuffs excreted in the faeces are a great loss in terms of feed utilization, particularly in ruminants.

Digestibility of a feed denotes the percentage of the whole feed or of any single nutrient in the feed which is not excreted in the faeces and is therefore assumed to be available to the animal for absorption from the gastrointestinal tract. There are objections to this assumption which will be discussed later. The digestibility is commonly expressed by the digestibility coefficient:

$$\text{Digestibility coefficient (\%)} = \frac{\text{Nutrient intake} - \text{nutrient in faeces}}{\text{nutrient intake}} \times 100$$

Evaluation of digestibility involves the determination of how much of a feedstuff or an individual nutrient is not degraded and absorbed while passing through the animal. This is an important facet of nutrient utilization.

MEASUREMENT OF DIGESTIBILITY

The digestion trial involves measuring the intake of a well-sampled feedstuff or ration given to the animal, and the total output of faecal excretion of animals being fed a certain amount of the feedstuff under evaluation.

Several techniques are used to collect faeces quantitatively uncontaminated by urine. The animals are kept in specially designed crates; in some types of crate the bottom is a metal grid through which faeces and urine pass, the faeces being received in a suitably placed container (see PLATE XI). Collections of faeces can be made from male animals using a harness and bag attached to the animal. This technique allows the animal more freedom and can be adapted for use under grazing conditions (see PLATE XIIA and later). For females a special device channels faeces into the bag while diverting the urine.

It is essential that the faeces collected represent quantitatively the undigested residue of the measured amount of food consumed, and not also faeces originating from some pre-experimental ration. The time required for feed residues to pass through the gastrointestinal tract is 1–3 days for monogastric animals and 8–10 days for ruminants. Therefore a preliminary period is needed for the digestive tract to become free of residues from previous feeds and for the animal to adapt to the test diet. The collection and analysis of faeces is then begun after this preliminary period. For pigs and horses preliminary and collection periods of 4–6 days each are commonly used; for ruminants these periods must be extended to 8–10 days. The ration to be tested is fed in constant daily amounts to minimize day-to-day variations in faecal output.

Samples of the food used and the faeces collected are analysed. A digestion trial involves a record of the nutrients consumed and their amounts voided in the faeces. These data make it possible to calculate the digestibility of a single nutrient, a feedstuff as a whole or a ration (see Table 13.1).

In digestibility trials several animals must be used and the results are averaged to minimize individual variability. Replication allows more opportunity for detecting errors of measurement.

The determination of the digestibility of roughage is a relatively easy matter because roughage can be given as the sole ingredient in the ration of ruminants. But when it is desired to determine the digestibility of concentrate feeds by ruminants it is necessary to feed them with roughage as a base feed because concentrate feeds alone do not furnish sufficient bulk. In this procedure the digestibility of the roughage has to be determined first, then in conjunction with the concentrate feed in question for a second test; a calculation is then made to obtain the digestibility of the concentrate alone. The coefficients of digestibility of the roughage are used to calculate how much of the excreta and their constituents should remain if the roughage was digested to the same extent as when fed alone. The remainder of the excreta is attributed to the concentrate feed (digestibility by difference). Determination of digestibility by difference is not very reliable where there is an associative effect between feeds (see later).

The determination of digestibility in poultry requires a special technique since the faeces and urine are excreted together from a single orifice, the cloaca. The nitrogenous compounds present in the faeces and urine can be

Table 13.1 *Digestibility of hay by a calf* (a calf was fed 10 kg hay per day and an average quantity of 15 kg wet faeces were excreted per day)

Constituent	Composition (%)		Quantity (kg)		Quantity digested (kg)	Coefficients of digestibility	Digestible nutrients in hay (%)
	Hay	Faeces	in 10 kg hay	in 15kg faeces			
Organic matter	9.5	18.6	7.95	2.79	5.16	64.9	51.6
Crude protein	9.7	2.6	0.97	0.39	0.58	59.8	5.8
Ether extract (fat)	2.5	0.8	0.25	0.12	0.13	52.0	1.3
Crude fibre	26.3	6.4	2.63	0.96	1.67	63.5	16.7
Nitrogen free extractives	41.0	8.8	4.10	1.32	2.78	67.8	27.8

separated chemically by determining uric acid and ammonia which represent the urinary output, whereas most faecal nitrogen is present as true protein. Separation of faecal and urinary nitrogen can be attained also by operative techniques such as a colostomy, in which a new opening is made for the large intestine to the outside of the body, or surgical exposure of the ureters so that urinary flow is diverted from the cloaca to a collecting device.

VALIDITY OF DIGESTIBILITY COEFFICIENTS: TRUE VERSUS APPARENT DIGESTIBILITY

There are two limitations to the validity of digestibility: (1) methane arising from fermentation of carbohydrates (see Chapter 5) is lost by eructation, but will be classed as being digested, thus leading to overestimation of digestible carbohydrates; (2) the coefficients of digestibility, usually determined by subtraction of the nutrients excreted in the faeces from those ingested in the diet (see Table 13.1), do not always reflect exactly the availability of the nutrients. The faeces are not only undigested feed residues, but also include metabolic products derived from body tissues (sloughed intestinal cells, digestive juices) and intestinal microorganisms (see Chapter 7). Consequently, the balance between the diet less the faeces is defined as apparently digested food, while true digestibility is the balance between the diet and the feed residues in the faeces exclusive of metabolic products. Nitrogenous compounds, lipids and minerals of metabolic origin are admixed with the faeces, but there is no secretion of carbohydrates into the gut. The coefficients are calculated as follows:

Apparent digestibility of nitrogen =

$$\frac{\text{nitrogen intake} - \text{nitrogen in faeces}}{\text{nitrogen intake}} \times 100$$

True digestibility of nitrogen =

$$\frac{\text{Nitrogen intake} - (\text{nitrogen in faeces} - \text{metabolic nitrogen})}{\text{nitrogen intake}} \times 100$$

Digestibility coefficients are not calculated for minerals. As discussed in Chapter 10, large and variable portions of the minerals absorbed are secreted into the gut, thus values of digestibility coefficients of minerals would be meaningless. The actual available proportion of minerals can be estimated by use of isotopes (see Chapter 10).

Apparent digestibility coefficients for protein and fat, excluding protein and fat of metabolic origin, are always lower than the coefficients of true digestibility. The apparent digestibility coefficients are sufficiently accurate for most practical needs, and are given in tables on the nutritive value of feedstuffs.

The amount of metabolic nitrogen excreted in the faeces can be estimated or calculated as follows. The amount is governed by the total feed intake, but is relatively constant at differing protein intakes. Thus, the metabolic nitrogen can be determined during a special collection period in which a nitrogen-free diet is fed but containing the same amount of dry matter as the ration under test. Since animals cannot survive on a protein-free diet it is a common method to place animals on diets with minimal amounts of nitrogen so that a regression line can be extrapolated to zero intake and the excretion of metabolic nitrogen thus calculated.

Routine calculations of the amount of metabolic nitrogen excreted can be made on the basis of the following: for animals consuming rations low in indigestible matter (e.g. rat, dog, pig and man), the metabolic faecal nitrogen averages about 0.20–0.25 g per 100 g of dry matter consumed, and for ruminant animals the metabolic nitrogen averages about 0.5–0.6 g per 100 g of dietary dry matter. The latter value is equivalent to about 4% of dietary protein and thus the coefficient of apparent digestibility for protein is negative in ruminant diets that contain less than 4% protein.

FACTORS AFFECTING DIGESTIBILITY

Different species of animals digest the same nutrients with different efficiency (see Chapter 3). The greatest differences are between monogastric species and ruminants. The variations are largest in the case of roughage; however, swine digest most concentrate feedstuffs as well as ruminants and a few of the latter feedstuffs even more fully. The desirability of determining separate digestibility coefficients for each species of animal is apparent. Due to the economy of sheep digestion trials, coefficients obtained with sheep are commonly used in computing rations for cattle. But digestion values determined for sheep and cattle are not always identical. Sheep have higher digestion coefficients than cattle at digestibilities above 66%, which probably reflects the lower metabolic losses in sheep. Below 66% digestibility, cattle tend to have higher digestibility than sheep, which reflects the greater capacity of cattle for digestion of fibre.

Effect of Food Composition on Digestibility

Digestibility of food is closely related to the chemical composition, and a food like cereal grains which varies relatively little in composition from one sample to another shows little variation in digestibility. Fresh and conserved herbages are much less constant in composition and vary more in digestibility. Crude fibre tends to depress digestibility chiefly by protecting constituents of the foods from attack by digestive or microbial enzymes, in monogastric species to a greater extent than in ruminants as expressed in the following regression equations:

for ruminants:

$$Y = 90 - 0.85X$$

for horses and pigs:

$$Y = 90 - 1.60X$$

for poultry:

$$Y = 90 - 2.30X$$

X = percentage crude fibre in dry matter, Y = coefficient of digestibility of the organic matter.

The stage of maturity of forage plants influences their digestibility; as the plant matures, its cell wall content increases, the soluble cell content decreases and the plant becomes less digestible.

Effect of Ration Composition

The digestibility of a feedstuff is influenced not only by its composition, but also by the composition of other feedstuffs ingested with it. *Associative effects* can be attributed to the fact that two feeds together constitute a less (or more) balanced diet than a certain feed given alone. For example, cellulose digestion in ruminants is much reduced if there is a lack of nitrogenous substances or essential minerals (see Chapter 5); or the digestibility of crude fibre is lowered when there is a large proportion of easily digestible carbohydrates, because the microorganisms utilize the more readily available carbohydrates instead of attacking the resistant cellulose.

Level of Feeding

High plans of nutrition result in lower digestion. An increase in the quantity of a ration eaten by an animal causes a faster rate of passage of digesta, thus allowing less time for digestion and absorption. When ruminants are fed roughage alone, increasing the feeding level reduces digestibility to a small extent, but the influence becomes larger (by units of intake) as the proportion of concentrate in the total ration increases. A small depressant effect on the digestibility is exerted by high levels of feed consumed by rapidly growing pigs. When feed intakes are reduced below the maintenance level (see Chapter 17), animals become more efficient in digesting feed and metabolizing nutrients as well (see Chapter 16).

Preparation of Feeds

Grinding grain or seed does not usually increase digestibility for animals that masticate their food thoroughly, but seeds that escape mastication may be excreted largely undigested. Unbroken seed coats resist the action of digestive enzymes. Sheep masticate their feed effectively, so there is no advantage in grinding grain for them, except in the case of very hard seeds. Cattle chew cereal grains less thoroughly and thus they digest them somewhat better when they are ground or crushed. Such mechanical pretreatment of grains is necessary for pigs.

Unlike grains, roughage is chewed by all ruminants sufficiently to break it up, so that digestive juices can penetrate. There is no advantage in grinding or chopping hay; on the contrary, the fine-grinding of hay lowers its digestibility, because the ground hay passes too rapidly through the gut.

Pelleting of hays (grinding, pressing and extruding of feedstuffs: size of pellets = 6–20 mm in diameter and in length) may even decrease fibre digestion, also because of increased rate of passage. The increased efficiency of hay pellets is largely due to improved acceptability to ruminants.

Products of very low feeding value such as cereal straws, cotton straw, seed hulls, etc., can be made suitable as feedstuffs for ruminants by treatment with sodium hydroxide or other chemicals. Lignin is removed by this treatment; the cell wall structure is destroyed and thus the digestibility of the remaining constituents is considerably improved.

The digestibility of some specific feedstuffs may be improved by heating, e.g. potatoes for pigs and poultry or soybean meal and cottonseed meal for monogastric species and ruminants. The reasons for the benefit arising from heat treatment of starch and proteins are noted in Chapters 4, 7 and 8, as well as the causes of the decrease in the digestibility of proteins due to overheating (see Chapter 7).

CHEMICAL AND BIOLOGICAL PROCEDURES FOR ESTIMATING DIGESTIBILITY

The proximate system of food analysis (Weende method; see Chapter 1) does not give a direct estimate of the digestibility or of nutritive value of feedstuffs, since the digestibility of plant feedstuffs, particularly roughage, is governed not only by the proportions of the components but also by the morphological structure. Thus, even the effort expended in providing regression equations between compositional parameters and digestibility, does not lead to reliable prediction of the digestibility.

For replacing the less useful Weende system van Soest used detergents to separate feedstuffs into three fractions based on nutritional availability. These fractions are (1) mostly available, (2) incompletely available, and (3) unavailable (see Figure 13.1).

A sample of feedstuff is divided by boiling with neutral detergent into soluble cell contents (NDS = neutral detergent solubles) and insoluble cell-wall constituents (NDF = neutral detergent fibre) which include the digestible fibrous fraction and the indigestible fraction. NDS contains lipids, sugars, starch, protein and organic acids; pectin, normally a cell-wall component, is

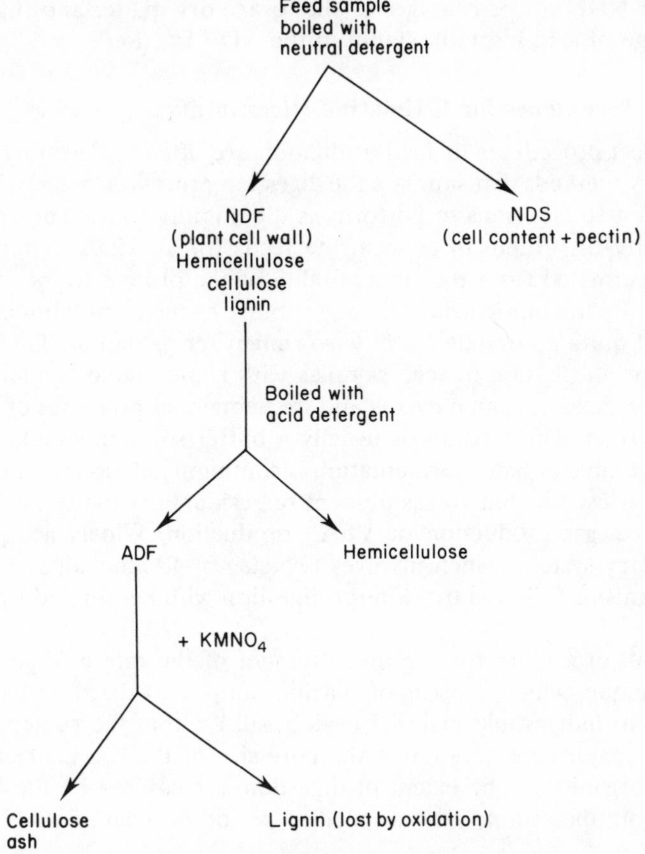

Figure 13.1. Basic scheme of forage analysis using detergents according to Van Soest—partition of feedstuffs into fractions of different nutritive availability (after Van Soest, 1982). Abbreviations are explained in the text.

also included as it has a very high nutritive availability. The NDF is boiled with acid detergent, thereby hemicellulose is hydrolysed and a residue, the acid detergent fibre (ADF) is obtained. ADF contains cellulose and the least digestible non-carbohydrate fraction (lignin, cutin, silica and insoluble non-protein nitrogen). Treatment of ADF with permanganate oxidizes lignin; thus cellulose and ash are left behind and the value for cellulose is obtained upon ignition.

Considering the full availability of the cell content (NDS) and the effect of lignin in reducing the availability of cell-wall constituents (NDF), the following regression equation for estimating digestibility of forages has been devised by van Soest:

$$\text{Digestibility} = 0.98\ \text{NDS} + (1.473 - 0.789 \log_{10}[\text{lignin}])\ \text{NDF}$$

NDS and NDF are percentages of the forage dry matter and [lignin] is the percentage of acid-insoluble lignin in the ADF fraction.

In Vitro Techniques for Estimating Digestibility

Chemical procedures of feed evaluation are often replaced by enzymatic laboratory methods that simulate the digestion process. The enzymatic methods are not so laborious to perform as digestibility trials. The approximate digestibility of protein in concentrate feeds for all kinds of farm animals may be estimated from the susceptibility of the protein to incubation with pepsin + hydrochloric acid. The digestibility of feeds for ruminants can be measured quite accurately by *in vitro* rumen fermentation. The latter procedure, i.e. incubation of feed samples with rumen liquor under anaerobic conditions, reflects rumen events and the sequential processes of the rumen digestive tract. The medium is usually a buffer solution simulating rumen saliva. The time of batch fermentation is commonly 48 hours. The end-point procedures used include measurement of residual dry matter, cellulose disappearance, gas production or VFA* production. Widely accepted is the Tilley–Terry system, which involves two stages: 48-hour digestion with rumen organisms followed by 48-hour digestion with pepsin and hydrochloric acid.

A useful procedure for the measurement of the rate of digestion is the bag technique, which consists of placing samples of dried herbage in bags made of an indigestible material (nylon, silk) within the rumen through a fistula. A major consideration is the pore size of the bag material to allow entry of organisms. The extent of digestion is measured by the loss of dry matter from the sample after a specified period of incubation.

THE USE OF INDICATORS FOR DETERMINING DIGESTIBILITY

In some circumstances it is impracticable to perform laborious digestibility trials. Digestibility can be determined by use of indicators without measurement of feed intake and faeces output. Indicators may be natural constituents of the food or added to it. The following criteria for an ideal indicator have been proposed: (1) it should be totally indigestible and unabsorbable, (2) its determination should be sensitive and easy, (3) it should pass through the tract at a uniform rate, also similar to that of the ration under test, (4) the presence of an added marker in the feed should not in any way influence digestion. A number of internal indicators have been proposed such as lignin, acid-insoluble ash and chromogen; and as added indicators chromium(III) oxide (Cr_2O_3), ferric oxide and oxides of rare earths. Chromium(III) oxide is the indicator most commonly used in digestibility studies (see later). With the use of radioactive metal oxides, only minute amounts of tracers are needed and physical rather than chemical methods of faecal analysis can

* VFA = volatile fatty acids.

be used. Use of magnesium ferrite has recently been suggested by Neumark *et al.* (1974); this compound can be easily determined by comparison of its weight in a magnetic field and under normal conditions.

If the concentrations of the indicator substance in food and small samples of faeces are determined and if the percentages of the nutrients in feed and faeces are known, digestibility can be calculated from the ratio between these concentrations. For instance, if the feed contains 100 g but the faeces only 25 g of protein per unit of the indicator, the difference, amounting to 75% of the protein, has been digested. The digestibility of each nutrient can be found by means of the following equation without measuring either the food intake or the faeces output:

$$\text{Digestibility} = 100 - \left(100 \times \frac{\text{percentage indicator in feed}}{\text{percentage indicator in faeces}} \times \frac{\text{percentage nutrient in faeces}}{\text{percentage nutrient in feed}}\right)$$

The use of indicators is of great importance for the determination of digestibility and feed intake in grazing animals. Exact data cannot be obtained by the simplest method consisting of a normal indoor digestibility trial using cut herbage from the field. The reason is that the animals graze selectively preferring young plants to old, and leaf to stem, and the herbage they actually consume differs considerably from a cut sample. Feed intake of grazing animals and digestibility of the herbage consumed can be determined by the simultaneous use of two indicators, namely a naturally occurring indicator to measure digestibility (formula given above) and an external indicator such as chromium(III) oxide to measure dry matter intake, the calculation being made as follows:

$$\text{Dry matter intake (g per day)} = \frac{\text{units of indicator per g dry faeces} \times \text{g dry matter in faeces per day}}{\text{units of indicator per g dry matter of herbage}}$$

It is recommended to administer chromium(III) oxide to pasturing animals in factory-produced capsules or impregnated on Kraft paper in order to avoid its incomplete recovery in faeces as a result of incomplete mixing of food and indicator during passage through the gastrointestinal tract.

DETERMINATION OF DIGESTIBILITY IN THE VARIOUS SECTIONS OF THE GASTROINTESTINAL TRACT

A knowledge of the partition of digestion within the gastrointestinal tract is essential for an understanding of the amounts and proportions of the particular nutrients absorbed. Study of digestibility of feeds in various sections of the tract is also important since the nutritive value of food constituents will depend not only on the extent to which it is digested but also on the site of digestion; for example, utilization of glucose in ruminants differs when it is digested in the rumen or in the intestine (see Chapter 5). Markers such as ^{91}Y (yttrium) may be used for these purposes in different ways, or as a single dose for estimating mean retention times, or by continuous input either by infusion or by homogeneous mixing into the diet.

When it is desired to determine absorption from intestinal segments of chickens, the animals are fed a diet mixed with the marker and killed after several days. Then the intestine is divided into various segments and the contents of each are analysed for the nutrient under study and the marker. The percentage of the net absorption on passage from one segment to the other is given by

$$\text{Net absorption (\%)} = 100\,\frac{(N/R)_1 - (N/R)_2}{(N/R)_1}$$

where $(N/R)_1$ and $(N/R)_2$ are the nutrient/reference substance ratios in the consecutive intestinal segments, respectively. The rate of nutrient absorption by any intestinal segment in units per weight can be calculated by

$$\text{Daily net absorption} = \Delta(N/R)\times R_{\mathrm{m}}$$

where $\Delta(N/R)$ is the difference in N/R between the beginning and the end of the intestinal segment, and R_{m} is the daily intake of the reference substance.

The flow-rate of feed in any intestinal segment can be calculated by

$$\text{Flow-rate (min)} = 1440\,\frac{R_1}{R_{\mathrm{m}}}$$

where R_1 is total isotope content in the particular intestinal segment and R_{m} is the total intake of the indicator (1440 = number of minutes in 24 hours).

The following two results, obtained by Hurwitz *et al.* using this technique, demonstrate the usefulness of labelled reference substances for the determination of the intestinal site of absorption: (1) Calcium absorption in laying hens, which is increased during shell formation, occurs mainly in the proximal intestine. The calcium/^{91}yttrium ratios can be properly used for estimating the net absorption of calcium since this marker moves in the intestine at a similar rate to calcium. (2) The main site of protein absorption in broilers is the jejunum, and the duodenum in poults.

Since the collection of digesta from various parts of the gastrointestinal tract of slaughtered big animals is very cumbersome, techniques have been devised to study the progress of digestion within various sections of the tract in pigs and ruminants. The surgical establishment of cannulae at various points along the gastrointestinal tract permits sampling of the content of the respective sections. Two types of cannulae have been devised; simple 'T-piece' cannulae are used for discontinuous spot sampling of digesta, in conjunction with non-absorbed markers, while re-entrant cannulae (see Figure 13.2) divert the entire flow of digesta outside the animal and allow the direct measurement of the total flow of nutrients during periods of 24 hours or more.

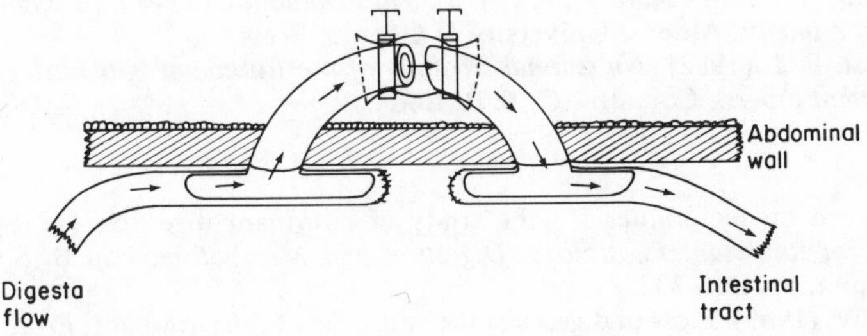

Figure 13.2. Diagrammatic representation of digesta flow through Ash re-entrant cannula (after Macrae, 1975, p. 261). *Reproduced by permission of the University of New England Publishing Unit.*

Simple cannulae consist of two parts: an internal flange ingested into the walls of abomasum, rumen or intestine and a straight barrel exteriorized through the wall and capped with a screw-topped stopper. Removal of the stopper allows samples of digesta to be taken. Administration of markers and evaluation of results are made as described above for the determination of absorption of nutrients from intestinal segments of killed chickens. However, when sampling through a cannula it is difficult to obtain samples containing solid particles and dissolved substances in the same proportions as in the digesta flowing past the cannula. Similarly, any single marker may not be present in a sample in the same concentration as in the digesta flowing past the cannula. This problem can be overcome by the use of two markers, one soluble to indicate the movement of soluble substances, while a solid marker indicates the movement of the larger particles.

The principle of the re-entrant cannula is shown in Figure 13.2. Both ends of the two cannulae are connected by a tube running outside the animal. The tube is opened for collecting digesta from the proximal part. After measuring the volume of the digesta, they are sampled and returned to the distal fistula. The re-entrant cannula allows the flow of digesta to be measured directly and hence the digestibility to be calculated as in regular digestibility trials. The accuracy of the collection procedure can be controlled by the use of markers.

The great importance of the site of digestion of proteins and soluble carbohydrates for their nutritive action in ruminants is emphasized in Chapters 5 and 8. Most of our knowledge about the partition of digestion and absorption between stomach (rumen), small and large intestine has been provided by the use of animals equipped with fistulas

BIBLIOGRAPHY

Books

Boorman, K. N., and Freeman, B. M. (1976). *Digestion in the Fowl.* Edinburgh: British Poultry Science Ltd.

Schneider, B. H., and Flatt, W. P. (1973). *The Evaluation of Feeds through Digestibility Experiments.* Athens: University of Georgia Press.

van Soest, P. J. (1982). *Nutritional Ecology of the Ruminant (Chemistry of Forages and Plant Fibers).* Corvallis: O. & B. Books.

Articles

Symposium on techniques for the study of ruminant digestion (1975). In *Symposia on Ruminant Physiology, Digestion and Metabolism,* Fourth Symposium, Australia, pp. 261–332.

Ash, R. W. (1969). Flow of digesta in the intestine of the ruminant. *Proc. Nutr. Soc.,* **28**, 110–114.

Barnes, R. F. (1973). Laboratory methods of evaluating feeding value of herbages. In *Chemistry and Biochemistry of Herbage* (G. W. Butler and R. W. Bailey, eds), Vol. 3, pp 179–214. London and New York: Academic Press

Faichney, G. J. (1980). The use of markers to measure digestive flow from the stomach of sheep fed once daily. *J. Agri. Sci.,* **94**, 313–318.

Hurwitz, S., and Bar, A. (1965). The absorption of calcium and phosphorus along the gastrointestinal tract of the laying fowl as influenced by dietary calcium and egg shell formation. *J. Nutr.,* **86**, 433–438.

Kolb, A. T., and Luckey, T. D. (1972). Markers in nutrition. *Nutr. Abstr. Rev.,* **42**, 814–845.

Macrae, J. G. (1975). The use of re-entrant cannulae to partition digestive function within the gastro-intestinal tract of ruminants. In *Symposia on Ruminant Physiology, Digestion and Metabolism.* Fourth Symposium, Australia, pp. 260–276.

Mertens, D. R., and Ely, L. O. (1982). Relationship of rate and extent of forage utilization. *J. Anim. Sci.,* **54**, 895–905.

Neumark, H., Halevi, A., Amir, S., and Yerushalmi, S. (1974). Assay and use of magnesium ferrite as a reference in absorption trials with cattle. *J. Dairy Sci.,* **58**, 1476–1481.

Osborne, D. F., and Terry, R. A. (1977). In vitro technique for the evaluation of ruminant feeds. *Proc. Nutr. Soc.,* **36**, 219–225.

Pigden, W. J., Balch, C. G., and Graham, M. (1980). *Standardization of Analytical Methodology for Feeds.* Ottawa: International Development Research Centre.

Sutton, J. D., and Oldham, J. D. (1977). Feed evaluation by measurement of sites of digestion in cannulated ruminants. *Proc. Nutr. Soc.,* **36**, 203–209.

Tyrrel, H. F., and Moe, P. W. (1975). Effect of intake on digestive efficiency. *J. Dairy Sci.,* **58**, 1151–1163.

van Soest, P. J. (1967). Development of a comprehensive system of feed analysis and its application to forages. *J. Anim. Sci.,* **26**, 119–128.

Wainman, F. W. (1977). Digestibility and balance in ruminants. *Proc. Nutr. Soc.,* **36**, 195–202.

CHAPTER 14

Energy Metabolism

Meeting the energy requirements of animals is the major cost associated with feeding animals, and the efficiency of utilization of energy is, from a quantitative and economic standpoint, the primary consideration. The animal derives energy by partial or complete oxidation of carbohydrates, fats and proteins ingested and absorbed from the diet or from breakdown of glycogen, fat or protein stored in the body.

Even in a non-productive state, animals need energy for sustaining the body, keeping a stable body temperature, and for maintaining muscular activity. Domestic animals require additional energy to support the work of production: growth and fattening, pregnancy and lactation.

PARTITION OF FOOD ENERGY WITHIN THE ANIMAL

The manner in which energy is partitioned into various fractions in terms of animal utilization is shown in Figure 14.1.

Gross Energy

Gross energy (GE) is defined as the energy liberated as heat when a food, faeces or animal tissue is fully oxidized by burning a sample completely in a bomb calorimeter (see PLATE XIIB). A bomb calorimeter consists of a strong metal chamber resting in an insulated tank of water. Oxygen is admitted under pressure. The heat produced by the oxidation is calculated from the rise in temperature of the surrounding water. Not all the gross energy value of a food is available to animals because of the losses of energy that occur during digestion and metabolism.

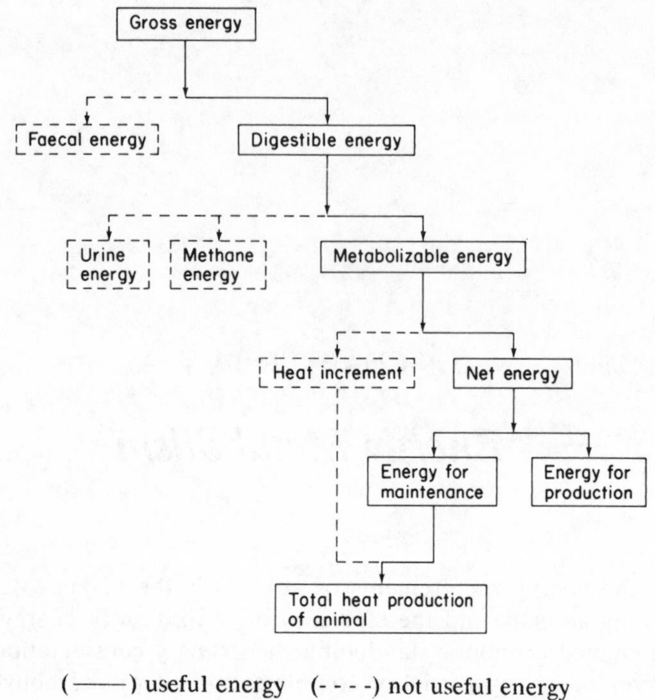

(——) useful energy (- - - -) not useful energy

Figure 14.1. The partition of food energy within the animal.

Digestible Energy

Digestible energy (DE) is the energy of the food (GE) less the energy of the faeces. By determining the heat of combustion of food and faeces collected in a digestibility experiment (see Chapter 13) one obtains the apparent digestible energy. It is called apparent digestible energy as distinct from the truly digestible energy because the faecal energy includes that of undigested feed as well as that of metabolic products derived from body tissues (see Chapter 13).

Energy lost in the faeces accounts for the single largest loss of ingested energy. In ruminants the losses are 40–50% in the case of roughage and 20–30% in the case of concentrates. In horses faecal losses account for about 40% of the energy ingested and in pigs for 20%.

Metabolizable Energy (ME)

ME is defined as DE less the energy lost in urine and in combustible gases leaving the digestive tract, chiefly methane. Loss of energy in gases reaches appreciable proportions only in ruminants (see Chapters 3 and 5). In other words ME is the ingested gross energy minus faecal energy, minus urinary energy, minus energy in the gaseous products of digestion. ME is that portion of food energy that is available for metabolic processes in the animal. Therefore ME provides a satisfactory measure of the nutritive value of feedstuffs (see Chapter 16). To measure ME for ruminants it is necessary to collect faeces, urine and methane. Faeces and urine are collected in digestibility trials from animals placed in metabolism cages provided with a device for collecting urine. For measuring methane produced a respiration chamber is needed (see Chapter 15). When no respiration chamber is available, methane losses can be calculated as 8% of the GE intake or from the amount of digested carbohydrates (see Chapter 5). Methane contains 31.8 kJ per gram.

Metabolizability is defined as the ME of a feedstuff divided by the gross energy (GE). ME may be calculated by multiplying DE by 0.82. This is only an approximation as the ME/DE ratio varies considerably with the nature of the diet.

Because birds void urinary and faecal losses together, the ME values for poultry can be determined by standard digestibility procedures. ME is commonly used to evaluate feedstuffs for poultry (see Chapter 16). Losses of combustible gases occurring as a result of fermentation in the caecum and large intestine of monogastric species are negligible.

The urine loss of energy results from the excretion of incompletely oxidized nitrogenous compounds associated with protein metabolism, primarily urea, in mammals and uric acid in birds (see Chapter 3). Urinary losses are quite stable in a given animal species and run in the order of 2–3% of the GE intake in pigs and of 4–5% in cattle, although they reflect differences in the diets, particularly when excess protein is fed. The energy value of each gram of nitrogen excreted as urea is 23 kJ and as uric acid is 28 kJ. For this reason each gram of urinary nitrogen excreted by ruminants accounts for 31 kJ, in pigs 28 kJ and in poultry 34 kJ.

Whereas the energy value of carbohydrates of fats absorbed by animals equals their heats of combustion determined in the calorimeter, the energy value of absorbed protein is lower than its heat of combustion because of the energy lost in nitrogen-containing constituents excreted with the urine. In the case of protein a subtraction of 5.3 kJ or 1.25 kcal has to be made from the GE content of 1 g of protein (23.7 kJ or 5.65 kcal) for the energy lost in the urine. In the calorimeter, protein, i.e. amino acids, are completely oxidized to carbon dioxide, water and gaseous nitrogen; for example, alanine is oxidized according to the equation

$$4\,CH_3.CH(NH_2).COOH + 15\,O_2 \longrightarrow 12\,CO_2 + 14\,H_2O + 2\,N_2$$

but the oxidation of alanine occurring in the animal body stops at the intermediate stage of urea

$$4\,CH_3.CH(NH_2).COOH + 12\,O_2 \longrightarrow 10\,CO_2 + 10\,H_2O + 2\,CO(NH_2)_2$$

In human nutrition ME is generally obtained by considering only the small faecal and urinary losses. The so-called physiological fuel values of human foods having a similar significance as metabolizable energy are calculated from the analysis of a food for carbohydrate, protein and fat using the values of an available energy content of 16.8 kJ (4 kcal) for 1 g of protein or carbohydrate and of 37.8 kJ (9 kcal) for 1 g of fat.

The DE and ME contents are affected by the amount of feed consumed, because the more an animal eats the faster is the passage of food through the gastrointestinal tract. The additional losses in faeces caused by increasing intake are partially compensated by reduction in losses of energy in urine and methane. The overall effect of increasing intake on reduction of ME is more marked with low-quality feeds and then this reduction reaches in ruminants 10% by doubling the intake.

Metabolizable energy values of some typical foods are given in Table 14.1. The ME values for some feedstuffs which are digested to much the same extent by ruminants and non-ruminants are greater for non-ruminants than for ruminants since the energy of nutrients is always utilized less efficiently when they are fermented in the rumen than when they are digested in the intestine.

Table 14.1 Energy values for some typical feeds (expressed as MJ/kg dry matter) (after McDonald *et al.*, 1981, and Church, 1979)

Animal	Feed	Gross energy	Loss of energy			DE	ME
			Faeces	Urine	Methane		
Fowl	Maize	18.4	2.2				16.2
	Wheat	18.1	2.8				15.3
	Barley	18.2	4.9				13.3
Pigs	Maize	18.9	1.6	0.4		17.3	16.9
	Barley	17.5	2.8	0.5		14.7	14.2
Ruminants	Maize	18.9	2.8	0.8	1.3	16.1	14.0
	Barley	18.3	4.1	0.8	1.1	14.2	12.3
	Lucerne hay	18.3	8.2	1.0	1.3	10.1	7.8
	Grass hay	17.9	7.6	0.5	1.4	10.3	8.4
	Maize silage	18.9	6.0	0.8	1.3	12.9	10.8

Net Energy (NE) and Heat Increment (HI)

'Net energy' is obtained from ME by subtraction of HI, i.e. NE = ME − HI; thus NE differs from ME by the amount of heat lost as a result of the chemical and physical processes associated with digestion and metabolism, i.e. the heat increment (HI). The NE of a feedstuff is that portion of its energy content that is completely useful for the body as it is available to the animal for maintenance and productive purposes.

The portion of NE used for maintenance is the energy expended to sustain the life processes of an animal; it serves for muscular work needed for minimal movement, maintenance and repair of tissues (including turnover; see Chapter 2) and to keep up the temperature of the body in a cold environment (see Chapter 17). The other portion of NE (fed above maintenance needs for production) is the energy retained in tissue gain of growing or fattening animals or in milk or eggs produced, i.e. the caloric value of animal products. NE is the part of GE completely useful to the body. Methods for determination of NE will be described in the following chapter.

Heat production of an animal consuming food is composed of HI and heat used for maintenance. The energy needed for maintenance of a fasting animal (termed basic metabolic rate) is covered by breakdown of body reserves, mainly body fat, and the energy equivalent of the decomposed body constituents is dissipated as heat (see Chapter 17).

If a fasting animal is given food, within a few hours its heat production will increase to about the level represented by basal metabolism. The HI increases with the amount of food consumed and may be used by animals in a cold environment to warm the body; otherwise the HI must be considered, like the energy of the excreta, as a loss of the energy contained in the food. The main causes of this heat increment (HI is also called specific dynamic effect) are the following: (1) The work of nutrient metabolism. The energy liberated by oxidative reactions occurring inside the tissue is never fully utilized to the benefit of the animal, since a part of it is lost as heat as a result of incomplete transfer of energy. As will be shown later in this chapter, if, for example, glucose is oxidized, almost 70% of the energy liberated is captured as ATP and the remainder is inevitably converted to heat. (2) Heat arises from the work of digestion, i.e. from mastication of the food and it propulsion through the tract. (3) A part of HI chiefly in ruminants consists of the heat of fermentation. (4) Heat production results from work of excretion by the kidney. (5) Heat production also results from increased muscular activity of various organs due to metabolism of nutrients.

As is seen in Table 14.2, the HI and consequently the efficiency of feedstuffs vary with species of animals, type of nutrients, character of the diet and also according to the specific purposes for which the food is used. It is very interesting that the HI of protein is considerably higher than that of the other nutrients; this is due to the following: (1) synthesis of urea that is excreted; (2) the energy cost of concentrating and excreting the waste products via the kidney; (3) metabolism of the carbon skeleton of amino acids (see Chapter 7). In cases of amino acid imbalance, these reactions occur to a greater extent, and then the HI is even higher.

The HI of feeding in ruminants is greater and more persistent in ruminants than in monogastric species. Apart from the higher energy costs of

Table 14.2 Heat increment of feeding (in percentage of ME) (after Church and Pond, *Basic Animal Nutrition and Feeding*, 1974, p. 87)

Nutrient	Species		
	Pig	Sheep	Cattle
Fat	9	29	35
Carbohydrate	17	32	37
Protein	26	54	52
Mixed rations	10–40	35–70	35–70

eating, ruminating and of the work of the digestion and fermentation, the higher HI in ruminants is related to the large amounts of volatile fatty acids formed in the rumen. These acids, the major energy sources for ruminants, provide a larger HI than glucose (see Chapter 5). Increasing proportions of acetic acid in the volatile fatty acid mixture present in the rumen content give rise to HI. Thus the efficiency of utilization of volatile fatty acids for fattening declines with increasing proportions of acetate. This would explain the inverse relationship between the fibrousness of a food and the capacity of its ME to promote body gain (see Chapter 18).

According to recent measurements, the losses of energy as heat (HI) in ruminants are quite high and account in the case of cereal grains for 50% of the ME and in the case of forages for as much as 60% (Webster, 1981).

The great differences in the efficiency of ME for various functions in ruminants are of greatest importance for the evaluation of diets (see later). The following values for the conversion of ME into NE result from comprehensive metabolism experiments (see Chapter 15) carried out by Moe and Tyrell (1974):

Cold stress	100%
Maintenance	70–80%
Lactation	60–70%
Growth	40–60%
Pregnancy	10–25%

The considerable ranges in the values for some functions reflect the influence of different types of diets. However, in monogastric species such as the growing pig, the efficiency of utilization of ME is very similar for both maintenance and body growth.

BIOCHEMICAL INTERPRETATION OF ENERGY UTILIZATION

Understanding of the efficiency of energy transformations occurring in the whole body of an animal has been made possible by the study of the energetics involved in biochemical reactions proceeding at the cellular level. The central position of ATP in catabolic and anabolic reactions occurring in the cells has been elucidated in Chapter 2. ATP formed during the breakdown of carbohydrates, fats and proteins can be used as an energy source for synthesis and growth, muscular contraction, glandular secretion, active absorption (see Chapter 3) and for other activities. While discussing many of the metabolic reactions in the preceding chapters, it has been noted that not all the energy in foods is transferred to ATP; instead, a large proportion of this energy becomes heat. On average, about 55% of the energy in foods becomes heat during ATP formation. Then still more energy becomes heat as energy is transferred from ATP to the functional systems of the cell. Thus, no more than about 25% of all the energy from the food is finally utilized for the functions of the animal. Accordingly, the net energy provided by high-quality concentrate feeds accounts also for about 25% of the GE.

Because of the paramount importance of ATP in fulfilling energetic functions in the body (for directing biosynthetic reactions, for muscle contraction), it is important to know how many moles of ATP can be produced in the animal body by oxidation of various nutrients absorbed from the food or liberated from body substances. The net yields of ATP produced from some substrates given in Table 14.3 reflect the efficiency of these substrates as energy sources.

Table 14.3 The caloric equivalent of deriving one high-energy phosphate bond from oxidation of different nutrients

Nutrient	Heat of combustion (kJ/mol)	ATP yield obtained from 1 mol of nutrient	Energy required for formation of 1 mol of ATP	
			kJ	Relative
Glucose	2816	38	74.0	100
Stearic acid	11342	146	77.9	105
Acetic acid	816	10	87.4	118
Propionic acid	1536	18	85.3	115
Butyric acid	2193	27	84.1	110
Protein	2143*	22†	94.9	128

* kJ per 100 g

† mol ATP/mol amino acid (molecular weight = about 100)

The complete oxidation of 1 mol of glucose (heat of combustion = 2816 kJ) by an animal yields 38 mol of ATP, so that the synthesis of 1 mol of ATP requires 74 kJ (see Table 14.3). Because of the biochemical pathways followed, ATP yields are some 5% lower for energy in fat and 10–25% lower for energy in volatile fatty acids and in protein than for energy in glucose. The lower yield of ATP from the oxidation of amino acids is due to the expense of urea synthesis and the work done by the kidney in excreting the urea formed (see Chapter 7). Volatile fatty acids are the major energy-yielding substrates in ruminants (see Chapter 5). Two mol of acetic acid formed by fermentation of 1 mol of glucose yield 20 mol of ATP upon their oxidation in the tissues of ruminants, whereas 38 mol of ATP are obtained from the aerobic oxidation of glucose which takes place in monogastric species. Moreover, the microbial conversion of 1 mol of glucose into 2 mol of volatile fatty acids yields 4 mol of ATP which are available for microbial growth and for the benefit of the host animal. In summary, 24 mol of ATP are obtained in ruminants from the complete oxidation of 1 mol of glucose via acetic acid.

Likewise, the costs of ATP needed for biosynthetic purposes can be calculated. The conversion of nutrients into protein and fat follows known metabolic pathways. During some parts of these pathways ATP is being produced, while in other parts often considerable amounts of ATP are being utilized. The efficiency of a biosynthetic process is the heat of combustion of the product divided by the sum of the heat of combustion of the nutrients used, firstly to provide the energy to drive the synthesis and secondly to provide the molecules used in the synthesis. The efficiency of some biosynthetic processes is given in Table 14.4. In the biosynthesis of glycogen from glucose this efficiency is 94%. The overall reaction can be formulated as

Efficiency = 100 × [(calorific value of 1 mol glycogen)/(calorific value of n mol glucose obtained from hydrolysis of glycogen + joules required to synthesize $2(n - 1)$ mol ATP)]

If digested fat is incorporated into the body without breakdown the process is very efficient (98%). If digested fat is first broken down to acetyl-CoA molecules which are used for resynthesis of fat (see Chapter 6), the efficiency is about 90%. The efficiency of the conversion of glucose into body or egg fat is considerably lower, and in consequence 25% of the energy present in glucose is lost as heat. When protein is used as precursor of fat, the efficiency is again lower, about 60%, for reasons mentioned above.

The theoretical energy costs for protein synthesis in monogastric species can be calculated as follows. The total number of moles of ATP needed to synthesize one peptide bond (also for activation and coding of one amino acid) is 5. One mol of amino acid is equal to about 100 g, and its caloric value is 2.4 MJ. ATP is produced by oxidation of glucose. The complete oxidation of 1 mol of glucose (heat of combustion = 2.8 MJ) by the animal is considered to yield 38 mol of ATP, so that the synthesis of 1 mol of ATP requires 75 kJ. Since 5 mol of ATP can be produced by oxidizing 5 × 75 kJ (0.38 MJ) glucose, the energetic efficiency of synthesizing 2.4 MJ protein from 2.4 MJ amino acid + 0.38 MJ glucose is [2.4/(2.4 + 0.38)] × 100 = 87%.

Table 14.4 Energy efficiency of biosynthetic processes occurring in monogastric species

Synthesis			Energy efficiency*
Fat synthesis			
Food fat	⟶	body fat	90–98
Glucose	⟶	body fat	75–80
Protein	⟶	body fat	50–60
Carbohydrate synthesis			
Lactic acid	⟶	glucose	92
Glucose	⟶	glycogen	94
Glucose	⟶	lactose	94
Protein synthesis			
Amino acids	⟶	protein	87

* (Calorific value of product formed)/(calorific value of precursors of synthesis).

In ruminants the efficiency of protein synthesis is only about 79%, mainly due to the fact that volatile fatty acids as an energy source are less efficient than glucose (see Chapter 8).

BIBLIOGRAPHY

Books

Blaxter, K. L. (1967). *The Energy Metabolism of Ruminants*, 2nd edn. London: Hutchinson.

Brody, S. (1945). *Bioenergetics of Growth.* New York: Reinhold.

Kleiber, M. (1961). *The Fire of Life.* New York: John Wiley.

Monteith, J. L., and Mount, L. E. (1974). *Heat Loss from Animals and Man.* London: Butterworths.

National Research Council (1981). *Nutritional Energetics of Domestic Animals*, 2nd edn. Washington, DC: National Academy Press.

Articles

European Association for Animal Production (EAPP) Symposium on Energy Metabolism of Farm Animals:

*3rd Symposium*1965, Scotland (K. L. Blaxter, ed.). EAPP No. 11. London: Academic Press.

4th Symposium 1967, Poland (K. L. Blaxter *et al.*, eds). EAPP No. 12. Newcastle: Oriel Press.

5th Symposium 1970, Switzerland (A. Schürch and C. Wenk, eds). EAPP No. 13. Zurich: Juris-Verlag.

6th Symposium 1974, Germany (K. H. Menke *et al.*, eds). EAPP No. 14. Hohenheim: Universität.

7th Symposium 1976, France (M. Vermorel, ed.). EAPP No. 19. Clermont-Ferrand: G. de Sussac.

8th Symposium 1979, Cambridge (L. E. Mount, ed.). EAPP No. 26. London: Butterworths.

9th Symposium 1982, Norway (A. Ekern *et al.*, eds). EAPP No. 29. Norway: Agricultural University.

Baldwin, R. L., and Bywater, A. C. (1984). Nutritional energetics of animals. *Annu. Rev. Nutr.*, **4**, 101–114.

Baldwin, R. L., and Smith, N. E. (1979). Regulation of energy metabolism in ruminants. In *Adv. Nutr. Res.*, **2**, 1–27.

Graham, N. (1976). Specific dynamic action. In *Energy Metabolism of Farm Animals, 7th Symposium* (see above), pp. 113–116.

Hegsted, D. M. (1974). Energy needs and energy utilization. *Nutr. Rev.*, **32**, 33–38.

Moe, P. W. and Tyrell, H. F. (1974). Observations on the efficiency of utilization of metabolizable energy for meat and milk production. In: *Nutrition Conference for Feed Manufacturers*, Vol. **7**, 27–35. London: Butterworths.

Moore, T. (1977). The calorie as a unit of nutritional energy. *World Rev. Nutr. Diet.*, **26**, 1–25.

Webster, A. J. F. (1979). Energy metabolism and requirements. In *Digestive Physiology and Nutrition of Ruminants* (D. C. Church, ed.), Vol. 2, pp. 210–229. Corvallis: O. and B. Books.

Webster, A. J. F. (1981). The energetic efficiency of metabolism. *Proc. Nutr. Soc.*, **40**, 121–128.

Webster, A. J. F., Osuji, P., and Weekes, A. (1976). Origins of the heat increment in sheep. In *Energy Metabolism of Farm Animals, 7th Symposium* (see above), pp. 45–48.

Wilhelm, C. M. (1963). The specific dynamic action of food. *Physiol. Rev.*, **15**, 202–220.

CHAPTER 15

*The Measurement of the Net Energy of Feedstuffs**

The most suitable estimation of the feeding value of feedstuffs or diets is the measurement of their net energy value. The determination of net energy is especially important for the evaluation of feeds and planning of diets for ruminants, since these animals consume feeds of low and high fibre content and derive very different amounts of net energy from feeds of similar content of metabolizable energy. Since net energy (NE) is the difference between the metabolizable energy (ME) and the heat increment (HI) of a feed, the estimation of ME and HI determines NE (see Chapter 14).

In order to determine net energy values of feeds, it is necessary to measure, in addition to metabolizable energy (see Chapter 14), either the energy stored in the animal (including the energy stored in animal products like milk) or the heat production. The heat produced may originate from the oxidation of food constituents occurring in the body or from oxidation of fat and protein liberated from body tissues; the heat originating from the breakdown of body constituents is used mainly for covering maintenance requirements (in underfed or fasting animals; see Chapter 17).

Heat production from an animal can be measured directly in an animal calorimeter or calculated from the respiratory exchange of an animal (indirect calorimetry). The application of calorimetry enables the determination of the energy balance of an animal, which in turn enables the estimation of the requirements for maintenance and for productive functions as well (see following chapters).

The principles of direct and indirect calorimetry and those underlying the measurement of energy retention are outlined below.

---

* In collaboration with A. Arieli.

DIRECT CALORIMETRY

Direct calorimetry is simple in theory, but difficult in practice. Calorimeters for measuring the heat production of an animal are based on the same general principle as the bomb calorimeter, in that the heat evolved is used to increase the temperature of a surrounding medium. The animal calorimeter is an airtight insulated chamber, in which oxygen is supplied by flow of air. An animal loses heat to its environment in two ways, usually called sensible and evaporative heat. Sensible heat is lost from the body surface to the environment by radiation, convection and conduction. Similarly, evaporative heat is lost from the surfaces of the skin and the respiratory tract due to the vapour difference between the animal surface and the surroundings. Evaporation of water forms an important fraction of the total heat loss of a live subject and is measured by recording the volume of air drawn through the chamber and its moisture content on entry and exit. Water evaporation takes up heat from body surfaces. The quantity of this latent heat of evaporation is 2.5 kJ/g at 0 °C and 2.4 kJ/g at 40 °C.

Heat loss from animals can be measured directly by using heat sink or thermal gradient calorimeters. In the *heat sink calorimeter* sensible heat loss from an animal is measured as a rise in temperature in an absorbing medium which may be the airstream ventilating the chamber or water circulating outside the walls. The weight of water circulated per unit time multiplied by the rise in its temperature gives the sensible heat loss. In the *gradient layer calorimeter* the heat-emitting subject is completely surrounded by a uniform layer of insulating material; the difference in temperature between the inner and outer surface of this layer will be proportional to the rate of sensible heat loss of the animal. The outer surface is maintained at a constant temperature with a water jacket; the temperature gradient is measured electrically with thermocouples which line the inner and outer surfaces of the wall.

The heat increment of the food under investigation is measured by feeding at two levels of intake, first at a low level and then at a higher level, or by fasting the animal and then feeding. Two levels are necessary because a part of the animal's heat production is contributed by its basal metabolism. An increase in food intake causes total heat production to rise, but the basal metabolism remains the same. The increase in heat production can thus be attributed to the heat increment of the additional food given.

INDIRECT CALORIMETRY BY MEASUREMENT OF RESPIRATORY EXCHANGE

Because the animal body ultimately derives all its energy from oxidation, the magnitude of energy metabolism can be estimated from the exchange of respiratory gases; i.e. the ratio of the volumes of carbon dioxide produced to oxygen consumed, or the respiratory quotient (RQ).

Since, under the same conditions of temperature and pressure the same number of moles of different gases occupy the same volume, the molar ratio may be replaced by the volume ratio. The numerical value of the RQ is dependent on the chemical nature of the substance being oxidized within the body.

The main substances oxidized in the body, and whose energy is thus converted into heat, are carbohydrates, fats and proteins.

The overall oxidation of glucose, the form in which carbohydrates are catabolized, takes place according to the following equation:

$$C_6H_{12}O_6 + 6\,O_2 \longrightarrow 6\,CO_2 + 6\,H_2O + 2.82\text{ MJ}$$

and that of a typical fat, tripalmitin, as follows:

$$C_3H_5(OOC.C_{15}H_{31})_3 + 72.5\,O_2 \longrightarrow 51\,CO_2 + 49\,H_2O + 32.04\text{ MJ}$$

The RQ of carbohydrate is 1, and that of tripalmitin is 0.7. Since the carbohydrate molecule contains hydrogen and oxygen in the proportion to form water, oxygen is required only for the oxidation of carbon. One molecule of carbon dioxide is formed for each molecule of oxygen consumed:

$$\text{RQ} = \frac{6\text{ vols . carbon dioxide}}{6\text{ vols . oxygen}} = \frac{6\text{ mol carbon dioxide}}{6\text{ mol oxygen}} = 1$$

Fats require more oxygen than carbohydrates for combustion since the fat molecule contains a low ratio of oxygen to carbon and hydrogen. Thus, for a fat such as tripalmitin,

$$\text{RQ} = \frac{51\text{ vol . }CO_2}{72.5\text{ vol . }O_2} = 0.70$$

Calculation of heat production from gaseous exchange is relatively easy. The utilization of 1 litre of oxygen used for the oxidation of glucose leads to the production of 21 kJ of heat: 2820/(6 × 22.4) = 21 kJ; 2820 kJ are liberated by the oxidation of 1 mol of glucose; 1 gram molecule of a gas occupies 22.41 litres. The caloric equivalent of 1 litre of oxygen for oxidation of fat, calculated analogously, is 19.6 kJ. Caloric equivalents for 1 litre of carbon dioxide produced are: carbohydrate, 20.9 kJ; and fat, 27.6 kJ.

If mixtures of carbohydrates and fats are catabolized, the RQ varies between 0.7 and 1.0. The caloric equivalent of oxygen or carbon dioxide will also vary with the RQ, as seen from the data in Table 15.1. If the RQ for a mixture of fats and carbohydrates is known, the proportions of both substrates oxidized can be found from the data plotted in Figure 15.1. The caloric equivalent of oxygen is linearly related to the RQ: i.e. kJ/1l oxygen consumed = 15.962 + 5.155 RQ.

Table 15.1 Heats of combustion of carbohydrates, fats and protein, and respiratory quotients and caloric equivalents of oxygen consumption

Substrate	Heat of combustion (kJ/g)	Respiratory quotient	kJ per litre oxygen consumed
Carbohydrate	16.74	1.0	21.12
Fat	37.66	0.70	19.61
Protein	22.59	0.80	18.62
Mixtures (carbohydrate/fat ratio)			
15 : 85	34.52	0.751	19.83
30 : 70	31.38	0.795	20.06
40 : 60	29.29	0.825	20.21
50 : 50	27.20	0.853	20.36
60 : 40	25.10	0.882	20.50
70 : 30	23.01	0.912	20.66
85 : 15	19.87	0.956	20.87

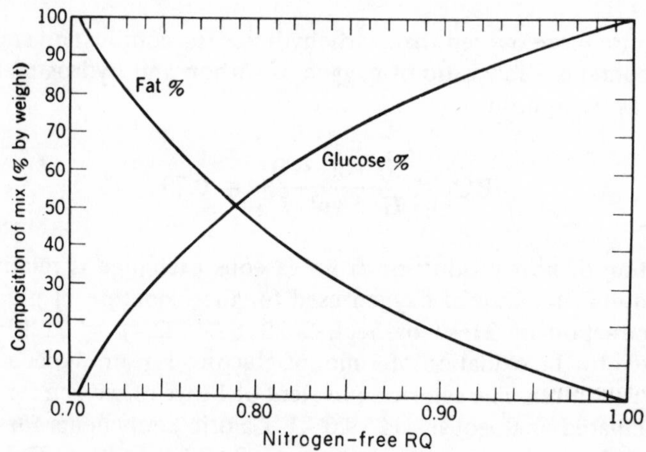

Figure 15.1. Calculation of the proportions of glucose and fat oxidized, from the respiratory quotient (nitrogen-free RQ) (after Kleiber, *The Fire of Life*, 1961, p. 86). New York and London: John Wiley and Sons. *Reproduced by permission of the publisher.*

Nutrients oxidized in the body generally include protein, and the relative contribution of protein to the oxidative metabolism has to be considered when dealing with the estimation of the heat of combustion of food from the exchange of respiratory gases. The quantity of protein catabolized can be estimated from the output of nitrogen in the urine, 0.16 g of urinary nitrogen being excreted for each gram of protein. This calculation is based on the assumption that all nitrogen excreted in the urine originates from catabolism of amino acids, the carbon skeleton of which is oxidized within the body (see Chapter 7).

In the complete oxidation of protein taking place in the bomb calorimeter, the end-products are carbon dioxide, water and nitrogen (formula, see Chapter 14). The heat of combustion of protein, i.e. the heat produced when it is completely oxidized, varies according to the amino acid composition, averaging 22 kJ/g.

Animals cannot oxidize protein, i.e. amino acids, completely. Considering urea as the principal nitrogenous excretory product in mammals, the average amount of heat produced by oxidation of 1 g of protein, as occurring in the animal body, is 18.0 kJ, the RQ is 0.83, and the caloric equivalent of 1 litre of oxygen used for the oxidation of protein is 18.8 kJ.

In order to calculate the heat production from respiratory exchange one has to subtract from the total amounts of oxygen used and carbon dioxide produced those amounts related to protein metabolism. For each gram of protein oxidized, 0.77 litre of carbon dioxide is produced and 0.96 litre of oxygen used. On the assumption that all urinary nitrogen is derived from protein and that protein contains 16% of nitrogen, the amounts of oxygen and carbon dioxide (in litres) associated with protein metabolism are calculated by multiplying grams of urinary nitrogen by 5.91 and 4.76, respectively. Subtraction of these volumes from the total leaves the non-protein respiratory exchange. Then, from graphs or tables showing the caloric values of the oxygen used or the carbon dioxide produced at the non-protein RQ (see Figure 15.1 and Table 15.1), the heat production from carbohydrate and fat metabolism is obtained. An example of the calculation of heat production from respiratory exchange is shown in Table 15.2. No account was taken of methane production because none was to be expected in the young animal (40 kg).

The determination and calculation of heat production by respiratory exchange and nitrogen balance is laborious and has certain limitations, particularly in ruminants due to incomplete oxidation of carbohydrates leading to methane formation (see Chapter 5) and other limitations to be mentioned later. The amount of methane produced can be determined experimentally (see later) or estimated by empirical calculation based on the amount of digestible carbohydrates ingested (see Chapters 5 and 14). Heat production calculated from respiratory exchange has to be corrected by the deduction of 2.1 kJ for each litre of methane. The following equation has been developed to predict heat production (see Flatt and Moe, 1969):

$$H\,(\text{KJ}) = 16.180\,O_2 + 5.02\,CO_2 + 2.17\,CH_4 - 5.99\,N$$

where O_2, CO_2 and CH_4 refer to gaseous exchange (litres) and N refers to urinary nitrogen (g).

In practice, the following two simplified procedures can be used for indirect determination of heat production.

(1) One procedure is based on the determination of the total RQ instead of the non-protein RQ. The contributions of nitrogen excreted in urine and of methane formed to the calculation of heat production (according to the last equation) are small and if they are neglected the error in estimating heat production would seldom surpass 2%.

(2) A further simplification is to measure oxygen consumption alone and to assume an intermediate respiratory quotient (0.82) and assign to oxygen a caloric equivalent of 20 kJ/litre. Since the caloric equivalent varies only from 19.7 kJ/litre at an RQ of 0.7 to 21.0 kJ/litre at an RQ of 1 (a range of about 6%), the error introduced by the preceding assumption is no more than 3.5%.

Table 15.2 Calculation of the heat production of a calf from oxygen consumption, carbon dioxide production and urinary nitrogen excretion (after Blaxter *et al.*, 1955, *J. Agric. Sci.*, **45**, 10, adapted by McDonald *et al.*, Animal Nutrition, 3rd. ed., 1981, Longman).

Results of the experiment (per 24 hours)		
Oxygen consumed		392.0 l
Carbon dioxide produced		310.7 l
Nitrogen excreted in urine		14.8 g
Heat from protein metabolism		
Protein oxidized	(14.8×6.25)	92.5 g
Heat produced	(92.5×18.0)	1665 kJ
Oxygen used	(92.5×0.96)	88.8 l
Carbon dioxide produced	(92.5×0.77)	71.2 l
Heat from carbohydrate and fat metabolism		
Oxygen used	(392.0−88.8)	303.2 l
Carbon dioxide produced	(310.7−71.2)	239.5 l
Non-protein respiratory quotient		0.79
Thermal equivalent of oxygen when RQ = 0.79		20.0 kJ
Heat produced	(303.2×20.0)	6064 kJ
Total heat produced	(1665 + 6064)	7729 kJ

Interpretation of the Respiratory Quotient

Although oxygen consumption is a valid measure of the rate of energy metabolism, the volume of carbon dioxide eliminated is influenced by factors other than metabolism in the tissues. Also, there may exist differences between the amounts of carbon dioxide produced and that exhaled during a certain period of time. In ruminants, anaerobic fermentation in the rumen produces large amounts of carbon dioxide and methane, and this carbon dioxide cannot be distinguished from that produced in body tissues by oxidative metabolism. Also, under various other circumstances the RQ measured by exchange of respiratory gases may lead to erroneous interpretation. For example, lactic acid which accumulates during strenuous muscle exercise is buffered with bicarbonate in the plasma. The carbon dioxide which is displaced from plasma bicarbonate is eliminated along with that of metabolic origin.

The metabolic interconversion of feedstuffs may also alter the RQ. Respiratory quotients considerably higher than 1 may be obtained when carbohydrate is being converted into fat because oxygen-poor fat is being formed from oxygen-rich glucose (see Chapter 6). Respiratory quotients of about 1.4 were found in rapidly fattening pigs and in geese stuffed with grain. On the other hand, an RQ of less than 0.7 has been observed during starvation, particularly in hibernating animals. It has been suggested that this phenomenon may indicate synthesis of carbohydrate from fat. The RQ is lowered in the metabolic disorder of ruminants, known as ketosis, when fatty acids are not completely oxidized to carbon dioxide and water, and carbon and hydrogen are excreted as ketone bodies (see Chapter 24). It is noteworthy that the RQ in birds may be below 0.7 since uric acid formation leads to a lower RQ than the formation of urea.

In conclusion, the RQ may provide valuable information about the metabolic processes in the body. However, because of uncertainty in the interpretation of the RQ it is questionable whether its determination always leads to a more accurate estimation of caloric production than the use of oxygen consumption as the sole parameter for indirect calorimetry.

Measurement of Respiratory Exchange

To measure oxygen consumption and carbon dioxide production per unit of time, it is necessary either to confine the animal to a temperature-controlled chamber where changes in the ventilating air can be determined, or else to connect the animal by face-mask or tracheal cannula to some measuring apparatus.

In both types of respiration chambers—the closed and open circuit types—the central feature is an airtight container for the animal which incorporates devices for feeding and watering of the animal and the collection of faeces and urine.

In the *closed circuit system* (Figure 15.2(b)) the same air is continuously circulated from the chamber through carbon dioxide and water absorbers and back to the chamber. The absorbers are weighed before and after use to determine the quantities of carbon dioxide and water produced.

The consumption of oxygen by the animal results in a fall in the pressure of the gas mixture in the chamber. Oxygen in the circulating air is renewed by a spirometer, by means of which the volume added is recorded. Thus the consumption of oxygen can be measured by changes in pressure of the closed system or by adding measured amounts of oxygen to replace that removed, so maintaining pressure and volume constant.

Methane is determined in the residual air. The air sample is drawn over platinized kaolin at red heat; the methane is thus oxidized:

$$CH_4 + 2\ O_2 \longrightarrow CO_2 + 2\ H_2O$$

and determined from the carbon dioxide formed.

Closed systems are of high accuracy and particularly suitable for work on small animals; they are less practicable for large animals in view of the large amounts of absorbents required for recycling air in a closed system.

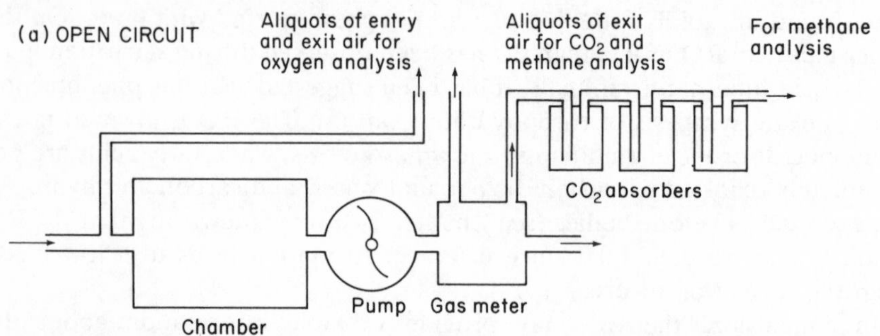

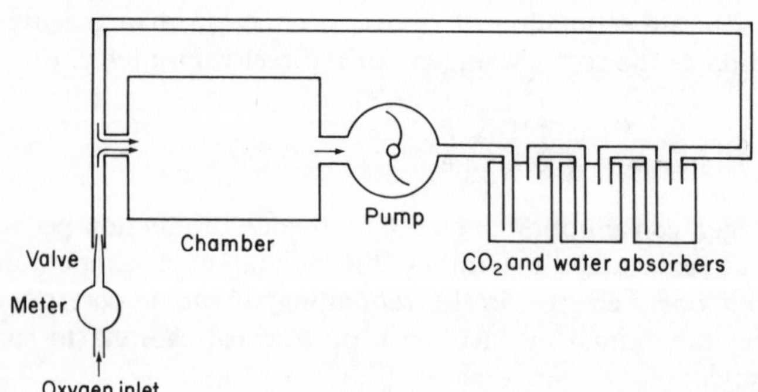

Figure 15.2. Diagrams of respiration chambers: (a) open-circuit type, (b) closed-circuit type (after McDonald *et al., Animal Nutrition*, 3rd. ed. 1981, p. 212). Reproduced by permission of Longman Group Ltd.

The other type of apparatus, the *open circuit system* (Figure 15.2(a)), differs from the closed circuit type in that outside air is continuously drawn through the chamber in which an animal is enclosed, and then discharged. During its passage the air increases in carbon dioxide and methane content and decreases in oxygen content. The volume and composition of the air (also its water content) entering and leaving the chamber determine the respiration balance. The net oxygen consumption and carbon dioxide production are calculated by integrating gas flows and compositions over time. Use of the open circuit system had been limited for many years because of the large number of high-precision gas analyses that are required. However, the recent development of automatic sensing and integration devices in combination with a computer have overcome the serious labour problems and have made it possible to obtain precise measurements of relatively undisturbed animals. Oxygen is determined by its paramagnetic effect and carbon dioxide and methane by infrared absorption.

Additional Techniques for Measuring Respiratory Exchange

Equipment is also in use which combines the features of a respiration chamber and an animal calorimeter so that direct measurements of heat loss and gaseous exchange can be carried out simultaneously.

Various systems have been developed to measure the volume of expired air, respiratory gas exchange and the energy requirement of grazing animals. The animals are equipped with a face-mask or a cannula inserted into the trachea. The animal inhales outdoor air from nostrils or mouth and expired air is measured by a portable gas meter which is combined with the tracheal cannula and strapped to the back of the animal. A tube from the gas meter to a bag is also attached to the animal and collects a small aliquot of respiratory gases for analysis (see Figure 15.3).

Determination of Energy Balance

Energy is stored by growing and fattening animals, mainly as protein and fat; carbohydrate reserves of the body are small, less than 1% of the body weight (see Chapter 4), and negligible. Energy retention can be determined by measurement of carbon and nitrogen balance or by direct calorimetry. According to the direct method, energy retained is determined by deduction of the sum of the heat of combustion of the faeces, urine and methane plus the heat produced by the animal from the energy content of the food ingested (see Table 15.3).

For the determination of gain or loss of protein and body fat by carbon and nitrogen balance, inputs and outputs of these elements have to be determined. Determinations are made of nitrogen and carbon in the food, faeces and urine, and of carbon in the gaseous output (carbon dioxide and methane). By measuring the amounts of carbon and nitrogen entering and

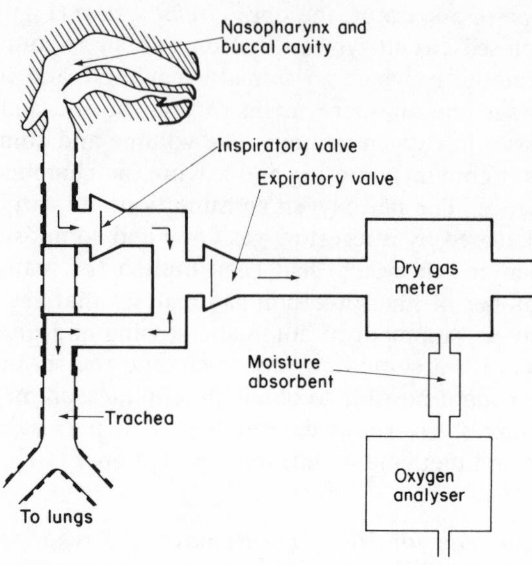

Figure 15.3. Diagram of the tracheal cannula technique and gas metering system after Young and Webster, 1963, adapted by Van Soest, P. J., *Nutritional Ecology of the Ruminant*, 1982. p. 298. Corvallis: O & B Books, Inc. *Reproduced by permission of the author.*

leaving the body, by difference the amounts of protein and fat gained or lost from the body may be calculated. Finally, the energy retained can be calculated by multiplying the quantities of protein and fat stored by average calorific values for fat and protein (see later). The quantity of protein stored is calculated by multiplying the nitrogen balance by 6.25. Protein is assumed to contain 16% nitrogen and 51.2% carbon; thus the amount of carbon stored as protein can be calculated. The remainder of the carbon is stored as fat which contains 74.6% of carbon. Fat storage is therefore calculated by dividing the carbon balance, minus the carbon stored as protein, by 0.746. The caloric equivalents of 1 g of protein and 1 g of fat stored are the following: 23.6 kJ/g protein and 39.9 kJ/g fat. A regression equation has been developed for calculating the amount of energy stored from carbon and nitrogen balances:

$$\text{Energy stored (kJ)} = 51.8\ \text{C (g)} - 19.40\ \text{N (g)}$$

As an example of the calculation of energy balance, data obtained during a classical experiment carried out by H. P. Armsby with his respiration calorimeter are given in Table 15.3. The data presented in the left part of Table 15.3 were obtained by direct heat estimation and those in the right part of this Table by measurement of carbon/nitrogen balance (see also Table 16.5). It should be noted that after accounting for all energy losses a balance

Table 15.3 Energy, carbon and nitrogen balance figures determined on bullock by Armsby, adapted by C. Tyler, *Animal Nutrition*, 2nd ed. (1964), pp. 134 and 138. London: Chapman and Hall

Items	Amount		Daily energy balance		Carbon/nitrogen balance			
					Nitrogen		Carbon	
			Gain (MJ)	Loss (MJ)	Gain (g)	Loss (g)	Gain (g)	Loss (g)
Feed	6998 g	timothy hay	116.02		56.4		2831.7	
	400 g	linseed meal	7.57		21.9		172.6	
Excreta	16.619 g	faeces		59.59		33.5		1428.7
	3357 g	urine		5.06		32.4		124.2
	37 g	brushings		0.37		1.3		8.0
	142 g	methane		7.93				106.6
	4730 g	carbon dioxide						1290.2
Heat emitted				48.09				
Gain by body				2.54		11.1		46.6
Total			123.59	123.59	78.3	78.3	3004.3	3004.3

of 2.54 MJ was left as a net gain of the animal from the feed ingested. The large loss of energy as heat represents approximately 40% of the total energy intake.

Armsby used the data from the nitrogen/carbon balance to compute the heat loss in order to compare the values thus obtained with those determined by direct calorimetry. Gains of 11.1 g of nitrogen and 46.6 g of carbon correspond to gains of 66.6 g of protein and 15.2 g of fat (calculated on the basis of average nitrogen and carbon contents of protein and carbon content of fat as given above). Considering 23.85 kJ as the energy value of each gram of stored protein and 39.75 kJ as that of each gram of fat, the energy content of the body gain is 2.19 MJ. The metabolizable energy is calculated by subtracting all energy losses (as faeces, urine plus methane = 72.96 MJ, but *not* the heat emitted) from the gross energy of the food (123.59 MJ) and is therefore 50.63 MJ. Subtracting the stored energy 2.19 MJ from the metabolizable energy, a balance of 48.44 MJ is obtained, which represents the energy lost as heat. This computed value for heat emitted agrees well with the direct measurement of 48.09 MJ. The error is +0.7%.

OTHER METHODS FOR DETERMINING ENERGY RETENTION

The measurement of energy retention in growing and fattening animals by direct or indirect calorimetry requires elaborate apparatus, and during the experimental period the animals are confined to a respiration chamber under unnatural conditions. Thus, other methods to measure energy retention have been proposed. The most obvious, measurement of the live weight gain, does not provide accurate estimates of energy retention since the calorific value of body gains or losses can vary widely according to changing proportions of bone, muscle and fat and the amounts of water present in gut or bladder. Energy retention in the form of milk or eggs is easily measured, but energy retention in these products is accompanied by changes of energy stores in the body, particularly in lactating cows (see Chapter 21). In the *comparative slaughter technique*, energy retention is measured as the difference in energy content of the carcass of animals slaughtered before and after a feeding period. The method is useful for chickens (see Chapter 18), but expensive and laborious when applied to large animals. Thus alternative methods have been proposed for estimating body composition of living animals, such as densitometry, and dilution techniques for determination of the water content of the body or assessing the concentration of potassium. The principles or these methods are outlined in Chapter 1.

Apart from methods based on the determination of the body composition, additional techniques have been proposed for the measurement of total energy metabolism in unconfined animals, such as measurement of the heart rate and isotope dilution methods.

Heart rate is easily measured by radiotelemetred electrocardiograms. A good correlation between heart rate, which is easily measured, and oxygen consumption exists in individual animals. Therefore this method involves individual calibration in the laboratory of the heart rate versus oxygen consumption under the experimental conditions such as activity and temperature.

One radioisotopic method uses water ($^{2}H_2^{18}O$) that is doubly labelled and depends on the fact that the oxygen in respiratory carbon dioxide is in isotopic equilibrium with the oxygen in the body water. The turnover rate for hydrogen in body water is slower than that of oxygen which is lost in water and carbon dioxide as well. The difference between these turnover rates is proportional to carbon dioxide expired and can be used as a measure of energy metabolism.

BIBLIOGRAPHY

Further descriptions of apparatus used for measurement of the net energy value of feedstuffs are given in books and some articles listed in the Bibliography of Chapter 14.

Blaxter, K. L. (1971). Methods of measuring the energy metabolism and interpretation of results obtained. *Fed. Proc.*, **30**, 1436–1443.

Brockway, J. M. (1978). Escape from the chamber: alternative methods in large animal calorimetry. *Proc. Nutr. Soc.*, **37**, 5–12.

Farrell, D. J. (1974). General principles and assumptions of calorimetry. In *Energy Requirements of Poultry*, pp. 1–24. *Poultry Science Symposium Series* No. 9. Edinburgh: British Poultry Science Ltd.

Flatt, W. P. (1969). Methods of calorimetry (B) indirect. In *Nutrition of Animals of Agricultural Importance.* (D. Cuthbertson, ed.), Part 1, pp. 491–520. Oxford: Pergamon Press.

Pullar, I. D. (1969). Methods of calorimetry (A) direct. In *Nutrition of Animals of Agricultural Importance* (D. Cuthbertson, ed.), Part 1, pp. 471–490. Oxford: Pergamon Press.

Van Es, A. J. H. (1966). Labor-saving methods for energy-balance experiments with cattle, description of equipment and methods used. *Neth. J. Agr. Sci.*, **14**, 32–46.

Young, B. A., and Webster, M. E. D. (1963). A technique for the estimation of energy expenditure in sheep. *Austr. J. Agr. Res.*, **14**, 867–873.

CHAPTER 16

Methods for Evaluating Feeding Stuffs[*]

Scientific rationing of animals is based on (1) the assessment of their requirements for energy and nutrients, and (2) the nutritive value of the feedstuffs. Requirements of animals and the ability of feedstuffs to provide nutrients must be expressed in the same terms. The methods to design diets for animals involve firstly a tabulation of their daily needs for energy and nutrients according to species, body weight, age, sex and type and level of production (meat, milk, eggs, etc.) and secondly a tabulation of the energy values of feedstuffs and their contents of available nutrients. The computation of a suitable ration from the feedstuffs available is then a matter of arithmetic calculation or of linear programming techniques (see Appendix).

Standards for requirements of nutrients and energy are under constant review and revision, and are published every few years by official government committees such as the NRC (National Research Council, USA) or the ARC (Agricultural Research Council, Great Britain). In ruminant nutrition first consideration is given to the supply of energy and protein. Energy is quantitatively the most important item in the diet of ruminants. Great losses of energy are involved in the utilization of feedstuffs by ruminants, the losses varying widely with the types of feedstuffs and kinds of production (see above). Therefore different systems for expressing the energy value of foods for ruminants have been proposed and will be discussed below. Diets devised to meet the energy and protein requirements and found to be deficient in minerals or vitamins can be rectified simply by adding concentrate sources of minerals or vitamins such as vitamin A.

Nowadays rationing of pigs and poultry is also based on supply of energy, particularly in order to assure the correct relationship between energy content to content of essential nutrients such a amino acids, vitamins and minerals (see later).

---

* In collaboration with the late Dr. H. Neumark.

MEASURES OF FOOD ENERGY FOR RUMINANTS

The importance of food evaluation was recognized first by Thaer in 1810, who expressed the value of all feeding stuffs in terms of hay (hay equivalents); for example, the nutritive value of 1 kg of potatoes equals that of 0.5 kg of hay. Some later workers based the evaluation of feedstuffs on their chemical composition. Real advance came by expressing feed values in terms of digestible nutrients. Modern approaches tend to express the total effect of a food in a single value by determining its energy content. Feed evaluation systems are based on three energy forms: digestible, metabolic and net energy. Net energy is the only actually useful form of energy efficient in the animal body and most systems of feed evaluation at present in use are based on net energy, although its experimental determination is somewhat tedious (see later).

Total Digestible Nutrients (T.D.N. According to Morrison)

In this simplest scheme of food evaluation, animal requirements and the value of foods in meeting these requirements are expressed in terms of the weight of digestible material in the food. In this system the content of TDN is calculated for each as follows:

$$\text{TDN (per 100 kg)} = \text{kg dig.protein} + \text{kg dig.crude fibre} + \text{kg dig.N-free extractives} + 2.25 \times \text{kg dig.crude fat}$$

The digestible crude fat is multiplied by 2.25 because of the comparatively higher energy value of crude fat. This system takes account of the faecal losses of energy which in all species, even in ruminants, are the most important sources of energy losses; but losses in combustible gases, heat and urine are not considered. This is a strong limitation to the usefulness of TDN for evaluating feeds for ruminants. The losses in methane and heat are relatively larger per unit TDN for roughage than for concentrates; for example, 1 kg of TDN in low-grade roughage contains only 50% of the net energy present in 1 kg of TDN in maize grain. Thus, low-quality feeds are overestimated by the TDN system.

The TDN system can be used for feeding pigs and horses and it is also reliable for comparative evaluation of ruminant rations of similar composition. Estimates of TDN requirements for maintenance and for production of milk and meat have been made under practical conditions and therefore this simple system of formulation of rations enables adequate feeding, providing that feedstuffs of very poor or very high quality are not used. It is often preferred to express TDN in calories of digestible energy (DE). DE can be readily determined by digestibility experiments and measuring the energy value of food and faeces (see Chapter 14). DE can also be calculated by multiplying TDN by a factor of 4.4. It was found experimentally that the average caloric value of 1 g of TDN is 4.4 kcal (= 17.5 kJ).

Kellner's Starch Equivalent

The classical method of Kellner developed at the beginning of this century in Germany is a net energy system, since the production value of feeds is measured by their utilization for fat deposition in adult animals relative to the fat-producing power of 1 kg of starch. Kellner's system was based on the determination of the carbon–nitrogen balance by respiration experiments. An adult bullock was given a ration representing the maintenance requirement, then this ration was supplemented with the feedstuff to be tested. An example of the evaluation of the results of carbon–nitrogen balance determination is given in Table 15.3. Gain of fat and protein were calculated. The small amount of protein stored in adult animals in addition to fat may be expressed as the isodynamic amount of fat, i.e. as the weight of fat supplying the same amount of energy that the protein contains.

One kg of starch fed in excess of maintenance requirements produced 248 g of body fat or, since 1 g of fat is equivalent to 9.5 kcal (39.9 kJ), the net energy value of 1 kg of starch for fattening is 2.36 Mcal or 9.9 MJ. Kellner expressed the fat-producing power of feedstuffs, i.e. their energy value, in terms of the number of kg starch that produce the same amount of fat as 100 kg of the respective feed. This value is called the starch value or starch equivalent; for example the starch value of wheat bran, 71.3, is the number of kg of starch producing as much fat as 100 kg of wheat bran. Starch values of typical feedstuffs have been determined by carbon–nitrogen balance experiments. For rationing diets, starch values are computed from their content in digestible nutrients. Kellner determined the actual fat-producing power of isolated nutrients typical of the proximal constituents of feedstuffs; the results are summarized in Table 16.1.

It appears from Table 16.1 that the starch values of starch, crude fibre and nitrogen-free extractives are equal. The fat-producing power of protein and

Table 16.1 Energy efficiency and starch equivalents of pure nutrients (after experiments of Kellner with fattening cattle)

Digestible nutrient*	Fat deposited		Starch equivalent factor
	g	MJ	
Starch	250	9.95	1.0
Crude fibre	250	9.95	1.0
Ether extract			
from oil seeds	600	23.89	2.4
from cereals	525	20.92	2.1
from roughage	474	18.87	1.9
Protein	235	9.31	0.9
Sugar	188	7.49	0.8

* 1 kg fed in addition to maintenance ration

sugars is lower since breakdown of protein to nitrogen-free substances and formation and excretion of urea need energy (see Chapter 7), and loss of energy is involved in the fermentation of sugars. The fat-producing power of the ether extract from oilseeds is considerably higher than that from cereals and roughage because the latter fractions contain a greater percentage of non-glyceride compounds such as waxes and pigments than the purer fat from oilseeds.

For the calculation of starch values of two typical feedstuffs as given in Table 16.2, the percentages of the digestible nutrients are multiplied by the respective starch equivalent factors (see Table 16.1). The arithmetic sum of these products is called the production value. In the case of oil cakes the calculated production value is identical with the starch value determined experimentally, but the starch value of oats or feedstuffs high in crude fibre is lower than the calculated production value. The actual starch value is obtained from the production value by multiplying by the 'value number'. The value number expresses the ratio between the starch value of a feedstuff and that of the pure nutrients contained in the feedstuff (the calculated production value). The value number for oil cakes is 100%, for oat grain 95% and for roughage less (e.g. wheat straw 30%).

Since value factors were determined by balance experiments for a limited number of feedstuffs only, Kellner recommended an alternative way of calculating starch values from the predicted production values: by correction on the basis of the crude fibre content of the feedstuffs. The production value of a roughage would be reduced by 0.58 units for every 1% crude fibre present in this roughage. For example: the calculated production value of a hay containing 28.4% crude fibre is 59; for 28.4% crude fibre present in this hay (28.4 × 0.58) 16.5 units are deducted, thus the corrected starch value is 42.5 (59 − 16.5). If the crude fibre content is less than 16%, units lower than 0.58 will be deducted for every 1% of crude fibre.

Kellner related the difference between calculated production value and actual starch value to the energy cost of digesting crude fibre. It is now known that the work of digestion of crude fibre is quite small and the effect of crude fibre is better explained by its influence on fermentation in the rumen. As the fibre content of the ration rises, the ratio of acetic to propionic acid in the rumen content increases and the energy of such acid mixture is utilized less efficiently for body fat synthesis (see Chapters 5 and 18).

The starch equivalent system suffers from the same weakness as other net energy systems, namely that the starch value of a ration is not constant at different levels of feeding, but decreases with increasing levels (see later), and the starch value differs considerably for different productive purposes, even at the same level of feeding. The efficiency of 1 starch equivalent for different functions is given in Table 16.3. The difference in the utilization of energy for maintenance and fat synthesis is stressed in Chapter 18. Kellner recognized that net energy values differed according to the function of the food, but instead of using starch values of different magnitude according

Table 16.2 Examples for calculation of starch value

	Oat grains				Coconut oil meal			
	Digestible nutrients (%)		Starch equivalent factor		Digestible nutrients (%)		Starch equivalent factor	
Protein	8.0	×	0.94	= 7.5	16.7	×	0.94	= 15.7
Fat	4.8	×	2.12	= 8.5	8.2	×	2.41	= 19.8
Crude fibre	2.6	×	1.0	= 2.6	9.3	×	1.0	= 9.3
Nitrogen-free extract	44.8	×	1.0	= 44.0	32.1	×	1.0	= 32.1
Production value				63.4				76.9
Value factor			0.95				1.0	
Starch value				60.2				76.9

Table 16.3 Efficiency of 1 kg starch value for different functions (after Schürch, 1969)

Function of food	1 kg starch value equivalent to
Maintenance	15.7 MJ
Fat production	9.9 MJ = 248 g fat
Meat production	17.1 MJ = 723 g dry meat
Milk production	11.7 MJ = 4 kg milk (3.5% fat)
Work production	5.2 MJ = 534 000 mkg

These calculations were based on the following energy values of production:
1 g fat = 9.5 kcal = 39.7 kJ;
1 g meat (dry and fat-free) = 5.7 kcal = 23.8 kJ;
1 g milk (3.5% fat) = 0.7 kcal = 2.9 kJ;
100 m-kg work = 2.3 kcal = 9.6 kJ

to the function of the food, he expressed energy values for feedstuffs and requirements for all functions in starch equivalents for fattening.

Kellner and later authors recommended the use of the starch equivalent system also for other species of animals. Data about the efficiency of starch for fat production in various species are given in Table 16.4. The great efficiency of pigs for fat formation deserves interest.

Table 16.4 Efficiency of conversion of starch to body fat in different species of adult animals (after Schürch, 1969)

Species of animals	1 kg digestible starch equivalent to		
	Metabolic energy (MJ)	Net energy (MJ)	Fat retained (g)
Cattle	15.7	9.9	248
Sheep	15.3	10.5	264
Pig	17.5	14.6	367
Hen	17.5	10.0	252

Armsby's Net Energy Values

Armsby, working at about the same time as Kellner, employed very similar ways of approach. Both Kellner and Armsby carried out difference experiments in which the net energy value of the food was determined as the increase in energy retention resulting from an increase in food intake. Kellner compared two levels above maintenance and measured energy values of foods for fattening. Armsby, however, compared two levels below maintenance—the higher level was close to maintenance—and calculated the net energy value of the food by relating the addition of more food to the resultant saving in body tissues. It will be stressed in Chapter 18 that the utilization of metabolic energy for maintenance is markedly higher than for fattening, and thus evaluation of the same feedstuffs according Kellner and Armsby leads to different results. Table 16.5 illustrates the measurement of the net energy of hay fed to a mature steer in an animal calorimeter. An increase of 1.83 kg of hay decreased the steer's loss of body substance by 8.49 MJ. One kg of hay thus contained 4.64 MJ of net energy.

Table 16.5 Determination of the energy value of hay (after Armsby) adapted by C. Tyler, *Animal Nutrition* (1964)

	Hay intake (kg/day)	Metabolizable energy (MJ/day)	Heat production (MJ/day)	Energy balance (MJ/day)
Trial 1	4.633	39.93	41.06	−1.12
Trial 2	2.801	24.19	33.74	−9.62
Difference (1−2)	1.832	15.74	7.22	8.48
Values per kg hay	1	8.64	4.00	4.64

Armsby expresses the net energy content of feeds in therms (1 therm = 1 Mcal = 1000 kcal) in order to avoid large figures. Forbes, continuing Armsby's work, stressed the influence of the kind of production and used different energy units for the different functions of ruminant animals as follows:

1 kg starch equivalent for maintenance = 2356 kcal

1 kg starch equivalent for fat production = 3100 kcal

1 kg starch equivalent for milk production = 3050 kcal

Scandinavian Food Unit System

This system was developed in Sweden by Niels Hansson for feeding milking cows. It resembles the starch equivalent system of Kellner, but the principle difference between both systems is in the energy evaluation of protein, since conversion of food protein into milk protein or meat protein is a more efficient process than synthesis of body fat. The amino acids that are retained in the body cannot give rise to a loss of energy by, for example, deamination of amino acids, or formation and excretion or urea.

The value of feeds is not expressed in energy terms, but relative to the value of a common feedstuff, barley. One kg of barley is the standard feed unit: 3 litres of milk (3.5% fat content) are produced by 1 kg of barley. The same factors are used for the calculation of feed units as for calculation of the starch value, with one exception: percentages of digestible protein are multiplied by the factor 1.43 instead of 0.94, the factor used by Kellner (see Table 16.2). The factor 1.43, according to Hansson, is the ratio of the energy content of protein to the energy content in carbohydrates: 5.7 : 4.0 = 1.43.

Feeds are evaluated in group feeding trials with a large number of lactating cows for a few months. The food to be tested is substituted for barley or another feedstuff of known value; the control and experimental rations are compared with another concerning the milk yield they promote.

Mollgard's System

A decade later than Hansson in Sweden, Mollgard developed his feed unit in Denmark. He based his system on energy-balance experiments with milk cows. He expressed his values in terms of net energy for fattening, NK_F. Mollgard recognized, like Forbes (see above), the great influence of function of food on the efficiency of energy utilization. For fat production in the body of cattle, 1000 NK_F produce 1000 kcal of fat but only 840 NK_F and 820 NK_F, respectively, are needed to give 1000 kcal in the form of milk or to cover a maintenance requirement of 1000 kcal. These values for energy requirements are correct if the production quotient k, i.e. the ratio of protein/energy (as NK_F) to total energy (as NK_F) is 0.2 in rations for milk production and 0.1 in rations for maintenance or fattening.

East German System of Nehring

This system is an improved modification of Kellner's system; the use of value factors for the individual feeds as prescribed by Kellner is no longer needed. Nehring determined by respiration experiments the gain of fat in adult animals obtained by addition of various feedstuffs to their maintenance rations. On the basis of these results multiple regression equations were computed relating the content of digestible nutrients in feeds to their energy potential, i.e. fat production in adult animals. The following equations were found for the energy values of concentrate feeds in cattle and pigs:

$$\text{Cattle:} \quad y = 1.78x_1 + 7.04x_2 + 2.37x_3 + 2.13x_4 \; (\pm\, 3.7\%)$$

$$\text{Pigs:} \quad y = 2.14x_1 + 7.71x_2 + 0.01x_3 + 3.27x_4 \; (\pm\, 5.8\%)$$

y designates (kcal) net energy. The new energy feed unit called NE_F (net energy for fattening) equals the energy value of 1 kcal of net energy produced in adult animals. x_1–x_4 are g digestible nutrients in 1 kg of food: x_1 = digestible crude protein, x_2 = digestible crude fat, x_3 = digestible crude fibre, x_4 = digestible nitrogen-free extract. The constant values 1.78 and 7.04 indicate kcal of fat produced by 1 g of digestible protein, fat, etc. Thus, the amount of fat produced can be calculated on the basis of the amount of digestible nutrients present in concentrate feeds without use of value factors which refer to individual feeds. It may be noted that nitrogen-free extract was utilized 50% more efficiently by pigs than by cattle.

Nehring's experiments were extended to the determination of the fat-producing power of feed mixtures containing concentrates and roughage (in the case of ruminants). The following regression equations were computed relating the contents of digestible nutrients present in all kinds of feedstuffs to their potential for fat formation:

Cattle: $y = 1.71x_1 + 7.52x_2 + 2.01\,(x_3 + x_4)$

Pigs: $y = 2.56x_1 + 8.54x_2 + 2.96\,(x_3 + x_4)$

Chickens: $y = 2.58x_1 + 7.99x_2 + 3.19\,(x_3 + x_4)$

Good agreement between NE_F values calculated and those obtained experimentally was found with all kinds of rations for monogastric species but not always with ruminants: greater standard deviations found with ruminants may be attributed to interaction in the rumen between concentrate feeds and roughage. Therefore, NE_F values calculated for ruminant diets outside the range of 67–80% digestible energy have to be corrected using correction factors that vary according to energy digestibility, for example 0.82 for 50% energy digestibility.

Tables of energy values for all feedstuffs and for energy needs for different performances of various animal species were established by Nehring. These tables contain the following parameters: digestibility of energy, energy feed units (EF/kg feed), energy concentration (EF/kg feed dry matter) and protein-energy quotient (PEQ = g digestible protein in 1 kg EF). For practical feed calculations EF, the energy feed unit, is introduced, which is a multiple of 1 kcal of NE_F. The EF for cattle equals 2.5 kcal of NE_F and for monogastric species 3.5 kcal of NE_F.

Blaxter's System Based on the Content of Metabolizable Energy in Feeds

Kellner recognized that net energy values of a feedstuff differ according to whether the feedstuff was used for fattening, maintenance or milk production. Instead of giving each food three different values, the animal requirements were determined for the different purposes, but all expressed in starch values, i.e. in terms of net energy for fattening.

According to the system developed by Blaxter in Britain, energy values of feeds are expressed in terms of metabolizable energy (ME) and the energy requirements of animals as net energy (NE). Allowances of net energy are given separately for each main function of the animals, i.e. in terms of the energy the animals expend in maintaining themselves, or retain in growth or secrete in milk.

A practical advantage of the system is that metabolizable energy contents of feedstuffs can be determined quite easily by digestion trials (see Chapter 14) whereas the determination of the net energy of feedstuffs needs time-consuming and complicated experiments with respiration chambers or animal calorimeters (see Chapter 15). The metabolic energy contents of a large number of feedstuffs determined mainly with sheep are listed in MAFF (1975), whereas the values for net energy have so far been determined in only a limited number of samples.

These two sets of data—ME concentrations in feedstuffs and energy allowances expressed in NE—are linked together by considering the efficiency with which ME present in various types of rations is converted into NE when serving different animal functions. Data concerning the efficiency of utilization of ME for maintenance, for growth and fattening and for milk production are presented in Table 16.6, particularly in their dependence on the quality of the diets. Below this table regression equations predicting efficiency of utilization of ME for the different functions are given. The terms k_m and k_f express the net efficiency with which increments of ME below and above maintenance are retained in the body, and k_l is the respective efficiency for milk production. The efficiency for maintenance (k_m) is on average higher than for lactation (k_l) or for growth and fattening (k_f). Moreover, the values of k_m and k_l are less influenced by the energy concentration than the value of k_f (see Table 16.6 and Figure 21.6).

Table 16.6 Efficiency of utilization of metabolizable energy by ruminants (according to ARC, 1980, p. 81).

	Metabolizability (q)			
Function	0.40	0.50	0.60	0.70
Maintenance (k_m)	0.643	0.678	0.714	0.750
Growth and fattening (k_f)	0.318	0.396	0.474	0.552
Lactation (k_l)	0.560	0.595	0.630	0.665

q (metabolizability) $= \frac{\text{ME}}{\text{GE}}$ (ME = metabolizable energy, GE = gross energy)

Regression equations:
$$k_m = 0.35q + 0.503$$
$$k_f = 0.78q + 0.006$$
$$k_l = 0.35q + 0.420$$

The decline of k_f with increasing crude fibre content of the diet appears from the equation $k_f = 0.650 - 0.084C$, where C is the crude fibre content of the dry matter of the diet in g/kg. The use of single values of k_m (0.72) and of k_l (0.62) for all types of diets involves little error, but k_f varies from 0.30 to 0.60. k_f values for roughage and silage are very dependent on ME content. For pelleted diets and mixtures of cereal grains and roughage, k_f varies much less with changes in ME content. It should be noted that efficiency of utilization of ME for maintenance falls as the protein content of the diet increases, but efficiency for production increases with increasing protein content.

It has been shown that as the level of feeding—particularly of forage diets-rises, the proportional loss of energy in the faeces increases (see Chapter 13) and the metabolizability of energy decreases. Thus Blaxter applied a correction for feeding level using the equation:

$$q_L = q_m + (L-1)(0.20[q_m - 0.623])$$

q_L is metabolizability of the gross energy at any feeding level L and q_m is that determined at maintenance. L is a multiple of maintenance. Since the intake effect is comparatively small, this correction has not been accepted by recent modifications of the Blaxter system such as that to be outlined below.

Use of Blaxter's System for the Formulation of Rations for Meat Animals

According to Blaxter's system, (1) the ME value of a ration is calculated by adding up the contribution of individual feeding stuffs, (2) NE requirements for meat animals consist of the energy used for maintenance plus the energy content of the gain. The principles of measurement of each of these values are outlined in Chapters 14 and 15. In practice, these values are estimated by means of equations (based on experimental results) which express dependence of the energy requirements on live weight and live weight gain (for further explanation see the following example for use of Blaxter's system). To compare these requirements given as NE with the feeding value of the ration (in ME), the data on requirements are converted into ME by use of the factors given in Table 16.6.

To simplify the calculations, data on ME requirements for maintenance and growth are tabulated in ARC, 1980. These estimates are given separately for bulls, bullocks and heifers since bulls have a higher fasting metabolism than heifers and bullocks and there are differences in the composition of gains which arise from differences in mature size and sex.

In order to illustrate the use of Blaxter's system, an example for prediction of live weight gain from ME intake (after ARC, 1975) is given here.

A 250 kg steer receives the following daily rations:

Food	Dry matter (kg/day)	Metabolizable energy (MJ/day)
4.1 kg hay	3.6	32.1
1.7 kg barley	1.4	17.9
Total intake	5.0	50.0

The concentration of metabolizable energy in dry matter is $\frac{50}{5} = 10$ MJ/kg.

Allowance for Maintenance. According to tabulated values, 31 MJ of ME are needed for maintenance of a steer of 250 kg. This requirement can also be calculated from the amount of net energy required for the fasting metabolism (FM) together with the efficiency (k_m) with which dietary ME is used to satisfy this requirement. FM is estimated by means of the regression equation:

$$\text{FM (MJ/day)} = 5.67 + 0.6\,W$$

where W is live weight in kg. For estimating ME—using the constant factor $k_m = 0.72$ (see above) and including a 0.05 safety margin—this equation becomes

$$M_m = 8.3 + 0.091\,W$$

where M_m is the ME requirement for maintenance. Fasting metabolism, thus maintenance requirement, is a simple function of live weight.

Calculation of the ME Available for Live Weight Gain. The ME available for live weight gain (MEP) can be found by deducting the ME allowance for maintenance (M_m) from the total ME of the ration: 50 − 31 = 19 MJ.

The efficiency with which MEP will be utilized for producing body gain depends on the energy concentration of the ration (M/D*), and in this example of 10 MJ/1kg, $k_f = 0.0435$. Allowing for a 0.05 safety margin, the net energy that will be used for growth (E_g) can be calculated as follows:

$$E_g(\text{MJ}) = \frac{\text{MEP} \times 0.0435(M/D)}{1.05} = \frac{19 \times 0.0435 \times 10}{1.05} = 7.9$$

The net energy requirement for gain (E_g) is the energy content of the gain.

The live weight gain (LWG) that can be achieved from the retained energy (E_g) is dependent on the energy value of the gain (EV_g) which in turn is related to the live weight and maturity of the animal. The energy value of 1 kg of the gain may be calculated using the following equation:

$$\text{EV}_g\ (\text{MJ/kg}) = 6.28 + 0.3\,E_g + 0.188\,W$$

Since

$$E_g = \text{LWG} \times \text{EV}_g,$$

then

$$E_g = \frac{\text{LWG}\ (6.28 + 0.188\,W)}{(1 - 0.3\,\text{LWG})}$$

* D = dry matter.

LWG is obtained using the equation

$$\text{LWG} = \frac{E_g}{(6.28 + 0.3E_g + 0.0188\ W)}$$

Accordingly, the live weight gain predicted for a 250 kg steer from a net energy (E_g) of 7.9 MJ is 0.60 kg/day.

Blaxter's system not only enables calculation of the live weight gain of an animal that consumes a certain amount of a specific diet. It can also be operated in reverse, to formulate a ration to give a desired rate of growth.

Use of Metabolizable Energy System for Dairy Cows. The ME system for dairy cattle involves separate calculation of the maintenance and production allowances. Constant values for utilization of ME are used ($k_m = 0.72$ and $k_l = 0.62$) because the efficiency of the ME for both purposes varies only slightly with the ME concentration in the diet (see above) and a limited range of feeds is given to dairy cows.

As an example, ME allowance for a cow weighing 650 kg and producing 20 kg of milk of 36 g/kg BF and 85 g/kg SNF (BF = butter fat content, g/kg, and SNF = non-fat solids content, g/kg) will be calculated (MAFF, 1975). To calculate the ME requirement for maintenance of the cow the same regression may be used as for growing cattle:

$$M_m = 8.3 + 0.091\ W$$

ME allowance for maintenance of the cow: $M_m = 63$ MJ.

The energy value of the milk secreted is calculated from the equation (see Chapter 21):

$$\text{EV}_l = 0.0386\,\text{BF} + 0.0205\,\text{SNF} - 0.236$$

where EV_l = energy value of the milk secreted in MJ/kg. The ME requirement for milk production (M_l) will be given by $\text{EV}_l/0.62$, which becomes $M_l = 1.694\ \text{EV}_l$ MJ/kg milk (with inclusion of 0.05 safety). The ME requirement for the production of 1 kg of milk is 4.9 MJ. A milk yield of 20 kg requires 20 × 4.9 = 98 MJ ME. Thus total ME allowance = 63 + 98 + 161 MJ/day.

Calculations of rations for dairy cows are often complicated by changes in their body weight. Body tissue has an energy value of 20 MJ/kg; it can be mobilized under certain physiological conditions (see Chapter 21) and used for milk production with an efficiency of 0.82. Thus each kg of tissue mobilized will allow the secretion of 20 × 0.82 = 16.4 MJ as milk, equivalent to a dietary ME of (16.4 × 1.05)/0.62 = 28 MJ, or to 5.2 of kg of milk (4% fat).

ME is used for body gain in the lactating dairy cow with an efficiency of 0.62. A gain in weight of 1 kg increases the animal's requirement for dietary ME by 20/0.62 = 32.3 MJ.

California Net Energy System

The NRC (1976) bases the energy requirements for growth in beef cattle on a system developed by Lofgreen and Garret (1968) in California. This system expresses NE requirements and NE content of feed in two different values: NE_m used for maintenance and NE_g used for the production of weight gain. The similar system of Blaxter is based on the different efficiencies of ME for each purpose. Values for NE_m and NE_g have been deduced from experiments with cattle fattened under commercial conditions. The NE_g value of a feed is determined by feeding at two levels and measuring the energy deposition resulting from the increase in feed intake. The energy gain is determined from measurements of specific gravity on carcasses as slaughter (see Chapter 1). Fasting metabolism (NE_m) is predicted by measuring live weight changes of animals fed two levels of food and extrapolating the energy retention to zero energy intake (see Chapter 17). Basal metabolism thus predicted is $0.332\ \text{MJ}.W^{0.75}$ which is accepted by the NRC (1976) as the NE_m requirement for all classes of beef cattle. Since heifers deposit more fat per unit of body weight than steers, separate equations are used to relate weight gain to NE_g (in MJ):

$$\text{For steers: } NE_g = (0.2205\ G + 0.0286\ G^2)\,W^{0.75}$$

$$\text{For heifers: } NE_g = (0.2344\ G + 0.0529\ G^2)\,W^{0.75}$$

G = daily weight gain.

Comprehensive tables of values for NE_m and NE_g in feedstuffs have been published by the NRC (1976) in spite of the amount of work required to evaluate each single feed. Since the partial efficiency of energy utilization for maintenance is higher than the partial efficiency of the energy used for the production and storage of fat and protein, the NE_m value for a feedstuff or a ration is always higher than its NE_g value. The major advantage of separate net energy requirements for maintenance and gain is that animal requirements stated in this way do not vary when different roughage/concentrate ratios are fed—contrary to Blaxter's system and all other energy systems based on a single unit because of their tendency to under-evaluate roughage in relation to concentrates.

In order to illustrate the use of the Californian system for prediction of the live weight gain which may be achieved by intake of a certain ration, the following example will be given (according to Church, *Digestive Physiology and Nutrition of Ruminants*, Vol. 2, 1971):

(1) *Ration composition.* The following amounts of energy would be supplied by 1.8 kg of alfalfa hay and 5.4 kg of barley.

	Dry matter (kg)	NE_m (MJ)	NE_g (MJ)
Alfalfa hay	1.8	9.2	4.3
Barley	5.4	48.1	31.6
	7.2	57.3	35.9

This ration contains 7.95 MJ of NE_m (57.3 : 7.2) and 4.98 MJ NE_g (35.9 : 7.2) per kg of dry matter.

(2) NE_m. The maintenance requirement accepted according to the California system in 0.332 MJ. $W^{0.75}$ for NE_m; thus a 300 kg steer requires for maintenance 23.22 MJ contained in 2.92 kg of the ration (23.22 : 7.95 = 2.92).

(3) NE_g. Since 2.92 kg of the ration are needed for maintenance, 4.28 are available for gain (7.2 − 2.92 = 4.28). Thus 4.28 kg ration are equivalent to 21.3 MJ of NE_g. According to the equation given above [$NE_g = (0.2205\ g + 0.0286\ G^2)W^{0.75}$], 21.3 MJ of NE_g are retained in 1.15 kg body gain of a 300 kg steer.

Flatt's System for Rationing Dairy Cows

Since energy is used with similar degrees of efficiency for maintenance and milk production in lactating animals, and in order to simplify the formulation of rations, a single feed unit (NE_l) is adequate to calculate rations for both maintenance and milk production. The NE_l unit is based on the caloric value of milk. The energy values of feeds are listed in terms of NE_l units. The same unit is used for expressing the requirements for maintenance, milk production and body weight change. Regression analysis data obtained by energy balance trials with lactating cows indicated that the maintenance requirement is 0.336 MJ. $W^{0.75}$. The net energy requirement for a lactating cow is the sum of the energy content of milk produced (calculated by formulae based on milk fat and total solids as given above in this chapter and in Chapter 21) plus the maintenance requirement, and with consideration of body weight changes. As a means of compensating for the depression of digestibility that occurs with rations high in roughage (see Chapter 13), an increase of 3% of the feed has been proposed for each 10 kg of milk produced above 20 kg per day.

Energy Systems for Ruminants: Summary

The main goals of feeding systems are (1) formulation of rations which permit certain performances, i.e. production of certain amounts of milk or meat, (2) estimation of animal response if feed intake and feed quality are known, and (3) replacement of a single feedstuff in the ration without altering the nutritive value of the ration. Most feeding systems are based on the evaluation of net energy. The earlier standards use the same unit for expressing the energy value of feeds for different functions. The newer systems recognize different efficiencies of ME for maintenance and growth and assign two energy values to each feedstuff—one for maintenance and one for body gain. The new systems take account of the type of diet in view of the influence of the energy concentration in the whole ration on the metabolizability of energy; however, according to the earlier methods, the energy value of a ration is calculated by summing the energy values of the ingredients. Even in the new systems, requirements for lactation and maintenance can be listed in terms of the same unit because of the similar rates of efficiency of ME for both purposes (see Chapter 21).

The new energy systems incorporate various new findings and become more precise, but also more complex. Nowadays the need for simplicity of calculation is diminished by the increasing availability of computers.

EVALUATION OF ENERGY OF FEEDSTUFFS FOR POULTRY AND PIGS

The losses of energy involved in the utilization of food by monogastric species are much smaller than the respective losses occurring in ruminants; for instance, the losses of energy due to the production of combustible gases are very small in monogastric species and can be ignored. However, nowadays, feeding of poultry and pigs is based on the energy evaluation of their diets; this arises mainly for the following reasons:

(1) It is common practice to feed poultry, and sometimes growing pigs, ad libitum or under conditions of permanent access to food. The amount of food consumed daily by these animals is to a large extent controlled by the energy content of the diet. It follows that dietary nutrients (amino acids, minerals and vitamins) are required in some specific ratio to the energy content of the diets; conventionally these requirements are expressed in terms of dietary concentrations such as grams per kg of diet, when energy density is specified.

(2) Increasing the energy density of diets by the addition of fat improves the rate and efficiency of body gain in chickens as well as growing pigs (see Chapters 6 and 18).

(3) Correct evaluation of various kinds of cereal grains that differ in their energy content is only possible by estimating the content of available energy. The energy content of cereal grains declines gradually in the order maize > milo > wheat > barley > oats.

(4) The desired lean/fat ratio of pig carcasses may be obtained by controlling the daily energy intake.

The energy system commonly used in poultry nutrition is that of ME. The ME content of poultry feedstuffs is easily determined since in birds faeces and urine are combined excreta. DE and ME systems are accepted in pig nutrition: the determination of DE is simple, and the ME in feedstuffs for pigs may be calculated from the ratio ME/DE, which fluctuates between 0.91 and 0.97 according to the diet; this ratio is 0.96 for the mostly used diets based on cereals (see later).

Net energy for poultry nutrition, also called 'productive energy', would be the most logical measurement of the energy available for maintenance and production also in monogastric species, but the determination of NE is complicated and as yet has only been carried out for a few feedstuffs. The net availability of ME varies not only according to the function of the diet in the body but also with the different dietary constituents. In growing chicks the ratio NE/ME for fat is 0.90, for carbohydrates 0.75 and for protein 0.60 (De Groote, 1974).

Estimates of the requirements for ME in chicks and pigs may be made by two different approaches. According to the functional approach, the requirements are determined for each of the major processes contributing to energy needs such as those for maintenance and protein and fat deposition in growing animals or those for egg production and those needed in pregnancy and lactation of sows. According to the second approach, responses to variation in dietary energy intake are related to changes in the performance of pigs. Changes in body weight gain or carcass lean content may be the basis on which energy requirements can be predicted.

The differences in energy cost of deposition of protein and fat in the animal body deserves greatest interest. The efficiency of ME for fat deposition (k_F) is much higher than that for protein deposition (k_M). k_F in growing pigs ranges from 0.62 to 0.92 with an average of 0.74; k_M ranges from 0.35 to 0.80, averaging 0.56. The ME cost of deposition of 1 g of fat in pigs is 53.8 kJ (caloric value of 1 g fat is 39.8 kJ and cost of synthesis is 14 kJ); the ME cost of deposition of 1 g of protein is 43.9 kJ (caloric value is 23.8 kJ and cost of synthesis 20 kJ) (Kilianowski, 1972).

Determination of Metabolizable Energy in Feedstuffs for Poultry and Pigs

ME in feedstuffs for poultry and DE in feedstuffs for pigs are determined by measuring food consumption and output of excreta by animals kept in metabolism cages (see Chapter 13) and calorimetric determination of GE in food and faeces. The use of an inert indicator such as chromium(III) oxide eliminates the need for measuring food intake and faeces output. (Hill, 1958). Instead, the ratios of indicator to GE in pooled samples of feed and excreta are measured. Since most individual feedstuffs are imbalanced with respect to nutrient content and impalatable when fed alone, a *diet replacement assay* has been developed (Sibbald, 1979). This procedure involves the feeding of two diets: a reference diet composed partly of glucose and partly of feedstuffs or only of feedstuffs, and a similar diet in which a proportion of the reference diet is replaced by the test material.

In a recently recommended *rapid bioassay*, first the alimentary tract of chicks was emptied by starving for 40 hours, then the test diet was fed and the collection period could be shortened to 24 hours.

The procedures for ME determination described here fail to take into account the metabolic faecal energy and the endogenous urinary energy (see Chapter 3). A more precise estimate of ME, i.e. true ME (TME) instead of apparent ME (AME), can be obtained by correcting ME for nitrogen gained or lost from the body. For pigs a correction factor of 29.5 kJ and for poultry such a factor of 34.4 kJ can be used for each g of nitrogen above or below nitrogen equilibrium. This correction is added to ME for animals in negative nitrogen balance and subtracted when the animal is in positive nitrogen balance.

For a quick prediction of the ME of poultry feedstuffs or diets, regression equations have been developed which include the positive and negative contributions of the chemical constituents:

$$\text{ME} = (\text{carbohydrate} \times \text{DC} \times \text{GE}) + (\text{fat} \times \text{DC} \times \text{GE}) + (\text{protein} \times \text{DC} \times \text{GE}) - k$$

where DC and GE are the digestibility coefficients and gross energy of the respective constituents; k reflects the negative influence of crude fibre. Satisfactory agreement has been obtained between ME values estimated by such formulae and those determined biologically.

Determination of Net Energy in Poultry

Productive energy (= net energy) can be determined by measurement of the energy gain attained in growing chicks fed with the test diet. The diet is fed to two sets of chicks to allow for different rates of growth. The proportions used for production and maintenance are calculated from simultaneous equations:

$$WM + G = FX$$

W = average chick weight at end of experimental period
M = maintenance requirement of chick
G = gain in carcass energy during feeding period
F = food intake
X = productive energy value of diet per unit weight

Because the measurements involved in the determination of productive energy are difficult to obtain with precision, the productive energy value of a certain feedstuff is less reproducible than the ME value.

Two procedures for calculating the values of net energy have been described. One was proposed by De Groote (1974), who calculated NE values from those of ME, and took into consideration the different metabolic efficiencies of carbohydrates, fat and protein (see above). The principle underlying the second system, developed by Nehring (1973) for feedstuffs for pigs and poultry, has been given above.

BIBLIOGRAPHY

Measurement of food energy for ruminants

Books

Ministry of Agriculture, Fisheries and Food (1975). *Energy Allowances and Feeding Systems for Ruminants. Technical Bulletin.* (MAFF). London: HMSO.

Agricultural Research Council (1980). *The Nutrient Requirements of Ruminant Livestock*, 2nd edn. Slough: Commonwealth Agricultural Bureau.

National Research Council (1982). Nutrient requirements of dairy cattle. No. 3. Washington, DC: National Academy of Sciences.

National Research Council (1984). Nutrient requirements of beef cattle. No. 4. Washington, DC: National Academy of Sciences.

National Research Council (1985). Nutrient requirements of sheep. No. 5. Washington, DC: National Academy of Sciences.

Articles

Blaxter, K. L. (1974). Metabolizable energy and feeding systems for ruminants. In *Nutrition Conference for Feed Manufacturers*, No. 7, pp. 3–26. London: Butterworths.

Blaxter, K. L., and Boyne, A. W. (1978). The estimation of the nutritive value of feeds as energy sources for ruminants and the derivation of feeding systems. *J. Agr. Sci.*, **90**, 47–68.

Blaxter, K. L. (1979). Further developments of the metabolizable energy system for ruminants. In *Recent Advances in Animal Nutrition—1979* (W. Haresign and D. Lewis, eds), pp. 79–92. London: Butterworths.

Flatt, W. P., and Moe, P. W. (1974). Nutritional requirements of lactating animals. In *Lactation* (L. L. Larson and V. R. Smith, eds), Vol. 3, pp. 311–349. New York and London: Academic Press.

Garrett, W. N. (1979). Relationships among diet metabolizable energy utilization and net energy values of feedstuffs. *J. Anim. Sci.*, 1403– 1409.

Graham, N. McC. (1982). Energy feeding standards. A methodological problem. Proc. 9th Symp. on Energy Metabolism, Norway (as above), pp. 108–111.

Knox, K. L., and Handley, T. M. (1973). The California net energy system: theory and application. *J. Anim. Sci.*, **37**, 190–199.

Kromann, R. P. (1973). Evaluation of net energy systems. *J. Anim. Sci.*, **37**, 200–212.

Lofgreen, G. P., and Garrett, W. N. (1968). A system for expressing net energy requirements and feed values for growing and finishing beef cattle. *J. Anim. Sci.*, **27**, 793–806.

Moe, P. W., and Tyrell, H. F. (1973). The rationale of various energy systems for ruminants. *J. Anim. Sci.*, **37**, 183–189.

Moe, P. W., and Tyrell, H. F. (1974). Observations on the efficiency of utilization of metabolizable energy for meat and milk production. *Nutrition Conference for Feed Manufacturers*, No. 7, pp. 37–78. London: Butterworths.

Nehring, K., and Haenlein, G. F. W. (1973). Feed evaluation and ration calculation based on net energy. *J. Anim. Sci.*, **36**, 949–964.

Schürch, A. (1969). The future of animal production. *Mitt. Tierhaltung*, **123**, 1–10.

Van Es, A. J. H., Vermorel, M., and Bickel, H. (1978). Feed evaluation for ruminants: new energy systems in The Netherlands, France and Switzerland. *Livestock Prod. Sci.*, **5**, 327–371.

Measures of food energy for poultry and pigs

Books

Agricultural Research Council (1984). *The Nutrient Requirements of Poultry.* Slough: Commonwealth Agricultural Bureau.

Agricultural Research Council (1981). *The Nutrient Requirements of Pigs.* Slough: Commonwealth Agricultural Bureau.

National Research Council (1977). Nutrient requirements of poultry. *Nutrient Requirements for Domestic Animals*, No. 1. Washington, DC: National Academy of Sciences.

National Research Council (1979). Nutrient requirements of pigs. *Nutrient Requirements for Domestic Animals*, No. 2. Washington, DC: National Academy of Sciences.

Morris, T. R., and Freeman, B. M. (eds) (1974). Energy requirements of poultry. *Poultry Science Symposium Series* No. 9. Edinburgh: British Poultry Science Ltd.

Articles

De Groote, G. (1974). Utilization of metabolizable energy. In *Energy Requirements of Poultry* (T. R. Morris and B. M. Freeman, eds), *Poultry Science Symposium Series*, No. 9, pp. 113–134. Edinburgh: British Poultry Science Ltd.

Fisher, C. (1982). Energy evaluation of poultry rations. *Recent Advances in Animal Nutrition—1982* (W. Haresign, ed.), pp. 113–140. London: Butterworths.

Hill, F. W., and Anderson, D. L. (1958). Comparison of metabolizable energy and productive energy determinations with growing chicks. *J. Nutr.*, **64**, 587–603.

Kielanowski, J. (1972). Energy requirements of the growing pig. In *Pig Production* (D. J. A. Cole, ed.), pp. 183–202. London: Butterworth.

Sibbald, I. R. (1979). Metabolizable energy evaluation of poultry diets. In *Recent Advances in Animal Nutrition—1979* (W. Haresign and D. Lewis, eds), pp. 35–50. London: Butterworths.

Sibbald, I. R. (1985). The true metabolizable energy bioassay as a method for estimating bioavailable energy in poultry feedstuffs. *Wld. Poult. Sci. J.*, **41**, 179–187.

Vohra, P., Wilson, W. O., and Stopes, T. D. (1975). Meeting the energy needs of poultry. *Proc. Nutr. Soc.*, **34**, 13–19.

Wiseman, J., and Cole, D. J. A. (1983). Predicting the energy content of pig feeds. In *Recent Advances in Animal Nutrition—1983* (W. Haresign, ed.), pp. 59–70. London: Butterworth.

Part IV

NUTRITIVE REQUIREMENTS FOR BODY PROCESSES AND PRODUCTIVE FUNCTIONS

CHAPTER 17

The Maintenance Requirement

Obviously a certain amount of energy is expended by non-producing, resting or sleeping animals. Energy is continuously required by the heart to pump blood, by the diaphragm for respiration, by the nervous system to maintain its own activity and muscle tone, for temperature regulation mainly but not only in warm-blooded animals, for the general metabolism of most tissues, for active absorption and transport of chemical compounds, for repair of damaged or worn tissues, including turnover (see Chapter 2) and for the production of hormones and enzymes.

Mammals and birds are warm-blooded animals. Most domestic animals are in these two classes and their relative constant body temperature usually exceeds that of the environment. As a result heat flows from the animal to its environment. Within a certain range of ambient temperatures the heat produced by the normal metabolism of a resting animal is minimal and suffices to cover this heat loss. This range is referred to as the *comfort zone* or the *thermoneutral zone.* The former term is self-explanatory, the latter infers that no physiological processes requiring the expenditure of a considerable amount of energy are activated in order to maintain normal body temperature. In the comfort zone, body temperature is physiologically regulated by the constriction or dilation of peripheral blood vessels and by some sweating. Very little energy is required for these processes. The low and the high limits of the comfort zone are referred to as the *lower* and *upper critical temperatures,* respectively. When the ambient temperature is below the lower critical point, body temperature is regulated by shivering. Above the upper critical point animals pant, i.e. they increase their rate of respiration, in addition to sweating. They also gradually lose their thermoregulatory ability. Shivering and panting require more energy than the regulation of peripheral blood flow and sweating. The consumption of feed increases the heat production of an animal. The amount of excess heat produced by consuming feed is its *heat increment* which is a function of the diet (see Chapter 14). The heat increment also serves to compensate the animal for heat lost to a cold environment. Consequently, the comfort zone of a fed animal is lower than that of a fasted one.

A fasted animal will shiver at a higher ambient temperature than a fed animal. To illustrate the point, a fasted calf will shiver at about 19 °C where a fed calf will do so only when the temperature drops to about 7 °C. The same is true for panting: a fasted animal will pant at a higher temperature than a fed animal. To generalize, fasting elevates and feeding reduces both the lower and the upper critical temperatures, i.e. they shift the comfort zone upwards and downwards, respectively. Many factors affect the heat exchange between the body and the environment: body size, shape, physical activity, endocrine function, insulation and behaviour. Subcutaneous fat, and to some extent the skin, insulate and tend to shift the comfort zone downwards, i.e. they increase the tolerance to low temperatures and reduce tolerance to heat. Hair, wool and feathers have complex effects because, in addition to being insulators, they may retain or repel water and radiation. Behaviour patterns like huddling, and brooding of the dam, effectively reduce the lower critical temperature of neonatal animals. In hot weather animals tend to reduce their insulation, limit their feed intake and physical activity, and seek shade and water.

BASAL METABOLIC RATE (ENERGY METABOLISM IN FASTING ANIMALS)

In a starving animal the energy required for the above-mentioned physiological processes essential to life is obtained by the breakdown of the body's constituents, first glycogen, then fat and protein. The energy expended in the fasting animal is represented by the fasting heat production, which is referred to as *basal metabolic rate* (BMR) or fasting catabolism. The first energy requirement of an animal is to receive enough energy in its diet to keep it alive. The energy required for *maintenance* is the amount needed to prevent any loss from the animal's body, i.e. to keep the animal in energy equilibrium. When expressed as net energy (NE), the BMR represents the maintenance requirement, but when expressed as metabolic energy (ME), the maintenance requirement includes BMR plus the heat increment of the diet supplied for covering BMR. Like heat production by animals in general, the BMR can be measured by direct or indirect calorimetry (see Chapter 15).

The BMR, which depends mainly on body characteristics including endocrine activity, age, size, etc., is often measured in animals and man for research or diagnostic purposes. The maintenance requirement, on the other hand, is measured for very practical reasons related to production in animals and weight control in humans.

Measurement of the BMR

In order to measure the BMR of an animal the following conditions must be observed:

(1) The animal must be fasting and its alimentary tract must be free of feed residues since digestion and absorption increase heat production. This is commonly referred to as the *post-absorptive state.*

(2) The animal must be resting.

(3) The ambient temperature must be in the thermoneutral zone.

(4) The animal must be in reasonably good body condition; poor condition may reduce the metabolic rate.

These conditions will now be discussed in detail.

Attainment of the Post-absorptive State

The excretion of feed residues takes longer in ruminants than in monogastric species. The excretion of feed residues can be monitored by observing the disappearance of dyed feed particles added to the diet before fasting. Results of such trials with cattle and pigs are presented in Figure 17.1. The graphs represent the number of particles excreted, expressed as a percentage of the total excreted, as a function of time. In pigs all the dyed particles of feed were excreted in about 48 hours. In normally fed cows about 80% of the dyed particles were excreted in 80 hours but thereafter excretion was

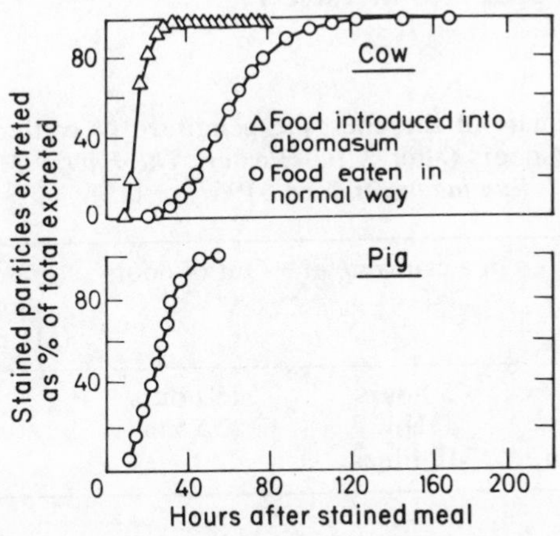

Figure 17.1. Passage of food through the digestive tract of cattle and of pigs (after Blaxter, K. L. *The Energy Metabolism of Ruminants*, 1967, p. 81). *Reproduced by permission of Hutchinson Edition and Sir K. Blaxter).*

slow, dyed particles possibly appearing in the faeces for 10 days. If, however, the feed is introduced directly into the abomasum, it is eliminated as rapidly as in monogastric animals. This indicates that the passage of feed from the rumen is slow and prolongs the excretion time. Therefore, it is customary to measure the BMR in ruminants only 10 days after the last meal.

Decline of methane production to very low values can also be used as an index of the attainment of a post-absorptive state since methane is produced only by fermentation processes in the rumen.

Another criterion of the attainment of the post-absorptive state is a non-protein respiratory quotient of about 0.7, which indicates that no carbohydrate is being oxidized and the energy is being obtained only from body reserves (see Chapter 15).

Rest

Standing requires more energy than lying down. The difference may be of the order of 15%. This difference may be smaller in the horse, which is better adapted to standing than ruminants, because of the structure of its ligaments. Perfect resting conditions may prevail only in brief measurements on human subjects. In ruminants it is commonly assumed that the animal will stand one-third of the time rather than lie down. The measurement of the BMR is more difficult in young animals because they tend to engage in spontaneous physical activity. This is reflected in their maintenance requirement (see later).

Freedom to move will further increase the energy expenditure of animals. A fasting bull with a BMR of 3.34 MJ/day may spend 15% more energy when free to move than when in an animal calorimeter. A breakdown of the added energy cost is presented in Table 17.1.

Table 17.1 Estimate of the energy expenditure of cattle in a calorimeter and out of doors (after K. L. Blaxter, *The Energy Metabolism of Ruminants*, 2nd. ed., 1967, p. 112)

Activity	In a calorimeter	Out of doors	Additional energy cost (MJ per 24 hours)
Standing	5 hours	15 hours	2.09
Walking	Nil	2.5 km	2.91
Changing position	10 times	12 times	0.10

Sleep reduces the BMR by about 7% in the dog, the cat, the horse and in man. In ruminants this is not the case because the nature of their sleep is different.

Thermoneutral Zone

To measure the BMR, the ambient temperature must be within the comfort zone so that the animal does not increase its heat production to keep warm (as in shivering) or increase its metabolic rate trying to keep cool (as in panting). It must be emphasized that the comfort zone may be reduced when humidity is high. At high temperatures humidity prevents cooling by reducing the evaporation of sweat. At low temperatures water vapour, i.e. humidity, increases the specific heat of air and thus heat loss by convection. Therefore, relative humidity in an animal calorimeter must not exceed about 50%. Typical thermoneutral zones for different species are listed in Table 17.2.

Table 17.2 Typical thermoneutral zones of several species (°C) (according to K. Nehring, *Animal Nutrition*, 7th ed., 1959)

Rat	28–29	Ovine	21–31
Dog (with long hair)	13–16	Bovine	5–20
Fowl	12–26	Caprine	10–20
Turkey poults	20–28	Swine	20–26
Calves	10–20	Man	28–32

Animal factors that may affect the BMR have been discussed above. Results obtained for the heat production of a dog, in the clipped and in the unclipped state, and at various temperatures, are presented in Figure 17.2. The comfort zone of the dog in the unclipped state was between 20 and 30 °C. Below 20 °C and above 30 °C the heat production of the unclipped dog increased. In the clipped state the lower critical temperature appears to be 30 °C, which is the upper critical temperature in the unclipped state. Thus, clipping shifted the lower critical temperature from 20 °C to about 30 °C. In sheep carrying a fleece with a fibre length of about 50 mm, the lower critical temperature may well be 9 °C compared with 28 °C for shorn sheep.

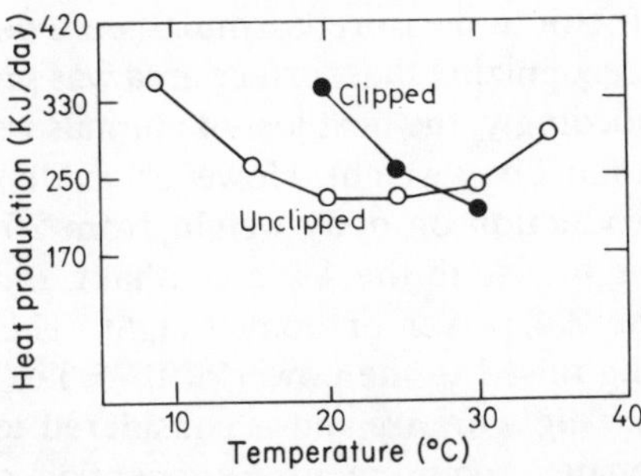

Figure 17.2. Heat production of a dog, clipped and unclipped at different ambient temperatures (adapted from Nehring, 1959).

Body Condition of the Animal

Fasting is a prerequisite of measuring the BMR of an animal. However, prolonged fasting will reduce the condition of the animal and its metabolic rate to a point where it cannot maintain its normal temperature. Data on the utilization of body fat and protein in a rabbit fasted for 18 days (Table 17.3) indicate that, during the first two days of the fast, with feed still being absorbed, protein was utilized. Thereafter, from day 3 to day 15 mainly fat was utilized. Finally, when body fat was almost exhausted, body protein was the major source of energy. The measurement of the BMR should be performed at that point in time when feed residues are nil and fat utilization is maximal. In the example in Table 17.3, this probably occurred on day 3 or 4.

Table 17.3 The daily utilization of body fat and protein in a rabbit fasted over 18 days (after K. Nehring, 1959, as above)

Days of fast	Live weight (g)	Utilization		
		Protein (g)	Fat (g)	MJ
1–2	2341	9.75	–	0.231
3–9		6.70	10.0	0.552
9–15		5.92	7.4	0.431
16–18	1388	13.27	1.0	0.353

Factors Affecting the BMR

Most of the energy required to maintain the BMR is used to keep body temperature at a constant level. Heat loss to the environment is proportional to the difference between body temperature and ambient temperature, and to body surface area. Since the body temperature of all mammals, regardless of their size and of the environment in which they live, is between 36 and 39 °C, body surface area is the major variable that determines heat loss. Since the surface area is difficult to measure, formulas were devised for computing it from body weight, recognizing that surface area was proportional to the 2/3 power of weight. Accordingly, the heat loss of animals would be proportional to the 2/3 power of their body weight. However, evidence derived from the regression of heat production on body weight from 26 mammalian species ranging from the 20-g mouse to the 4-ton elephant, indicates that heat loss is proportional to the 3/4 power of body weight. The body weight of an animal expressed in kg raised to the power of 0.75, i.e. $W^{0.75}$, is referred to as its *metabolic body weight* or *size* and is considered to be proportional to the BMR and maintenance requirement. Exponentials ranging from 0.60 to 0.82 have been proposed by different authors for the calculation of metabolic body weight, but values between 0.72 and 0.75 are most commonly used.

THE PRACTICAL SIGNIFICANCE OF THE BMR AND METABOLIC BODY SIZE

The BMRs of cattle at different body weights are presented in Table 17.4 and those of several species in Table 17.5. The values in these tables indicate that the total BMR of animals increases with body weight, both within and between species, but the increase is not proportional to body weight. Within one species, and especially among mature animals (cattle from 300 to 700 kg), the BMR per unit of surface area tends to increase with weight though the increase is not proportional to that of body weight. Clearly, the large decrease in BMR per unit of body weight with increasing size is the result of the *relatively* smaller surface area of large animals. The most striking fact is that the BMR per metabolic body size is remarkably constant both within and between species. The average BMR per kg of metabolic body weight

Table 17.4 The values of basal metabolic rate of cattle at different body weights (according to K. Nehring, 1959, as above)

Body weight (kg)	Basal metabolic rate (MJ/day)		
	Total	Per 100 kg weight	Per square metre
300	17.62	5.87	4.33
400	21.09	5.27	4.27
500	24.23	4.85	4.26
600	27.15	4.53	4.21
700	29.87	4.27	4.17

Table 17.5 The basal metabolic rate of adult animals of several species (according to McDonald *et al.*, *Animal Nutrition*, Third ed., 1981, p. 267)

Animal	Live weight (kg)	Basal metabolic rate (MJ/day)			
		Per animal	Per kg live weight	Per square metre	Per kg $W^{0.75}$
Cow	500	34.1	0.068	7.0	0.32
Pig	70	7.5	0.107	5.1	0.31
Man	70	7.1	0.101	3.9	0.29
Sheep	50	4.3	0.086	3.6	0.23
Fowl	2	0.60	0.300	–	0.36
Rat	0.3	0.12	0.400	3.6	0.30

for the species in Table 17.5 is 0.27 MJ. This value is a good approximation for calculating the total BMR of any warm-blooded mature animal from the equation:

$$\text{BMR} = 0.27\ W^{0.75}$$

where the BMR is expressed in MJ/day and W is the body weight of the animal expressed in kg (the equivalent in kcal is 70 $W^{0.75}$).

The concept of metabolic body size has been developed in order to reduce the BMR to a common denominator. However, the rate of many other physiological functions is also related to metabolic body size rather than to body weight. This is clearly demonstrated in Table 17.6. The maximal dry matter intake during lactation and the peak milk production of either of two greatly different species, such as cattle and rats, can be reasonably well predicted from those of the other species by using the ratio of their metabolic body size. This is also true with respect to many other physiological functions including growth. In recent years it has become common practice to express nearly all the rates of physiological functions (intake, output, deposition, mobilization, etc., of miscellaneous metabolites) relative to metabolic body size rather than to body weight.

Table 17.6 Body weight, metabolic body size, peak feed intake during lactation and peak lactation in a high-producing dairy cow and in the rat

Species	Body weight (kg)	Metabolic body size ($W^{0.75}$)	Peak dry matter intake during lactation (g/day)	Peak lactation (MJ/day)
Cattle	600	121.2	22 000	158.90
Rat	0.28	0.35	60	0.50
Cattle/rat ratio	2143	346	367	317

The most noteworthy difference in BMR among domestic animals occurs between cattle and sheep. The lower BMR of sheep (about 0.23 versus 0.32 MJ per kg $W^{0.73}$) is probably the result of the fact that sheep carry a fleece and are better insulated against both low and high temperatures. Similar differences have also been observed between breeds of sheep and cattle; for example, the difference in BMR found by Blaxter between Ayrshire and Angus steers amounts to 20%. Furthermore, there are quite considerable variations between individuals of the same breed (5–10% in cattle and sheep).

Age and BMR

Age is the major source of variation of the BMR per unit of metabolic body size in animals and man. As a rule, the BMR decreases with age. Factors contributing to this decline are (1) the proportion of energy derived from oxidation of protein instead of fat decreases with age (see Table 17.7), (2) a simultaneous decrease in turnover of protein with age (see Chapter 7).

Table 17.7 The fasting energy metabolism of a 14-day-old calf, and a 3-year-old steer (after K. L. Blaxter, *The Energy Metabolism of Ruminants*, 1967 as above, p. 93)

Animal	Weight (kg)	Total heat (MJ per 24 hours)	Proportion of heat derived from oxidation of	
			protein (%)	fat (%)
Calf	38	5.27	20.8	79.2
Adult steer	478	30.19	13.9	86.1

In early life metabolism is very high, much higher than the interspecies normal of 0.27 MJ/kg$^{0.75}$. The relatively high metabolism of young ruminants compared with that of the foetus is of considerable interest. Heat production of foetal tissues of ewes is about the same as that of mature ewes at rest. Within 24 hours of birth the metabolism of the lamb is three times as great. Afterwards the BMR declines gradually with age in sheep and also in cattle. Some values recently obtained for the BMR of cattle at ages ranging from one to 48 months are presented in Table 17.8.

Table 17.8 The basal metabolic rate of cattle from one month of age to 48 months

Age (months)	BMR (MJ/$W^{0.73}$)	Age (months)	BMR (MJ/$W^{0.73}$)
1	0.586	18	0.419
3	0.565	24	0.398
6	0.523	36	0.377
7	0.460	48	0.356

In sheep, as mentioned above, the BMR is generally lower than in cattle. It has been estimated that between the ages of one month and 48 months the BMR of sheep declines from 0.44 to 0.24 MJ/$W^{0.73}$. Young growing pigs, due to their higher activity and high rate of nitrogen turnover, show values around 0.58 to 0.033 MJ/$kg^{0.75}$. Because the BMR per unit of metabolic body size decreases with age, the maintenance requirement of growing animals increases to a lesser extent than that indicated by their metabolic body size.

In humans, the metabolic rate usually decreases after the age of 15 years; after 26 years of age the decrease may be 1% per year.

Sex, Hormones and the BMR

The BMR of males is higher than that of non-pregnant females of similar age and size. Castration greatly reduces the BMR in both males and females, though the effect is greater in males. This indicates that the sex hormones increase the BMR of both males and females. Thyroxine has an important role in the regulation of the BMR. In the past, thyroid activity was commonly estimated by measuring the oxygen consumption of human patients after a night's rest and before a meal. Administration of thyroxine increases the BMR and thyroxine antagonists decrease it (see Chapter 10).

THE PRACTICAL MAINTENANCE REQUIREMENTS OF ANIMALS

By definition, a non-producing animal in which no weight changes occur is using 100% of its ration for maintenance. Adult humans, apart from pregnant women and hard-working labourers, live on maintenance rations. Farm animals, however, are infrequently kept in a non-productive state, e.g. mature non-lactating and non-pregnant females. Some knowledge of maintenance is needed in order to formulate rations for productive animals by adding together requirements that are calculated separately for maintenance and for production. The maintenance portion forms a considerable part of the whole ration. Ruminants or pigs at high levels of production (meat or milk) use about 35% of the energy in their food for maintenance, egg-laying hens even more than 50%.

Principles of Methods for the Determination of Energy Requirements for Maintenance

Data on maintenance requirements for energy have mainly been obtained in the following ways:

(1) Determination of Basic Metabolic Rate

Measurement of heat expenditure of a fasting animal (BMR) by direct or indirect calorimetry theoretically provides an estimate of the amount of NE which the animal must obtain from its food in order to meet its requirement for maintenance. Determinations of fasting metabolism are usually performed on mature and non-producing animals under standardized conditions as described above. It is often difficult to translate values of BMR into practical recommendations for maintenance requirements. For adjusting BMR values the following factors should be considered:

(a) According to the definitions given above, the maintenance requirements (expressed in any other measure of food energy apart from NE, particularly as ME) comprise BMR plus the heat increment of the ingested feed, i.e. the energy needed to metabolize the maintenance ration *per se* (see Chapter 14). The resting heat production of an animal eating a maintenance ration is 25–40% above basal metabolism, depending on the efficiency of utilization of ME for maintenance, which is about 0.8 J/J in pigs and 0.63 to 0.73 in ruminants depending on the fibrousness of the diet.

(b) Since muscular activity of an animal confined to a calorimeter is minimal, an activity increment of 20–30% should be added for converting BMR values into NE requirement for maintenance.

(c) Additional energy is needed for compensating for losses occurring in animals kept at temperatures outside the comfort zone (see above).

(2) Determination of Maintenance Requirements of Fed Animals

Since it is not easy to derive from determinations of BMR accurate figures of the ME requirements for maintenance, it is preferable to determine ME requirements direct by experiments with fed instead of fasting animals and to obtain results directly applicable to the practical conditions.

(a) Direct Measurement by Feeding Trials. An easy way of measuring the maintenance requirement, used in early feeding standards, is to feed a diet of known energy value and to determine the amount of food required to hold adult animals at constant body weight over an extended period. In principle this method is accurate; however, measurements of live weight are imprecise because live weight is greatly affected by changes in the amounts of water retained in gut or bladder and it does not indicate changes in body composition which are essential for calculating energy balance.

A more reliable criterion for determination of the maintenance requirement than measurement of feed intake that holds the animal at constant weight, is the measurement of the feed intake that promotes energy equilibrium. The term maintenance metabolism is usually defined as the metabolism when the body is in energy equilibrium, i.e. when energy balance is zero because energy input equals output. Energy balance can be measured from calorimetric experiments.

(b) Estimation of Maintenance Requirements by Regression Methods. It is more usual to estimate maintenance requirements with producing animals

fed the same ration at two or more levels of intake and production than with mature animals maintaining constant weight. Therefore, the measurement involves at least two calorimetric trials. By means of regression of these data, an estimate is made of the energy required for maintenance by extrapolating intake to zero energy production.

Figure 17.3 presents schematic results of calorimetric trials performed to determine the maintenance requirement of an adult sheep. Intake in terms of metabolizable energy appears on the x-axis (abscissa) and energy balance on the y-axis (ordinate). In the example given in Figure 17.3 the maintenance requirement in terms of ME is 4600 kJ/day. In this Figure the slopes of the straight lines below and above maintenance are unequal; they represent the partial efficiencies of the utilization of ME for maintenance, and for growth or fattening, respectively, for the diet used. For all diets the former is greater than the latter (see Chapter 16); average values for maintenance are 0.82 (k_m) and for growth or fattening 0.64 (k_f). Therefore, when two calorimetric trials are performed to determine the maintenance requirement of an animal, the feeding level in both trials must be either below or above maintenance.

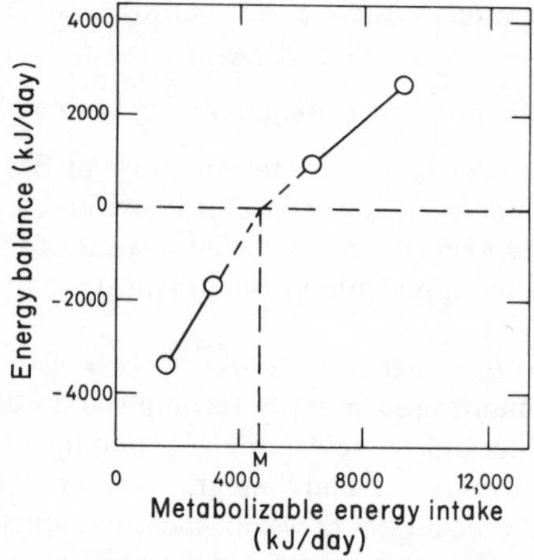

Figure 17.3. Measurement of the maintenance requirement by extrapolation (M = maintenance).

According to a modification of this method, changes in the energy content of animals are measured by means of a *slaughter technique* (see Chapters 1 and 18) without use of respiration chambers (Lofgreen and Garrett, 1968). By determining body composition and body weight before and after a feeding period, it is possible to calculate the net energy required for maintenance and gain. Heat produced (HP) daily at zero feed intake (composed of BMR and HP activity increment) is measured indirectly by deducting energy balance (EB) from ME intake: HP = ME − EB. If HP is determined at various levels it is possible to estimate it at zero feed intake by extrapolation.

Some Problems in Meeting Maintenance Requirements

Energy retention determined by the difference trial assume that the relationship between food intake and energy retention is rectilinear (see Figure 17.4). Actually, this relationship over the complete range of intake from zero to ad libitum is curvilinear. The convention adopted to accommodate this is an approximation with two straight lines (see Chapter 16).

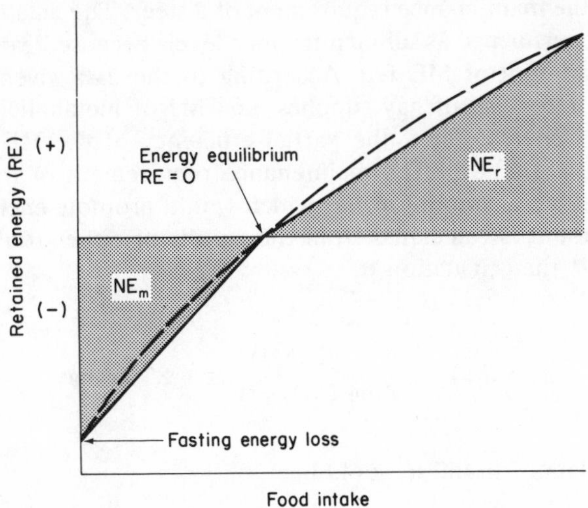

Figure 17.4. Relationship between retained energy (RE) and net energy (NE). The dashed line shows the curvilinearity between RE and food intake; the solid lines are linear aproximations (after National Research Council, *Nutritional Energetics of Domestic Animals*, 2nd edn, 1981, p. 23, as above, Chapter 14). *Reproduced by permission of National Academy Press.*

The determination of the maintenance requirements on producing animals has some limitations. Increments of metabolism which correlate with increments of production (such as increased blood circulation and respiration, gastrointestinal and metabolic activity) are excluded from the maintenance estimate. However, not excluded are changes of maintenance metabolism that are not correlated with the level of production but with the fact of production *per se*, e.g. due to the change in hormonal status from non-lactating to lactating. For the dairy cow such a lactation effect causes an increase in the maintenance requirement of lactating cows by 10–20% (Van Es, 1972; see Chapter 21).

Examples of the Determination of the Maintenance Requirement

Two classical examples of the determination of the maintenance requirement will be presented here. Both experiments were performed on fed animals: one study was carried out by Armsby, the other by Kellner (see Chapter 15).

Example 1

The results obtained in Armsby's difference trial for the determination of the energy value of a hay, presented in Table 16.5, can be used for the calculation of the maintenance requirement of a steer. The data indicate that the trials were performed at submaintenance levels because heat production exceeded the amount of ME fed. According to the data given in the last line of Table 16.5, 1 kg of hay supplies 8.64 MJ of metabolic energy and 4.64 MJ of net energy. Thus, the partial efficiency of the ME in the hay for maintenance is 53.8%. The maintenance requirement of the steer can be calculated: it is the amount of hay which would promote energy balance, i.e. 0. This amount is calculated from the results of either trial. From the results of trial 1 the calculation is:

$$4.633\ (\text{kg}) - \frac{-1.12\ (\text{MJ})}{4.64\ (\text{MJ/kg})} = 4.87\ \text{kg hay}$$

and from the data of trial 2 it would be

$$2.801\ (\text{kg}) - \frac{-9.62\ (\text{MJ})}{4.64\ (\text{MJ/kg})} = 4.87\ \text{kg hay}$$

To generalize, in terms of feeds (kg):

$$\text{Maintenance requirement} = \text{intake (kg)} - \frac{\text{energy balance (MJ)}}{\text{NE/feed (MJ/kg)}}$$

In terms of ME the maintenance requirement is 41.98 MJ, and in terms of NE 22.60 MJ.

Example 2

Kellner used a diet of known NE value. Each of three steers (684 kg average body weight) was fed a daily diet of 4.4 kg SE (starch equivalent value) per day and gained 0.41 kg fat per day, i.e. the energy balance was positive and amounted to 16.2 MJ/day. According to Table 16.3, 1 kg of SE fed above maintenance produces 0.248 kg of body fat equivalent to 9.9 MJ.

Thus, 0.41 kg of fat actually gained is equivalent to 16.2 MJ. However, 4.4 kg SE supplied 43.56 MJ daily. The maintenance requirement in terms of NE for fattening is the difference between the NE supplied and the NE gained, i.e.

$$43.6\,(\text{MJ}) - 16.2\,(\text{MJ}) = 27.4\,\text{MJ}$$

or in terms of SE, 27.4 MJ : 9.9 MJ = 2.75 kg SE.

The Effects of Physical Activity on the Maintenance Requirement

The physical activity of domestic animals in their usual environment always exceeds their activity in the calorimeter. Therefore, the theoretical maintenance requirement for energy is a minimal estimate of the practical one. In animals kept on the farm, the latter may be 20–30% greater in ruminants and pigs and 50% greater in poultry. In grazing ruminants the energy requirement for walking and grazing may be considerable.

The energy cost of locomotion is proportional to the body weight of the animal and is somewhat lower for large animals than for small animals. For walking on a horizontal surface, the cost of energy is about 1.7–2.5 J per m per kg body weight. For the vertical component of locomotion, i.e. for a change in altitude, the energy cost is about 12–20 times higher. The efficiency of conversion of body energy into the vertical component of locomotion is about 30–35%, similar to that for work (see Chapter 22). A sheep weighing 50 kg, walking 6 km daily and climbing to an altitude of 100 m, would expend about 1000 kJ/day on locomotion which would be 20% of its theoretical maintenance requirement. The additional energy cost of activity of a 500-kg steer getting all its food by grazing on the range would be between 4.7 and 16.1 MJ per 24 hours, an increase in heat expenditure of 7–23% above that of the confined animal. Because the calculation of the energy expenditure of grazing animals involves many factors, its prediction is difficult, but the energy cost of grazing must never be neglected.

The Energy Cost of Maintenance in Pigs and Poultry

An estimate of the BMR of pigs and fowl is presented in Table 17.5. Further data on maintenance requirements in poultry are presented in conjunction with their requirements for growth in Chapter 18 and of layers in Chapter 20.

THE MAINTENANCE REQUIREMENT FOR PROTEIN

All the proteins in the tissues and fluids of animals undergo continuous catabolic and anabolic changes. Since part of the catabolized protein (or nitrogen) is wasted, animals have a protein requirement for maintenance that is independent of the rate of production of protein-containing products (milk, eggs, wool and tissue). A fasted animal excretes a certain amount of protein and some non-protein nitrogenous products; their excretion is increased by feeding. Therefore, by analogy to energy, the maintenance requirement for protein consists of the amount of protein used by an animal to maintain its mass unchanged and by a further amount used to metabolize the maintenance requirement. It is interesting to note that protein turnover bears a relationship to fasting metabolic rate—the range of values in various species is between 14.5 and 24.8 g of protein per $kg^{0.75}$ (Garlick, 1980). The chemical nature of protein metabolism has been discussed in Chapter 7. The emphasis here is on the quantitative aspects of the maintenance requirements for protein.

The Theoretical Protein Requirement for Maintenance

On theoretical grounds, the maintenance requirement for protein is the sum of the amounts of nitrogen excreted in the urine and in the faeces by an animal fed a nitrogen-free diet which, however, supplies energy, minerals and vitamins at the maintenance level. The reason for using a nitrogen-free diet is to eliminate the presence of (a) undigested feed nitrogen in the faeces, and (b) digested feed nitrogen absorbed and excreted in the urine. After the nitrogen-free diet is offered, a certain length of time is required for the animal to reduce its nitrogen excretion to the minimum. Energy for maintenance must be supplied, because under fasting conditions body protein would be mobilized to provide energy and would elevate the excretion of nitrogen, especially in the urine. Thus, under ideal conditions for measuring the protein required for maintenance, urinary nitrogen excretion is minimal. This nitrogen is endogenous, i.e. its immediate precursors are components of tissue rather than feed (see Chapter 7). It has been experimentally shown that the minimum urinary nitrogen (MUN) excretion is proportional to metabolic body size (see above) rather than to body weight. Therefore, the relationship can be expressed by the equation:

$$\mathrm{MUN} = k.W^{0.73}$$

where W is body weight in kg, MUN is in mg and k is a factor which depends on the animal. The value of k in the above equation varies from 80 to 200 for different species and classes of animals. A mean value of 120 has been proposed for mature cattle and 190 for young calves.

The protein requirements to replace nitrogen loss in scurf (sweat, hair and other keratinous waste) and for wool production are also dependent on metabolic body size. For practical reasons, it is convenient to consider these as part of the maintenance requirement, although wool is a useful animal product rather than a waste product; the value of k in the above equation may be adjusted to include scurf and wool. Approximate increments of k for nitrogen lost in scurf and wool production are 20 and 50, respectively.

The faecal nitrogen excretion from a nitrogen-free diet supplying energy, minerals and vitamins for maintenance will not be minimal but will contain no undigested feed nitrogen. It will be endogenous, i.e. it will include only constituents derived from gastrointestinal epithelium, and some blood. It has been experimentally shown that the amount of faecal endogenous nitrogen (FEN), also called metabolic nitrogen (see Chapter 7), depends primarily on the amount of dry matter passing through the gastrointestinal tract. In order to simplify its estimation, it is usually expressed as a function of dry matter intake. The amount of FEN has been found to be 1–2 g/kg dry matter intake in monogastric species and 4–6 g/kg in mature ruminants. In ruminants the distinction between endogenous and undigested faecal nitrogen is blurred; because bacterial synthesis in the rumen and in the caecum most of the faecal nitrogen is bacterial nitrogen.

Ruminants often refuse to consume nitrogen-free diets; also, roughage, i.e. high-fibre diets, will increase the amount of endogenous faecal nitrogen per dry matter intake. This is consistent with the assertion that faecal endogenous nitrogen depends on the amount of dry matter passing through the tract, rather than on the amount consumed.

Determination of the Practical Protein Requirement for Maintenance

For practical purposes the maintenance requirements for protein can be determined by feeding an animal two (or more) levels of a diet with a limited protein content and extrapolating or interpolating to zero nitrogen balance. The important point in such a determination is not to feed an excess of protein under any circumstances, because the excess would be excreted as urea and produce an exaggerated estimate of the protein requirement for maintenance. Therefore, if a mature (non-growing) animal is used, it must be offered less protein than is sufficient for maintenance, and energy to supply maintenance. In growing animals, feeding levels above maintenance may be used, provided, as indicated above, that protein is the limiting factor.

Factorial Determination of the Protein Requirements

When the dietary protein requirement for maintenance is calculated rather than directly measured, it is computed from available estimates of MUN and FEN. In calculating the dietary protein requirement for any function, one should consider the efficiency of digestion (or absorption) and the efficiency of retention of digested protein (biological value; see Chapter 7).

Calculation of the Dietary Protein Requirement for Maintenance of a Cow Weighing 500 kg Consuming 10 kg Dry Matter Per Day

A value of 5 g FEN per kg dry matter intake and one of 0.12 g (120 mg) of MUN per kg metabolic body size will be used.

(1) The amount of nitrogen as FEN per day would be:

(10 kg dry matter per day) × (5g nitrogen per kg dry matter) = 50 g nitrogen

(2) The amount of nitrogen as MUN would be:

$$(0.12 \text{ g nitrogen kg}) \times (500^{0.73} \text{ kg}) = 11.2 \text{ g nitrogen}$$

(3) The total amount of nitrogen for both functions is 61.2 g, or, in terms of protein, 6.25 × 61.2 g = 383 g.

(4) Assuming that the biological value of digested protein is 0.70 and the true digestibility of protein is 0.80, the requirement for dietary protein is:

$$\frac{383 \text{ g}}{(0.70) \times (0.80)} = 683 \text{ g}$$

The general formula for the calculation is:

$$R = \frac{6.25[(\text{FEN} \times \text{DM}) + (\text{MUN} \times W^{0.73})]}{\text{BV} \times \text{TD}}$$

where

R	=	the maintenance requirement for protein (g/day)
FEN	=	faecal endogenous nitrogen (g/kg dry matter intake)
DM	=	dry matter intake (kg/day)
MUN	=	minimum urinary nitrogen (g per metabolic weight in kg)
W	=	body weight of animal (kg)
BV	=	biological value
TD	=	true digestibility

A modified formula for the calculation of the maintenance requirement for protein of cattle has been suggested by Swanson (1977):

$$\text{TCP} = \frac{U + F + S}{E_{\text{p}}}$$

where

TCP is total crude protein expressed as g/day

U represents the protein equivalent of the amount of (endogenous) nitrogen excreted in urine on a nitrogen-free diet and is 2.75 $W^{0.75}$

F is 6.25 × faecal endogenous nitrogen lost on a nitrogen-free diet and is calculated on the basis of faecal dry matter in kg/day (FDM) as F = 0.068 FDM.

S represents protein in scurf calculated as 0.2 $W^{0.6}$

E_p is a factor which converts feed protein to protein available for maintenance; its value is 0.45 in mature cattle and 0.75 for milk-fed calves (see above).

THE MAINTENANCE REQUIREMENT FOR MINERALS

Most of the constituents of the diet are efficiently reabsorbed from the gut or from the kidney and recycled; the efficiency of these processes tends to increase with respect to deficient animals. Well-known examples of efficiently retained minerals are calcium, iron, sodium and chloride. Probably, most of the trace minerals are also efficiently retained. However, there are losses of endogenous minerals in faeces and urine, and thus even non-producing animals require minerals for maintenance.

It is customary to determine the maintenance requirement for minerals by measuring their excretion under conditions of depletion, analogous to those used for protein (see above). The daily endogenous calcium excretion of a cow weighing 450 kg has been found to be 7.3 g. Assuming the true absorption rate of calcium in cattle is 45%, the maintenance requirement would be 7.3/0.45 = 16.2 g. Radioactive isotopes are often used to determine the endogenous mineral excretion (see Chapter 10).

It should be noted that in a state of deficiency of protein or energy, protein is broken down and the animal is unable to achieve sulphur and phosphorus balance because both are constituents of protein. Conversely, a deficiency of phosphorus or sulphur will induce negative nitrogen balance. Generally, the mineral requirements for maintenance are proportional to the BMR, i.e. to metabolic body size. Thus, the calcium for maintenance has been expressed as 1.3 mg per 4 kJ of fasting heat production.

REQUIREMENTS FOR WOOL PRODUCTION

As mentioned above, it is useful to include the requirement for wool production in the requirement for maintenance. A fine-wool sheep may produce a fleece of 4 kg per year; about 3 kg consists of actual wool fibres, the remaining 1 kg being wool wax, suint and water. Wool wax, secreted by the sebaceous glands, is a mixture of many esters of water-insoluble aliphatic alcohols and sterols with fatty acids having 8–26 carbon atoms. Suint, the secretion of sweat glands, is a mixture of inorganic salts, potassium soaps and potassium salts of lower fatty acids.

The growth rate of wool depends on the nutritional status of the sheep, but wool grows even under conditions of negative energy and nitrogen balance. As a rule, adequate nutrition increases both the length and the diameter of the wool fibre. The major limiting factor of wool production is the absorption rate of cystine (or cysteine) from the gastrointestinal tract. The wool fibre consists of keratin and related fibrous proteins with unusually high levels of cystine (7–13% of the nitrogen in cystine) and very little methionine (less than 0.6% of the nitrogen in methionine). Infusions of cystine into the abomasum or into the blood may double wool growth. Infusion of methionine increases wool growth by supplying sulphur for the synthesis of cystine, but it is toxic above a certain level. Protected proteins, i.e. proteins treated with formaldehyde, which escape bacterial hydrolysis in the rumen (see Chapter 8), also increase the production of wool by increasing the amino acid supply.

The energy requirement for wool is low; it does not exceed 5% of the basal metabolic rate. It appears that for most diets the supply of amino acids will be limiting for wool growth and any effect of dietary energy will be exerted primarily through its influence on amino acid metabolism and the association between energy intake and the synthesis of microbial protein.

Adequate copper nutrition is necessary for maintaining the quantity and quality of wool production (see Chapter 10). Wool becomes brittle and lacking in crimp because of zinc deficiency in sheep.

BIBLIOGRAPHY

Blaxter, K. L. (1976). Environmental factors and their influence on the nutrition of farm livestock. In *Nutrition and Climatic Environment* (D. Haresign and H. Swan, eds), pp. 1–16. London: Butterworths.

Close, W. H., and Mount, L. E. (1975). The rate of heat loss during fasting in the growing pig. *Br. J. Nutr.*, **34**, 279–290.

Ferrel, C. L., and Jenkins, T. G. (1984). Energy utilization by mature, nonpregnant, nonlactating cows of different type. *J. Anim. Sci.*, **58**, 234–243.

Garlick, P. J. (1980), see Bibliography to Chapter 7.

Graham, N. McC. (1967). The metabolic rate of fasting sheep in relation to total and lean body weight, and the estimation of maintenance requirements. *Austr. J. Agr. Res.*, **18**, 127–136.

Graham, N. McC., Searle, T. W., and Griffiths, D. A. (1974). Basal metabolic rate in lambs and young sheep. *Austr. J. Agr. Res.*, **23**, 957–971.

Lofgreen, G. P., and Garrett, W. N. (1968) see Bibliography to Chapter 16.

Mount, E. L. (1978). Heat transfer between animal and environment. *Proc. Nutr. Soc.*, **37**, 21–28.

Swanson, E. W. (1977). Factors for computing requirements of protein for maintenance of cattle. *J. Dairy Sci.*, **60**, 1583–1593.

Thonney, M. L. *et al.* (1976). Interspecies relationships between fasting heat production and body weight. A reevaluation of $W^{0.75}$. *J. Anim. Sci.*, **43**, 693–704.

Turner, H. G., and Taylor, C. S. (1983). Dynamic factors in models of energy utilization with particular reference to maintenance requirement of cattle. *World Rev. Nutr. Diet.*, **42**, 135–190.

Van Es, A. J. (1972). Maintenance. In *Handbook of Animal Nutrition* (W. Lenkeit and K. Breirem, eds), Vol. 2, pp. 1–54. Hamburg and Berlin: Paul Pasey.

Webster, A. J. F. (1978). Prediction of the energy requirements for growth in beef cattle. *World Rev. Nutr. Diet.*, **30**, 226–189.

Requirements for wool growth

Black, J. L., and Reis, P. J. (1979). *Physiological and Environmental Limitations in Wool Growth.* Armidale: University of New England Publications Unit.

Thomas, P. C., and Rook, J. A. F. (1983). Diet and wool growth. In *Nutritional Physiology of Farm Animals* (J. A. F. Rook and P. C. Thomas, eds), pp. 538–557. London and New York: Longman.

CHAPTER 18

The Nutritional Requirements for Production of Body Gain

INTRODUCTION

Rations for productive animals must supply the requirements for the energy and specific nutrients (proteins, minerals and vitamins) that are needed for maintenance, and for the formation of all types of animal products (meat, milk and eggs) and for performing work. The portion of the diet needed for production must provide not only the energy deposited in animal products (net energy) but an additional amount required for the conversion of feedstuffs (mostly of plant origin) to animal products, i.e. energy needed for digestion of the food, absorption of the digested nutrients, their transport to the sites where biosynthesis occurs and for the metabolic processes involved in the conversion of the absorbed nutrients into components of body tissues, milk or eggs.

The energy required for the conversion of food into animal products is dissipated as heat and is equal to the difference between metabolizable energy and net energy (see Chapter 14). The efficiency with which ME is utilized for productive purposes depends not only on the type of product (meat, milk or eggs) and on the animal species, but also on the nature of the food and the nutrients liberated by digestion. This may be clarified by the following example: if a sheep is given a maintenance ration, then, for production of body fat which equals 418 kJ, an additional 600 kJ of energy must be provided as glucose by infusion into the abomasum. The same amount of fat can be synthesized by addition of acetic acid; this requires the supply of 1250 kJ as acetic acid in addition to the maintenance ration. The conversion of these two nutrients into a fixed amount of body fat requires different amounts of energy. Data on the efficiency of conversion of energy into various animal products in ruminants and monogastric species will be presented in this and following chapters.

In addition to providing energy, dietary protein is most important for ensuring the satisfactory performance of farm animals. The efficiency of conversion of dietary protein into animal products varies with the type of production and is affected by the ratio of protein to energy consumed.

Although the supply of high-quality protein for humans is the main purpose of animal production, the fat content of animal products is comparatively high. The proportion of energy in fat (as a percentage of the total energy) is 44–49% in milk (with 3–4% milk fat), 61% in eggs, 65% in beef and 45% in broilers. A large portion of the fat in the animal body is inedible (see later). A high feed intake is involved in the production of fat because of its high energy value. Humans in affluent countries reject the consumption of animal products high in fat, partly because of an aversion to fat and partly because of the health risks presumed to be associated with excessive intake of animal fat (see Chapter 6).

GROWTH OF ANIMALS

The rapid growth of animals is economically important. However, in meat production the emphasis is on maximizing the mass of muscles and glands which are edible and minimizing that of fat. The growth rate of young animals raised for other productive purposes must be aimed at ensuring maximal efficiency in performing the respective functions at later periods of life (milk and egg production and fertility).

Organisms grow by hyperplasia—an increase in the number of cells—by hypertrophy—an increase in the size of existing cells, and by an increase of extracellular fluid. The maximal growth rate of cells, tissues and of the whole body depends on genetic factors. An animal fully expresses its genetic potential only if its requirements for energy, protein, minerals and vitamins are met.

Increase in live body weight is the most common measure of growth. However, real growth is characterized by an increase in weight of muscles, bones and the vital organs which means accretion of protein, minerals and water as the animal approaches maturity. With advancing age, weight gains consist of more fat and less water. Farmers distinguish between growth in young animals and fattening in older ones after approaching maturity. The proportions between different organs and between chemical constituents of the body change because of different rates of growth of its various parts. In ruminants and pigs the head and extremities develop early, the hindquarters and loin region very late.

The Role of Hormones in Regulation of Growth

Hormones secreted from the anterior lobe of the hypophysis, especially growth hormone (STH, somatropic hormone), participate in the regulation of growth. STH stimulates various biosynthetic processes: it influences the biosynthesis of protein in muscle cells directly, and indirectly by supplying amino acids to the sites of biosynthesis as it facilitates the transportation of amino acids and their absorption into the muscle cells. The activity of STH is manifested in increased nitrogen retention in the body. The secretion of this hormone is enhanced following feeding of protein. This reflects the great influence of feeding on growth. STH increases the growth of skeletal tissues as well.

Other hormones also have a positive effect on the biosynthetic processes involved in growth, e.g. androgens. Male sex hormones enhance protein biosynthesis, and at maturity the development of the massive musculature typical of males. The activity of thyroxine is also of great importance in the growing animal (see Chapter 10).

Insulin regulates protein balance in skeletal muscles by promoting protein synthesis and the transport of amino acids. High correlations have been found between body fat content and plasma insulin levels. Insulin stimulates lipogenesis and inhibits lipolysis. The use of anabolic agents, hormones and antibiotics as feed additives to increase the rate of gain of young animals and improve feed utilization will be discussed later.

EFFICIENCY OF FEED UTILIZATION FOR MEAT PRODUCTION

Information on the ratio between protein and fat in body gain of animals is necessary to determine their energy requirements since these two components vary considerably in their energy content (see Chapter 16). This ratio depends mainly on the age of the animal (see Chapter 1), but also on the level of nutrition. A surplus of energy in the ration which increases rate of gain, also increases fat content in the body gain above that expected for the age of the animal. The synthesis of muscle protein produces a higher rate of gain since muscle has a high moisture content (up to 75%). The energy value of muscle tissue is 5.0–5.9 kJ/g, even though the energy value of protein is 24 kJ/g.

Fat is deposited in the animal body free of water; in fact, it replaces tissue water. Adipose tissue is very low in moisture content. The deposition of 1.4 g of fat replaces 0.4 g water, and the actual gain is 1 g only; however, the accretion of energy is 54.6 kJ (the energy value of 1 g of fat is 37.8 kJ). Hence, the change in energy value of 1 g weight gain produced by fat is 11 times greater than that of 1 g added as muscle protein.

For the exact estimation of the energy utilization for deposition of protein and fat, the costs of protein and fat synthesis must be considered. From animal balance trials it appears that the cost of synthesizing protein is greater than that of synthesizing fat (Lindsay, 1983). For protein synthesis, efficiencies range from 0.40 to 0.70, whereas for fat they are about 0.70–0.75. The observed difference is surprising because both efficiencies estimated from the biochemical pathways are much higher, approximately 0.85 (see Chapter 14). The discrepancy between the efficiencies of protein synthesis determined in animal trials and those estimated from biochemical pathways may be caused by the fact that the latter ignores the energy cost of protein turnover and transport of metabolites. As indicated in Chapter 7, protein deposition is the net effect of protein synthesis and degradation; the latter also requires energy.

Fully mature animals store energy provided in excess of their maintenance requirement almost entirely as fat, but some protein is synthesized in the growth of the cellular matrix of adipose tissue, hypertrophy of muscles fibres, growth of hair and replacement of keratinized cells of the skin.

Although fat may be deposited in various organs of the body (even muscles contain small amounts of fat), it is stored in subcutaneous adipose tissue, i.e. under the skin, in the abdominal cavity and in connective tissue. The production of inedible fat stores cannot be prevented.

Thus, the production of fat meat requires a high feed intake owing to the high energy value of fat and the high dry matter content of the gain. The production of very fat meat is objectionable in most Western markets (see above). However, a minimum of fat deposition is inevitably associated with protein synthesis and small amounts of intermuscular and intramuscular fat are needed for the tenderness and palatability of meat.

It is an old objective of animal producers to limit the inedible portion, and particularly excess fat, in meat animals by breeding and by adequate feeding.

The organic components of the skeleton are also produced at the expense of offered feed. The inedible portion in various animals ranges between 40% and 50% of their weight.

The duration of the feeding process has a considerable effect on the efficiency of feed utilization. The higher the rate of gain is, the shorter the time required for the animal to reach slaughter weight or condition, and the smaller will be the proportion of feed used for maintenance. The comparison of feed required by 300 rabbits weighing 1 kg each to that of a 300 kg steer could serve as an example. Because of their smaller body weight rabbits require much more energy for maintenance than the steer. This shortcoming is compensated by the greater relative rate of gain of the rabbits. The steer requires 120 days to produce the same weight as that produced by the rabbits in 30 days. Feed requirement is 1 ton of hay in both cases. The nutritional factors affecting the efficiency of feed utilization for meat production will be discussed below.

RATE OF GAIN AND CHANGES IN BODY COMPOSITION

The growth rate of animals is commonly expressed in terms of the change

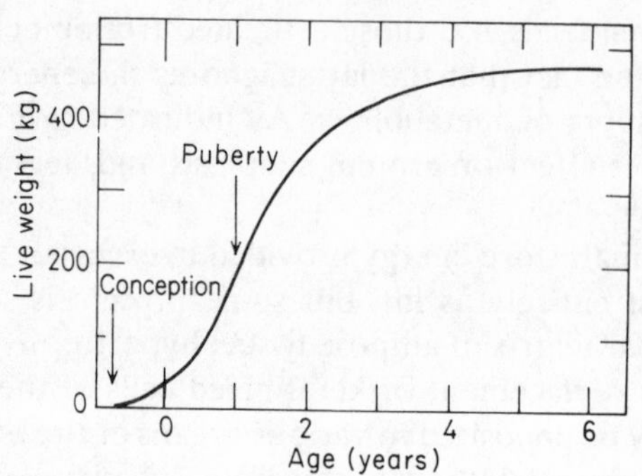

Figure 18.1. Typical growth curve for the dairy cow (from McDonald *et al.*, *Animal Nutrition*, 3rd edn, 1981, p. 278). *Reproduced by permission of Longman Group Ltd.*

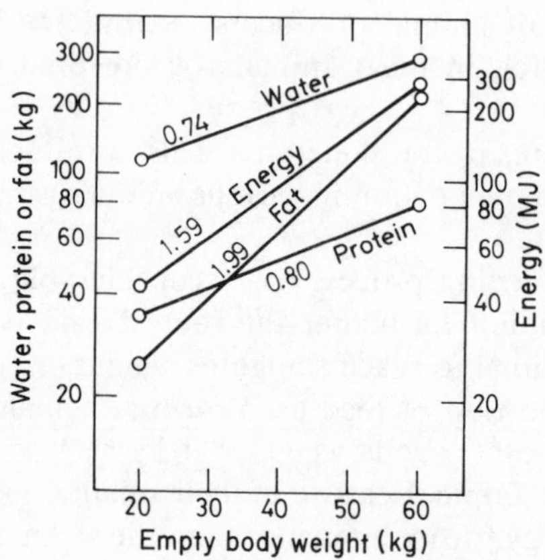

Figure 18.2. Growth coefficients for water, protein, fat and energy in the whole empty body of sheep (from McDonald, *et al.*, *Animal Nutrition*, 3rd edn, 1981, p. 279). *Reproduced by permission of the Longman Group Ltd.*

in their live weight within a given period of time. However, the growth curve of animals is not always linear and the weight of gut content, which in ruminants may reach 20%, introduces a considerable error. A typical growth curve which applies to most domestic animals is presented in Figure 18.1. Growth rate in the prenatal and post-natal periods accelerates. As the animal approaches sexual maturity growth rate is linear and decelerates before the animal attains mature weight. With most species of farm animals the inflection of the growth curve occurs at about 30% of the mature weight.

Different body parts and organs develop at a different rate, hence the proportions between them change during growth.

The relationship of the weight of a certain tissue, chemical component or organ to the weight of the entire body is best expressed by the equation: $Y = BX^a$ where Y is the weight of the component studied, X is the weight of the entire body, B is a constant, and a could be termed the coefficient of relative growth. On a logarithmic scale, $\log Y = \log B + a \log X$, the changes are expressed by straight lines. When the component in question increases at the same rate as the entire body, $a = 1$; $a > 1$ when the growth rate of the component is faster than that of the entire body. Figure 18.2 presents on a logarithmic scale the relative growth coefficients of water, protein, fat and energy in the body of sheep. The coefficients of protein and water are smaller than 1, indicating that their rate of growth is decreasing and the coefficient of fat and accordingly that of energy are greater than 1, indicating an increase in their rate of growth as the animal ages.

ENERGY UTILIZATION BY GROWING RUMINANTS

Some of the changes in body composition of various types of animals during growth, e.g. increase in fat content and decrease in protein and water, are presented in Table 1.2 and in Figure 1.2. As body weight increases from birth to slaughter, the most marked changes are a decrease in the proportion of bone and an increase in the proportion of depot fat; for example, subcutaneous fat increases in the carcass of British Frisian steers between the age of 12 days to 18 months from 1.1% to 7.1% and that of bones declines from 27% to 14%.

The quantitative differences in body composition during growth are very important because they affect the quality of meat and serve as a basis for calculating the energy requirements.

Calculation of Energy Requirements for Gain in Ruminants

Whereas the maintenance requirement is a function of body weight, live-weight gain and the ratio of protein to fat in the tissue gain are the main determinants of the energy required for production. Live-weight gain and its energy value are also related to body weight and energy content of the body as expressed in the equations for calculating live-weight gain (LWG) and its energy value (E_g). These equations, given in Chapter 16, serve as a basis for estimating energy requirements according to the British systems.

Regression equations for the calculation of the energy requirements for steers and heifers in relation to body-weight and the desired body weight gain were recommended by the California net energy system and are presented in Chapter 16. Data calculated on the basis of these equations show that heifers retain more energy per unit of gain than steers. This explains in part the faster rate of gain in males.

The following equations expressing the utilization of energy for maintenance and body gain (fat in mature animals and meat in growing ones) were developed by Blaxter (1968):

(a) For calculation of energy retention when the animal receives more energy than that required for maintenance:

$$R = k_f(M - \frac{F}{k_m})$$

(b) For calculation of loss of energy from the body when the animal receives less energy than that required for maintenance:

$$R = k_m(M - F)$$

where

R = the retention of energy (cal/day)
M = the intake of metabolizable energy (cal/day)
F = the fasting metabolism (cal/day)
k_m = the efficiency of utilization of ME for maintenance
k_f = the efficiency of utilization of the ME for growth and fattening.

The validity of these equations was examined in an experiment with sheep. Sixteen two-year-old sheep were given a standard maintenance diet which was then increased and kept constant at that level for 10 months. The results of this experiment, given in Table 18.1, indicate that energy retention is directly related to energy intake. Ingestion of feed at a level of twice the maintenance requirements increased energy retention to over 11 times the retention at maintenance feeding level (groups 1 and 3). Rates of protein deposition have a maximum at maintenance level of feeding; the limit of protein deposition is much lower than that of fat (see later).

The observed changes in body weight were compared to those expected by prediction from the equations relating food intake to the attainment of weight equilibrium (transformed equations (a) and (b); see Table 18.1).

Table 18.1 The effect of dietary energy supply on growth: weight of sheep and mean daily retention of energy, fat, flesh and wool following changes in diet (after Blaxter, 1963, p. 329)

Group	Change in daily intake of metabolizable energy (MJ)	Body weight, kg		Composition of gain or loss per day*			
		At the beginning	After 10 months	Energy value (MJ)	Wool (g)	Body fat (g)	Body protein including water (g)
1	+5.88	54.5	78.7 (observed) 81.6 (expected)	2.70	7.3	65.0	35.9
2	+2.94	49.9	64.3 (observed) 63.3 (expected)	1.83	6.2	43.6	34.2
3	0	48.6	50.6 (observed) 48.6 (expected)	0.24	4.6	3.8	−8.0
4	−2.94	43.8	33.2 (observed) 25.7 (expected)	−0.25	2.7	−7.6	−17.6

* During a 10-month period of constant feeding.

Good agreement was found, except for the underfed sheep which lost less body weight than expected. The latter may have adapted themselves to undernutrition by the reduction of the maintenance cost.

SOME FACTORS INFLUENCING ENERGY REQUIREMENT OF GROWING RUMINANTS

Additional factors influencing the energy requirement and feed efficiency in growing ruminants are age, rate of gain, sex, breed, ratio of energy to protein in the diet and the presence of growth stimulants. Energy retention under various feeding conditions has been determined in energy balance experiments during representative parts of the growth period (see Chapter 15) or in comparative slaughter experiments (see Chapter 1), in which initial body composition and changes in response to nutritional treatments were measured.

Influence of Age of Animals

The gross efficiency of conversion of feed to body substance declines considerably during growth because of the increasing fat content of body gain and increase in the maintenance requirement that had to be met during the whole growth period. According to classical experiments carried out by Ritzman and Colovos (1943), the efficiency of utilization of energy provided above the maintenance requirement declines as the animals grow and approach maturity. Growing dairy cattle under 200 kg body weight use metabolizable energy provided above maintenance with an efficiency of 64% and those weighing 300–600 kg with an efficiency of only 42%. It must be stressed that energy retention may be predicted on the basis of weight and rate of gain, but not on the basis of age.

Level of Feeding

The data presented in Table 18.2 indicate that the composition and energy content of weight gain change with the age of the animal and with its rate of gain which is enhanced by an increasing level of nutrition. Gains made rapidly contain more fat and are therefore richer in energy. Thus the requirement for weight gain is higher in fast growing animals, but the lower proportion of energy used for maintenance may offset this disadvantage.

Recent research of Byers (1982) indicates likewise that fat deposition increases with accelerated growth rate: body fat concentration increases from 23% to 31% for cattle (Hereford steers) growing at 0.4 and 1.0 kg/day.

The recent systems for feed evaluation (see Chapter 16) take account of the fact that the requirement for energy increases with the rate of gain as shown in Figure 18.3.

Table 18.2 Energy value (MJ/kg) of the weight gains of cattle growing at different rates (from Blaxter, *Energy Metabolism of Ruminants*, 1967, p. 170)

Age of animal (months)	Fast growth (about 1 kg/day)	Medium growth (about 0.6 kg/day)	Slow growth (about 0.4 kg/day)
	12.6	10.0	8.0
3	13.8	10.9	8.4
6	15.5	12.1	9.2
9	17.2	13.8	9.6
18	20.9	15.1	12.1
24	24.7	17.6	13.4

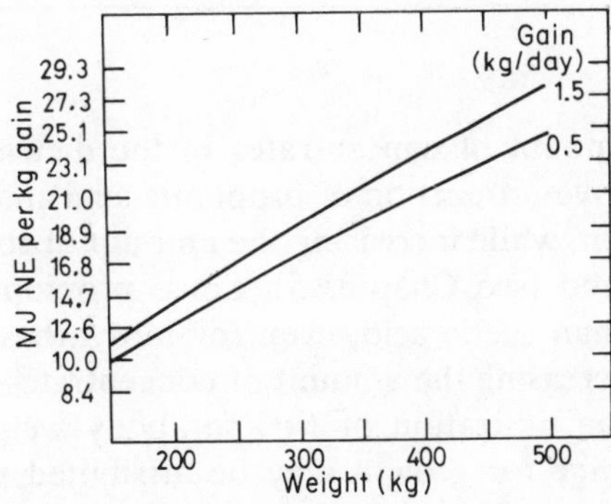

Figure 18.3. Effect of body weight and rate of gain on energy requirements (from Lofgreen and Garret, 1968, p. 793, adapted by Fox and Black, 1984, p. 729). *Reproduced by permission of the* Journal of Animal Sciences.

The increase in energy requirement with age or live weight increases the feed costs of meat animals at high weights. Therefore the marketing of meat animals must not be unduly delayed. The edible portion of meat animals increases with growth up to a certain limit. Therefore marketing time for maximum returns depends, among other factors, on feed costs, on one hand, and the increase in meat yield on the other.

Composition of the Diet

The efficiency of utilization of ME for growth of ruminant animals varies considerably with the type of diets. Results of feeding experiments carried out by Kay (1976) in England show that the efficiency with which the metabolizable energy of barley diets is converted into body energy is 75% while that of dried grass is only 30%. These coefficients relate to the whole energy intake, including maintenance, and were obtained from animals slaughtered at different ages (see Table 18.3).

Table 18.3 Net availability of ME [(energy retained)/(ME intake) × 100] for production of body gain at different live weights of steers given different diets (from Kay, 1976, p. 260)

	Live weight (kg)			
	365	450	520	585
Barley diet	77	77	75	84
Dried grass diet	36	30	28	30

Increasing the amount of concentrates in the diet is associated with an increase in the relative proportion of propionic acid and a decrease of acetic formed in the rumen, while increasing the amount of roughage is associated with an inverse trend (see Chapter 5). Since propionic acid is used with greater efficiency than acetic acid, even for biosynthesis of fat* (Chapter 6) it follows that increasing the amount of concentrates in the diet increases the efficiency of the utilization of ME for body weight gain. The lower efficiency of roughage for growth may be attributed to a lesser extent to losses involved in crude fibre digestion and the elimination of indigestible constituents. The experimental results presented in Figure 18.4 may be explained by the interrelationship between proportions of volatile fatty acids formed in the rumen and energy efficiency for growth. When cattle or sheep were fed diets containing maize grains or hay in various proportions between 0–100%, the utilization of metabolizable energy for fattening increased gradually from 29% to 61%, but the respective efficiency for maintenance was much higher than for fattening and remained almost constant (71–79%) in spite of the changes in the ratio of concentrates to roughage in the diet.

Although meat animals may be safely fed diets containing widely varying ratios of concentrates to roughage, because of the much more efficient utilization of concentrates for meat production it is common practice to use cereal grains or other starchy feeds in diets for growing ruminants and to increase the proportion of these feeds in the diet with advancing age and weight. It must be noted, however, that the inclusion of a small amount of roughage, preferably hay in the diet, enhances feed intake, efficiency and rate of gain, and prevents digestive disturbances (see Chapter 24) which

* This agrees with the view that the utilization of acetate for lipogenesis is dependent upon an adequate supply of glucose or glucose precursors (propionic acid).

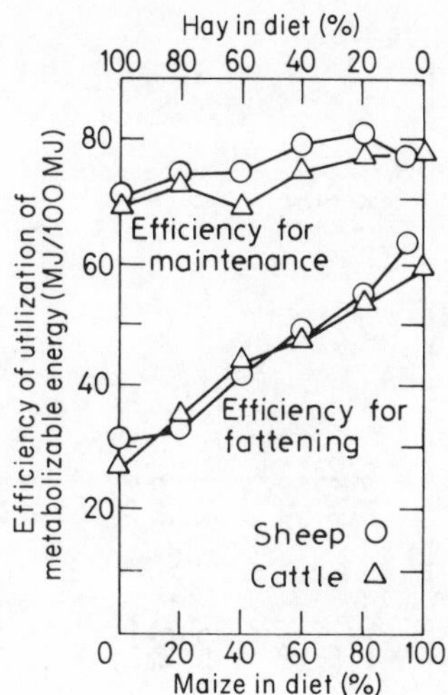

Figure 18.4. Efficiency of utilization of ME for fattening and maintenance in relation to diet composition (from Blaxter and Wainman, 1964, p. 123). *Reproduced by permission of Cambridge University Press.*

may arise on all-concentrate diets. Even so, with due care, all-concentrate diets have been successfully used in the fattening of ruminants, especially of lambs. In order to minimize digestive disturbances on all-concentrate diets, feed must be continuously available and mineral supplements of antacids (e.g. sodium bicarbonate) may be added to the diet. In practice, the availability and price of concentrates and roughage determine the composition of diets used for growth and fattening (see Fox and Black, 1984).

Influence of Sex and Breed

Rate of gain is influenced by sex and breed of animals. Table 18.4 presents data derived from an experiment in which bulls, steers and heifers of the Aberdeen Angus and Holstein breeds were fed the same diet ad libitum and slaughtered after several periods of growth. The criterion of efficiency is the amount of protein produced per unit of digestible energy ingested. Holstein cattle grew faster and matured later than the Angus, bulls grew faster than steers and steers faster than heifers, irrespective of breed. Rapidly growing, slowly maturing cattle are more efficient producers of protein than are slower growing, more early maturing cattle with a propensity to fatness.

Interaction of Protein and Energy

A shortage of protein or energy in the diet prevents the animal from using fully their potential for growth. As the rate of growth of muscles and bones is limited, excessive energy intake will be converted into fat, even in

Table 18.4 Production efficiency by bulls, steers and heifers of each of two breeds (after Reid, 1975, p. 20)

	Aberdeen Angus			Holsteins		
	Body weight (kg)		Protein (g)/DE (Mcal)	Body weight (kg)		Protein (g)/DE (Mcal)
	Initial	Final		Initial	Final	
0–93 days on feed						
Bull	234	327	7.8	210	319	11.6
Steer	234	324	6.1	201	293	9.2
Heifer	235	313	5.6	187	268	7.4
94–196 days on feed						
Bull	328	421	5.4	332	454	7.6
Steer	325	410	4.6	292	395	5.7
Heifer	292	387	4.3	276	368	5.1
196–400 days on feed						
Bull	456	565	2.7	465	668	3.5
Steer	463	607	2.2	420	562	3.3
Heifer	435	519	1.7	379	510	3.2

the young animal. A limit in the rate of retention of protein in the body exists even in very young calves fed large quantities of milk—a diet excessive in protein (see later). Protein and fat retention were measured in growing calves by nitrogen and carbon balance; the results, presented in Figure 18.5, show that with increasing food intake protein retention increases at an ever-diminishing rate whereas fat deposition increases at an every-increasing rate.

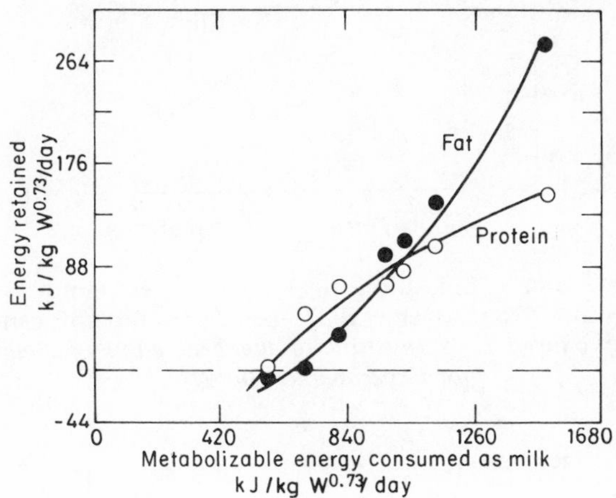

Figure 18.5. Retention of energy as fat and as protein by calves given different amounts of milk as single feed (from Blaxter, K. L., *Energy Metabolism of Ruminants*, 1967, p. 168). *Reproduced by permission of the editor and of Hutchinson Edition.*

The existence of a similar relationship between fat and protein gain in steers growing at various rates appears from the curves presented in Figure 18.6. Rates of protein growth increased slowly up to a certain limit, whereas rates of fat deposition increased considerably with increasing levels of nutrition and rates of gain (Byers, 1982).

A correct supply of protein is essential not only for pursuing the main purpose of animal growth, i.e. maximal muscle production, but also for ensuring optional energy utilization. The existence of a range of protein concentrations in the diet within which maximal energy utilization occurs may be exemplified by experiments with growing rats (see Figure 18.7). Maximal energy utilization was found when the diet contained between 18% and 30% protein, and it declined at higher and lower protein contents. Feeding of-low protein diets was associated with high proportions of fat in the gains. The decline in efficiency of energy when excessive protein was fed can largely be attributed to the energy cost involved in urea formation and excretion (see Chapter 7).

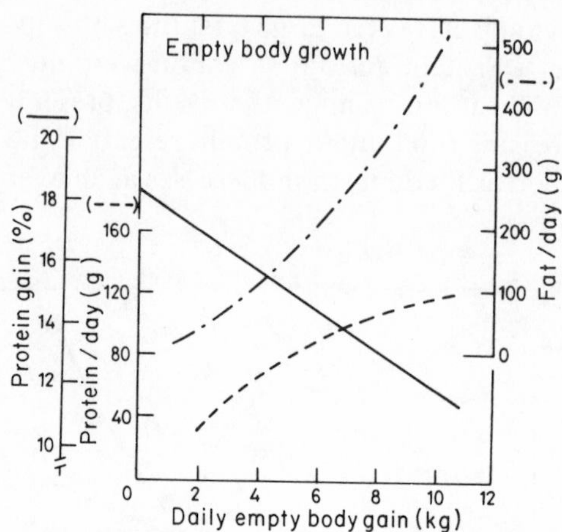

Figure 18.6. Rate and composition of growth in Hereford steers fed diets of forage and grain (from Byers, 1982, p. 2563). Protein gain (----); fat gain (·—·—·) *Reproduced by permission of the Federation of American Societies for Experimental Biology.*

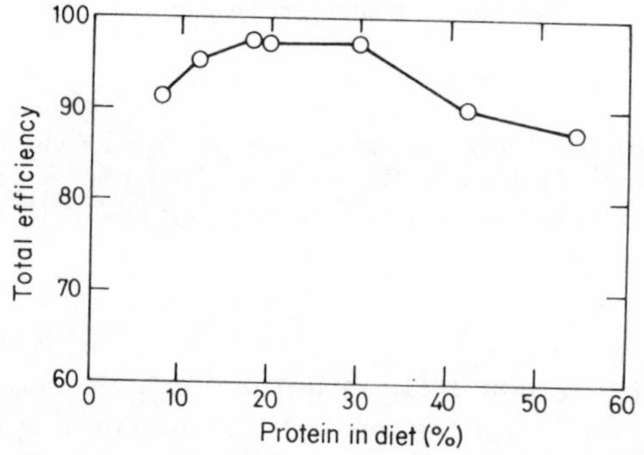

Figure 18.7. Total efficiency of energy utilization in rats given diets containing different proportions of protein (from Blaxter, K. L., *Energy Metabolism of Ruminants,* 1967, p. 250). Total efficiency = (NE/ME) × 100. *Reproduced by permission of the Author and Hutchinson Edition.*

USE OF GROWTH STIMULANTS

Growth-stimulating agents are used on a wide scale to improve the efficiency of meat production and to produce leaner meat. There are two major types of growth-promoting agents effective in ruminant animals: (1) hormone-like substances which increase the efficiency of utilization of absorbed nutrients, and (2) antibiotic-like substances, which act on the ruminal microflora and thereby modify the quantity and quality of the rumen fermentation products.

Action of Hormone-like Substances

The differences in growth rate and mature body size between male and female animals suggest that sex hormones play an important role in the control of growth. Several analogues of the female and male sex hormones have been used successfully to stimulate the growth of meat-producing animals. Oestrogen-like substances appear to stimulate the release of endogenous anabolic agents, growth hormone and insulin (see above). The mode of action of androgens is even less clear, they may act by reducing protein turnover and increase feed efficiency because protein turnover is an energetically expensive process (see Chapter 7).

Growth promoters available for cattle and sheep are oestradiol or synthetic oestrogens such as diethylstilboestrol, hexoestrol, or zeranol, prepared from corn mould, and male hormones such as purifed testosterone or synthetic trienbolone. These anabolic agents may be administered either as feed additives, oil-based injections or subcutaneous implants. The correct choice of anabolic agents depends on the species and sex of the recipient animal. The best responses are obtained when androgens are used in females, androgens combined with oestrogens in females and androgens combined with oestrogens in steers. Diethylstilboestrol was recognized as an effective growth promoter in steers, but several countries have banned its use because of presumed health hazards.

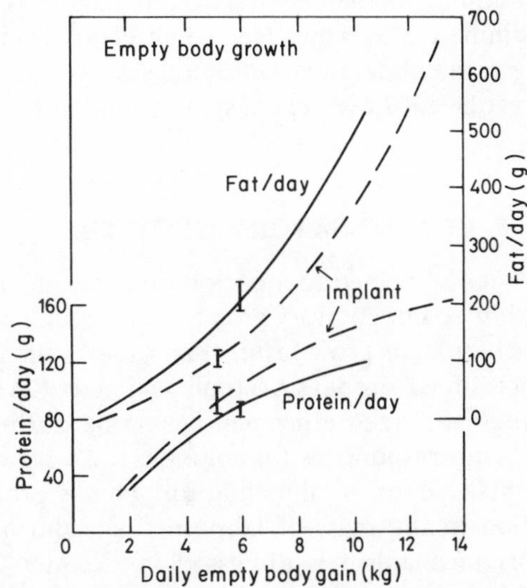

Figure 18.8. Effects of anabolic agents on growth of protein and fat in steers growing at various rates (from Byers, 1982, p. 2563). *Reproduced by permission of the Federation of American Societies for Experimental Biology.*

Synthetic oestrogens produce in castrated male animals an improvement of about 15% in growth rate and a 10% decrease in the amount of feed required per unit of gain.

A comparison of growth curves for cattle (Figure 18.8) illustrates the effect of anabolic agents in the diversion of energy from fat to protein growth; thus the use of anabolic agents increases the physiological limit for protein growth (see above).

Action of Antibiotics

Monensin, an antibiotic produced by *Streptomyces cinnamonensis*, has gained wide acceptance as a feed additive since it improves feed efficiency of ruminants by modifying rumen fermentation (see later). Cattle fed a monensin-containing diet gain slightly faster, consume about 6–7% less feed than control animals, and feed efficiency is increased by 7–8%

Monensin, like some other antibiotics, is a ionophore. The basic mode of action of ionophores is to modify the movement of ions across the membranes of microbial cells, thereby interfering with the uptake of ions. This leads to changes in the composition of the ruminal microflora and of rumen fermentation products. The most consistent effect of monensin is its ability to increase the molar proportion of propionic acid by 50% at the expense of acetate and butyrate (see Chapter 5), and propionate is more efficiently used by the tissues than acetate. The change in the microbial population and the decrease in the total number of microorganisms induced by monensin result in the depression of methanogenesis (see Chapter 5) and of available proteolytic and deaminative enzymes (see Chapter 8). Thereby the amount of protein escaping ruminal degradation increases and monensin exhibits a protein-sparing effect (see Rumsey *et al.*, symposium on Monensin in Cattle, 1984).

COMPENSATORY GROWTH

Animals that have been subjected to a period of undernutrition and thus to retarded growth, exhibit compensatory growth during a subsequent period of re-alimentation. Such animals grow faster than other animals of similar age or weight when liberal feed supplies become available. Several factors may contribute to the increased feed efficiency occurring during compensatory growth. The major factor responsible for compensatory growth is an increase in voluntary food intake upon re-alimentation. This is probably caused by the relative reduction in the mass of adipose tissue during the period of undernutrition. This mechanism is discussed in Chapter 23, which deals with the control of food intake. The increase in voluntary food intake may also be caused by an increase in space in the abdominal cavity which would reduce pressure on the gastrointestinal tract. The reduction in the size of the underfed animals reduces their maintenance requirement, and changes in intermediary metabolism have been shown to occur in underfed animals, increasing their net food efficiency. During re-alimentation, hypertrophy of the gastrointestinal tract occurs and the amount of digesta in the gut is greater than in animals of similar size which have not been underfed. This further increases the apparent efficiency of re-alimented animals to gain body weight.

Many feeding experiments were carried out in order to examine whether compensatory growth following feed restriction would enable a considerable saving in the amount of feed required. Figure 18.9 shows that calves fed at maintenance level from 6 to 12 months of age and then fed ad libitum attained almost normal weight during the second year of life. Nevertheless recent studies did not reveal economic benefits from the proposed system (compensatory growth following feed restriction) in intensive beef production.

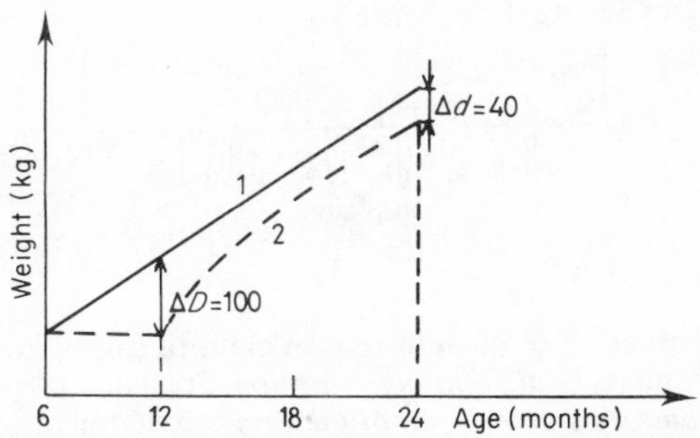

Figure 18.9. Compensatory growth. Increase of weight in calves fed according to requirements (1) and in calves fed for six months on maintenance level and re-alimentated later (2) (from Kolb and Gurtler, *Nutritive Physiology of Farm Animals*, 1971, p. 452). *Reproduced by permission of Gustav Fischer Verlag.*

PROTEIN REQUIREMENTS FOR GROWING RUMINANTS

Dietary protein is of greatest importance in its effects on biological and economic performance. Beef cattle are known for their low rate of conversion of dietary protein into protein for human consumption. Their efficiency is only one-third that of the high-producing dairy cow. Knowledge of the protein requirements of growing ruminants permits maximum efficiency and prevents wastage of energy. As shown above, the efficiency of energy utilization depends on the quantity of protein supplied.

Methods of Expressing Requirements for Protein

Practical protein requirements for growth (including maintenance) may be determined from the results of feeding trials in which several groups of animals are fed on rations supplying increasing levels of proteins. Such rations must be similar in ME, mineral and vitamin content. Above a certain level in the diet, protein produces diminishing returns. The protein level at which nitrogen retention or live weight cease to increase is usually designated as the requirement, although a slightly lower level may be more economical. An example of such a nitrogen-balance trial is presented in Figure 18.10. Calves weighing 50 kg received diets containing 93–230 g of digestible protein per day; maximum nitrogen retention was obtained with a minimum intake of 190 g of digestible crude protein per day.

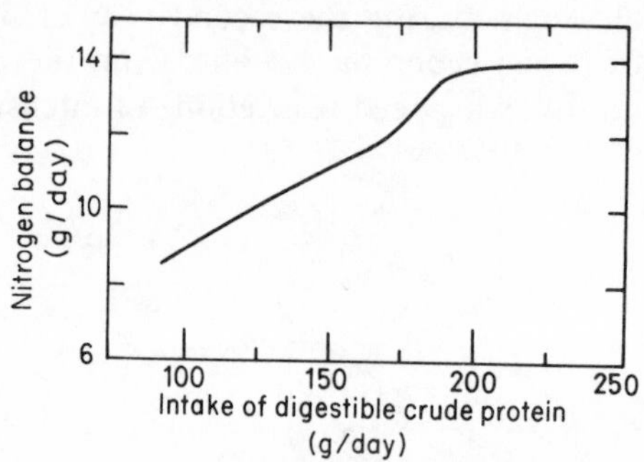

Figure 18.10. Determination of protein requirement of calves from their nitrogen balance (from McDonald, *et al.*, *Animal Nutrition*, 3rd edn, 1981, p. 286). *Reproduced by permission of the Longman Group.*

Protein allowances for growing ruminants may be calculated according to two factorial methods used for estimating the protein requirement for maintenance as explained in Chapter 17. The total protein requirement including that for body gains (G = nitrogen retention in body gain in g/day) is expressed in the following equation*:

$$R = \frac{6.25(\text{FEN} + \text{DM} + \text{MUN}) \times W^{0.73} + G}{\text{BV} \times \text{TD}}$$

Likewise, the modified factorial method of Swanson can be used for calculations of the protein requirement for growing ruminants. The respective equation is

$$\text{TCP} = \frac{U + F + S + G}{E_p}$$

* Definitions for the other abbreviations have been given in Chapter 17.

where G is protein deposited in body-weight gain. This author assumes that the protein content of gain is 19% in newborn calves and decreases to 16% as animals approach mature size.

In analogy to the equations proposed by Lofgreen for calculating the energy requirement for growing ruminants (see Chapter 16), the following equations for the requirements of digestible protein as functions of body weight and body weight gain were developed by Preston (1966).

For young cattle: $DP = 2.79\ W^{0.75}\ (1 + 1.905\ G)$

For young lambs: $DP = 2.79\ W^{0.75}\ (1 + 6.02\ G)$

where DP refers to grams of digestible protein per day, W is body weight in kg and G is daily gain in kg.

Since, according to recent research, digestible protein does not reflect quality in ruminants, a new method of calculating the protein requirement of ruminants has been developed by the ARC (UK). This system is based on an estimate of the total amino acid nitrogen absorbed from the small intestine and the efficiency of its utilization for tissue synthesis and maintenance. Apart from small amounts of nitrogen in endogenous secretions (see Chapter 7), the amino acids absorbed from the intestine are derived from microbial protein produced in the rumen and also from dietary protein which has escaped degradation in the rumen (undegradable protein, UDP). Degradable protein (RDP) is provided in the diet and the microorganisms can utilize it to form microbial protein.

The protein required for maintenance and productive purposes may be expressed as the sum of RDP and UDP. The extent of synthesis of microbial protein depends upon the energy available to the microorganisms. Thus the RDP need (g/day) is $6.25 \times 1.25\ M_E$* a factor developed by assuming that 1 kg of organic dry matter is equal to 19 MJ of DE and 18% of the energy is lost as methane and urine.

The degradability of proteins in the rumen varies considerably among feeds. The method generally used for measuring the extent of degradability of protein feedstuffs is the incubation of a sample in a bag suspended in the rumen of a fistulated animal (see Chapter 13).

The theoretical background of this method is sound; however, the values of several assumed coefficients are rather variable and it remains to be seen whether this method will gain acceptance.

Factors Affecting Protein Requirements

As explained in the preceding section on energy requirement, a number of

* M_E = Metabolizable energy requirement ((MJ/day). Factorial method for calculation of UDP is given in the ARC publication (1980), p. 138 (see Bibliography for Chapter 18).

factors affect the ability of growing ruminants to synthesize tissue proteins, e.g. genotype, sex, body weight, rate of growth and supply of energy.

The great influence of genotype and sex on protein accretion is indicated by the following facts: conventional early maturing growing steers (Hereford or Aberdeen Angus) retain only 12–15% of the total energy gained as protein. However, protein requirements are relatively higher for growing bulls of late maturing breeds (Simmental or Charolais) which retain 35–45% of their energy gain as protein (Geay, 1984).

Protein requirements seem to be less important than energy requirements for early maturing, growing ruminants. The rate of protein accretion depends on energy intake, as illustrated in Figure 18.11. Protein retention increases with energy intake until protein supply becomes limiting; it increases with higher protein levels P_2 and P_3. For practical purposes one must recall that high-energy diets increase protein deposition, but promote also fat deposition.

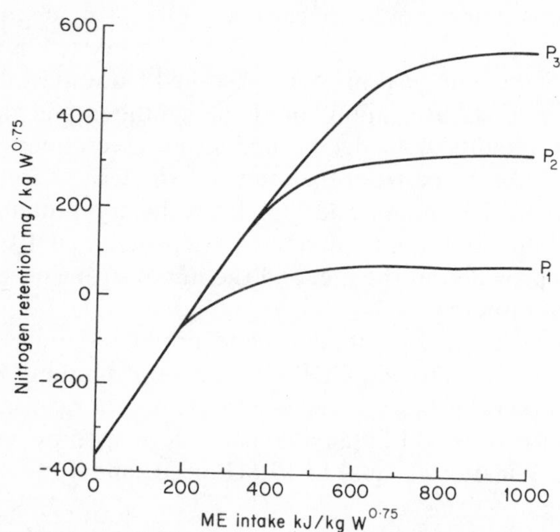

Figure 18.11. Effects of energy intake on protein retention with three levels of protein intake (from Orskov, E. R., *Protein Nutrition in Ruminants*, 1982, p. 106). *Reproduced by permission of Academic Press and of the Author.*

When energy is insufficient nitrogen balance may become negative even if sufficient protein is fed. On the other hand, an excess of protein is wasteful. The protein content of live-weight gain may not be increased by feeding a surplus of protein. This is suggested by the results of recent experiments, presented in Table 18.5.

In summary, beef animals may be efficiently fed by supplying the required amounts of protein and energy and by maintaining the desirable ratio of protein to energy in the diet.

Table 18.5 Composition of the gain of steers and heifers fed different levels of protein (after Garrett, 1977, p. 116)

	Protein in diet (%)	Composition of the gain (%)		
		Protein	Water	Fat
Steers	11.7	12.4	38.9	45.8
	19.4	11.9	36.8	48.6
Heifers	13.4	8.7	24.5	64.8
	22.7	8.9	25.4	63.6

For growth and fattening, up to 25% of the dietary nitrogen may be supplied to ruminants by simple non-protein nitrogen compounds, particularly urea (see Chapter 8). The substitution of urea usually reduces the cost of the diet. Since urea contains neither available energy nor sulphur, appropriate carbohydrates and minerals must be provided together with the urea to ensure its efficient utilization. Starch is the most appropriate carbohydrate for the utilization of non-protein nitrogen; therefore, cereal grains or other starchy feeds are preferable to molasses.

MINERAL AND VITAMIN REQUIREMENTS OF GROWING RUMINANTS

The determination of feeding standards, i.e. the quantitative requirements for minerals and vitamins, is considerably more difficult than that for energy or protein. A deficiency of energy or protein rapidly reduces the growth rate of animals. For minerals and vitamins the criteria for adequate dietary levels are much less clear-cut. With certain diets deficient in minerals or vitamins the deficiency may be latent and normal growth may continue for some time. Considerable bodily reserves of certain minerals and vitamins, e.g. calcium, phosphorus, copper and vitamin A, may be present at the beginning of an experiment and may complicate the interpretation of growth experiments and even that of balance trials. The metabolism, storage and the interactions of minerals are discussed in Chapter 10.

There are two general approaches for the estimation of mineral requirements: the *factorial method* and *growth and balance trials.*

The factorial method is not unlike that used for estimating the protein requirement for growth (see above). This is a quasi-theoretical approach although it relies on previously obtained information from actual experiments. The factorial method is based on the assumption that the *net* requirement of the growing animal for a specific mineral is the sum of the endogenous loss (in the faeces and in the urine) and the amount retained by the animal.

The *gross* or *dietary* requirement is calculated from the net requirement by correcting it for the availability of the mineral which is always lower than 1.0 (100%). Thus:

$$\text{Dietary requirements} = \frac{\text{Endogenous excretion} + \text{retention}}{\text{Availability}}$$

For example, a heifer of 300 kg liveweight gaining 0.5 kg/day may have an endogenous loss of calcium of 5 g/day and may retain 6 g of calcium per day. If the availability of calcium for this age and weight group were 40%, the gross requirement would be 11/0.40 = 27 g of calcium per day (after McDonald *et al.*, *Animal Nutrition*, 1981).

The values for endogenous loss, retention and the availability of the minerals must be obtained from actual experiments. Such values are not easily obtained and vary even among experiments performed under presumably similar conditions. Therefore, the theoretical factorial assessments of mineral requirements often do not agree with practical estimates.

Growth and balance trials, on the other hand, are practical experiments in which graded levels of a specific mineral are offered. Thus they reveal the minimum level required to produce maximum retention and growth. It is obvious that the diets used must be normal and adequate in all other respects. The results of such balance data may be used to produce regression equations to express the requirement for a certain mineral as a function of body weight and live-weight gain. For example, the phosphorus requirement of growing lambs has been expressed by the following equation:

$$P = 0.0194\, W(1 + 0.0171\, G)$$

where P is the daily requirement in g, W is body weight in kg, and G is the daily gain in g (Preston and Pfander, 1964).

Growth and balance trials are very useful for the determination of mineral requirements provided they are continued long enough. Since the storage of a mineral in excess or the depletion of a stored mineral during a trial may bring about the respective overestimation or underestimation of the requirement, it is useful to measure the storage and depletion of minerals in live animals by different techniques. For instance, radiography can be used to determine calcium accretion in bones, and liver biopsies to determine the storage of iron and copper. Blood analyses can detect deficiencies of certain minerals, especially trace elements, but the serum levels of calcium, sodium, potassium, magnesium and inorganic phosphate are reasonably well regulated and are affected only after a long period of deprivation.

The requirements for vitamins can be determined only by growth trials. Only fat-soluble vitamins are required by mature ruminants and these have been discussed in Chapter 11. Vitamin A can be stored in the liver in large amounts; thus, the determination of the requirement is not an exact procedure. Recent recommendations for vitamin A (or carotene) and vitamin D in young ruminants (ARC, 1980) exceed earlier ones by a factor of 2–3 in order to accommodate modern beef production systems.

NUTRITION OF PRERUMINANTS

Newborn ruminants subsist on milk during the first few weeks of life. Because of the reflex closure of the oesophageal groove, when young calves or lambs consume liquid feed, nearly all of it by-passes the rumen into the omasum and from there it reaches the abomasum. The digestion occurring in the gastrointestinal tract of preruminants is not unlike that of monogastric species (see Chapter 3). The tendency to market more milk and to reduce the milk consumption of stock has led to the use of milk substitutes in young ruminants from the age of two or three days. In order to simulate whole milk, most milk substitutes contain reconstituted dried skim milk, some homogenized vegetable or animal fat and fat-soluble vitamins.

Colostrum

The first milk product produced after parturition is colostrum; it greatly differs from milk produced later during lactation (see Table 18.6). The major difference, and probably the most important one, between normal milk and colostrum is that the latter contains a large proportion of albumin and globulins. Colostrum is richer than normal milk in most nutrients (including vitamins), apart from lactose. However, its major effect is to confer passive resistance on the newborn against pathogenic microorganisms. Immunoglobulins present in colostrum are absorded intact by pinocytosis (see Chapter 3), passing through the mucosa of the gut into the lymphatic system, and reach the circulation through the thoracic duct. The capacity of the newborn to absorb the antibodies intact from the gut contents declines rapidly and lasts for only about 12–24 hours after birth. Since placental transfer of antibodies to foetal tissues does not occur in ruminants, their neonates depend on colostrum as a source of antibodies. This passive immunity is necessary for the young until they develop active immunity. The colostrum of ruminants contains a trypsin inhibitor which protects the immunoglobulins from digestion. The globulin fraction in colostrum declines quite rapidly with successive milkings. Within three or four days after parturition milk loses its colostral properties and becomes normal.

Digestion in Preruminants

Proteins

One of the most important factors in the digestion of milk proteins is their slow release from the abomasum. Here, casein clots under the influence of

Table 18.6 Composition of colostrum (first 24 hours after calving) and of milk (after Roy, 1980, and Walker, 1979)

Component	Colostrum	Milk
Total solids (%)	21.9	12.5
Fat (%)	3.6	3.7
Protein (%)	14.3	3.3
Casein (%)	5.2	2.6
Albumin (%)	1.5	0.5
β-Lactoglobulin (%)	0.8	0.3
α-Lactalbumin (%)	0.27	0.13
Serum albumin (%)	0.13	0.04
Immunoglobulin (%)	5.5–6.8	0.09
Lactose	3.1	4.6
Ash	1.5	0.8
Calcium (%)	0.26	0.13
Phosphorus (%)	0.24	0.10
Magnesium (%)	0.04	0.01
Sodium (%)	0.07	0.06
Potassium (%)	0.14	0.16
Chloride (%)	0.12	0.10
Iron (mg per 100 g)	0.20	0.05
Copper (mg per 100 g)	0.06	0.02
Carotenoids (μg/g fat)	35	7
Vitamin A (μg/g fat)	45	8
Vitamin D (ng/g fat)	30	15
Vitamin E (μg/g fat)	125	20
Thiamine (μg per 100 g)	60	40
Riboflavin (μg per 100 g)	500	150
Niacin (μg per 100 g)	100	80

hydrochloric acid, pepsin and rennin (see Chapter 7). As the clot begins to break up, the partially digested casein and trapped lipids are slowly released into the small intestine, allowing sufficient time for their effective abomasal and intestinal digestion.

Milk proteins, unlike most plant proteins, are almost completely digestible and are efficiently utilized because their combined biological value is high. However, milk proteins may be adversely affected by improper heat treatment during the dehydration of either whole or skim milk. Overheated dehydrated milk products do not coagulate in the abomasum and their reduced abomasal retention time results in decreased digestibility of the protein. The passage of undigested protein or other indigestible constituents into the intestine predisposes to infection and may cause diarrhoea.

Various inexpensive non-milk proteins have been tried in milk substitutes, such as soybean products and fish meals. These non-milk protein sources may cause digestive disturbances and poor growth rates. However, according to Nitsan *et al.* (1972), a large proportion of the dietary protein may be substituted by properly toasted soybean protein, particularly when calves are started on this diet not from the beginning.

Carbohydrates

The supply of carbohydrates, which play a central role in foetal metabolism, is drastically reduced at birth and the newborn must adapt to fat as the major energy source in milk. The only carbohydrates that can be efficiently used by preruminants are lactose and its components galactose and glucose. The activity of intestinal lactase is high at birth and declines with age, but even at eight weeks of age it is 10 times higher in the calf than in mature animals (see Chapter 4).

Starch is poorly digested in the small intestine of preruminants and only small amounts of starchy feedstuffs (cereal grains, etc.) must be used in milk substitutes; undigested starch may lead to digestive upsets in the preruminant.

Fats

Pancreatic lipase together with pregastric esterase produced in the saliva are responsible for the high extent of digestion of milk fat. Salivary esterase is capable of hydrolysing only glycerides of short-chain fatty acids. Unless correctly emulsified, vegetable fats and animal fats such as tallow incorporated into milk substitutes are poorly digested and tend to cause diarrhoea. Adequate emulsification, which can be obtained by the use of lecithin (an emulsifier) and mechanical homogenization, ensures the extensive digestion of the fat.

There appears to be a critical level of polyunsaturated fatty acids above and below which increased susceptibility to infections occurs. Weakness of the legs and dull hair coat symptoms of essential fatty acid (EFA) deficiency are prevented by raising calves on a diet containing 0.11% of the energy as EFA, compared to a minimum requirement of 1% as EFA suggested for piglets. It is imperative to obtain the right balance between EFA concentration and vitamin E level, which should be between 1.5 and 2.5 mg vitamin per g linoleic acid (see Chapter 11).

Utilization of Feed by Preruminants

Energy

In the preruminant the utilization of liquid feed based on milk is much more efficient than in the ruminant fed grains and roughage. The digestibility of the dry matter of milk is about 95% compared with 82% for that of the most digestible plant materials. Energy losses as a result of fermentation are minimal in the milk-fed preruminant. The heat increment of milk is much smaller than that of plant feedstuffs. However, the specific energy requirement for maintenance, i.e. per unit of metabolic body weight, and the energy requirement per unit of body weight gain, are considerably higher in the very young animal than in the older ones. This is evident from a comparison of the requirements for ME of preruminant and ruminant animals in Table 18.7.

The high energy requirement for maintenance of young animals is the result of their greater voluntary activity (see Chapter 17) and intensive protein turnover (see Chapter 7). The energy requirements for body gain of very young animals are high since the lipid reserves of calves and lambs at birth are very low (2–3% of live weight); thus in the postnatal period there is a continued accretion of fat, protein and also minerals. The net efficiency of ME utilization for maintenance in preruminants is 0.80–0.85, considerably higher than that for body gain (0.65–0.70). In this respect, young animals are not unlike older ones.

Rumen development, as indicated mainly by papillary proliferation and an increase in rumen volume (see Chapter 3) and weight, occurs after the age of three weeks. However, rumen development depends on the presence of volatile fatty acids and therefore on the consumption of dry feed. Veal calves, which must be fed for rapid gains and a special type of carcass, may also be offered an all-milk liquid diet. Therefore, veal calves are nonruminants even at 8–10 weeks of age. However, heifer calves raised for dairy herd replacement, must be fed for moderate gains. This may be done by limiting their milk supply and offering them dry feed from birth. Such a feeding regime advances dry feed consumption and rumen development and is more economical. Bull calves of the dairy herd raised for beef should be fed milk more liberally than heifer calves, because rapid postweaning growth rate depends largely on rapid early preweaning weight gains.

Protein

The protein requirements of preruminants calculated on the basis of live weight are similar to those of ruminating animals. They may be estimated by the same factorial equation used for ruminating growing animals (see above and Chapter 17). The following factors in this equation are higher for preruminants than for ruminants: nitrogen content of gain (G), true digestibility (TD), biological value of the proteins (BV) and particularly the endogenous urinary nitrogen (MUN); the average MUN value for preruminants is 70 mg/kg body weight, whereas in ruminant calves it decreases from 58 mg nitrogen per kg body weight at 12 weeks to 25 mg at maturity. However, the influence of these factors on the protein requirement of preruminants is balanced by that of metabolic faecal nitrogen (FEN) which is considerably lower in preruminants than in ruminants. The value of *FEN* for the milk-fed

Table 18.7 Requirements of the preruminant and ruminant calf for metabolic energy (after Roy, 1980, p. 234)

	Live weight (kg)	Metabolizable energy (MJ)			
		Maintenance	Maintenance +0.5 kg gain per day	Maintenance +1 kg gain per day	Maintenance +1.5 kg gain per day
Preruminants (whole-milk diet)	50	8.5	15.0	23.4	30.9
	75	11.5	18.0	28.1	36.4
	100	14.3	20.8	32.5	41.6
		Maintenance (50% hay + 50% concentrates)	Maintenance +0.5 kg gain/d (15% hay + 85% concentrates)	Maintenance +1.0 kg gain/d (9% hay + 91% concentrates)	Maintenance +1.5 kg gain/d (100% concentrates)
Ruminants	50	12.1	18.8	27.6	38.9
	75	15.1	21.8	31.0	43.9
	100	18.4	25.1	35.1	47.7

calf is about 1.9 g/kg dry matter intake, whereas for young ruminants fed diets containing 5% and 30% crude fibre the values are 3.3 and 4.2 g/kg dry matter, respectively (see p. 365).

The growth rate of milk-fed preruminants is more likely to be limited by energy intake than by protein. The protein to energy ratio is higher in milk than that required in diets enabling the rapid growth of calves.

Minerals and Vitamins

Preruminant animals are susceptible to deficiencies of amino acids and B vitamins like monogastric species. They depend on their diet (colostrum or milk) to supply the needs for B vitamins. Characteristic symptoms of deficiencies of B vitamins may occur if milk substitutes are fed that are inadequately supplemented with these vitamins (see Chapter 12).

The great importance of the supply of vitamin A to young animals in colostrum or milk has been stressed in Chapters 11 and 21.

Since levels of iron, copper and magnesium in milk may be insufficient for rapid growth, the supplementation of these elements is desirable when milk only is fed for an extended period. The low concentration of pigment in veal reflects an iron deficiency due to an all-milk diet; however, pale veal is a market requirement (see Chapter 10).

NUTRITION OF GROWING PIGS

Energy Supply and Utilization in Pigs

The basic aim of feeding is the most efficient conversion of feedstuffs to lean body tissue without excessive deposition of fat. Composition of the body gain and rate of growth of pigs depends on the dietary regime, genetics and age of the animal. The utilization of dietary energy during the whole growth period of pigs is illustrated in Figure 18.12. The cumulative protein and lipid accretion from birth to 140 kg body weight and the total heat loss from the animal are indicated in this figure. It is obvious from the graph that as the animal gets older the energy retained as protein progressively decreases, whereas that retained as fat increases. The intake of ME must provide for maintenance (0.44 MJ/kg $W^{0.75}$) and for the cost of protein and lipid deposition. The efficiency of ME utilization for fat deposition in pigs is much higher than that for protein deposition, as outlined in Chapter 16. The partition of daily intake of ME by a 60-kg pig is given in Table 18.8; this is the live weight at which optimal energy utilization for the production of lean tissue is achieved. Even at this age (15 weeks) only about 9% of the dietary ME is retained as protein, but only two-thirds of the retained protein is in the muscle tissue; thus only about 6% of the dietary ME is retained as edible protein.

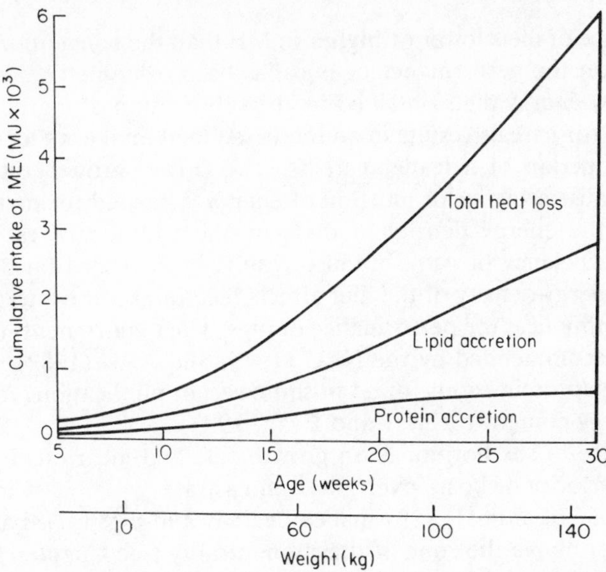

Figure 18.12. Partition of cumulative metabolizable energy during growth of pigs between heat loss and the accretion of protein and lipid (after Fowler, 1978, p. 76). *Reproduced by permission of Butterworths and of the Author.*

Table 18.8 Partition of daily intake of metabolizable energy by a 60-kg pig (after Kielanowski, 1972, p. 143)

	Protein deposition (100 g)	Lipid deposition (200 g)	Maintenance	Total	Percentage of intake
Retention (MJ)	2.3	8.0	–	10.3	38
Associated heat loss (MJ)	4.3	2.9	9.2	16.5	62
Total (MJ)	6.7	10.9	9.2	26.8	100

Diets fed to growing pigs are usually based on maize or other cereal grains and soybean meal and have an energy density of about 12.5 MJ ME per kg. Maximum growth rate and efficiency of feed utilization are achieved by allowing the pigs to eat such a diet to appetite. Certain types of pig can consume food ad libitum until slaughter at 45 kg body weight without undesirable deposition of fat. With other types of pig, however, feed intake must be restricted to produce a leaner carcass, even during earlier growth.

The influence of diets lower or higher in ME than the conventional cereal–soybean diet on the performance of pigs has been examined systematically. The use of low-energy diets obtained by the substitution of ingredients high in crude fibre for cereals results in an increased feed intake, reduced growth rate and production of a leaner carcass. The slower growth rate reduces efficiency because a higher proportion of energy is needed for maintenance.

Increasing the energy density of diets by the addition of fat improves the rate and efficiency of gain, but also results in increased fat deposition. Because the energy density of the diet affects feed intake, the energy/protein ratio greatly influences the performance of pigs. The requirements for energy and protein recommended by the NRC (1979) and ARC (1982) reflect the optimal energy/protein ratios found in some recent publications for diets of different energy content (Seerley and Evan, 1983).

The influence of sex hormones on growth rate and meat quality appears from the superiority of boars over gilts and castrates.

The efficiency of animal performance decreases in pigs raised at temperatures below or above the zone of thermoneutrality (see Chapter 17). Extra energy is needed to maintain body temperature in a cold environment and also for active thermoregulation at high temperature. Pigs have little coat and poor thermal insulation; consequently they respond more markedly than other species to change in temperature.

Protein Supply to Pigs

Requirements of protein and essential amino acids needed for optimal growth and carcass composition depend on the energy density of the diet and vary with stage of growth. Maximal carcass leanness requires greater intake of amino acids than maximal rate of weight gain. Excess protein intake improves meatiness of carcasses, but it usually reduces daily gain and feed efficiency (gain to feed ratio). Gilts and boars are leaner than castrates and require higher levels of protein and amino acids than castrates.

A great deal of attention has been given to the experimental determination of the level of individual amino acids at which the maximum response is achieved. Lysine is the first limiting amino acid in the usual pig diets based on cereals and soybean meal.

Supply of Minerals and Vitamins to Pigs

Improvements in conversion of energy and protein to lean body tissue achieved by genetic and nutritional means necessitated a reappraisal of mineral and vitamin requirements for pigs. Vitamin premixes designed to meet the deficiencies of vitamins in diets based on cereals and soybean meal have been developed; it became necessary to include vitamins E and K and selenium in these diets. The importance of zinc to prevent or cure parakeratosis and of copper for growth promotion have been stressed in Chapter 10.

NUTRITION OF GROWING POULTRY*

Supply and Utilization of Energy by Growing Chickens

As outlined in Chapter 16, chickens, like other growing animals, need energy for maintenance and for growth. Thus energy intake of chicks varies with the metabolic body weight (accountable for maintenance) and with the rate and composition of body gain (depending on age, sex and strain of the birds). The maintenance requirement is also influenced by environmental factors such as temperature (see Chapter 17). The ME content of broiler diets increases with age corresponding to the rise in body fat content with age. The ME content of starter diets fed from 0 to 3–4 weeks of age is generally between 13.1 and 13.8 MJ ME per kg: that fed from 3–4 weeks to 6–7 weeks of age ranges from 13.6 to 14.1 MJ ME per kg and from 6–8 weeks the energy content is increased from 13.8 to 14.4 MJ ME per kg. Likewise, the diet for egg-type strains of chickens is usually changed three times between hatching and start of laying. Replacement pullets must receive low-energy diets (from hatching to 10 weeks, about 10 MJ/kg) to prevent excess fat accumulation in the pullets at sexual maturity. For turkeys six diets are given between hatching and marketing at 20–24 weeks for males and at 17–20 weeks for females.

As stated in Chapter 16, chickens maintain constant energy intake when fed diets of different energy levels between 11.8 and 14.3 MJ/kg. Therefore, the relationship of change in dietary energy concentration to body weight gain, feed consumption and feed conversion efficiency was generally found to be linear (see Figure 18.13(a)–(c)). However, when the dietary energy concentration is below a certain limit, food consumption is determined by the physical capacity of the digestive tract. This limit is about 11.8 MJ ME per kg for chicks and varies with species and strains. In a classical experiment carried out by Hill (1954), groups of chicks were fed diets of gradually diminishing energy content; such diets were prepared by mixing standard diets with oat hulls. Intake of the diet with the lowest ME concentration (7.5 MJ/kg versus the control diet 13.2 MJ/kg) was followed by an increase of 25% in food intake but a decrease of 29% in energy intake, probably because of the physical limitations of the digestive tract.

On the other hand, favourable effects on performance of broilers and turkeys were shown by the use of high-energy diets prepared by supplementation with 2–5% fat (see Chapter 6). Supplemental fat increases the energy efficiency even more than would be expected from the additive effects of the dietary constituents. The term 'extracaloric effect' of fats is used to describe this favourable influence of fat on the performance of poultry. The extracaloric effect can be attributed to the following factors: (1) The presence of unsaturated fatty acids in a mixture with saturated fatty acids

* In collaboration with Mrs. Z. Nitsan and I. Nir.

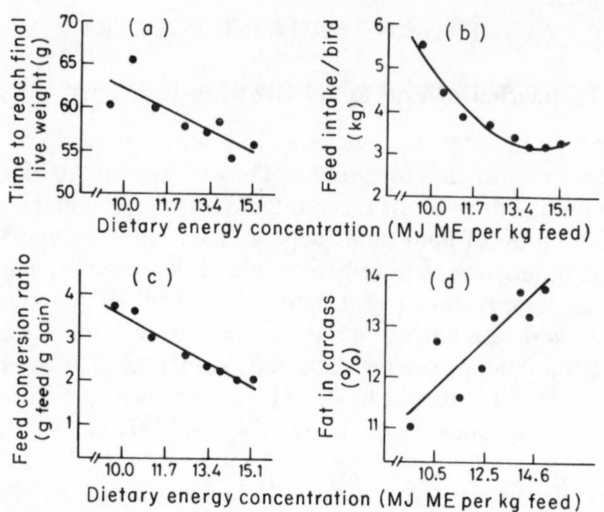

Figure 18.13. The relationship between dietary energy concentration and (a) growth performance, (b) food intake, (c) food conversion ration and (d) carcass fat content (after Farell, 1974, p. 31). *Reproduced by permission of* British Poultry Science.

improves the absorption of the latter and thereby increases energy utilization of dietary fat. (2) An important proportion of the extracaloric effect may be related to increased utilization of dietary carbohydrates. (3) Supplemental fat slows the rate of passage of food through the gastrointestinal tract and therefore the digestibility of the diets is raised by the addition of fat. As the energy concentration of the diet increases above about 14.3 MJ ME due to the addition of 10% fat, chickens do not adjust their energy intake exactly but consume somewhat more energy and this results in deposition of fat (Figure 18.13(d)). Whereas the presence of some muscular fat is desirable for the palatability of chicken meat, the accumulation of excess fat as abdominal fat is objectionable because of its visual appearance and is wasteful from the point of view of production economics.

When high-energy fat-supplemented diets are used, high feed efficiency and production of broilers or turkeys of desirable quality may be achieved by maintaining certain energy/protein ratios in the diet; estimates for optimal ratios are given in Table 18.9.

As the protein level of a diet is decreased and the energy/protein ratio widens, feed consumption is increased in order to overcome a protein deficiency stress, and, as a result of the overconsumption of energy, the efficiency of food conversion decreases and deposition of carcass fat increases.

When there is a relative excess of protein and the energy/protein ratio declines, the energy efficiency of the diet is reduced since a considerable amount of energy is required to synthesize uric acid.

Table 18.9 Estimates of optimal energy/protein ratios for broilers and turkeys (adapted from Lewis, 1978, p. 17)

	Energy (kJ) to protein (%) ratio
Broiler starter (1–2 weeks)	598
Broiler starter (2–3 weeks)	636
Broiler finisher (5–10 weeks)	628
Turkey starter (0–6 weeks)	418
Turkey grower (12–18 weeks)	628
Turkey finisher	941

Protein Supply to Chickens

The supply of dietary protein and of essential amino acids can be considered as one of the most important determinants (see Chapter 7) of growth of chickens and egg production (see Chapter 20). The amino acid content of the diet must be such that the growing chick receives sufficient of each essential amino acid and also some combination of dispensable amino acids for optimal synthesis and deposition of tissue proteins. The pattern of absorbed essential amino acids originating from dietary protein should simulate the amino acid pattern in the body gain; thus the requirements of the growing bird for amino acids are related to their content in organs and tissues and the relative growth of the latter. For such comparison the amino acid contents of proteins present in tissues, feathers, skin and eggs and also of proteins in a usual standard diet for chicks based on soybean meal and maize are given in Table 18.10. The amino acid pattern of the main protein source present in poultry diets varies from that of body proteins and does not contain some essential amino acids (sulphur-containing amino acids and lysine) in the required proportion. The quantities of the usual plant protein sources supplying adequate amounts of these limiting amino acids would be associated with a wasteful excess of the other amino acids. In order to avoid this and to ensure a proper balance of amino acids, the diets can be supplemented by protein feeds of animal sources (fish meal) or by synthetic amino acids (see Chapter 7).

As the conventional ingredients are well endowed with most of the essential amino acids, there remain only the following five amino acids to be considered during the formulation of diets for growing birds: lysine, tryptophan, arginine, cystine and methionine. As explained in Chapter 7, requirements for individual amino acids are influenced by interactions among essential amino acids themselves and between essential and non-essential amino acids, such as the possibilities of meeting the requirement for cystine by methionine or that for tyrosine by phenylalanine, but not the reverse. Furthermore, in Chapter 7 the harmful influence of an amino acid when ingested in considerable excess above its required amount on the utilization of certain others is discussed.

Table 18.10 Amino acid composition of tissue proteins, egg protein and dietary protein (in percentages of proteins)

	Chick tissues	Skin	Feathers	Egg	Mash based on soybean meal and maize
Arginine	6.8	6.9	7.3	6.4	6.7
Cystine	2.4	1.8	7.4	2.4	1.8
Glycine	8.2	10.5	6.8	3.8	5.0
Histidine	4.1	2.2	0.6	2.1	2.4
Isoleucine	3.9	3.4	6.4	8.0	5.1
Leucine	6.5	6.5	8.5	9.2	9.6
Lysine	9.9	5.7	1.6	7.2	4.9
Methionine	1.9	1.5	0.5	3.4	1.7
Phenylalanine	3.6	4.0	5.5	6.3	5.2
Threonine	3.4	3.9	4.7	4.9	4.1
Tryptophan	1.0	–	0.7	1.5	1.2
Tyrosine	3.1	3.8	2.5	–	–
Valine	4.4	4.3	8.9	6.7	5.1

Determination of the Requirements of Growing Birds for Protein and Essential Amino Acids

Measurements of these requirements may be carried out according to two principles: (1) experimental determination of the minimal amount of an essential amino acid needed for optimal growth; and (2) factorial summation of the requirements needed for maintenance, carcass growth and feather growth.

Growth Assay. The requirement of growing animals for essential amino acids is assessed by preparing a response curve recording some parameter, such as gain, in relation to graded inclusions in the diet of the particular amino acid. In devising the basal diet it is important to ensure that the supply of all other indispensable nutrients, particularly amino acids, is adequate but not excessive. The dietary level at which maximum growth performance or nitrogen utilization (measured by nitrogen balance) is achieved is generally recognized as the requirement for maintenance and growth together. Figure 18.14 illustrates an example of the application of this principle to the requirement of chicks for arginine. The data plotted semilogarithmically produce straight response lines up to the limit of response. The biological response to an increased supply of an essential nutrient has frequently been found to follow a logarithmic curve. The limit of response is well defined by the plateau and the minimal requirement is readily seen from the intersection of the response and plateau lines.

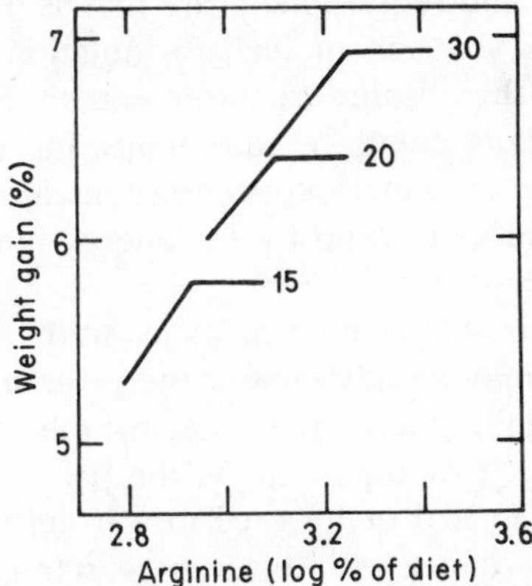

Figure 18.14. The relation of daily rate of weight gain of chicks to the logarithm of the arginine level in the diet (in per cent) at protein levels of 15%, 20% and 30% in the diet as indicated in the figure (after Almquist, H. J., *Proteins and Amino Acids in Animal Nutrition*, 5th edn, 1972, p. 22).

Factorial Method. According to the factorial methods developed by Fisher (1980) and Hurwitz (1978), the requirements for protein and essential amino acids are taken as the sum of the needs for maintenance and the accretion in carcass and feathers. The maintenance requirement is based on determination of the excretion of nitrogen from animals fed a nitrogen-free diet. The nitrogen required for maintenance serves to replace obligatory losses of nitrogen and depends on the metabolic body weight (see Chapter 17). The requirements for growth are deduced from the protein content or the amino acid composition of the carcass and feathers of chickens killed at different ages. The actual needs for amino acids are estimated by dividing their contents in gains of carcass and feathers by 0.85 since 85% is the average availability of proteins and amino acids in the conventional protein sources for poultry.

The amino acid composition of feather proteins differs from carcass proteins mainly by the high content of sulphur-containing amino acids in feather proteins. The amino acid requirements may also vary according to the relative growth of the organs and tissues in different avian species. For example, in the growing chicken the feathers constitute about 4% of the body weight and contain about 15% of the total body protein. In the growing gosling the respective values are about 7% and 25%.

Since the major part of the amino acids needed for maintenance is associated with replacement of tissue proteins, it can be assumed that the amino acid pattern for maintenance is very similar to that needed for tissue growth. While most (about 80%) of the energy consumed by growing chicks is diverted to maintenance, the major part (60–70%) of the amino acids consumed is directed to body gain.

The protein and amino acid requirements are met as outlined in absolute terms and cannot be used as such in feed formulation since poultry are fed ad libitum. Therefore the absolute requirements for protein and amino acids are divided by the daily energy requirements and expressed in g/MJ. The requirements in terms of dietary concentrations are then obtained by multiplying the amino acid/joule ratio by the energy concentration of the diet.

The expression of amino acid requirements as grams of amino acid per MJ metabolizable energy may be advisable if the protein content is altered to compensate for a change in energy requirement; then the amino acid requirement will also change. Correspondingly, the percentage requirements for amino acids should be raised or lowered in warmer or colder environments, respectively, in accordance with differences in feed or energy intake, so that a certain daily intake of the amino acid is assured.

Some Dietary Factors Influencing Amino Acid Requirements

(1) Dietary protein and amino acid concentrations are reduced with increasing age and body weight, while energy needs increase with age. The growth of meat-type strains is very rapid; therefore, their needs for adequate amino acid concentrations are more critical than those of egg-type strains or of adult birds. Since the required ratio between energy and protein increases continuously during the growth period, the diet has to be changed at various stages of growth. Thus the diet of broiler chickens is changed three times between hatching and marketing (7–9 weeks; see preceding section).

(2) Amino acid requirements are considerably higher for turkey pullets than for broiler chickens, since carcass protein and accordingly amino acid concentrations are higher in the carcass and feathers of turkeys than in those of chickens. The opposite is true for water-fowl (geese and ducks) since their body contains more fat and less protein than the chicken body.

(3) The importance of the energy/protein ratio in the diet has been stressed in the preceding section. If the protein supplied provides all essential amino acids in perfect balance, then the quantity of protein that needs to be supplied is minimal and the energy/protein ratio appears to be optimal. However, as the supply of amino acids deviates from the balance, a larger amount of protein is required and the energy/protein ratio may be reduced, leading to undesired fattening as explained above.

Protein Deposition in Growing Birds

Protein deposition is a most important factor in poultry productivity, and some relevant data determined by Fisher (1980) are summarized in Table 18.11.

During normal commercial growing periods the average protein deposition varies among species. The average daily rate of protein deposition is

Table 18.11 Average rates of protein deposition in poultry (according to Fisher, 1980)

Type (period)	Sex	P (g day)	$P/W^{0.75}$ (g)	% P at slaughter
Broiler	M	8.1	6.3	17
(t_0–t_{56})	F	5.8	5.6	16
Turkey	M	12.4	6.3	21
(t_0–t_{84})	F	9.2	5.8	22
Duck (t_0–t_{49})	M/F	9.4	6.2	14
Layers		5.1	3.0	12

Abbreviations: P = protein mass; % P = percentage protein content of body; t = time (days); t_{56} = 56 days of age.

highest in turkeys grown to t_{84}, followed by ducks (t_{49}) and broiler chickens (t_{56}). If these results are expressed on a metabolic weight basis, then a relative constant figure of 6 g protein per day per kg $W^{0.75}$ is found for all the growing birds. This result may be expected in view of the relatively small differences in body protein content and the fact that similar stages of maturity are involved. However, there is a gradual decline in protein deposition during growth. The average daily rate of protein deposition in broiler chickens expressed as a function of metabolic weight decreases from 12.1 g protein per kg $W^{0.75}$ during the first week of life to 5.1 g in 8-week-old chickens. In the laying hen the rate of daily protein deposition (as egg protein) per unit of metabolic weight is much lower, being only 3.0 g.

Supply of Minerals and Vitamins

Since growth of the skeleton is quite rapid in young chicks and poults, during the first four weeks of the life of the chicks the amount of bones increases 20-fold, but that of the whole body only 8-fold; thus the supply of adequate amounts of calcium and phosphorus in the correct ratio is essential. Excessive amounts of calcium, magnesium and phosphorus depress growth, and excess phosphorus also depresses food conversion. Deficiencies of phosphorus cause leg weakness and bone deformation, resembling perosis in external morphological appearance (see Chapter 10). Leg weakness in chicks and poults—an economic problem facing the broiler industry—is not the result of simple nutritional deficiencies (of phosphorus, vitamin D or manganese); genetic susceptibility, environmental stress and infectious disease are all important factors in its occurrence.

The balance of electrolytes (mequiv./g): (Na^+ + K^+ + Mg^{2+} versus Cl^-

$+ HPO_4^{2-} + SO_4^{2-}$) in the diet is of great importance in the growth of chicks. Growth of chicks is optimal in diets providing a sodium/chloride (w/w) ratio near to 1, and decreases sharply as this ratio decreases or increases.

The importance of manganese, zinc and selenium for growth of chicks has been stressed in Chapter 10 and that of vitamins in Chapters 11 and 12. The basic feedstuffs present in the diet contribute sufficient amounts of most trace minerals and of some B-vitamins to meet requirements without any further supplementation. However, it is necessary to supplement the diets with manganese, zinc, fat-soluble vitamins, riboflavin, pantothenic acid and vitamin B_{12} to meet the recommended requirements. Moreover, nicotinic acid and choline are added to vitamin concentrates for chicks at levels that provide margins of safety.

BIBLIOGRAPHY

Growth of animals and efficiency of animals for meat production

Books

Cole, D. J. A., and Lamming, J. E. (1975). *Meat.* London: Butterworths.
Lawrence, T. L. J. (1980). *Growth in Animals.* London: Butterworths.
Lodge, J. A., and Lamming, J. E. (eds) (1968). *Growth and Development of Mammals.* London: Butterworths.

Articles

Symposium on nutrition and growth (1976). *Proc. Nutr. Soc.*, **35**, 309–391.
Berg, R. T., and Walters, L. E. (1983). The meat animal: changes and challenges. *J. Anim. Sci.*, **57**, Suppl. 2, 123–146.
Christie, W. W. (1980). The effects of diet and other factors on the lipid composition of ruminant tissues and milk. *Prog. Lipid Res.*, **17**, 245–278.
Lindsay, D. B. (1983). Growth and fattening. In *Nutritional Physiology of Farm Animals* (J. A. F. Rook and P. C. Thomas, eds), pp. 261–313. London and New York: Longman.
Lister, D., Perry, B. N., and Wood, J. D. (1983). Meat production. In *Nutritional Physiology of Farm Animals* (J. A. F. Rook and P. C. Thomas, eds), pp. 476–537. London and New York: Longman.
Prior, R. (1982). Symposium on partitioning of nutrients between muscle and adipose tissue growth in the ruminant, particularly hormonal effects. *Fed. Proc.*, **41**, 2536–2566.
Reeds, P. J., Whale, K. W. J., and Haggarty, P. (1982). Energy cost of protein and fatty acid synthesis. *Proc. Nutr. Soc.*, **41**, 155–159.
Van Es, A. J. H. (1977). The energetics of fat deposition during growth. *Nutr. Metab.*, **21**, 88–104.
Van Es, A. J. H. (1980). Energy costs of protein deposition. In *Protein Deposition in Animals* (P. J. Buttery and D. B. Lindsay, eds), pp. 215–224. London: Butterworths.
Webster, A. F. J. (1980). The energetic efficiency of growth. *Livestock Product. Sci.*, **7**, 243–252.
Webster, A. F. J. (1986). Factors affecting the body composition of growing and adult animals. *Proc. Nutr. Soc.*, **45**, 45–53

The role of hormones in regulation of growth

Book

Lu, C. F., and Rendel, J. (1976). *Anabolic Agents in Animal Production.* Stuttgart: Georg Thieme.

Articles

Buttery, P. J., Vernon, B. G., and Pearson, J. T. (1978). Anabolic agents—some thoughts on their mode of action. *Proc. Nutr. Soc.*, **37**, 311–315.

Heitzman, R. J. (1978). The use of hormones to regulate the utilization of nutrients in farm animals: current farm practices. *Proc. Nutr. Soc.*, **37**, 289–293.

Trenkle, A. (1981). Endocrine regulation of energy metabolism in ruminants. *Fed. Proc.*, **40**, 2536–2542.

Young, V. R. (1980). Hormonal control of protein metabolism with particular reference to protein gain. In *Protein Deposition in Animals* (P. J. Buttery and D. B. Lindsay, eds), pp. 167–191. London: Butterworths.

Energy requirements of growing ruminants: Beef production

Books

Matsuchima, J. K. (1979). *Feeding Beef Cattle.* Berlin: Springer-Verlag.

Perry, T. W. (1980). *Beef Cattle. Feeding and Nutrition.* New York: Academic Press.

Preston, J. R., and Willis, M. B. (1974). *Intensive Beef Production.* Oxford: Pergamon Press.

Swan, H., and Broster, W. H. (1976). *Principles of Cattle Production.* London: Butterworths.

Wilson, P. N., and Brigstocke, T. D. A. (1981). *Improved Feeding of Cattle and Sheep.* London, Toronto, Sydney, New York: Granada.

Articles

Blaxter, K. L. (1968). The effect of dietary energy supply on growth. In *Growth and Development of Mammals* (J. A. Lodge and J. E. Lamming, eds), pp. 329–344. London: Butterworths.

Blaxter, K. L. and Wainman, F. W. (1964). Use of energy of different rations by sheep and cattle. *J. Agric. Sci.*, **63**, 113–128.

Byers, F. M. (1982). Nutritional factors affecting growth of muscle and adipose tissue in ruminants. *Fed. Proc.*, **41**, 2562–2566.

Fox, D. G., and Black, J. R. (1984). A system for predicting body composition and performance of growing cattle. *J. Anim. Sci.*, **58**, 725–739.

Garrett, W. N. (1980). Factors influencing energetic efficiency of beef production. *J. Anim. Sci.*, **51**, 1434–1440.

Geay, Y. (1984). Energy and protein utilization in growing cattle. *J. Anim. Sci.*, **58**, 766–778.

Kay, M. (1976). Meeting the energy and protein requirements of the growing animal. In: Principles of Cattle Production (H. Swan and W. H. Broster, eds), pp. 255–270.

Orskov, E. R. *et al.* (1979). Efficiency of utilization of volatile fatty acids for maintenance and energy retention by sheep. *Br. J. Nutr.*, **41**, 541–551.

Reid, J. T. (1975). Comparative efficiency of animals in the conversion of feedstuffs to human foods. Cornell Nutr. Conference for Feed Manufacturers, pp. 16–24. Ithaca: Departments of Animal Science, Cornell University.

Ritzman, E. G., and Colovos, N. F. (1943). Physiological requirements and utilization of protein and energy by growing dairy cattle. Univ. New Hampshire Tech. Bull. 80; quoted by Blaxter (1967), Energy Metabolism of Ruminants.

Rumsey, S. T. *et al.* (1984). Symposium on monensin in cattle. *J. Anim. Sci.*, **58**, 1461–1539.

Tyrrell, H. F., Reynolds, P. J., and Moe, P. W. (1979). Effect of diet on partial efficiency of acetate use for body tissue synthesis by mature cattle. *J. Anim. Sci.*, **48**, 598–606 and 1491–1500.

Van Es, A. J. H. (1976). Factors influencing the efficiency of energy utilization by beef and dairy cattle. In *Principles of Cattle Production* (H. Swan and W. H. Broster, eds), pp. 237–254. London: Butterworths.

Webster, A. J. F. (1978). Prediction of the energy requirements for growth in beef cattle. *World Rev. Nutr. Diet.*, **30**, 189–226.

Compensatory growth

Burton, J. H., Anderson, M., and Reid, J. T. (1974). Some biological aspects of partial starvation. The effect of weight loss and regrowth on body composition in sheep. *Br. J. Nutr.*, **32**, 515–527.

Folman, Y. *et al.* (1974). Compensatory growth of intensively raised bull calves. *J. Anim. Sci.*, **39**, 788–795.

Protein and mineral requirements for growing ruminants

Book

Owens, F. N. (1982). *Protein Requirements for Cattle, Symposium.* Oklahoma State University.

Articles

Garrett, W. N. (1977). Protein production by growing ruminants as influenced by dietary nitrogen and energy. In *Second International Symposium on Protein Metabolism and Nutrition*, pp. 115–118. Wageningen: Centre for Agricultural Publishing and Documentation.

Gunther, K. D. (1972). Growth and mineral metabolism. In *Handbook of Animal Nutrition* (W. Lenkeit and K. Breirem, eds), Vol. 2, pp. 463–489 and 585–592. Hamburg and Berlin: Paul Parey.

Kay, M. (1976). Meeting the energy and protein requirements of the growing animal. In *Principles of Cattle Production* (H. Swan and W. H. Broster, eds), pp. 255–270. London: Butterworths.

Lobley, G. E. *et al.* (1980). Whole body and tissue protein synthesis in cattle. *Br. J. Nutr.*, **43**, 491–502.

Preston, R. L. (1966). Protein requirements of growing–finishing cattle and lambs. *J. Nutr.*, **90**, 157–160.

Preston, R. L. and Pfander, W. H. (1964). Phosphorus metabolism in lambs fed varying phosphorus intakes. *J. Nutr.*, **83**, 369–378.

Rohr. K., Schafft, H., and Lebzien, F. (1983). Critical analysis of present protein allowances for growing ruminants. *4th International Symposium on Protein Metabolism and Nutrition*,Vol. 1, pp. 449–461. Paris: Institut National de la Recherche Agronomique.

Roy, J. M. B. *et al.* (1977). Calculation of the N-requirement of ruminants from nitrogen metabolism studies. In *Second International Symposium on Protein Metabolism and Nutrition*, pp. 126–129. Wageningen: Centre for Agricultural Publishing and Documentation.

Nutrition of preruminants

Book

Roy, J. H. B. (1980). *The Calf.* 4th edn. London: Butterworths.

Articles

Symposium on Development of the ruminant stomach (1974). In: *Digestion and Metabolism of the Ruminant,* 4th Internat. Symposium Australia (see above), pp. 1–59.

Faulkner, A. (1983). Foetal and neonatal metabolism. In *Nutritional Physiology of Farm Animals* (J. A. F. Rook and P. C. Thomas, eds), pp. 203–242.

Foley, I. A., and Otterby, D. E. (1978). Storage, composition and feeding value of surplus colostrum. *J. Dairy Sci.,* **61**, 1033–1060.

Nitsan, Z., Volcani, R., Hasdai, A., and Gordin, S. (1972). Soybean proteins substitute in milk replacers for suckling calves. *J. Dairy Sci.,* **55**, 811–821.

Noble, R. C., and Shand, J. H. (1982). Fatty acid metabolism in the neonatal ruminant. *Adv. Nutr. Res.,* **4**, 287–338.

Stobo, I. J. F. (1983). Milk replacers for calves. *Recent Advances in Animal Nutrition—1983,* pp. 113–140. London: Butterworths.

Thivend, P., Toullec, R., and Guilloteau (1979). Digestive adaptation in the preruminant. In *Digestive Physiology and Metabolism in Ruminants* (Y. Ruckebush and P. Thivend, eds), pp. 561–586. Lancaster: MTP Press.

Walker, D. M. (1979). Nutrition of preruminants. In *Digestive Physiology and Nutrition of Ruminants* (D. C. Church, ed.), 2nd edn, Vol. 2, pp. 258–280.

Nutrition of pigs

Books

Cole, D. J. A. (1972). *Pig Production.* London: Butterworths.

Cunha, T. J. (1977). *Swine Feeding and Nutrition.* New York: Academic Press.

Articles

Symposium on pig nutrition (1980). *Proc. Nutr. Soc.,* **39**, 149–211.

Baker, D. H., and Speer, V. C. (1983). Protein–amino acid nutrition of nonruminant animals with emphasis on the pig. *J. Anim. Sci.,* **57**, Suppl. 2, 284–299.

Close, W. H., and Fowler, V. R. (1982). Energy requirements of pigs. *Recent Advances in Animal Nutrition—1982,* pp. 159–174. London: Butterworths.

Fowler, V. R. (1976). Some aspects of energy utilization for the production of lean tissue in the pig. *Proc. Nutr. Soc.,* **35**, 75–79.

Fowler, V. R. (1978). Energy requirements of the growing pig. In *Recent Advances in Animal Nutrition—1978,* pp. 143–161. London: Butterworths.

Fuller, M. F. (1982). Protein requirements of pigs. In *Recent Advances in Animal Nutrition—1982,* pp. 175–186. London: Butterworths.

Kielanowski, J. (1972). Energy requirements of the growing pig. In *Pig Production* (D. J. R. Cole, ed.), pp. 183–203. London: Butterworths.

Low, A. G. (1980). Amino acid use by growing pigs. In *Recent Advances in Animal Nutrition—1980,* pp. 141–156. London: Butterworths.

Miller, E. R., and Kornegay, E. T. (1983). Mineral and vitamin nutrition of swine. *J. Anim. Sci.,* **57**, Suppl. 2, 315–329.

Seerley, R. W., and Ewan, R. C. (1983). An overview of energy utilization in swine nutrition. *J. Anim. Sci.,* **57**, Suppl. 2, 300–314.

Whittemore, C. T. (1986). Causes of variation in the body composition of growing pigs. *Proc. Nutr. Soc.,* **45**, 111–117.

Nutrition of growing birds (See also Bibliography to Chapter 16)

Bolton, W., and Blair, R. (1977). *Poultry Nutrition*, 2nd edn. Ministry of Agriculture, Fisheries and Food, HMSO.

Boorman, K. N., and Wilson, B. J. (1977). *Growth and Poultry Meat Production. Poultry Science Symposium*, No. 12. Edinburgh: British Poultry Science Ltd.

Mead, G. C., and Freeman, B. M. (1980). *Meat Quality in Poultry and Game Birds. Poultry Science Symposium*, No. 15. Edinburgh: British Poultry Science Ltd.

Scott, M. L., Nesheim, M. C., and Young, R. I. (1982). *Nutrition of the Chicken*, 3rd edn. Ithaca: M. L. Scott.

Articles

Symposium on advances in poultry nutrition (1975). *Proc. Nutr. Soc.*, **34**, 1–49.

Baker, D. H. (1977). Amino acid nutrition in the chick. *Adv. Nutr. Res.*, **1**, 299–336.

Combs, G. F. (1962). The interrelationships of dietary energy and protein in poultry nutrition. In *Nutrition of Pigs and Poultry* (J. T. Morgan and D. Lewis, eds), pp. 127–147. London: Butterworths.

Farrell, D. J. (1974). Effects of dietary concentration on utilization of energy by broiler chickens. *Br. Poult. Sci.*, **15**, 25–41.

Fisher, C. (1980). Protein deposition in poultry. In *Protein Deposition in Animals* (P. J. Buttery and D. B. Lindsay, eds), pp. 251–270. London, Butterworths.

Fisher, C. (1982). Energy evaluation of poultry rations. In *Recent Advances in Animal Nutrition—1982*, pp. 113–139. London: Butterworths.

Hill, F. W., and Dansky, L. M. (1954). Studies on the energy requirements of chickens. *Poult. Sci.*, **33**, 112–119 and 791–798.

Hurwitz, S. (1980). Protein requirement of poultry. *Proceedings 3rd EAAP Symposium on Protein Metabolism and Nutrition*, Vol. 2, pp. 697–706. Braunschweig, Germany.

Hurwitz, S., Sklan, D., and Bartov, I. (1978). New formal approaches to the determination of energy and amino acid requirements of chickens. *Poult. Sci.*, **57**, 197–205.

Leeson, S., and Summers, J. D. (1980). Feeding the replacement pullet. *Recent Advances in Animal Nutrition—1980*, pp. 203–212. London: Butterworths.

Lewis, D. (1978). Protein–energy interactions in broiler and turkey rations. In *Recent Advances in Animal Nutrition—1978*, pp. 17–30. London: Butterworths.

Lewis, D., and Annison, E. F. (1974). Protein and amino acid requirements of poultry, pigs and ruminants. In *Nutrition Conference for Feed Manufacturers*, Vol. 8, pp. 27–47. London: Butterworths.

McLeod, J. A. (1982). Nutritional factors influencing carcass fat in broilers. *World Poult. Sci.*, **38**, 194–200.

Mongin, P. (1981). Recent advances in dietary anion–cation balance: application in poultry. *Proc. Nutr. Soc.*, **40**, 295–306.

Moran, E. T. (1986). Variations in body composition of poultry. *Proc. Nutr. Soc.*, **45**, 101–109.

CHAPTER 19

Nutritional Requirements for Reproduction

Reproduction is of great economic importance in all the farm animals. In order to optimize the production of meat and milk, farm animals must produce offspring regularly. Especially in grazing stock, conception and parturition are often necessarily seasonal and related to the availability of pasture. Thus, adequate fertility, when applied to farm animals, implies regular conception and parturition at certain seasons of the year.

In both sexes, fertility depends on genetic traits and on a variety of environmental factors. An adequate supply of nutrients is essential for normal reproduction. Undernutrition delays the maturation of the reproductive system and may impair its function after puberty. Nutritional inadequacies are reflected in the decreased secretion of the pituitary hormones involved in reproductive function, such as the follicle-stimulating hormone (FSH), luteinizing hormone (LH) and prolactin. The secretion rates of FSH and LH are controlled by a gonadotrophic releasing hormone (GnRH) which originates in the hypothalamus. Undernutrition and in particular hypoglycaemia lead to the reduction or inhibition of GnRH secretion.

An excess of energy may also impair fertility. Obese animals tend to be infertile, often because of endocrine disturbances.

Onset of Puberty and Sexual Development

The onset of puberty in ruminants is largely determined by body weight or size. Therefore, in ruminants, an insufficient supply of energy to growing animals, reduces growth rate and delays the sexual maturity of both sexes.

In studies carried out at Cornell University, wide differences in the level of nutrition of heifer calves influenced the age of maturity. Heifers fed a ration supplying 146% of the recommended level of total digestible nutrients (TDN), had their first oestrus at 9 months of age, compared with 11 months for calves fed according to recommendations and 20 months for those fed only 62% of the recommended level. Nevertheless, the overfeeding of heifer calves must be avoided because it reduces their subsequent milk production and fertility.

Diets high in concentrates or diets supplemented with monensin, which increase the level of propionic acid in the rumen (see Chapters 5 and 18), may advance puberty in ruminants.

In contrast to ruminants, age rather than body size is the primary determinant of puberty in gilts. A high plane of nutrition does not advance sexual maturity in pigs.

INFLUENCE OF NUTRITION ON REPRODUCTION IN FEMALE ANIMALS

Level of Nutrition

Following puberty, the level of nutrition continues to affect reproduction. Undernutrition of sexually mature females may prevent ovulation or fertilization, or increase the incidence of early embryonic mortality

Because of economic considerations, it is desirable that females conceive soon after parturition. Normally, therefore, breeding and conception must occur during lactation; cows are often bred shortly after the peak of lactation. In most females, milk production induces a temporary negative energy balance (see Chapter 21) which is not conducive to successful reproduction. In dairy breeds of cattle, sheep and goats, high milk production tends to interfere with reproduction. Therefore, in dairy cattle, the conception rate of heifers is usually greater than that of cows, but that of both classes fluctuates widely depending, among other factors, on their body condition and level of nutrition. In heifers, adequate nutrition permits a conception rate of 65–75% compared with 50% or less when the level of nutrition is low, the diet is unbalanced or the heifers are too fat. Conception rate in cows is even more variable than in heifers.

Regular reproduction depends on a reasonably early resumption of the ovarian cycle after calving and on successful conception. Both are affected by the level of nutrition before and after parturition. To a limited extent, a high level of nutrition before parturition may compensate a female for a low level of nutrition thereafter and vice versa. However, as mentioned already, obesity may also interfere with reproduction; therefore body condition at parturition must preferably be moderate.

In many experiments, the early resumption of the ovarian cycle and high conception rate have been related to adequate body condition and positive energy balance. In dairy breeds, energy balance appears to be the more important variable, probably because the output of energy in milk is relatively much greater than in other breeds. In beef breeds, body condition appears to be more important. If the energy supply during lactation is limited, beef cows tend to reduce their milk production and maintain body stores more efficiently than dairy cows. However, suckling *per se*, rather than being milked by hand or mechanically, delays the resumption of the ovarian cycle in both cows and ewes because of its effect on the central nervous system unrelated to nutrition. This is one reason for not permitting suckling in dairy breeds.

Overfeeding cows in late lactation or during the dry period may lead to fatty infiltration of the liver and impaired liver function. These changes are usually associated with the fat-cow syndrome (see Chapter 24) and may be related to nutritional deficiencies or imbalances of an unknown nature. The condition has also been associated with impaired reproductive performance.

In litter-bearing species, e.g. sheep and pig, the level of nutrition affects ovulation rate and thus fertility. Ewes or gilts in poor body condition have low ovulation rates; a high level of energy fed for a few weeks before mating increases their ovulation rate and the number of lambs or piglets in the litter. This practice is referred to as *flushing.* Flushing has no effect on the ovulation rate of animals well fed before mating.

The Effect of Protein

Low levels of protein in the diet delay puberty in both females and males. A shortage of protein in the diet diminishes the fertility of females after puberty. A deficiency of protein reduces food consumption (see Chapter 23) and is therefore frequently associated with deficiencies of other nutrients. In young females, diets adequate in protein for growth are adequate for reproduction, i.e. for conception and for early embryonic development.

An excess of protein in the diet added in the form of soybean meal (but not fish meal) is associated with a reduced conception rate in dairy cows. This effect is compounded by age and by high environmental temperatures and may not be due to protein *per se* (see below).

Minerals and Vitamins

All diets deficient in minerals or vitamins may result in impaired reproductive function. However, reproductive problems have been associated more often with deficiencies of phosphorus, calcium, manganese, copper, zinc and cobalt, and with those of the fat-soluble vitamins A and E. In proven deficiency conditions, the respective supplements have improved conception rate or fertility. A deficiency of phosphorus, common in ruminants grazing on herbage deficient in this element, delays puberty, suppresses oestrus and greatly reduces the calf or lamb crop, though the effect is often complicated by other deficiencies and low food intake (see Chapters 10 and 21).

The retention of the placenta following parturition in ruminants may indirectly delay conception by causing uterine infections. The retention of the placenta is associated with nutritional deficiencies of several factors: selenium, copper and the fat-soluble vitamins A and E.

A deficiency of vitamin A or β-carotene is associated with many disorders including reproductive dysfunction in both males and females (see Chapter 11). Recently, a specific β-carotene requirement for reproduction in the bovine has been suggested by some authors, as distinct from a requirement for retinol (vitamin A); however, the evidence presented so far is not conclusive (see Ascarelli *et al.*, 1985).

Several compounds that may be present in the diet, and certain nutrients in excess, may also interfere with normal reproduction in ruminants, especially by reducing conception rate: goitrogens and plant oestrogens or isoflavones, nitrates and thyroprotein (see Chapter 10), and excessive levels of sodium, phosphorus, calcium and fluorine. Soybean meal, the most common protein supplement for all farm animals, contains a weak goitrogenic compound and also some genistein, an isoflavone. The effect of the former may be counteracted by iodine; the latter may contribute to infertility when high levels of soybean meal are included in the diet.

INFLUENCE OF NUTRITION ON REPRODUCTION IN THE MALE

The maintenance requirement of males for energy and protein is greater than that of females. However, no additional amounts of feed are required for semen production and reproductive activity. The average ejaculate of a bull, including spermatozoa and accessory secretions, contains only 0.5 g of dry matter. After prolonged undernutrition the semen quality of bulls, rams and boars was adversely affected and so was the composition of seminal plasma. In the latter, changes included a reduction in the level of fructose (the source of energy for sperm motility) and of citric acid. Improved nutrition reverses these changes.

A deficiency of vitamin A causes the degeneration of the germinal epithelium of the testes which reduces or eliminates spermatogenesis. A severe and prolonged deficiency of vitamin A causes the complete failure of reproduction in both males and females (see above). The zinc requirement for testicular growth and normal sperm production is higher than that for normal growth. The testicular tissue and spermatozoa have a high content of zinc.

NUTRITION OF PREGNANT FEMALES AND ITS INFLUENCE ON FOETAL DEVELOPMENT

Pregnant females must be offered adequate amounts of energy and other nutrients in order to enable the satisfactory development of the foetus(es), uterine tissue gain, and to increase the dam's bodily reserves needed for milk production following parturition. Foetal growth is accompanied by the formation of foetal membranes and by a considerable enlargement of the uterus proper. The growth of the placenta and the increase in uterine fluids occur in early and mid-pregnancy. By the end of the second trimester of pregnancy, placental development is virtually complete. On the other hand, 80% of foetal growth occurs in the last trimester and concurrently it becomes necessary to make provision in the ration for foetal growth and maternal tissue gain (see Figure 19.1). Thus from conception to parturition, the requirements for pregnancy increase exponentially.

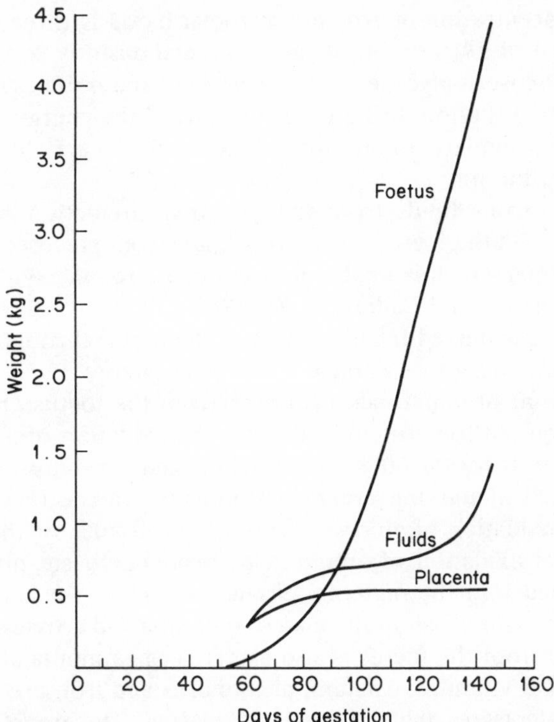

Figure 19.1. Growth of the sheep foetus and associated placenta and fluids (from Robinson, 1983, p. 148). *By permission of Butterworths and the author.*

Foetal Nutrition

Following fertilization the zygote is free in the uterus and absorbs nutrients directly from the surrounding fluids. Soon implantation occurs (the foetus becomes attached to the uterine wall) and the placenta develops. Following implantation, the foetal membranes permit the embryo to obtain nutrients from the maternal circulation across the placenta and to get rid of waste products. The interchange of metabolites includes that of oxygen and carbon dioxide and takes place in the capillaries of the placenta. In addition to its role in the transport and exchange of metabolites between the dam and the foetus, the placenta is a metabolically active organ which grows and also synthesizes hormones and therefore requires energy and other nutrients.

Metabolism in the Foetus

Glucose, derived from the maternal circulation, is the major energy source of the foetus. In ruminants, pigs and horses a portion of the glucose taken up by the placenta is converted to fructose which is transferred to the foetal blood. The concentration of fructose in foetal blood is three to four times higher than that of glucose, but its rate of metabolism is very limited. The foetus stores glucose as glycogen in the liver, lung and muscles mainly during the latter half of gestation; this glycogen is part of the energy reserve of the newborn. It is particularly important in species which carry little or no body fat at birth, e.g. the pig.

The foetus obtains lipids from the maternal circulation or synthesizes them. In ruminants the foetus synthesizes lipids from glucose, in contrast to the mature animal which is unable to use glucose for the synthesis of fatty acids (see Chapter 6, and Ballard *et al.*, 1969).

Amino acids transferred from the dam to the foetus through the placenta are used for protein synthesis and as a source of energy. Especially in ruminants, the amount of amino acids transferred to the foetus greatly exceeds the amount required for protein synthesis. The oxidation of amino acids in the foetus accounts for 20–60% of its total oxygen consumption; the higher value is obtained in starving pregnant ruminants. In the foetus of the fed ruminant, the oxidation of glucose accounts for 50–70% of the oxygen uptake. The foetal oxidation of amino acids produces waste nitrogen which must be excreted into the maternal circulation. Thus, the level of urea in foetal plasma is higher than in the maternal plasma and crosses the placenta. Waste nitrogen from the foetus is also excreted as ammonia and glutamate.

The levels of B vitamins, calcium, phosphorus and iron are several times higher in foetal plasma than in maternal plasma. The transfer of oxygen through the placenta is efficient because foetal haemoglobin has a higher affinity for oxygen than maternal haemoglobin.

Nutrient Requirements in Pregnancy

Estimates of energy and nutrient requirements can be arrived at factorially by combining the estimates of maternal and foetal requirements for maintenance and growth. Growth of the maternal organism during pregnancy occurs in (1) the products of conception associated with the foetus, and (2) the non-reproductive organs such as the gastrointestinal tract including the liver. In females bred before reaching mature size, the additional needs for growth must also be met. Determination of the rate of accretion of tissue in the products of conception in late pregnancy and division of these data by the coefficients for the efficiency of utilization of the respective nutrients provide an estimate of the dietary needs during the gestation period. The amounts of tissue deposited in the uterus and its energy and nitrogen content were determined by weighing and analysing foetuses and uteri from animals slaughtered at various stages of pregnancy. The results for cows are given in Table 19.1 and for sows in Figure 19.2. As indicated in Table 19.1, the amount of energy deposited in the uterus even in late gestation is small in comparison to the maintenance requirements of the dam. The dry matter in the products of conception consists largely of protein. The amounts of protein, calcium and phosphorus in the uterus rise considerably during late gestation. Relatively little tissue is deposited in the mammary gland. Even at the end of gestation the amount of protein deposited in the mammary gland of the cow does not exceed 45 g/day (see Table 19.1).

Table 19.1 Daily deposition of energy, protein, calcium and phosphorus in the uterus of cows at different stages of pregnancy and the estimated daily requirement for maintenance (from McDonald *et al.*, *Animal Nutrition*, Third ed. 1981, p. 306)

Days after conception	Deposited in uterus (per day)				Deposited in mammary gland
	Energy (kJ)	Protein (g)	Calcium (g)	Phosphorus (g)	Protein (g/day)
100	170	5	–	–	–
150	420	14	0.1	–	–
200	980	34	0.7	0.6	7
250	2 340	83	3.2	2.7	22
280	3 930	144	8.0	7.4	44
(Approx. net daily requirement for maintenance of 450-kg cow)	(35 000)	(70)	(8)	(12)	

The apparent efficiency of deposition of ME in the conceptus of ewes and cows is very low (10–25%), much lower than the respective efficiencies for maintenance and other productive processes (see Chapters 15, 17, 18 and 21). The apparent inefficiency stems from ignoring the sizable cost of maintenance of the foetus and the products of conception. Moreover, energy utilization for foetal growth is less efficient in ruminants than in pigs because the foetus requires energy in the form of glucose, which is produced in the ruminant dam by gluconeogenesis—a process of low energetic efficiency (see Chapter 5). The amount of energy stored in the conceptus is comparatively low, but a large amount of energy is spent in the processes involved in pregnancy; therefore, at the end of gestation the energy requirement of

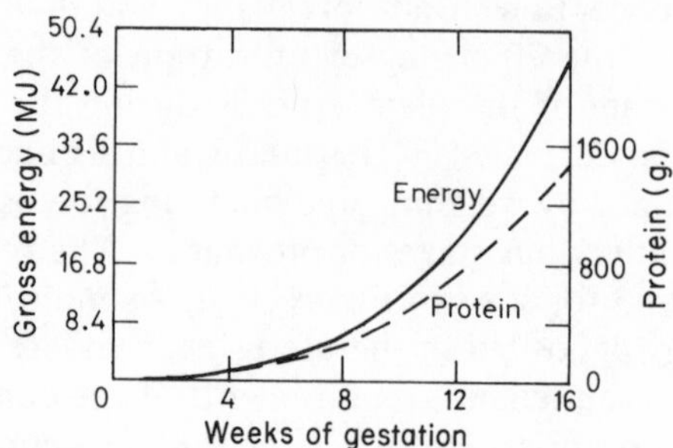

Figure 19.2. Gross energy and protein in products of conception at different stages of growth of the sow. *By permission from* Animal Nutrition, *7th edition, p.476, by L. A. Maynard* et al. *Copyright 1979 McGraw-Hill Book Co.*

the pregnant cow is about 50% higher than the maintenance requirement. In ewes bearing one foetus, the energy requirement in late gestation is also about 50% higher than maintenance, and in twin-bearing ewes 100% higher.

The efficiency of energy utilization for pregnancy is much greater in pigs than in ruminants. Although pigs deliver litters of 4–12 piglets, the increase in the amount of energy required for pregnancy is comparatively small. At birth, piglets are much less mature than calves or lambs; the heat increment of gestation increases with maturity.

An example of the factorial estimation of the energy required by pregnant cows is given in Table 19.2; the energy values of deposited tissue and the energy cost of its synthesis (i.e. the heat increment of gestation) may be added to the estimated maintenance requirement.

The overconditioning of pregnant animals must be avoided. In pregnant ruminants obesity is associated with the fatty infiltration of the liver and metabolic disturbances before and after parturition: Ketosis in dairy cows and pregnancy toxaemia in ewes, especially those bearing twins (see Chapter 24). The fat content of the liver in obese cows may exceed 20%; such cows require more services per conception and have considerably longer intervals from calving to conception.

Requirement for Protein and Other Nutrients

With advancing gestation, the requirement for protein increases even more than that for energy (see above and Table 19.1). Protein requirements for pregnancy may be estimated by adding to the maternal protein requirement that of the uterus and the mammary gland and using a suitable factor to express the efficiency of utilization of protein. An example of such a calculation is presented in Table 19.3 (see also Chapter 21).

Factorial estimates of pregnancy requirements have been calculated from the endogenous losses and deposition in the uterus of the minerals calcium, phosphorus, magnesium, sodium and chlorine.

Table 19.2 Estimated daily energy requirement for pregnancy in a cow (after Lodge, p. 168)

Days after conception	Body weight* (kg)	Fasting metabolism† (MJ)	ME for maintenance‡ (MJ)	Energy of uterine gains (MJ)	Heat increment of gestation (MJ)	Total ME (MJ)
0	500	33.2	45.1	–	–	45.1
100	505	33.3	45.3	0.2	2.4	48.0
150	510	33.4	45.5	0.4	4.0	49.7
200	520	34.0	46.2	1.0	7.0	53.9
250	545	35.2	47.9	2.3	11.0	61.4
280	570	36.2	49.2	3.9	14.8	68.1

* Calf of 40 kg at birth.

† 0.35 MJ/kg $W^{0.73}$.

‡ Fasting metabolism × 1.36.

Table 19.3 Estimated daily protein requirements for pregnancy in a cow (after Lodge, 1972, p. 174)

Stage of pregnancy, days	Body weight* (kg)	Nitrogen for maintenance† (g)	Nitrogen for pregnancy (g)	Total net protein (g)	Available protein requirement (g)
0	500	13.1	0	32	120
150	510	13.2	2.5	98	140
200	520	13.4	6.5	124	180
250	545	13.8	17.0	193	280
280	570	14.3	30.0	277	400

* Calf of 40 kg at birth

† 0.14 g/kg $W^{0.73}$

Deficiency Effects in Pregnant Animals

As a rule, maternal undernutrition, particularly a protein deficiency, affects the weight of the dam more than that of the newborn. Usually the foetus has a high priority for nutrients. However, the protection of the foetus is not absolute: a deficiency of vitamin A which is without effect on the ewe may produce serious symptoms of deficiency in the lamb. On the other hand, a deficiency of iron leads to anaemia in the dam, and while foetal haemoglobin levels may appear normal this may predispose the newborn to anaemia by minimizing the storage of iron in the liver. Neonatal anaemia is common in piglets (see Chapter 10).

As mentioned above, overconditioning ewes during pregnancy may cause pregnancy toxaemia, especially in twin-bearing animals. A frank deficiency of energy in late gestation may also precipitate the same disease (see Chapter 24).

Foetal malformations may be associated with a dietary deficiency of manganese, zinc or copper and also with that of vitamins of the B complex and vitamin A.

BIBLIOGRAPHY

Symposium on nutrition and the foetus (1977). *Proc. Nutr. Soc.*, **36**, 1–40.

Ascarelli, I. *et al.*, see Bibliography to Chapter 11.

Ballard, F. J., Hanson, W., and Kronfeld, D. S. (1969). Gluconeogenesis and lipogenesis in tissues from ruminant and non-ruminant animals. *Fed. Proc.*, **28**, 218–231.

Basset, J. M. (1986). Nutrition of the conceptus: aspects of its regulation. *Proc. Nutr. Soc.*, **45**, 1–10.

Dunn, T. G., and Kaltenbach, C. C. (1980). Nutrition and the postpartum interval of the ewe, sow and cow. *J. Anim. Sci.*, **81**, Suppl. 2, 29–39.

Faulkner, A. (1983). Foetal and neonatal metabolism. In *Nutritional Physiology of Farm Animals* (J. A. F. Rook and P. C. Thomas, eds), pp. 202–242. London and New York: Longman.

Lamond, D. R. (1970). The influence of under-nutrition on reproduction in the cow. *Anim. Breed. Abstr.*, **38**, 359–373.

Lenglands, J. P., and Sutherland, H. A. M. (1968). An estimate of the nutrients utilized for pregnancy by merino sheep. *Br. J. Nutr.*, **22**, 217–227.

Lodge, G. A. (1972). Energy and nutrition requirements for pregnancy. In *Handbook of Animal Nutrition* (W. Lenkeit and K. Breirem, eds), pp. 157–189. Hamburg and Berlin: Paul Parey.

Lodge, G. A., and Heaney, D. P. (1977). Energy cost of pregnancy in single and twin-bearing ewes. *Can. J. Anim. Sci.*, **51**, 479–489.

Moe, P. W., and Tyrrel, H. P. (1972). Metabolizable energy requirements of pregnant dairy cows: *J. Dairy Sci.*, **55**, 480–483.

Munro, H. N., Pilistine, S. J., and Fant, M. E. (1983). The placenta in nutrition. *Annu. Rev. Nutr.*, **3**, 97–124.

Rattray, P. V. (1977). Nutrition and reproductive efficiency. In *Reproduction of Domestic Animals* (H. H. Cole and P. T. Cupps, eds), pp. 553–576. New York: Academic Press.

Rattray, P. V., Garrett, W. N. *et al.* (1974). Efficiency of utilization of metabolizable energy and the energy requirement for pregnancy in sheep. *J. Anim. Sci.*, **38**, 383–393.

Robinson, J. J. (1983). Nutrient requirement of the breeding ewe. In *Recent Advances in Animal Nutrition—1983*, pp. 143–162. London: Butterworths.

CHAPTER 20

The Nutritional Requirements for Egg Production

The egg with completed shell is the sole reproductive product of birds. Consequently, the avian embryo is enclosed in the egg, which contains all the nutrients for the production of a viable embryo, unlike the mammal in which a continuous supply of nutrients is available from the mother.

COMPOSITION OF THE EGG

Data on the relative distribution of the three basic parts of the egg (yolk, egg-white and shell) are given in Table 20.1, and data on the chemical composition of the egg and its fractions in Table 20.2.

Table 20.1 Relative composition of the egg

Components of the egg	Weight (g)	Percentage of whole egg
Shell	6.1	10.5
White	33.9	58.5
Yolk	18.0	31.0
Total	58.0	100.0

Table 20.2 Percentage composition of the egg and its components

Compound	Shell (membranes)	White	Yolk	Total egg (without shell)
Water	1.5	88.5	49.0	73.6
Protein	4.2	10.5	16.7	12.8
Lipid	0	0	31.6	11.8
Other organic compounds	0	0.5	1.1	1.0
Inorganic ions	94.3*	0.5	1.6	0.8

*Mostly calcium carbonate

The chemical composition of the hen's egg is almost constant apart from the composition of the lipid fraction, which is somewhat influenced by the composition of the fat ingested, and the concentrations of some trace elements and vitamins, the latter changing with their intake with the food.

The partition of protein between the fractions of the egg is as follows: the egg contains 12.8% protein on average, 41.9% as yolk protein, 53.6% as egg-white protein, 2.1% as membrane protein (in both shell membranes) and 2.4% as eggshell protein. The yolk contains nearly all the lipid material of the egg. The considerable difference in the water content between yolk and egg-white may be noted: the egg-white contains 88% water and the yolk 49%.

Composition of the Egg-yolk

The yellow egg-yolk which supplies the nutrients for the developing embryo is a complex mixture of water, lipids and protein and contains micronutrients such as minerals and vitamins (see Table 20.3).

Most of the lipids exist in lipoprotein form as lipovitellin and lipovitellinin, and these phosphate-rich compounds are often complexed with calcium and iron. Lipovitellin is composed of 11% phosphoprotein (phosvitin), 46% triglycerides, 23% phospholipids and 4% cholesterol. The phosphoproteins contain phosphoric acid esterified to the hydroxyl group of serine or threonine, e.g.:

$$\begin{array}{l} CH_2O-PO_3H_2 \\ | \\ CHNH_2 \\ | \\ COOH \end{array} \qquad \text{Phosphorylserine}$$

Most of the free proteins in yolk (livetins) are clearly identical to blood proteins such as serum albumin and serum globulin. Vitamins and trace minerals present in egg-yolk are of great importance in embryogenesis. The yellow colour typical of egg yolk is due to carotenoid pigments originating from the diet (see Chapter 11).

Table 20.3 Constituents of egg yolk (after Parkinson, 1966, p. 104)

		Percentage of egg-yolk solids
Proteins	Livetins	4–10
Phosphoproteins	Vitellin	4–15
	Vitellinin	8–9
	Phosvitin	5–6
Lipoproteins	Lipovitellin	16–18
	Lipovitellinin	12–13
Lipids	Triglycerides	46
	Phospholipids	20
	Sterols (mainly cholesterol)	3
Carbohydrates		2
Mineral constituents		2
Vitamins		Traces

Table 20.4 Composition of egg-white (after Smith, K. W., 1978, p. 84)

	Percentage of egg-white solids
Glycoproteins	87.4
Ovalbumin	56.8
Conalbumin	13.7
Ovomucoid	11.6
Ovomucin	3.15
Ovoglycoprotein	0.55
Ovomacroglobulin	0.55
Ovoinhibitors	0.10
Avidin	0.05
Flavoprotein	0.85
Protein	
Lysozyme	3.50
Unidentified proteins	9.10

Composition of Egg-white

Egg-white, sticky and viscous, consists almost exclusively of protein (1 part) and water (8 parts). Egg albumen is made up of a number of different proteins—mostly glycoproteins, which are macromolecules composed of polypeptide chains to which sugar molecules are attached (see Table 20.4). Glycoproteins are responsible for the gel-like qualities of egg-white.

The major egg-white protein, *ovalbumin*, is rich in essential amino acids, particularly methionine, and provides protein reserve for embryonic growth. The function of *flavoprotein* is the transfer of riboflavin to the embryo; *conalbumin* serves for the transfer of iron to the embryo. The carbohydrate-free protein *lysozyme* is an enzyme capable of dissolving some bacteria and presents a chemical defence mechanism against bacterial invasion through the eggshell. *Avidin* is able to combine with biotin and renders it unavailable as a vitamin (see Chapter 12). In this way it can act as an antibacterial agent. *Ovomucoid*, which contains high carbohydrate levels, is an enzyme inhibitor of proteases; in addition to ovomucoid, another stabilizer of egg-white is the element magnesium, which retards liquification of the gelatinous white and is related to the maintenance of egg quality in storage.

BIOSYNTHESIS OF THE COMPONENTS OF EGG-YOLK AND EGG-WHITE

Yolk proteins and lipids are formed in the liver under the influence of oestrogens; they are subsequently transported to the ovary and deposited in developing follicles. The rampant increase in plasma lipid levels associated with the onset of laying is striking. While hepatic lipogenesis in mature laying hens may be 10 times that in the livers of immature females, plasma-free fatty acids commonly increase 10-fold prior to egg laying. As soon as egg laying commences, these levels fall dramatically to levels slightly about basal. Such decreases represent the drain by yolk deposition in the face of accelerated hepatic lipogenesis. In addition, the fatty acid composition of the total lipids in liver and blood is altered in laying hens towards the composition of yolk fatty acids; the proportions of palmitic and oleic acids are increased and those of stearic and linoleic acids are decreased.

Unlike the yolk proteins, which are formed in the liver and deposited in the follicles, the proteins of egg-white are formed in the oviduct tissues (see Figure 20.1(a,b)). In the oviduct, egg-white and egg-shell are produced, a process lasting about 24 hours. When the yolk (oocyte) is released from the ovarian follicles, it passes into the oviduct. The shed yolk enclosed in its own vitellin membrane is picked up by the funnel-shaped mouth of the oviduct (infundibulum). Then the yolk is carried to the magnum, the first part of the oviduct, where it is surrounded by layers of albumen; the albumen is produced in the well-developed glandular elements of the magnum. Egg-white protein synthesis requires elevated ovarian steroid hormone levels

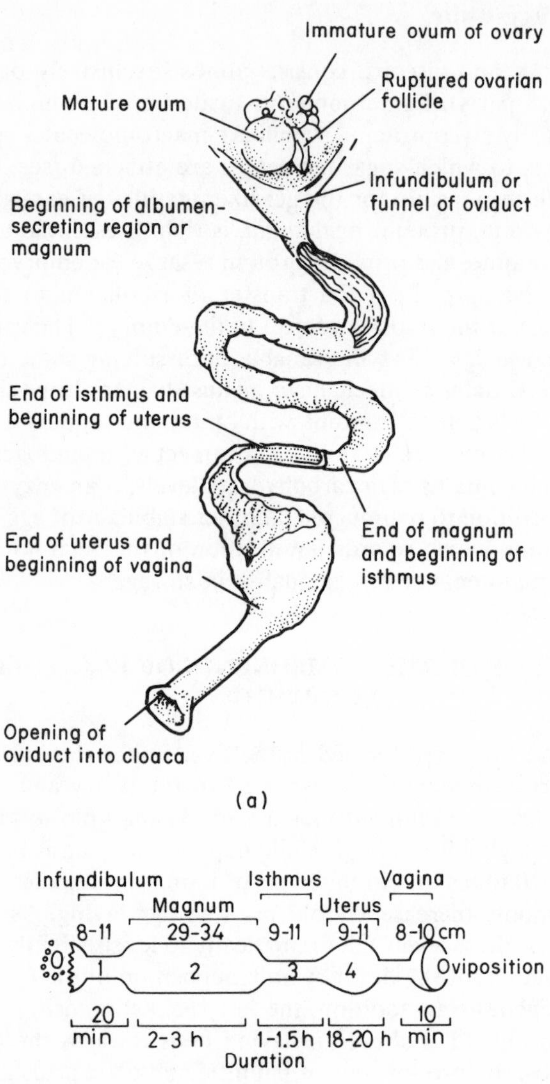

Figure 20.1. (a) Reproductive tract of the laying hen (after Sturkie, P. D. *Avian Physiology*, 3rd edn, 1976, p. 304). *Reproduced by permission of Springer Verlag.* (b) Schema of events leading to ovulation, oviductal passage and oviposition of the avian egg (after Hazelwood, R. L. In *Dynamic Biochemistry of Animal Production* (Riis, P. M., ed.), 1983, p. 412). The length of the avian structures and the average time spent by any one ovum are indicated, *Reproduced by permission of Elsevier Science Publishers B. V.*

in the tissues of the magnum. After 2–3 hours in the magnum, sequential peristaltic-like contractions move the egg with its thick albumen layer into the isthmus where during the next 1.5 hours some water is pumped into the egg-white and two shell membranes are formed (in tubular gland cells of the epithel of the isthmus) and laid down. These membranes are characterized by their high content of keratin (see Chapter 7) and hydroxyproline.

The yolk, with its coating of egg white, all enclosed in membranes, now passes to the shell gland (uterus). The egg remains in the shell gland for approximately 20 hours; here during the first 5 hours the egg 'plumps' due to the absorption of water and salts, and then during the last 15 hours the deposition of calcium carbonate takes place.

SHELL FORMATION

The beginning of shell formation occurs in the isthmus where small calcium carbonate crystals are attached to the surface of the outer shell membrane; an active process of calcification continues within the uterus, the small calcium carbonate crystals being used as seeding sites. External to the shell membrane is formed first a solid mammillary layer of calcium carbonate, then the more exterior, spongy pallisade layer, and finally the outermost surface crystal layer which will be coated with a proteinaceous cuticular coat.

Since the shell gland does not store significant amounts of calcium, the shell gland derives it from the circulating blood. During the 10 days before a pullet starts to lay, calcium levels in blood rise under the influence of oestrogen, from about 10 mg per 100 ml to about 25 mg. This rise being in the protein-bound calcium fraction can be attributed to the formation of a phosphoprotein complex (phosvitin; see above) which binds large quantities of calcium. Ionized calcium remains at 5–6 mg%, a value similar to that found in non-laying birds and mammals. As the shell formation proceeds in the hen the plasma level of total calcium decreases by about 20% during shell formation and remains so for an additional 10 hours after which it increases towards normal.

The transfer of calcium ions across the mucosa of the eggshell is thought to involve a carrier. A calcium-binding protein, CaBP, probably identical to intestinal CaBP, is present in the uterine mucosa of laying hens but not in immature birds. The exact role of CaBP in calcium translocation in the uterus has not yet been elucidated.

The rate of calcium turnover in the laying hen appears to be so large that the bird is incapable of absorbing sufficient calcium from the feed to cope with the needs for shell formation. Therefore it is necessary for it to lay down in the bones a readily mobilizable store of calcium before it reaches maturity and during its non-laying days so that on laying days it can add to the calcium absorbed from the food and thus produce a normal shell. As the pullet reaches sexual maturity, oestrogen released from the maturing ovary, acting in synergysm with the androgens, induces the formation of medullary bone in the narrow cavity, especially in the long bones of the skeleton. In the laying hen, medullary bone is in a dynamic state, being continuously deposited and broken down. Bone dissolution during shell formation may be stimulated by increased levels of parathyroid hormone invoked as a result of the depression of calcium in the blood (see Chapters 10 and 11).

All of the calcium secreted into the uterine lumen during shell formation is derived from the blood; the blood calcium is obtained from feed and from the bones. The relative importance of these two sources of eggshell calcium depends on the concentration of calcium in the diet. If the dietary calcium level is 3.6% (usual level in laying rations), 80% of the eggshell calcium is derived from the food and 20% from the bones. If the dietary concentration is 1.9%, bone supplies 30–40% of the shell calcium, and on calcium-deficient diets the skeleton may be the principal source. A low dietary intake of calcium in the laying fowl results in a reduction of egg yield and shell thickness, presumably to reduce the calcium loss from the body. The ability of the egg to withstand handling and transport depends on the thickness and strength of the eggshell and many methods have been devised to measure eggshell quality.

Carbonate Deposition

The eggshell contains about 60% carbonate. Carbon dioxide exists in the blood as dissolved gas (CO_2), as carbonic acid (H_2CO_3) and as bicarbonate (HCO_3^-), whilst the actual carbonate concentration (CO_3^{2-}) is negligible. The deposition of calcium carbonate in the forming shell is probably maintained through continuous secretion of bicarbonate from the shell gland rather than by secretion of carbonate within the shell gland mucosa. Carbon dioxide is hydrated under the influence of the zinc-containing enzyme carbonic anhydrase (see Chapter 10):

$$CO_2 + H_2O \rightleftharpoons H_2CO_3$$

Carbonic acid, dissolved in water, dissociates inside the mucosa cells as follows:

$$H_2CO_3 \longrightarrow H^+ + HCO_3^-$$

Bicarbonate is secreted into the lumen and equilibrates with hydrogen ions and carbonate:

$$HCO_3^- \longrightarrow H^+ + CO_3^{2-}$$

Carbonate and calcium ions are then precipitated on the shell membranes (see above).

Eggshell calcification results in the production of hydrogen ions: the formation of bicarbonate from carbonic acid is accompanied by release of hydrogen ions which must be buffered in the shell gland fluid and cells or released into the blood. Metabolic acidosis can be counteracted by increasing the respiratory rate and urinary acid excretion. Bone phosphate may also play a role as H^+ acceptor because it is mobilized as PO_4^{3-} and excreted as HPO_4^{2-} or $H_2PO_4^-$. This may contribute to the increased phosphorus content in the droppings of laying hens. Failure to buffer the hydrogen ions formed during calcification may interfere with eggshell formation. Environmental factors impairing eggshell quality are high ambient temperature reducing carbon dioxide solubility in blood leading to loss of carbon dioxide, and pesticide residues in feedstuffs inhibiting carbonic anhydrase.

NUTRIENT REQUIREMENTS OF LAYING HENS

The turnover of material during reproduction in poultry is enormous. During one year a hen may lay about 280 eggs, a total of 16 kg, a mass about 10 times its own body weight, including about 2 kg of protein, 1.7 kg of fat and 0.5 kg of calcium. This information may be used as the basis for factorial calculation of the nutrient requirements of layers. Besides the large nutritive demands for the formation of eggs as such, there are additional requirements for the production of eggs that will hatch and yield viable chicks, particularly concerning their requirements for essential nutrients, vitamins and trace elements.

Energy Requirements

The food ingested has to provide energy for maintenance, egg production and gain of body weight. A hen weighing 2 kg has a fasting metabolism of 0.36 MJ/kg $W^{0.73}$ or 0.60 MJ/day. The requirement for 70% egg production rate (= number of eggs/number of days; 56 g = average weight of eggs) may be calculated as follows. The energy value of 1 kg of egg substance = 6.7 MJ, that of 1 egg (56 g) = 0.375 MJ. Metabolizable energy (ME) is utilized for maintenance and egg production with an average efficiency of 80%. The requirement of ME for both these functions is 1.08 MJ/day and for 1 g body gain per day is 0.014 MJ.

Maintenance energy requirements are influenced by environmental temperature and by feathering. The effect of temperature is greater in unfeathered birds than in well-feathered birds, and maintenance energy requirements change with temperature by approximately 9.2 and 8.4 kJ/kg per day per °C for white and brown egg strains, respectively.

In agricultural practice, laying hens, like growing birds (see Chapter 18), are usually fed to appetite; their voluntary intake of ME is roughly commensurate with the calculated energy requirements (given above). The usual ME concentration in laying mash is 10–12 MJ/kg (i.e. 11.5–13.5 MJ/kg/DM). An increase or decrease of 1% in energy concentration (greater than 12 MJ or below 10 MJ) causes a corresponding increase or decrease in energy intake of only 0.5%. Use of diets containing less than 10 MJ may lead to a reduction in egg production; diets containing more than 12 MJ may increase body fat deposition, but not the number of eggs laid.

Protein Requirements

Protein deposition is a key element in the productivity of growing and laying birds; in birds receiving adequate energy the major determinant of protein deposition is the dietary supply of protein and essential amino acids. The requirements of laying hens for amino acids are comparatively high, as indicated by the fact that egg protein stands at the top of all protein sources in terms of amino acid composition (see Chapter 7 and Table 18.10). Protein and amino acid requirements of laying hens, like those of growing birds, are usually expressed as concentrations in the laying mash; these dietary concentrations depend primarily on the energy content of the mash. The ratio of energy to protein is important in the efficiency of utilization of the diet for laying hens and for growing birds as well (see Chapter 18). Metabolizable energy to protein ratios calculated by dividing the ME in kcal/kg or kJ/kg by percentage of protein have been assessed in relation to optimum performance.

The levels of protein and essential amino acids that ensure maximal productivity of laying hens can be investigated in the same two ways used for growing chicks—the empirical and causal ones (see Chapter 18).

Empirical Investigations

Numerous early studies on the protein requirement of laying hens indicated that, for an optimum production rate of about 75–80%, a mature white Leghorn hen required 17.5–18.5 g of protein per day. Because of the usual decrease in laying with age, higher protein levels are required during the earlier production period when higher rates of lay are prevalent than in the later period with declining production rate (phase feeding). The decline in protein requirement is not parallel to the decline in egg production, but less, as with the decrease in laying, egg weight increases with age.

For the determination of the requirements of essential amino acids, graded levels of the respective amino acid are added to a basal ration deficient in that amino acid; the response of groups of laying hens to different levels of amino acid intake is examined and response curves are fitted until a plateau is reached representing maximum output.

Causal Investigations

Factorial calculation of protein requirements of laying hens may be made by summing the requirements needed for maintenance, egg formation and body and feather gain (over the laying period). Requirements for maintenance and body gain are calculated as described in the section of protein requirements for growing chicks (see Chapter 18). Requirements for egg production are estimated by multiplying the protein content of eggs by an efficiency factor; values for this factor are based on the results of feeding experiments and defined as the ratio of protein amino acids in the egg to their amounts fed for egg production. Such values given in the literature vary between 0.55 and 0.85. Nowadays, a factor of 0.85 has been accepted for practical purposes. Thus for lysine, which is present in the whole egg at a level of 7.9 g/kg, the requirement for dietary available lysine is 7.9/0.85 = 9.5 g/kg of egg. The methionine requirements of laying hens were summarized by Combs by means of a factorial equation:

$$\text{Methionine requirement (mg/day)} = 5E + 50\,\text{BW} + 6.2\,\Delta\text{BW}$$

where E = production (g/day), BW = body weight, ΔBW = body weight gain.

Hurwitz and Bornstein (1973, 1977; see Smith, 1978) developed an alternative procedure for the factorial calculation of the protein and amino acid requirements of laying hens. These authors sum the requirements for maintenance, egg formation and body gain and use 0.85 as the coefficient for the conversion of fed amino acids into egg or body protein. The amino acid composition of the various fractions of egg protein are listed in a table in order to calculate the amounts of amino acids required for the synthesis of the egg proteins. Yolk and egg albumin, the major protein of egg-white, are synthesized continuously during 10 and 4 hours, respectively, so that the amino acids required for their synthesis can be derived directly from feed. However, the ovoglycoproteins and membrane proteins, forming together 14% of all egg proteins, are synthesized within 1 hour, which must involve the breakdown of tissue proteins in order to derive the necessary amino acids from body stores. Methionine is the limiting amino acid in ovomucoid. The methionine content of tissue proteins accounts only for half its content in ovomucoid. The formation of 1 g of ovomucoid requires the breakdown of 2 g of tissue proteins. Thus for the synthesis of 1 g of ovomucoid it is necessary to supply the quantities of amino acids present in 2 g of body tissues. Good agreement has been obtained for the egg output calculated according to the Hurwitz–Bornstein procedure with that found experimentally.

When protein intake is reduced below the required level or a particular amino acid is limiting, the amino acid composition of the eggs remains unchanged, but the total amount of eggs laid and feed efficiency decline subsequently. A considerable increase in the efficiency of laying diets can be achieved by supplementation with essential amino acids, as shown in experiments carried out by Scott in the early sixties. A rise of the methionine content of the usual maize–soybean diet from 0.29% to 0.33% (the calculated methionine requirement) improved feed efficiency, increased egg size and raised egg production rate (from 68% to 74%). However, dietary excesses of one or more of the essential amino acids can have detrimental effects on performance.

Supply of Minerals

Due to the large demand for calcium for eggshell formation, the calcium requirement of laying hens is many times that of non-laying hens. The formation of an eggshell containing 2.0–2.2 g of calcium requires an intake of 3.5–4.0 g of calcium since 50–60% of the ingested calcium is absorbed.

Under conditions of dietary adequacy the hen adjusts to its calcium needs by changing its food intake of the calcium-containing mash. Voluntary food consumption is about 25% greater on egg-forming days than on non-egg-forming days. But this effect is diminished when calcium is offered separately from the food, e.g. in the form of oyster shell; then the calcium supplement is consumed preferentially at times when it is required. Adaptation of intake of the mash to the calcium needs is not a desirable practice; thus provision of a separate calcium source (oyster shell or calcium grit) is now widespread. The effects of feeding calcium-deficient diets to hens were mentioned before.

The eggshell contains only small amounts of phosphorus: 20 mg compared with 120 mg in the egg content. These needs and those for many other functions can be met at very low dietary levels (about 0.3 g phosphorus per day). Excess levels of phosphorus lead to a reduction in shell thickness since apparently phosphate ions exert an inhibitory action on calcium phosphate precipitation.

The maintenance of acid–base balance depending mainly on the balance between sodium, potassium and chloride is important for the provision of bicarbonate needed for eggshell formation (see above). On the basis of the estimated requirements of sodium, potassium and chloride, supplementation of conventional layer rations with sodium chloride, but not with potassium chloride, is essential for optimum egg production and beneficial for counteracting cannibalism and feather picking (see Chapter 10).

The importance of iron, manganese, zinc and iodine in the nutrition of laying hens is noted in Chapter 10. Manganese and zinc deficiencies in the diet adversely affect egg productivity, hatchability and the viability of chicks after hatching, and a deficiency of manganese also increases the incidence of thin-shelled eggs.

Supply of Vitamins and Other Minor Nutrients

Vitamins must be supplied in adequate quantities not only to ensure that egg production is not disturbed (in the case of vitamin D also calcification of eggshell), but also that hatchability is good and that the chicks produced are healthy. For vitamins A, D and most of the B vitamins the quantities needed for maximum hatchability are considerably greater than those for egg production alone.

Whereas the content of the major nutrients and most minerals in the egg is not influenced by the diet, there are considerable effects in the case of manganese and iodine and vitamins, particularly A, D and some of the B group. An increase in the content of many vitamins in the egg is beneficial for hatchability and viability of the chick and improves the value of the eggs for human consumption.

The hen requires about 1% of linoleic acid in the diet to meet its needs for most production purposes. A deficiency of linoleic acid results in small egg size, depressed egg production and early mortality of the embryo during incubation. Conventional sources of linoleic acid are maize grains or soybean soapstock (see Chapter 6).

For the use of *carotenoids* for the pigmentation of egg yolks see Chapter 11.

BIBLIOGRAPHY

Formation and composition of the egg

Books

Bell, D. J., and Freeman, B. M. (1971). *Physiology and Biochemistry of the Domestic Fowl*, Vol. 3, Chapters 50–58. London and New York: Academic Press.

Carter, T. C. (1968). *Egg Quality. A Study of the Hen's Egg.* Edinburgh: Oliver and Boyd.

Freeman, B. M., and Lake, P. E. (1972). *Egg Formation and Production. Poultry Science Symposium*, No. 8. Edinburgh: British Poultry Science Ltd.

Articles

Gilbert, A. B. (1980). Controlling factors in the synthesis of egg proteins. In *Protein Deposition in Animals* (P. J. Buttery and D. B. Lindsay, eds), pp. 85–100. London: Butterworths.

Gilbert, A. B., and Pearson, R. A. (1983). Egg formation in poultry. In *Nutritional Physiology of Farm Animals* (J. A. F. Rook and P. C. Thomas, eds), pp. 243–260. London and New York: Longman.

Hazelwood, R. L. (1983). Adaptation of metabolism to various conditions: egg production in fowl. In *Dynamic Biochemistry of Animal Production* (P. M. Riis, ed.), pp. 389–430. Amsterdam: Elsevier.

Parkinson, T. L. (1966). The chemical composition of eggs. *J. Sci. Food Agr.*, **17**, 101–111.

Sturkie, P. D. (1976). Reproduction in the female and egg formation. In *Avian Physiology* (P. D. Sturkie, ed.), pp. 302–330. New York: Springer.

Nutrient requirements for laying hens

Books and articles dealing also with requirements of energy and nutrients for laying hens are listed in the Bibliography for Chapters 16 and 18. Literature concerning the supply of minerals for layers is given in the Bibliography for Chapter 10.

Bornstein, S., Hurwitz, S., and Lev, Y. (1979). The amino acid and energy requirements of broiler breeder hens. *Poult. Sci.*, **58**, 104–116.

Cherry, J. A. (1979). Adaptation in food intake after changes in dietary energy. In *Food Intake Regulation in Poultry* (K. W. Boorman and B. M. Freeman, eds), pp 77–86. Edinburgh: British Poultry Science Ltd.

Fisher, C. (1983). Egg production. In *Nutritional Physiology of Farm Animals* (J. A. F. Rook and P. C. Thomas, eds), pp. 623–638. London and New York: Longman.

Guilbert, A. B. *et al.* (1981). The egg laying response of the domestic hen to variation in dietary calcium. *Br. Poult. Sci.*, **22**, 537–548.

Hurwitz, S., and Bornstein, S. (1977). The protein and amino acid requirements of laying hens. *Poult. Sci.*, **56**, 969–978.

Menge, H. (1968). Linoleic acid requirement of the hen for reproduction. *J. Nutr.*, **95**, 578–582.

Pearson, R. A. (1982). Influence of nutritional factors on hatchability. *Recent Advances in Animal Nutrition—1982*, pp. 141–156. London: Butterworths.

Smith, K. W. (1978). The amino acid requirement of laying hens. *World Poult. Sci.*, **34**, 81–96 and 129–136.

Snetsinger, D. C., and Zimmerman, R. A. (1974). Limiting the energy intake of laying hens. In *Energy Requirements of Poultry* (T. R. Morris and B. M. Freeman, eds), pp. 185–189. Edinburgh: British Poultry Science Ltd.

CHAPTER 21

The Nutritional Requirements for Milk Production

THE BIOSYNTHESIS OF MILK

The weight of the mammary gland is only 5–7% that of the body and the gland is noted for every intensive metabolic activity. Due to selection and the improvement of management, especially that of nutrition, in a single lactation the dairy cow may produce an amount of milk sufficient for raising approximately 50 calves. A cow weighing 600 kg and having an annual production of 10 000 kg secretes in the milk an amount of dry matter which may exceed eight times that in her body.

The milk is produced in secretory cells arranged in *alveoli* (or acini) and is drained through a system of arborizing ducts towards the body surface; in the cow, ewe and goat these form a well-defined organ referred to as the *mammary gland* or udder. During sexual maturation the mammary gland develops under the influence of oestrogen, progesterone, adrenal corticoids, prolactin, growth hormone and possibly thyroxine; during pregnancy, hormones produced by the placenta, e.g. placental lactogen, reinforce the effects of ovarian and hypophyseal hormones. Only following parturition is milk secreted under the influence of adrenal corticoids, thyroxine, growth hormones, prolactin and insulin. The cessation of progesterone secretion at the time of parturition is a necessary condition for milk production since progesterone inhibits the secretion of hormones from the anterior lobe of the pituitary gland.

After milk production and secretion have begun they are also maintained by suckling and milking which stimulate the secretion of prolactin and other hormones active in the mechanism of milk secretion. Suckling and milking and other stimuli act through neuroendocrine mechanisms to promote milk production. During the first days of parturition the composition of milk is rather different from that produced in later lactation and is called *colostrum.* The composition of colostrum and its significance to the nutrition and survival of young is discussed in Chapter 18.

A pair of major arteries maintain the blood supply to the mammary gland. The production of 1 litre of milk requires the passage of approximately 500 litres of blood through the mammary gland. During lactation the mammary arteries are enlarged and blood flow through the mammary gland is very rapid; in a high producing cow it may reach 20 litres/min. Enzyme systems in the mammary gland convert substances carried in the blood to those secreted in the milk. Most of the precursors of milk carried by the blood to the udder undergo substantial chemical changes; only a few substances pass into the milk unchanged.

Different methods have been developed to study the biosynthesis of milk constituents in the body and especially in the mammary gland. The simplest method to study mammary gland metabolism is to determine *extraction rate* by measuring the change in the concentration of a substance entering and leaving the udder (arteriovenous difference techniques). A decrease in concentration would indicate that the udder has retained part of that substance. If, for example, the level of glucose, expressed as mg per 100 ml, is 46 in the mammary artery and 31 in the vein, one would assume that the udder has used 46 − 31 = 15 mg per 100 ml. The extraction rate of glucose would be 15/46 = 0.326 or 32.6%. The measurement of the rate of blood flow through the udder permits the calculation of the absolute amount of substances used by the udder per unit of time if the extraction rate is also measured. The amount of a substance used per unit of time is its *uptake.* Udders have been removed from slaughtered animals and perfused in isolation with blood or semisynthetic perfusates to study milk biosynthesis. Bovine udders have been kept functional for up to 8 hours and goat udders for up to 24 hours. The injection of compounds labelled by radioactive isotopes (tracers) is a powerful tool in studying milk biosynthesis. By this method chemical pathways, i.e. precursor–product relationships, can be established relatively easily in great detail. For example, an injection of a labelled precursor can be made directly into one quarter of the udder or elsewhere in the body. A higher activity of the product in the injected quarter indicates that synthesis occurs in the udder, whereas an even distribution of the label in all four quarters is evidence that synthesis occurs elsewhere in the body. Milk biosynthesis is also often studied in mammary slices, explants, and homogenates including particulate and cell-free preparations. Mammary arterial concentrations and extraction rates of several metabolites are presented in Table 21.1. Large differences in extraction rates of several metabolites are obvious. The most important pathways of milk biosynthesis will now be described, they are briefly summarized in Figure 21.1.

Table 21.1 Arterial concentrations of milk precursors and their extraction rates in the udder of goats (according to Linzell, 1974, adapted by Rook and Thomas, 1983)

Substance	Arterial concentration (mg/litre)	Extraction (%)
Blood		
Oxygen O_2 (vol./litre)	119	45
Glucose	445	33
Acetate	89	63
Lactate	67	30
Plasma		
3-Hydroxybutyrate	58	57
Triglycerides	219	40
Long-chain free fatty acids	87	3
Phospolipids	1600	4
Sterols	1040	0
Glycerol	3.4	7
Essential amino acids* (range)	2.7–27.9	37–72
Non-essential amino acids†	2.8–68.5	0–58

* In decreasing order of extraction rate: Met, Phe, Leu, Thr, Lys, Arg, Ile, His, Val.

† In decreasing order of extraction rate: Glu, Tyr, Asp-NH_2, Pro, Orn, Asp, Ala, Glu-NH_2, Gly, Cit, Ser.

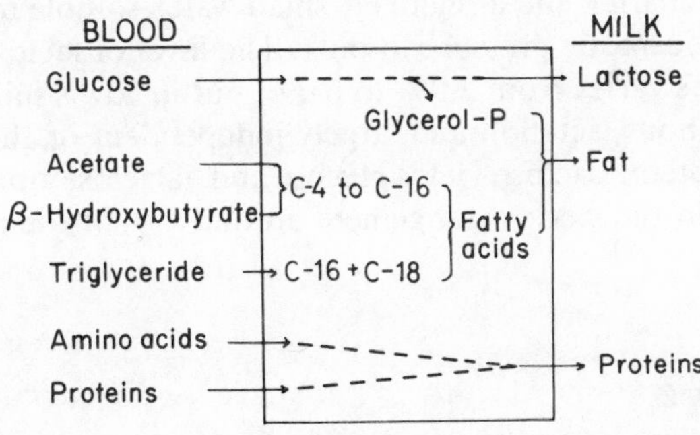

Figure 21.1 An outline of ruminant mammary metabolism.

Milk Proteins

Ninety-five per cent of the nitrogen in milk is in true proteins. The rest is in simple compounds which pass directly from blood to milk; urea, creatine and ammonia. Casein, which is not entirely uniform, represents the bulk of the protein in the milk of all non-human species: 82–86% in ruminants, 52–80% in monogastric species and 40% in humans. Only in colostrum do the levels of albumin and globulin exceed that of casein, though the absolute level of the latter in colostrum is higher than in milk produced later in lactation (see Chapter 18). Casein contains phosphate bound to the hydroxyamino residues (serine and threonine), and calcium. Although milk protein as a whole is rich in essential amino acids, casein is relatively poor in methionine.

About 90% of the bulk of the milk proteins in ruminants is synthesized in the udder from amino acids extracted from blood; these are casein, lactalbumin and lactoglobulin. The rest, including albumin, pseudoglobulin and *κ-casein* pass directly from blood to milk. Immunoglobulins, which act as antibodies and are abundant in colostrum, also pass directly from blood to milk.

The extraction of the essential amino acids in the udder (see Table 21.1) is about double that of the non-essential ones. Evidently, the udder synthesizes part of the non-essential amino acids from glucose, volatile fatty acids and essential amino acids (see Chapter 7). The energy required for the formation of peptide bonds in the milk protein is derived from ATP. Five mol of ATP are required for every peptide bond. ATP is produced in the udder mainly by the oxidation of glucose (see Chapter 14). Seven g of glucose are required to supply the energy to form 30 g of protein, which is approximately the amount found in 1 litre of milk (see Chapter 8).*

Lactose

Lactose, a disaccharide and a relatively small water-soluble molecule, produces most of the osmotic pressure in milk. The level of lactose in the milk of different species varies from 2.0% to 6.6%, but in cow's milk it is practically fixed throughout lactation and largely independent of the diet, unlike that of fat and protein. Lactose yields glucose and galactose upon hydrolysis; these differ only in the steric arrangement around C-4 of the molecule (see Chapter 4).

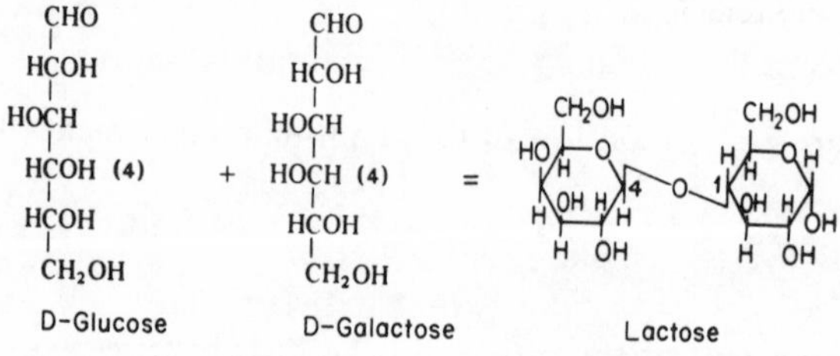

CHO–HCOH–HOCH–HCOH (4)–HCOH–CH_2OH (D-Glucose) + CHO–HCOH–HOCH–HOCH (4)–HCOH–CH_2OH (D-Galactose) = Lactose

* In the udder ATP is mainly produced by aerobic oxidation of glucose through the pentose phosphate pathway and in the rumen by anaerobic glycolysis (see Chapters 5 and 8).

Studies with ^{14}C indicate that blood glucose is the major precursor of both moieties of the lactose molecule. Glycerol is normally a hydrolytic product of triglycerides; studies with [^{14}C]glycerol indicate that it is a precursor of glucose, and hence of galactose and lactose. Because of low circulating levels it is of minor importance as a glucose source.

The conversion of glucose into galactose is catalysed by the enzyme *uridine diphosphate glucose epimerase* and occurs in four steps:

$$\text{Glucose} + \text{ATP} \longrightarrow \text{glucose-6 Ⓟ} \quad (1)$$

$$\text{Glucose-6 Ⓟ} \longrightarrow \text{glucose-1 Ⓟ} \quad (2)$$

$$\underset{}{\text{Glucose-1-P}} + \underset{\text{Uridine triphosphate}}{\text{UTP}} \longrightarrow \text{UDP-glucose} + \underset{\text{pyrophosphate}}{\text{P-P}} \quad (3)$$

$$\text{UDP-Glucose} \xrightarrow[\text{Epimerase}]{} \text{UDP-galactose} \quad (4)$$

The production of lactose is catalysed by the enzyme *galactosyltransferase* in a further step:

$$\text{UDP-galactose} + \text{glucose} \xrightarrow[\text{Galactosyltransferase}]{} \text{lactose} + \text{UDP} \quad (5)$$

Step (5) in the above sequence occurs only in the presence of another milk protein, *α-lactalbumin*. Galactosyltransferase and α-lactalbumin associate reversibly in a 1 : 1 molar ratio to form the *lactose synthetase* complex. This is a rare example of two proteins catalysing together one specific reaction.

Sixty per cent or more of the glucose extracted by the udder is used for lactose synthesis. Some of the rest may serve as a precursor of glycerol for the synthesis of fat, but most of it is oxidized and provides energy for the synthesis and secretion of milk constituents. Because of the high extraction rate of glucose in the udder its level in mammary venous blood is considerably lower than in other veins.

Milk Fat

Unlike in blood, where most of the lipids are bound to phosphorus (see Table 21.1), in milk 98% of the lipid is triglycerides, i.e. neutral fat. The source and synthesis of milk fat have been thoroughly studied. Fat, being insoluble in water, does not contribute to the osmotic pressure of milk and is present in milk in an emulsion. The fat content of milk is rather variable and so is its fatty acid composition.

The milk fat of ruminants is characterized by the high content of butyric acid (9–13 mol%) and other short-chain fatty acids in the range of 6–14 carbon atoms, which are practically absent in the milk fat of other species, and by the relatively low content of the polyunsaturated C_{18} essential fatty acids (see Chapter 6). In ruminants, most of the unsaturated fatty acids in the diet undergo hydrogenation in the rumen (see Chapter 6); thus, both body and milk fat of ruminants are highly saturated.

Fatty Acid Synthesis

The fatty acids in milk have two main sources. (1) One of these sources is triglyceride circulating in blood as lipoprotein (see Chapter 6) derived from dietary fat or, in negative energy balance as is often the case, from body fat as well. Lipoprotein is a source of milk fatty acids in all mammalian species. (2) In herbivores, especially in ruminants, milk fatty acids are also synthesized in the udder from fermentation products: acetate and 3-hydroxybutyrate. In ruminants acetate is directly absorbed from the rumen and 3-hydroxybutyrate is produced in the rumen wall from butyric acid (see Chapter 5). Some acetate is of endogenous origin.

There is good evidence for the direct passage of triglycerides from blood lipoproteins to milk. Nevertheless, during the passage of the triglycerides into the secretory tissue, fatty acid and glycerol are split by the enzyme *lipoprotein lipase* bound to the endothelial surface of the capillary wall. Sophisticated studies on goats, comparing the use of either [^{3}H]glycerol and [^{14}C]palmitic acid, or doubly labelled tri[^{14}C]palmityl-[^{3}H]glyceride produced virtually identical patterns of incorporation into milk fat. Following passage through the capillary walls, glycerol and fatty acids recombine and triglyceride is secreted into milk. Studies with labelled acetate and 3-hydroxybutyrate indicate that only fatty acids containing 16 carbon atoms or less may be synthesized wholly or partly from these precursors. It appears that a four-carbon fragment derived from 3-hydroxybutyrate or possibly longer fragments derived from dietary fat are elongated; elongation always occurs by the addition of two-carbon fragments derived from acetate. It should be noted that the production of milk fatty acids from triglycerides, acetate and 3-hydroxybutyrate is consistent with their high extraction rate in the udder (see Table 21.1). The saturated C_{18} acids (stearic acid) are derived exclusively from plasma lipids.

On the average about 50% by weight of the fatty acids in milk are derived from blood lipoproteins which in turn originate from adipose tissue or from the diet. A small portion of stearic (18 : 0) and palmitic (16 : 0) acids are desaturated in the udder to oleic (18 : 1) and palmitoleic (16 : 1) acids, respectively. Fatty acids having fewer than 16 carbons are not desaturated. The very small amounts of diene and triene acids present in milk are obtained only from the diet since they cannot be synthesized in the body.

The biosynthesis of fatty acids from acetate occurs as follows: first, acetate is activated in the udder and converted to acetyl-CoA. By the addition of the enzyme *acetyl-CoA carboxylase* malonyl-CoA is produced according to the following reaction:

$$CH_3.CO.CoA + ATP + HCO_3^- \xrightarrow[\text{Mg}^{2+},\ \text{Mn}^{2+}]{\text{Biotin}} {}^-OOC.CH_2.CO.CoA + ADP + P_i$$

In the second step acetyl-CoA (or butyryl-CoA or hexanoyl-CoA) is elongated by a *fatty acid synthetase*; the overall summary of this step is:

$$n\,\text{malonyl-CoA} + \text{acetyl-CoA} + 2n\,(NADPH + H^+) \longrightarrow CH_3.(CH_2)_{2n}.CO.CoA + (n + 1)CoA + 2n\,NADP^+ + n\,CO_2 + (n - 1)\,H_2O$$

The chain length of the fatty acids produced varies between 4 and 16 carbons; thus, in the above reaction, n may be any (whole) number from 1 to 7. The actual length of the chain is determined by species and enzyme specificity. As indicated above, 2 mol of NADPH are required for every mol of malonyl-CoA. The udder obtains NADPH (and ATP) by the oxidation of glucose through the pentose pathway (see Chapters 4 and 6).

As mentioned in Chapter 5, ruminants absorb little glucose; they produce most of it biosynthetically and must use it economically. Accordingly, in contrast to monogastric species which use glucose in the udder to produce NADPH, ATP and fatty acids via acetyl-CoA, in the ruminant there is no enzyme system in the udder to convert glucose into fatty acids. In fact, in the udder of the sow, milk fatty acids up to C_{18} are synthesized from blood glucose and acetate, or also obtained from plasma lipids.

Triglyceride Structure and Synthesis

Thereare 20 major fatty acids in bovine milk and more than 400 different fatty acids have been identified in milk lipids; therefore a very large variety of different triglycerides in milk fat is possible. Nevertheless, fatty acids are not randomly esterified among the three positions on the glycerol molecule. A list of the major fatty acids in bovine milk fat (96% by weight) and their distribution in the triglyceride molecule appear in Table 21.2. The figures indicate that 1,2-dipalmityl-3-butyryl glyceride is the most frequent triglyceride. The most common pathway for synthesis of triglycerides in mammary tissues is the glycerol-phosphate pathway described in Chapter 6. The major fatty acid entering position 3 of the glyceride is butyric acid. Thus the scarcity of short-chain fatty acids may limit the rate of milk fat production. The source of glycerol for milk fat synthesis in the udder is the hydrolysis of blood lipids (lipoproteins) and glucose in approximately equal parts.

Table 21.2 The major fatty acids of bovine milk and their positional distribution in the triglyceride molecule (after Mepham, 1983, p. 75)

Fatty acid	Wt% of total	Mol% of total	Mol% in position		
			1	2	3
4:0	3.3	8.49	–	–	8.49
6:0	1.6	3.62	–	0.24	3.38
10:0	3.0	2.26	0.39	0.61	1.26
12:0	3.1	6.10	2.55	3.23	0.31
14:0	9.5	9.44	2.73	4.91	1.80
16:0	26.3	23.24	11.01	10.48	1.74
16:1	2.3	2.06	0.74	0.95	0.37
18:0	14.6	11.63	5.71	5.26	0.66
18:1	29.8	23.92	9.97	6.27	7.68
18:2	2.4	1.95	0.44	0.91	0.60

Phospholipids and Cholesterol

Phospholipids are minor but essential components of the milk fat globule. They appear to be synthesized in the mammary gland. The cholesterol of milk lipids is partly synthesized in the udder and partly obtained from blood.

Inorganic Constituents of Milk

The inorganic constituents of bovine milk are listed in Tables 18.6 and 21.8. These pass into milk selectively. The calcium and phosphorus content of milk exceeds that of serum by an order of magnitude (calcium 13 times and phosphorus 10 times), but in milk both elements are mainly bound to casein. Milk differs from serum in ionic composition and is more like intracellular fluid; it contains five times as much potassium and three times as much magnesium, but only one-seventh as much sodium and one-third as much chloride as serum (see Chapter 10).

Under normal condition the levels of essential trace elements in milk vary considerably. If the diet is deficient in a trace element its level in milk is usually depressed. On the other hand, the effect of an excess in the diet varies greatly among elements. Copper, iron and zinc levels in milk respond poorly to dietary supplementation, those of manganese and cobalt respond more and the level of iodine in milk may reach levels which render milk unfit for consumption. However, the mammary gland is able to block the entrance of toxic levels of the essential elements selenium and fluorine and the non-essential cadmium and mercury.

Colostrum may contain twice the amount of minerals as milk (see Chapter 18). Potassium is lower in colostrum than in milk, but calcium, magnesium, phosphorus and chlorine are higher. The level of iron in colostrum may exceed that in milk by a factor up to 15.

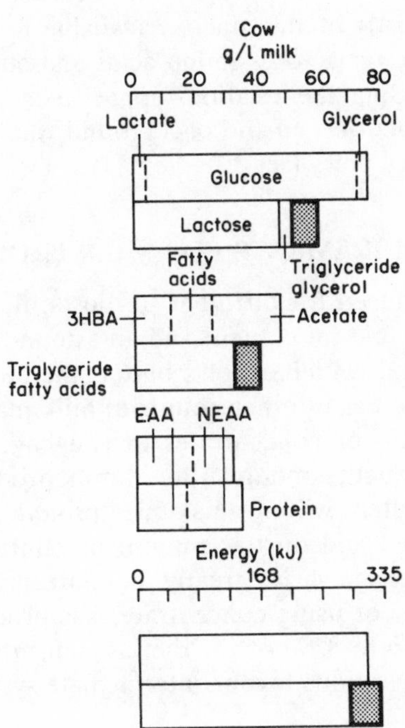

Figure 21.2. The utilization of precursors for milk production in the cow (after Rook and Thomas, 1983, p. 345). *Reproduced by permission of the Longman Group.* The upper areas represent the extraction rate from the blood, the lower areas the uptake by the mammary gland (EAA = essential amino acids; NEAA = non-essential amino acids; 3 HBA = 3-hydroxybutrate; stippled area = oxidized).

Summary

From the blood passing through the lactating udder, one- to two-thirds of all the amino acids, glucose, lactate, acetate, 3-hydroxybutyrate and triglycerides are extracted. The only exception are a few non-essential amino acids (evidently readily synthesized in the udder). The extracted substances are the major precursors of milk protein, lactose and fat. Data in Figure 21.2 indicate the high efficiency of milk biosynthesis. About 80% of the energy of the extracted organic substances is recovered in milk and only 20% is oxidized. Of the oxidized portion 30–50% is derived from glucose, 20–30% from acetate and the rest from long-chain fatty acids and amino acids.

In contrast to the ruminant animal as a whole which meets only 7–10% of its energy requirement from glucose, the contribution of glucose to the oxidative metabolism of the udder is, as mentioned, much higher (30–

50%). The contribution of acetate accounts for 20–30% of the oxidative metabolism in the whole ruminant animal and in the udder as well.

The importance of glucose supply to the lactating animal cannot be overemphasized; two-thirds of the glucose available to the animal is used by the udder as a source of lactose, amino acids and energy. This is rather important in understanding the relationship between the composition of the diet and milk production and the background for ketosis—a common metabolic disease of dairy cows (see Chapter 24).

ENERGY REQUIREMENTS FOR MILK PRODUCTION

Worldwide, milk is produced for different purposes: fluid milk for human consumption with a variable fat content and an astounding variety of dairy products. In New Zealand, which exports butter, the fat content of milk is of great importance, whereas in other countries milk mainly supplies high-quality protein. The ratio of concentrates to roughage in the diet has a significant effect on the quality and quantity of milk produced. A diet which will maximize the fat content will not maximize protein and vice versa. Depending on available land and on the amount of rainfall, the relative use of concentrates and roughage varies greatly. In contrast to New Zealand, in arid regions the necessity of using concentrates has produced viable intensive systems supplying up to 85% of the diet as concentrates. In temperate regions economic considerations favour intermediate systems.

Efficiency of Milk Production

For the economic production of milk the quantity of energy available for milk production is very important. Much more feed is required for the supply of energy needs for the dairy cow than for all other nutrients combined.

After calving, a cow requires energy for maintenance and for milk production. Requirements for milk production are expressed per unit of milk of a given fat content, since its energy value follows the fat percentage closely (see later). When pregnant, the cow needs energy for the foetus and a young cow also needs some for growth. The latter two needs have been discussed in Chapters 18 and 19.

The level and efficiency of milk production are not only influenced by the genetic ability of the animal and current nutritional status, but also by the plan of nutrition the animal received during growth. Cattle grown on high levels produce less milk. Underfed heifers generally perform as well as those grown on recommended allowances.

Economically and biologically it is convenient to consider the gross efficiency of milk production. In milk production, efficiency is the ratio between the energy in milk (output) and that in the feed consumed (input) during a given period, assuming no change in body reserves. It is convenient to express the energy in the feed as metabolizable energy (ME). Thus

$$\text{Gross efficiency of milk production} = \frac{\text{Energy in milk}}{\text{ME for maintenance + ME for milk}}$$

This ratio indicates how much milk energy is produced from 1 unit of feed energy. Gross efficiency measures productivity and is most meaningful when expressed on an annual basis.

The net efficiency of milk production is the ratio of energy in milk to the ME in feed excluding the maintenance requirement ('above' maintenance) and assuming no changes in body reserves. The net efficiency of milk production on balanced diets varies little and is a characteristic of this function.

Composition and Energy Content of Milk

Milk yield and its composition and energy content, which are closely related, determine the requirements for lactation. The content of the solid components of milk varies inversely with their contribution to osmotic pressure. Fat varies most, protein varies less and minerals and lactose vary least. Because of the great variation in fat content and its high caloric value, the energy value of milk depends mainly on its fat content. Milk fat within cows varies greatly during milking, between milkings, over the lactation cycle, seasonally and with the climate. Large variation exists between breeds and between cows; milk fat is the most heritable of all the productive traits studied. Regressions of the milk constituents versus energy in milk are presented in Figure 21.3; the position of the major breeds is indicated.

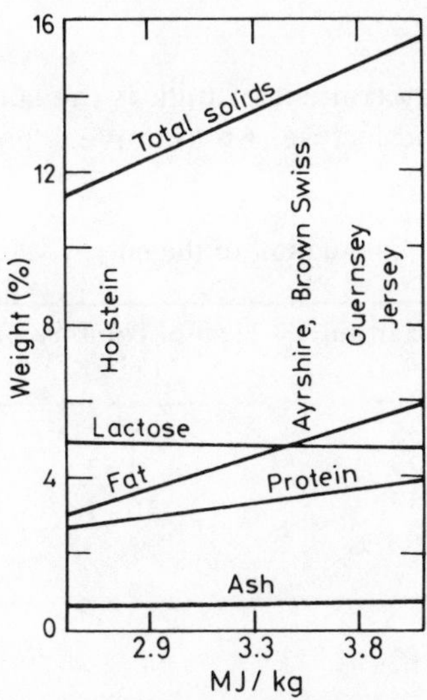

Figure 21.3. The relationship between milk constituents and the energy content of milk of the major breeds.

Between breeds and between cows the fat and protein content of milk are positively correlated. Nevertheless selection for fat and protein content in opposite directions is possible. Environmental effects, mainly those related to diet, usually cause a negative correlation between milk fat and protein content; this is discussed in more detail below.

Variation in Milk Composition Due to Factors Other Than Diet

The content of milk fat increases sharply during milking. Incomplete milking leaves milk with a high fat content in the udder. Unequal intervals between milkings reduce production and milk fat content. Increasing the frequency of milking increases milk production but decreases milk fat content. Yield and composition of milk change during the course of the lactation period (see later). For three to four weeks after calving the fat and the protein content of milk decrease, thereafter they gradually increase.

A seasonal effect on milk composition is noticed in many regions. Milk fat is usually lower in the summer, especially in warm climates. When dealing with seasonal changes in milk composition, the effects of diet, weather and day length may be easily confounded. Nevertheless, there is good evidence that hot weather reduces the fat content of milk.

Infection of the mammary gland (mastitis) alters the permeability of secretory tissue and impairs the secretory activity. Characteristically, there is a fall in milk yield, lactose and potassium content, and an increase in sodium and chloride content.

The Energy Value of Milk

The combustible energy content of milk is the sum of the heats of combustion of fat, protein and lactose. An example is given in Table 21.3.

Table 21.3 Calculation of the energy value of milk

Component	Content in milk (g/kg)	Heat of combustion (kJ/g)	Total (MJ/kg milk)
Fat	35	38.92	1.33
Protein	29	24.52	0.71
Lactose	49	16.54	0.81
		Total	2.85

Since milk fat is high in energy, rather variable and most often determined, regressions of milk energy on milk fat are used to calculate milk energy content. The most common formula is that of Gaines, proposed in 1922. He devised the following equation for expressing milk of different fat content in terms of milk with 4% fat, commonly called *4% fat-corrected milk* (4% FCM):

$$4\%\ \text{milk} = 0.4 + (0.15 \times F)$$

where F is the percentage fat of milk. Thus, for milk with 3.0% fat the conversion factor is 0.85 [0.4 + (0.15 × 3.0)]. Accordingly, 30.0 kg of milk with 3.0% fat is the equivalent of 25.5 kg of 4% FCM (0.85 × 30). Often 'FCM' is used without specifying the '4%'. A simple regression equation for estimating the energy value of milk is:

$$\text{Milk energy (MJ/kg)} = 0.0406\ F + 1.509$$

where F is the percentage fat content of milk. Both the above formulae disregard variation in milk protein. Protein in milk is easily determined, but it is easier to determine total solids by drying a sample at 100 °C; if milk fat is known the difference between total solids and milk fat is referred to as solids–non-fat (SNF). Since SNF includes protein, lactose and minerals and the latter two hardly vary, differences in SNF reflect differences in protein content. The following multiple regression equation may be used to obtain energy in milk reasonably exactly:

$$\text{Milk energy (MJ/kg)} = 0.386\ F + 0.0205\ \text{SNF} - 0.236$$

where F and SNF are percentage values of fat and solids–non-fat, respectively. Thus, for calculating the energy content of cow's milk with a fat content below 2.5%, this formula based on both fat and SNF is more accurate than Gaines' formula.

The mean milk composition of several domestic species and man is presented in Table 21.4. Whereas cattle and goat milk are reasonably close in composition, sheep milk is considerably higher in fat, protein and minerals. This greatly increases the requirement of sheep for milk production (see later).

Table 21.4 The percentage composition of the milk of some domestic species and humans

Species	Fat	Protein	Lactose	Calcium	Phosphorus	Energy (MJ/kg)
Cattle	3.5	2.9	4.9	0.12	0.10	2.86
Goat	4.5	2.9	4.1	0.13	0.11	3.12
Sheep	7.4	5.5	4.8	0.20	0.16	4.97
Horse	1.9	2.5	6.2	0.10	0.06	2.37
Sow	8.3	5.4	5.0	0.25	0.15	5.27
Human	3.8	1.0	7.0	0.04	0.01	2.86

Nature of the Lactation Cycle

The yield and composition of milk change during the course of the lactation cycle. A typical lactation curve of a high-producing dairy cow is presented in Figure 21.4.

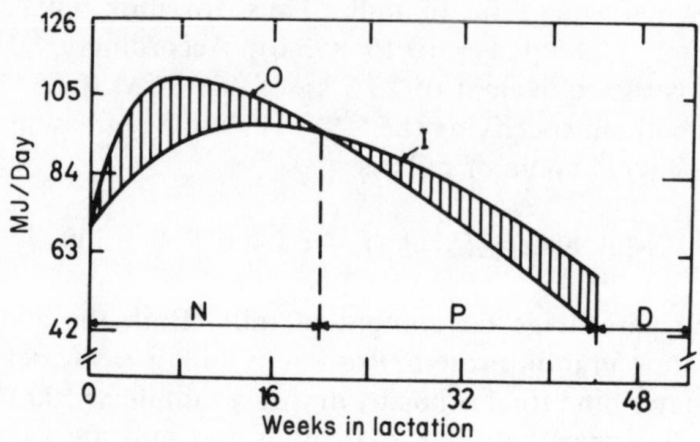

Figure 21.4. Schematic representation of the energy balance during the course of lactation. I is the input of net energy above maintenance. O is the total output of energy in milk, N is the period of negative energy balance, P the period of positive energy balance and D is the dry period.

Daily milk yield increases after calving for about 6–8 weeks, reaches a peak and then declines. Ideally the cow is bred at 11–12 weeks after calving and conceives; gestation is about 40 weeks (282 days) and the interval between two successive calving dates is about one year. For a large part of the time the cows are simultaneously lactating and pregnant. In order to maximize production in the following lactation, a dry period must precede calving; milking is stopped about 50–60 days before the expected calving date. Adequate nutrition during the ascending phase of lactation is rather important because there is a good correlation between peak daily yield and annual production. However, as indicated in Figure 21.4, in early lactation following calving, the intake of net energy *above maintenance* is usually lower than the output of energy in milk. Thus, calving is followed by a period of negative energy balance, i.e. the cow uses body tissue, mainly fat, to produce milk, so that the cow is commonly seen to lose an appreciable amount of body weight at this time. Normally, at a certain point, the daily intake of net energy above maintenance overtakes the daily output of energy in milk and a period of positive energy balance follows in which the cow builds up its reserves of fat. In Figure 21.4 N and P represent the periods of negative and positive energy balance, and the vertical distance at any point in time between the curves O (output) and I (input over maintenance) represents the daily deficit or surplus of energy, respectively. The period of negative energy balance is usually longer than the phase of increasing milk production, since the voluntary intake of the cow rises more slowly after parturition and the maximum may not be reached until many weeks after maximum milk yield.

Efficiency of Utilization of Energy in Early Lactation

The effect of negative energy balance on the net efficiency of milk production has been determined by Flatt (1965) in respiration trials with lactating cows. The following estimates have been established: feed ME to milk = 0.644; body fat to milk = 0.824; feed ME to body fat during lactation = 0.747; feed ME to body fat during the dry period = 0.587. Thus, the deposition of fat during lactation is 27% more efficient than in the dry period (0.747/0.587 = 1.273). Obviously, in order to maximize the efficiency over the entire cycle, the cow should regain the tissue lost after calving *before* the dry period rather than in the dry period. In order to establish the relationship between overall net efficiency and negative energy balance, the efficiency of converting ME into milk through body fat must be known. The latter is the product of two values: the efficiency of conversion of *feed ME to body fat* and that of *body fat to milk.* Two values are given for the former. The indirect efficiency of conversion of ME to milk through body fat is 0.824 × 0.747 = 0.616, if the fat is formed *during lactation,* and 0.824 × 0.587 = 0.483 if it is formed *during the dry period.* Both these values are lower than 0.644, the value for the direct conversion of ME to milk. The higher value, 0.616, is only 5% lower than that for the direct conversion of feed ME to milk.

The Partition of Energy

This is a term coined by Blaxter to describe energy exchange in cows. At any time during lactation part of the ME of the feed is used for milk production and the rest is used to produce body fat or to prevent its loss; loosely speaking the latter part is used for 'fattening'. Partitioning more energy for milk production is synonymous with increasing milk production at the expense of body fat.

The rate of energy intake after parturition can be somewhat raised by an increase in the concentrate/roughage ratio; thereby the disparity between output and input in early lactation would be reduced. However, it is an idealistic but unrealistic aim that a cow could consume during the whole of lactation the exact amount of feed required in order to resist body weight changes and to maximize the net efficiency of milk production.

Indirect Measures of Energy Balance

Milk production and the continuously changing energy balance during lactation are important indicators of the adequacy of nutrition. Excessive negative energy balance reduces conception rate (see Chapter 19) and increases the incidence of ketosis, a common metabolic disease of dairy cows (see Chapter 24). In order to determine exactly how much fat a cow is losing, respiration trial must be performed. In practice this is unrealistic. In applied research and farm practice, relative fat loss may be estimated in several ways. Changes in body weight are a crude way of estimating fat loss. Gross body weight in the lactating cow consists of 15–30% fat; the rest is water, protein, minerals, gut-fill, urine in the bladder and milk in the udder. The variation of these makes any extrapolation of changes in body weight to body fat very unreliable. Also fat lost from adipose tissue is temporarily replaced by water. In early lactation fat is lost rapidly, but gut-fill greatly increases and compensates for the change in body weight. Experienced farmers can estimate the relative body condition with remarkable accuracy by visual and manual inspection of cattle. High correlation between expert body condition scoring and in vivo measurements of subcutaneous fat by an ultrasonic device (see Chapter 1) confirm the reliability of both methods.

Principles of Expressing Feeding Standards for Milk Production

The evaluation of feeds for production has been presented in detail in Chapters 14–16. Feeding lactating cows is indeed complicated because the efficiencies of conversion of energy for maintenance, lactation, growth, fat deposition and pregnancy are not identical. In theory it is desirable to formulate an adequate ration for a dairy cow by considering each and every function. In reality, however, the cow stores and mobilizes energy during the cycle, and rationing exactly in the short term is less critical than in the long term. Nevertheless, it is necessary to measure the energy value of feeds for lactating dairy cows according to an objective yardstick in order to formulate least-cost diets.

Feeding standards for dairy cows based on digestible, metabolic and net energy have been described in Chapter 16; their merits and limitations will be pointed out below.

The Classical Feeding Standards for Dairy Cattle

The Total Digestible Nutrient (TDN) System. The TDN system is simple and it has the advantage that TDN values of many feeds have been determined. Its major limitation is that it grossly overestimates the energy value of feeds high in fibre. Though derived from work done in Germany in the nineteenth century, the TDN system has been popular mainly in the USA and in South America. The 1978 edition of *Nutrient Requirements of Dairy Cattle* of the USA National Research Council (NRC) includes a complete feeding standard based on TDN.

Kellner's Starch Equivalent (SE) System (see Chapter 16). Kellner developed his SE system by measuring fat deposition in mature animals and created a specific net energy system for fattening cattle, but modified SE values have been commonly used in feeding standards for dairy cattle in all European countries and in the British Commonwealth. The latter adjusted the values for the difference between the low efficiency of ME for fattening and the relatively high one for milk production. Nevertheless, a system developed for fattening but applied to lactation would *underestimate* feeds high in fibre. In Britain this was corrected by increasing the value of roughage by 20%; later the correction was dropped for straw but retained for hay.

Hansson's Feed Unit (FU) System. Hansson departed from the SE system because he realized that the value of protein in supporting lactation is considerably higher than for fattening. The energy value he assigned to protein was 50% higher than that in the SE system. This represented an improvement. Hansson advocated replacement trials with dairy cows; thus his system is in principle one of net energy for lactation. In practice, Hansson applied corrections to Kellner's SE values; therefore his tables also tend to underestimate the value of roughage. Nehring compared the feed requirements for producing 1 kg of 4% FCM according to five systems: the three systems mentioned above, Mollgard's system (see Chapter 16) and Armsby's system. Agreement was nearly perfect among the four systems; Armsby's system underestimated the requirement by about 23%. Armsby's net energy values of feeds were established at submaintenance levels and evidently overestimate their worth under normal conditions.

Modern Feeding Standards

Relatively recently, two largely similar feeding standards have been developed, by Blaxter in the UK and by Flatt in the USA. Both emphasize the differences with which ME is used for different functions. The estimates of Flatt for the efficiencies of conversion of ME to NE are presented in Table 21.5. The last value in the table is the efficiency of conversion of body substance to milk, this value may serve to calculate the efficiency of indirectly converting ME to milk via body fat.

The variation of the coefficients in Table 21.5 is mainly due to the effects of fibre in the diet. The effects of the diet on the production of milk, its composition and the efficiency of production are discussed in another section in this chapter.

The modern feeding standards have been described in detail in Chapter 16. Briefly, Blaxter's standard expresses feed values in ME and calculates requirements for maintenance, growth and lactation by applying different conversion factors to each function. A simplified version of the same system, devised by the ARC (1980) is also described. In Flatt's system for dairy cattle, feed energy values are expressed as NE for lactation (NE_L); these

Table 21.5 The efficiency of the direct conversion of metabolizable energy (ME) to net energy (NE) for different purposes and the efficiency of conversion of body substance to milk in cattle (after Flatt, *et al.*, 1972)

Function	Efficiency (NE/ME)
Maintenance	0.65–0.74
Walking	0.20–0.30
Foetus formation	0.10–0.20
Growth*	0.40–0.55
Fat deposition†	0.58–0.75
Milk production	0.60–0.70
Milk from body tissue	0.82 (NE/NE)

* After Lofgreen and Garrett, 1968 (see bibliography to Chapter 16).

† Low value in dry period, high value during lactation.

may also be used for maintenance. Flatt built his system on the results of many respiration trials with dairy cows, over a large range of production and body weights. His recommendations are summarized in Table 21.6.

Table 21.6 Summary of the energy requirements of dairy cows (after Flatt *et al.*, 1968, and Moe and Tyrrell, 1974¶)

	ME*	NE_L†
Maintenance (MJ/kg $W^{0.75}$)	0.475–0.502	0.305
Milk production (MJ/kg 4%FCM)	4.85–5.10	3.10
Pregnancy, last 75 days (MJ/day)	30.5	18.8
Tissue gain during lactation (MJ/kg)‡	33.4	20.9
Tissue loss during lactation (MJ/kg)‡	−33.4	−20.9
Tissue gain in dry period (MJ/kg)‡	41.8	25.9

* Requirement for ME, provided the energy content of the diet is between 8.4 and 12.5 MJ/kg dry matter.

† NE_L is based on the caloric value of milk (see Chapter 16) and can also be calculated from DE or ME by regression.

‡ Assuming that the energy value of 1 kg live-weight change is 25.1 MJ.

¶ See bibliography to Chapter 16

The most important findings about metabolic efficiency of lactating cows are probably three. (1) The mean net efficiency of conversion of metabolizable energy to milk is 0.64 and is relatively independent of level of feeding and composition of diets. (2) Lactating cows deposit body tissue, especially fat, more efficiently than dry cows. The importance of both findings has already been discussed in this chapter. (3) Further, the maintenance requirement per unit of metabolic weight is in lactating cows about 10–20% higher than in dry cows (see Chapter 17).

The Effects of Diet on the Production of Milk, its Composition and the Efficiency of Milk Production*

The Metabolizability† or the Ratio of Concentrates to Roughage in the Diet

As pointed out in this chapter, milk is economically produced on diets ranging in the content of concentrates from 0% to 85%. Nevertheless, the ratio of concentrates to roughage in the diet affects the level of production, milk composition and the efficiency of production. These are due to drastic changes in voluntary feed intake and in rumen fermentation which occur with an increased level of starch and a decreased level of fibre in the diet (see Chapter 5).

Metabolizability and Feed Intake. On an all-roughage diet, gut-fill limits feed intake, and a cow of high production potential is unable to consume enough feed to express it. The regulation of feed intake is discussed in Chapter 23. As stated above, there is a positive correlation between feed intake and milk production. In the last four decades reliable studies have shown that raising the proportion of high-starch concentrates (mainly cereal grains) in the diet at the expense of roughage increases feed intake and milk production, especially in dairy cows with high genetic merit.

The Effect of Metabolizability on Milk Fat Content. Increasing the metabolizability of a diet by adding starch at the expense of fibre or cell wall may severely reduce the fat content of milk by changing rumen fermentation (see Chapter 5). Some of the changes are schematically presented in Figure 21.5. The increase in the starch content of the diet increases the concentration of total VFA in rumen fluid, mainly by elevating the molar concentration

---

* Principles for design of feeding experiments with lactating cows are given in the Introduction to Part III.

† Metabolizability is defined in Chapter 16 as the ME of a diet divided by the gross energy, ME/GE, and increases with increasing proportion of concentrate in the diet at the expense of roughage.

of propionic acid and reducing the *relative* proportion of acetic acid. The net result is a gradual decrease in the molar ratio of acetic to propionic acid from about 4 to 1 and a non-linear decrease in the milk fat content from a level which is normal for the breed to one-third of the normal value of very low fibre diets (1–2%, rather than 3–4%).

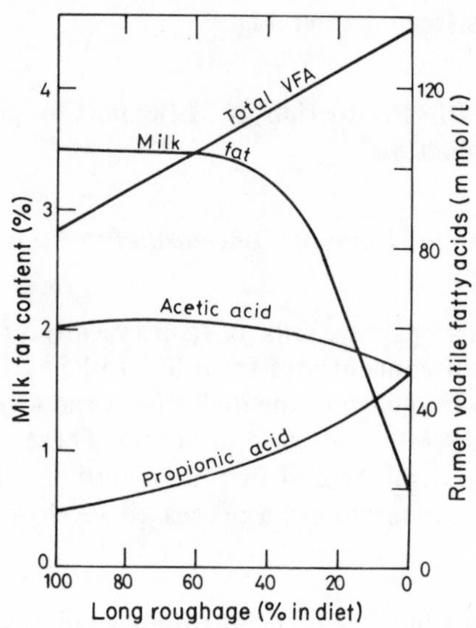

Figure 21.5. The generalized relationship between the proportion of long roughage in the diet, volatile fatty acid concentrations in rumen fluid and milk fat content.

The drop in milk fat content is probably caused by (1) a relative shortage of absorbed acetic acid which is a major precursor of milk fat, and (2) an inhibition of the mobilization of body fat produced by an excess of propionic acid (see Chapter 6). Although the point of the precipitous drop in milk fat and the rate of the depression vary greatly among breeds and cows, and are affected by environmental factors including ingredients in the diet *unrelated* to fibre, the shape of the curve in Figure 21.5 is generally valid.

Soluble sugars, when introduced into the diet as isolated material or as a part of natural feeds, such as fodder beet or whey, tend to promote butyric acid production in the rumen and to increase milk-fat content; butyric acid increases milk-fat secretion to a greater extent than acetic acid.

Inclusion of an adequate amount of fibrous material in the diet is critical for the maintenance of fat content at a satisfactory level, and a dietary crude fibre content of approximately 160 g/kg is often recommended (Thomas and Rook, 1983).

Other factors may depress milk fat. Fine chopping or grinding of hay depresses milk fat because reduced particle size increases its rate of passage from the rumen and reduces rumination and salivation. Reduced salivation promotes a lower rumen pH because saliva is mildly alkaline. The thermo-mechanical treatment of cereal grains, e.g. steam-flaking of sorghum, greatly increases the milk fat depressing effect by destroying the crystalline structure of starch and increasing its fermentability.

Alkaline minerals are often effective in counteracting the depression of milk fat because they elevate the pH of rumen fluid, cellulolytic activity and the ratio of acetic acid to propionic acid. Sodium or potassium bicarbonate, magnesium oxide and calcium hydroxide may be used singly or in combination. The substitution of cereal grains for roughage in the diet concomitantly reduces the alkaline mineral content of the diet because, as a rule, roughage supplies considerably more calcium, potassium, magnesium and sodium than cereal grains; this is often overlooked. Alkaline minerals which elevate the pH of rumen fluid are referred to as 'buffers' or 'antacids'.

The term *low-fat milk syndrome* is often used for the depression of milk fat just described. Although the depression reduces the energy content of milk and its market value, it is not a disease.

High-concentrate diets slightly increase the true protein content of milk, probably because propionic acid is a precursor of amino acids in the udder (see Chapter 8).

Fat in the Diet of Lactating Cows. The dietary fat requirement of ruminants is about 1% and is readily available in practical diets. However, in order to increase the metabolizability of the diet, fat is the only practical alternative to starch. Fat in the diet up to about 8% increases milk production, but its effect on milk fat content depends largely on the method of incorporating the fat into the diet. Fat is hydrolysed in the rumen and the long-chain fatty acids liberated severely reduce cellulolytic activity. Thus the ratio of acetic acid to propionic acid may be reduced and a depression of milk fat content may occur with an increase in milk production. Unsaturation of the long-chain fatty acids increases the depression. Highly unsaturated fish oils are known to cause dramatic decreases in milk fat content. On the other hand, when fat is incorporated into the diet in the form of unextracted oil seeds, e.g. soybeans or cottonseed, milk production tends to increase and milk fat production rises at least in proportion to milk. The results of an illustrative experiment are presented in Table 21.7.

The addition of fat to the diet in any form tends to reduce the protein content of milk (see Table 21.7).

Changes in the fatty acid composition of the dietary fats are always reflected in milk fat because part of the dietary long-chain fatty acids are absorbed and carried by the blood to the udder. Thus, the fatty acid composition of milk may be manipulated by changing the dietary fat. A procedure for raising the content of fat and of polyunsaturated fatty acids in milk by feeding 'protected lipids' has been described in Chapter 6.

Table 21.7 The effects of two methods of incorporating soybean oil in the diet on daily milk yield and composition (after Steele *et al.*, 1971, p. 43)

Ingredients*	Low-fat control	High fat	
Starch	1.76	–	–
Soybeans	–	3.95	3.32
Soybean meal	3.32	–	–
Soybean oil	–	–	0.63
Milk yield (kg)	18.60	21.0	23.0
Milk fat yield (g)	704	785	549
Milk fat content (%)	3.78	3.74	2.39
Milk protein (%)	3.31	2.96	2.96

* kg/day in the variable portion of the diet.

Metabolizability and the Net Efficiency of Lactation. At least from Kellner's time, it is well known that the net efficiency of the ME for growth and fattening increases rapidly with metabolizability (see Chapter 18). The relationship between metabolizability and the net efficiency of milk production is less striking; in field experiments with dairy cows tissue changes and the low-fat milk syndrome mask this effect. Nevertheless, many recent studies in Europe and in the US have shown that in the absence of changes in body tissue the net efficiency of milk production increases with the metabolizability of the diet. The changes in the net efficiencies of ME for growth and fattening, lactation and maintenance, as estimated by the Agricultural Research Council (UK) in 1980 are shown in Figure 21.6. Clearly, the net efficiency for growth and fattening is lower than for lactation and maintenance and its rate of increase with metabolizability is the highest. Although there is a concensus on the increase in the net efficiency for lactation with increasing metabolizability, there is no agreement on its linearity. Some claim that the net efficiency of lactation levels off when metabolizability exceeds 0.60 and that it declines at higher levels.

The reasons for this relationship are complex. Increased metabolizability usually implies higher levels of starch in the diet, elevated propionic acid production and greater efficiency of conversion for all purposes (see Chapter 5). A shift in starch digestion from low-efficiency ruminal fermentation to high-efficiency hydrolysis in the small intestine probably contributes to increased net efficiency. However, at very high levels of rapidly fermentable carbohydrates in the diet, ruminants suffer from various degrees of lactic acidosis in the subclinical or clinical form (see Chapter 24). Briefly, lactic acid accumulates in the rumen, pH drops and chronic systemic acidosis may set in and impair the net efficiency of lactation.

The implications of changing the metabolizability of the diets must be

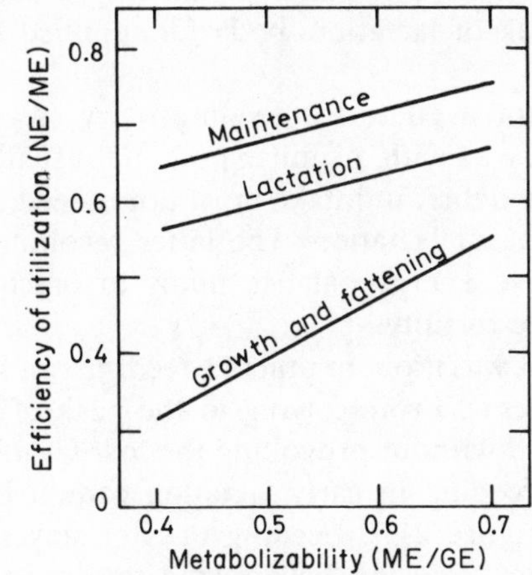

Figure 21.6. The efficiency of utilization (NE/ME) for growth and fattening, lactation and maintenance as a function of metabolizability (ME/gross energy) estimated by the Agricultural Research Council (UK), 1980. Adapted from Table 16.6.

properly weighed. To a certain extent, higher metabolizability increases feed intake and in cows with a high potential production the increased intake of energy promotes production. On the other hand, the changes in the volatile fatty acid composition may affect the partition of energy in the cow and promote fat deposition at the expense of milk production. Although these two opposing tendencies have been well established in relatively short-term experiments, the long-term effects of diets with varying metabolizability are not sufficiently clear.

Feeding for High Milk Production

The traits conducive to high overall productivity are (1) the ability to secrete large amounts of milk, (2) high voluntary feed intake, and (3) the ability to produce milk at the expense of body fat. A limited possibility for a trade-off exists between the second and the third. In the long run voluntary feed intake depends on the amount of fat lost in early lactation, but during the peak of lactation the correlation between intake and fat loss is negative.

Flatt reported on the variation in the ability to mobilize body fat in early lactation. Among 12 cows in early lactation, the mean loss of body fat was 29.0 MJ/day, but one cow lost up to 84 MJ/day. This cow, Lorna, became famous in the annals of dairy nutrition. Lorna lost in early lactation 150 kg of body fat. Using the coefficients of Flatt, and a value of 38.6 MJ/kg for energy in fat, it was calculated that Lorna produced 1558 kg of 4% FCM from body fat. Her 305-day production was 8768 kg of 4% FCM, and temporarily body fat supplied 18% of the energy used for annual milk production. At the peak of lactation, body fat supplied not less than 57% of the energy in milk.

Lorna's history is not a guide for feeding dairy cows; her feed intake was dangerously low. Cows with a similar potential usually go into negative energy balance, but the deficit in intake must not exceed 20% of the energy requirement at the peak of lactation. The latter recommendation is rather lenient because negative energy balance tends to precipitate ketosis (see Chapter 24) and reduce fertility.

Because of individual variation, in practice feeding can seldom be adjusted to individual requirements. From calving to the peak of lactation, feed intake must be maximized without provoking the low-fat milk syndrome. The results of inadequate feeding in early lactation cannot be corrected later. This is illustrated in Figure 21.7. Feeding in later stages must restore the energy lost in early lactation and be directed towards the next cycle. As pointed out above, restoration of body fat is most efficient while the cow is producing milk.

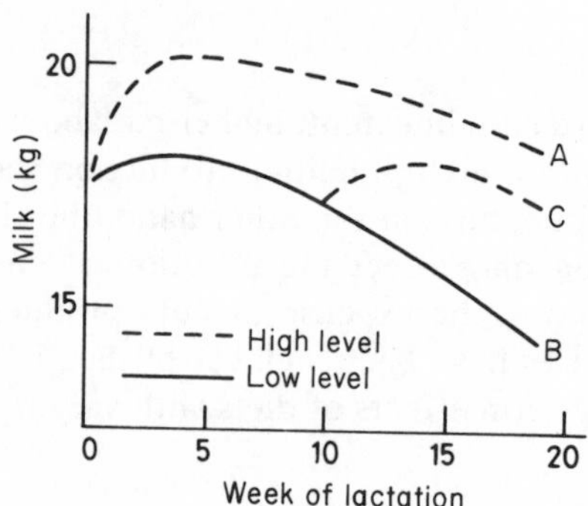

Figure 21.7. The influence of level of feeding upon milk production in later lactation (after Broster, *et al.*, 1969). Adapted by P. Van Soest, *Nutritional Ecology of the Ruminant* (1982), p. 306. Corvallis: O. & B. Books. Group A is a control group maintained continuously on a high plane of nutrition. Group B is maintained continuously on a low plane of nutrition. Group C was maintained for 10 weeks at low plane of nutrition and received from the tenth week the same amount of feed as Group A. *Reproduced by permission of author and publisher.*

Techniques of feeding dairy cows are beyond the scope of this book. It may briefly be noted that they differ greatly. Diets with all the ingredients thoroughly mixed ('total mixed rations') fed ad libitum minimize fluctuations in rumen activity. With this technique the metabolizability of the diet must be adjusted for groups of cows at different stages of lactation. At the other end of the spectrum are techniques using individual, continuous, preprogrammed allotment of concentrates through computer-controlled systems which identify cows electronically. Combinations of the above system are possible.

Mistakes to Avoid in Feeding for High Production

The general trend has been to abandon individual feeding according to daily requirements and to adopt group feeding. This reduces expenditure, but probably aggravates the two most common mistakes in feeding dairy cattle: underfeeding high producers, especially during early lactation, and overfeeding dry cows. The latter mistake produces the 'fat-cow syndrome' which is an increased incidence of periparturient disease and may impair appetite after calving and reproductive performance (see Chapter 19).

Effect of Level of Feeding on Efficiency of Milk Production

An increase in the amount of energy given to a cow in excess of that required by the standards results in increase in the yield of milk, and in an increase of live-weight gain. The responsiveness of milk yield, i.e. live-weight gain or loss, varies with the genetic potential, stage of lactation (see above) and nutritional history. A high-producing cow will respond with more milk for each extra unit of food compared to a low-producing cow; for the latter the responsiveness in live weight will be higher (see Figure 21.8).

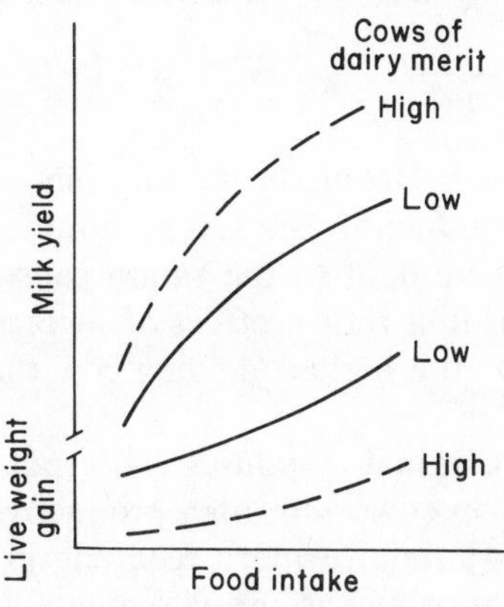

Figure 21.8. A schematic representation of the relationship of food to milk yield and live-weight gain in the cow: (----), cow of high milk potential; (——), cow of low milk yield potential (after Broster, 1976, p. 273). *Reproduced by permission of Dr. W. H. Broster.*

The response of milk secretion to addition of energy supplied to cows of different yield groups in excess of the requirement was examined by Blaxter (1967). His experiments were performed during the whole lactation period. The results of these trials are shown in Figure 21.9, and indicate that the response of milk secretion to additional feed increases with the milk yield; this in accordance with the model offered in Figure 21.9. After Blaxter, the response per MJ of ME increased from 0.061 kg when the daily yield was 10 kg to 0.172 kg when it was 25 kg.

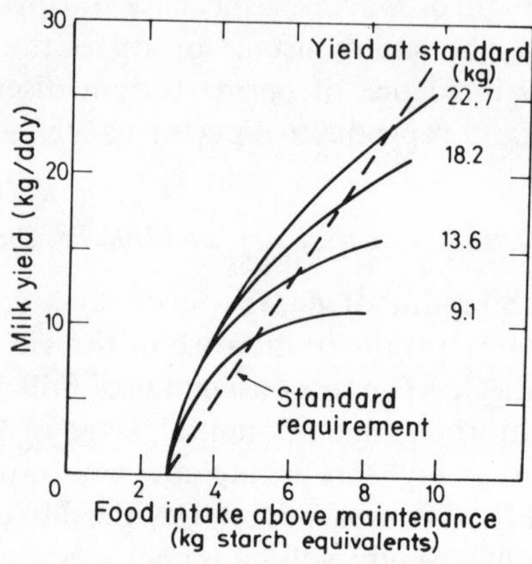

Figure 21.9. Milk production of cows in relation to their food intake during lactation (after Blaxter, K. L., *The Energy Metabolism of Ruminants*, 1967, p. 174). *Reproduced by permission of Hutchinson Edition and Sir K. Blaxter.*

Furthermore, it results from the findings shown in Figure 21.9 that at every level milk of production the law of diminishing returns is followed, i.e. each successive increment to the ration causes a progressively smaller response. The diminishing return effect of increasing intake on milk yield is less pronounced for the higher yielding cow than for the lower yielding animal.

The fact that the marginal responses (milk per additional units of feed) are apparently higher in cows with high productive capacity does not mean that high-yielding cows require more feed energy per unit of milk. It has been ascertained by calorimetric measurements with high-producing cows carried out by Flatt in the US, Van Es in The Netherlands and Schiemann in East Germany that the net efficiency of milk production from ME is practically independent of the level of production.

The high level of feeding required for lactating cows causes some depression in the apparent digestibility of the diets (see Chapter 13), but this effect tends to be somewhat balanced by decreased proportional losses of energy as methane and urine (see Chapter 14). Therefore, with increasing intake the real ME or NE of diets is affected much less than the TDN or DE value of the same diets.

PROTEIN REQUIREMENTS FOR MILK PRODUCTION

For persistent lactation at a high level it is essential to supply an adequate level of protein in the diet. In milk, about 25% of the dry matter and the same proportion of the energy are in the protein fraction. For comparison, in the gain of cattle, the proportion of energy in the protein fraction declines from 25% at 250 kg live weight to 14% at 550 kg. This suggests that the protein content of a lactation diet must be higher than that of a fattening diet. In addition, a high-producing cow secretes 1000–1500 g of protein per day whereas a rapidly growing calf or steer deposits only 250 g of protein per day. Indeed, intensive lactation diets contain roughly 50% more protein than fattening diets.

Protein Deficiency in Lactation

An inadequate level of protein in the diet reduces milk production. Among all the essential nutrients, the supply of protein to the high-producing dairy cow is probably the most critical in the short term. The ability of the dairy cow to mobilize protein from bodily reserves is very limited. A cow of 500 kg body weight cannot mobilize during one year more than about 8–9 kg of protein, mainly from muscle, to support lactation. This is the amount of protein in 250–300 kg of milk. In terms of the *energy equivalent*, the ability to mobilize protein is smaller by an order of magnitude than the ability to mobilize fat. If, as explained above, in early lactation the high producer may incur a significant deficit of energy and restore it later, protein consumption cannot be delayed. In addition, the underfeeding of protein gradually reduces voluntary feed intake.

Protein deficiency depletes muscle and liver protein, reduces the level of blood albumin and probably predisposes the cow to certain infectious and metabolic diseases; in early lactation it increases the weight loss and in pregnancy it reduces the growth rate of the foetus. Low levels of urea in the blood indicate efficient protein utilization, but a very low level (less than 2.5 mmol/litre) is a sign of protein deficiency.

Protein Quality and the Requirement for Lactation

Because of the assumption that microbial nitrogen metabolism in the rumen (see Chapter 8) compensates the host both for a deficiency in true protein and poor-quality protein, little attention had been paid to the quality of protein until about 1950. Although, in the UK in 1925, the suggestion was made to assign to non-protein nitrogen (NPN) in the diet a lower value than to true protein, after 1940, for three decades, many attempts were made, mainly in the USA, to replace plant nitrogen in ruminant diets to all classes of stock, including lactating cows, by synthetic NPN, nearly always urea (see Chapter 8). NPN can be utilized by the rumen flora only in low-protein diets where the protein content is lower than 13%, and then up to 30% of the dietary protein can be replaced by NPN. Since milk production requires relatively high levels of protein in the diet, the supply of true protein nitrogen for lactating cows is more important than for other classes of stock. At the same time it was also shown that proteins from various sources, e.g. cotton and soybean, differ in their ability to support milk production (the latter is superior). Recently, high-quality fish meal protein of very low degradability in the rumen has supported milk production at a higher level than plant proteins. The differences in the quality of true proteins for ruminants are partly due to their amino acid composition, but differences in their degradability in the rumen (see Chapters 8 and 18) may be more important. Briefly, when the microbial requirement for nitrogen has been satisfied, proteins having low degradability are superior because they increase the amount of *escape* protein (often incorrectly termed 'by-pass' protein) which reaches the abomasum and is available for digestion and absorption by the host.

Rationing Protein for Lactation

Protein requirements for lactating cows are calculated according to the same principles as those for growing ruminants (see Chapter 18). The crude protein (CP) requirement of the lactating dairy cow can be estimated by the factorial method as suggested by the National Research Council (USA) publication in 1978. The factorial approach for estimating requirements in terms of CP has been explained in detail for maintenance (see Chapter 17) and for growth (see Chapter 18). Expanded for lactation the factorial formula is:

$$\mathrm{TCP} = \frac{U + F + S + G + C}{E_{\mathrm{p1}}} + \frac{L}{E_{\mathrm{p2}}}$$

The definitions of TCP, U, F and S have been given in Chapters 17 and 18. G is protein in body-weight gain and the value suggested for cows is 0.16 g per g of live-weight gain. C is protein deposited in the products of conception (foetus, placenta, fluids and uterus); at 60 days prepartum it is estimated as $C = 1.136\ W^{0.7}$, where W is body weight in kg. L is the protein required for synthesis of milk protein and is the sum of milk protein and the added loss of protein in faeces (0.068 × faecal dry matter) due to the increased feed consumption for lactation. The percentage of protein in milk (P) may be estimated from the percentage of fat (F) by the formula:

$$P = 1.9 + 0.4\,F$$

E_p represents the product of two values: (1) the fraction of CP absorbed as amino acids, and (2) the mean efficiency of conversion of the absorbed amino acids to protein used for the functions which appear in the numerator. For (1) the NRC adopted a value of 0.75 for all functions in cattle weighing 100 kg or more. For (2) the value of 0.60 was adopted for all functions other than lactation (to calculate E_{p1}) and 0.70 for lactation (to calculate E_{p2}). Thus, in the factorial formula E_{p1} is 0.45 (0.75 × 0.60) and E_{p2} is 0.53 (0.75 × 0.70). It should be noted that in the above formula the value of the second term, L/E_{p2}, the CP requirement for lactation, is greater than the first, which represents maintenance, weight gain and pregnancy combined, whenever milk production exceeds roughly 13 kg/day.

The principles of a new system for computing the protein requirement of ruminants (suggested by the ARC) have been outlined in Chapter 18. The system seems to be adequate for high-producing cows because of the distinction between microbial and undegradable protein; the proportion between both forms of protein formed in the rumen varies with the type of the protein source ingested by the animals. The demand of high-yielding cows for amino acids exceeds the supply originating from microbial protein that reaches the small intestine. In such cases the amino acid supply to the intestine must be made up with undegraded protein originating from concentrate feeds.

Further Aspects of Protein Requirements for Lactation

The response of the lactating cow to increasing levels of protein in the diet obeys the law of diminishing returns, i.e. each successive increment to the ration causes a progressively smaller response. The schematic response to dietary protein at different levels of intake (and production) is shown in Figure 21.10. To a limited extent the level of protein in the diet will compensate for shortage of energy and vice versa.

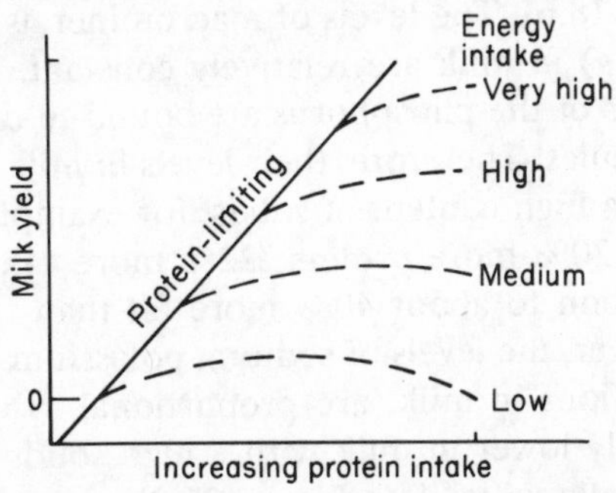

Figure 21.10. The response of lactating cows to protein intake at different levels of energy intake and productivity (after *Nutrient Requirements of Ruminant Livestock*. ARC, 1980, p. 159). *Reproduced by permission of the Commonwealth Agricultural Bureaux.*

Among cows, the recommendations of dietary protein content vary with milk production because the protein/energy ratio for lactation is higher than for maintenance. However, at the same level of milk yield, a cow before the peak of lactation requires a higher level of protein in the diet than after the peak. The reason for this is that before the peak of lactation feed intake is likely to be lower and the negative energy deficit higher than after the peak. By the same token, with advancing lactation the energy balance increases and the requirement for protein decreases.

An excess of protein in the diet is undesirable for several reasons. As suggested in Figure 21.10, it may depress production; it is also used as a source of energy very inefficiently. Recent studies have shown that increasing the dietary protein in early lactation, by using soybean meal as a supplement, reduced the conception rate of dairy cows (see Chapter 19). The mode of action is still unknown. Excessive levels of high-quality protein in the diet, especially of the undegradable type, may provoke ketosis (see Chapter 24). On the other hand, in high-starch, low-fibre diets provoking the low-fat milk syndrome, soybean meal may act as an antacid (see above) and increase the milk fat content. The latter effect may be due to an increase in the level of ammonia from degraded protein or to potassium which is high in soybeans (see Table 21.9). The use of soybean meal as a mere antacid is of course wasteful.

Rumen microbial protein is slightly deficient in methionine. Free amino acids added to ruminant diets are metabolized in the rumen; however, methionine hydroxy analogue (see Chapter 7) added at low levels to the diet (about 0.15%) tends to increase milk fat content and occasionally total milk production.

MINERAL REQUIREMENTS FOR MILK PRODUCTION

The general subject of minerals in nutrition has been dealt with in Chapter 10. Here, only the specific requirements and problems of lactating cows will be treated. The minerals of major importance in milk are shown in Table 21.8 (see also Table 18.6). The levels of macrominerals (present in amounts greater than 0.1 g/kg) in milk are relatively constant. Most of the calcium and about two-thirds of the phosphorus are bound to casein; sulphur is part of the protein molecules. Therefore, their levels in milk vary with that of the proteins. Milk with a high content of solids; for example, milk of the Jersey cow contains about 20% more protein, 22% more calcium and 12% more phosphorus in addition to about 40% more fat than that of the Holstein–Friesian cow. However, the levels of sodium, potassium and chloride, largely in the form of free ions in milk, are proportional to the water content of milk and thus slightly lower in milk with a high solids content. As a result total ash in milk hardly varies (see Figure 21.3).

A deficiency of any of the macrominerals in the diet may become the limiting factor in milk production even though traditionally only salt (sodium chloride), calcium and phosphorus have been added to lactation rations.

Table 21.8 Median concentrations of minerals in milk

Macrominerals (g/kg)		Microminerals (mg/kg)	
Ca	1.20	Fe	0.3–0.5
P	1.00	Cu	0.05–0.12
Mg	0.12	Zn	3.0–7.0
Na	0.55	Mn	0.02–0.13*
K	1.40	Mo	0.08–0.70*
Cl	1.00	I	0.02–2.00*
S	0.30		
			(mcg/kg)
		Co	0.5–6.0*

* Maximum value obtainable by supplementation.

The problems of macrominerals in lactation diets can be explained only in conjunction with information on the macromineral content of feeds. The latter values for roughage and concentrates derived from the two major botanical families are presented in Table 21.9.

Table 21.9 Average contents of macrominerals in cereal and legume roughage and concentrates expressed as percentage of the requirement in the diet of a high-producing dairy cow*†

Element	Roughages‡		Concentrates¶		Requirement (%)
	Cereal	Legume	Cereal	Legume	
Calcium	67	197	10	33	0.60
Phosphorus	68	68	90	140	0.40
Selenium	90	120	85	115	0.20
Magnesium	105	200	75	90	0.20
Potassium	220	248	55	184	0.80
Sodium	67	111	22	22	0.18
Chlorine	139	139	29	18	0.28

* Computed from *US–Canadian Tables of Feed Composition*, National Research Council, Washington, DC, 1982.

† Requirement as percentage of dry matter for highest level of production, according to *Nutrient Requirements of Dairy Cows*, National Research Council, Washington, DC, 1978.

‡ Fair to good hays.

¶ Unextracted seeds.

Sodium and Chloride

None of the four major classes of feeds listed in Table 21.9 supply adequate amounts of all the macrominerals, and grasses (cereals) supply less than legumes. Sodium is the most deficient element in feeds; its supplementation is a sine qua non of ruminant diets. Because of high variability, even leguminous roughage may be deficient in sodium. Common salt is widely used to supply the required sodium. Chloride is nearly adequate in roughage-based diets, but in high-concentrate diets the chloride in added salt may indeed be required. This amends the classical view which holds that sodium rather than chloride is the important component of salt.

Calcium and Phosphorus

Generally, all roughage is poor in phosphorus, and grasses may also be poor in calcium. The phosphorus and calcium supply in lactation must be considered together. In early lactation about 20% of the calcium and phosphorus in the skeleton may be normally mobilized. This is induced by endocrine mechanisms (see Chapter 10). A negative balance of either calcium or phosphorus will demineralize bone, but with different results. When calcium balance is negative, bone is promptly demineralized and the level of calcium in blood is maintained. There is no dramatic effect on physiological functions though in severe calcium deficiency milk production may be reduced, milk fever induced (see Chapter 24) and the demineralization of bone may cause spontaneous fractures. When phosphorus in the diet is the limiting factor, a reduction of inorganic phosphorus in plasma occurs and metabolism is severely disturbed; feed intake and milk production decline and reproductive functions are inhibited (see Chapter 19). Therefore, a negative calcium balance may be regarded as part of the lactation cycle; a phosphorus deficiency, however, is distinctly pathological.

In later lactation and in the dry period, there is a complimentary repletion of bone mineral and therefore a positive calcium balance; then it is possible to meet from the diet the needs for calcium of the cow for maintenance, milk secretion and growth of the foetus as well.

On roughage, especially when grazed or cut on phoshorus-depleted soils, phosphorus deficiency is notorious and has long plagued ruminants of all classes. Although the evidence is overwhelming, it is often confused by deficiencies of protein, vitamin A, calcium and possibly microminerals. Because of the mass of evidence on this point the recommendations for phosphorus have escalated with time. Also the purportedly direct effect of phosphorus deficiency on the fertility of beef cows in extensive systems, especially in South Africa, has led to the belief that phosphorus cures infertility in cows. This has often brought about the oversupplementation of phosphorus, mainly in high-concentrate diets which are quite adequate in phosphorus. The excessive supplementation of phosphorus is wasteful and may be harmful (see Chapter 10).

Normally, the relative absorption of phosphorus is constant for a certain source and the total amount absorbed increases with its level in the diet; the relative absorption of calcium, however, varies greatly with endocrine status and the total amount absorbed in mature ruminants is limited. In order to calculate the requirements for lactation, it has been assumed that the availability of calcium and phosphorus are 45% and 55%, respectively (NRC), but more recent estimates indicate a higher availability of 0.68 for calcium and of 0.58 for phosphorus (ARC, 1980).

In sharp contrast to phosphorus, mature ruminants can tolerate a level of calcium in the diet which is at least five times higher than the recommended level. The excess is excreted in the faeces. Since calcium offered as calcium hydroxide is an efficient antacid in the rumen, is active in increasing the digestibility of starch in the lower gut and is well tolerated by ruminants, the recommended levels of calcium are now on the increase. In Israel, in high-concentrate systems with a mean annual production ranging between 7000 and 10 000 kg of milk, calcium is added to the diet to supply 1.0–1.6%. This exceeds the NRC requirements by 67–167%. In the dry period, a few weeks before the expected calving date, the calcium supply must be severely limited to reduce the incidence of parturient hypocalcaemia (milk fever), a common metabolic disease of cows (see Chapter 24).

Potassium

The level of potassium in milk is higher than that of any other mineral. In the body nearly all the potassium is an integral part of soft tissue cells, i.e. none is stored. Roughage is high in potassium and so are seeds of legumes (see Table 21.9); the classical view holds that a potassium deficiency does not occur. Nevertheless, the recent shift to intensive systems with high levels of cereal grains in the diet may produce a marginal potassium deficiency in dairy cows. Weight loss and reduced feed intake are the non-specific signs of a marginal deficiency in dairy cows. A severe deficiency is rare indeed; it causes a deranged appetite (pica), lower plasma and milk potassium, dehydration indicated by higher haematocrit values and reduced milk production. It is preventable by supplementation with potassium salts.

Magnesium

The level of magnesium in milk is low and, depending on the preceding magnesium storage, bone may supply variable amounts in early lactation. Nevertheless, in sharp contrast to calcium depletion, following a fairly long magnesium depletion period, an often fatal disease in all classes of ruminants (hypomagnesaemia) is well known (see Chapter 24).

Sulphur

A sulphur deficiency in dairy cows mimics a protein deficiency because sulphur is primarily a component of the protein molecule. A deficiency occurs on diets high in NPN unsupplemented with sulphur and on crops low in proteins (see Chapter 10). Sulphur deficiencies are also related to sulphur deficiency in soils. A sulphur deficiency promotes lactic fermentation and affects the rumen microflora before it affects the host. The sulphur content of a diet is near the optimum when the sulphur/nitrogen ratio is 1 : 12.

Microminerals

The microminerals (trace minerals) have been discussed in Chapter 10. In many advanced systems the routine supplementation of microminerals (iron, zinc, manganese, copper, iodine, cobalt and selenium) has been adopted because all the ingredients of lactation diets may come from different unknown sources.

VITAMIN REQUIREMENTS FOR MILK PRODUCTION

The subject of vitamins has been reviewed in Chapter 11. Only points important in milk production are mentioned here.

Water-soluble Vitamins

All the vitamins of the B complex (see Chapter 12) are synthesized by the ruminal microflora in sufficient amounts for lactation; normally none are required by the cow. An exception to this rule are cattle suffering from lactic acidosis (see Chapter 24). In this complex metabolic disease, the biosynthesis of some vitamins of the B complex is reduced to the point where their injection produces a positive response.

By definition ascorbic acid (vitamin C) is not a vitamin in ruminants since they produce it in their tissues.

Fat-soluble Vitamins

Carotene and vitamin A in milk vary greatly with their level in the diet and the amounts stored in the liver and in body fat. They also vary with breed, as explained in Chapter 11. The requirement proper for vitamin A is strictly a function of body size; fed above the level required to maintain the cow's health its level in milk sharply increases, but this has no effect on milk production. Usually recommendations are liberal because synthetic vitamin A is inexpensive.

The supply of vitamin A in the liver may be sufficient to maintain a cow's health for several weeks, and the release of vitamin A from the liver is proportional to storage. Therefore, a vitamin A deficiency in all classes of stock develops slowly. An exception to the rule is the newborn calf, which is very sensitive to vitamin A deficiency. Since *in utero* it depends entirely on the dam and after birth on colostrum, an adequate supply of vitamin A in the dry period is probably more important than during lactation. Especially with the current trend of moderation in feeding dry cows in order to prevent periparturient problems, to the vitamin A status of dry cows must be paid special attention. The routine supplementation of vitamin A is justified because its level in feeds is extremely variable. A large excess of vitamin A is very toxic (see Chapter 11).

There is no evidence for a vitamin D requirement for lactation. The amount synthesized in the body by solar ultraviolet radiation appears to be sufficient (see Chapter 10). Its level in milk is not augmented by supplementing the diet of the cow. There is storage of vitamin D in the liver and a daily supply is unnecessary. The therapeutic use of vitamin D and some of its more active derivatives in the prevention of parturient hypocalcaemia have received considerable attention (see Chapter 24).

Vitamin E is essential for all classes of ruminants, but because of its prevalence in common feeds deficiencies in cows have not been reported. Vitamin E reduces the incidence of oxidized flavour in milk and prevents muscular dystrophy in calves and lambs in conjunction with selenium (see Chapters 10 and 11). However, the rate of passage of vitamin E into milk is less than 2% and it is expensive. Therefore, muscular dystrohpy in calves and lambs is often prevented by direct injections of vitamin E and selenium into young stock rather than by supplementing the dams. There is no information about the Vitamin E requirement of lactating cows. The same is true for vitamin K, which is prevalent in feeds and synthesized in the rumen.

REQUIREMENTS OF LACTATING GOATS AND SHEEP

The principles for feeding lactating goats and sheep are similar to those for feeding cows. As in cattle, milk composition varies with breed, stage of lactation and feeding. The composition and energy content of goat and sheep milk are described in Table 21.4. Since the net efficiency of conversion of ME to milk is about the same in all three species, the energy requirement for milk production must only be adjusted to the energy content of milk which is 9% higher in goats, and 73% higher in sheep, than in cattle. It is reasonably well established that sheep have a lower phosphorus requirement than cattle.

Ewes lactate for about 12–20 weeks. The peak of lactation is 2–3 weeks after lambing. In the ewe, as in some of the primitive breeds of cattle, milk production is greatly stimulated by the offspring; ewes suckling twins produce more milk than ewes suckling a single lamb, and in ewes raised for milk production suckling following milking also greatly increases total production. The correct prepartum feeding of ewes is more critical than that of cows; both underfeeding and overfeeding may precipitate pregnancy disease, a severe metabolic disturbance nearly limited to ewes carrying more than one foetus (see Chapter 24). Ewes are also very sensitive to underfeeding during lactation. It is important to realize that, relative to metabolic weight, milk production in goats and ewes may be just as high as in dairy cows.

REQUIREMENTS OF LACTATING SOWS

The length of the lactation period in sows is between 6 and 8 weeks, and milk production peaks between the third and fifth weeks of lactation. The average milk yield of a typical 8-week lactation is 300–400 kg; it varies with breed, age and litter size. Daily yield increases with the number of piglets suckled, increasing by 0.9–1.0 kg per piglet for litters of less than eight piglets and 0.7–0.8 kg for litters of 9–12. Sow's milk is more concentrated than that of the cow (see Table 21.4). Trends in the composition of milk throughout lactation are similar to those of cows, except in fat content which is highest in midlactation.

Energy requirements for lactating sows are calculated factorially by summing those for maintenance, milk production and body-weight change. It is usual to express energy requirements for swine as digestible energy (see Chapter 18). The efficiency of utilization of digestible energy for maintenance is 0.75 and for lactation 0.70.

Whereas foetal and postnatal growth are influenced little by energy intake during pregnancy, body-weight gain during pregnancy is important to serve as an energy reserve for being converted to milk. The intake of adequate amounts of energy during pregnancy and the restricted intake during subsequent lactation lead to increased production of milk of high fat content and thus enable the satisfactory rearing of piglets.

The protein requirement of lactating sows is the sum of their maintenance requirement and that required for milk production. The lactating sow is an efficient converter of protein into milk. Dietary protein is the sole source of essential amino acids. The high protein levels usually recommended for the diets of lactating sows are required only because of inadequate protein quality. For example, the biological value of barley protein can be raised from 0.56 to 0.72 by supplementation with L-lysine.

BIBLIOGRAPHY

Books

Falkoner, T. R. (1971). *Lactation.* London: Butterworths.

Larson, B. L., and Smith, V. R. (1974–1978). *Lactation. A Comprehensive Treatise,* 4 vols. New York: Academic Press.

Mepham, T. B. (1976). *The Secretion of Milk: Studies in Biology,* No. 60. London: Edward Arnold.

Mepham, T. B. (1983). *Biochemistry of Lactation.* Amsterdam: Elsevier.

Peaker, M. (1977). *Comparative Aspects of Lactation.* London: Academic Press.

Biosynthesis of milk

Articles

Clark, J. H., Spires, H. R., and Davis, C. L. (1978). Uptake and metabolism of nitrogenous components by the lactating mammary gland. *Fed. Proc.*, **37**, 1233–1238.

Ebner, K. E., and Schanbacher, F. L. (1974). Biochemistry of lactose and related carbohydrates. In *Lactation* (B. L. Larsen and V. H. Smith, eds), Vol. 2, pp. 77–113. New York: Academic Press.

Larson, B. L. (1979). Biosynthesis and secretion of milk proteins, a review. *J. Dairy Res.*, **46**, 161–174.

Linzell, J. L. (1974). Mammary blood flow and substrate uptake. In *Lactation* (B. L. Larson and V. R. Smith, eds), Vol. 1, pp. 143–225. New York: Academic Press.

Moore, J. H., and Christie, W. W. (1980). Lipid metabolism in the mammary gland of ruminant animals. *Prog. Lipid Res.*, **17**, 347–396.

Peaker, M., and Faulkner, A. (1983). Soluble milk constituents. *Proc. Nutr. Soc.*, **42**, 419–425.

Rook, J. A. F., and Thomas, P. C. (1983). Milk secretion and its nutritional regulation. In *Nutritional Physiology of Farm Animals*, (J. A. F. Rook and P. C. Thomas, eds), pp. 314–368. London and New York: Longman.

Thomas, C. P. (1983). Milk protein. *Proc. Nutr. Soc.*, **42**, 407–418.

Milk composition

Jenness, R. (1974). The composition of milk. In *Lactation* (B. L. Larson and V. R. Smith, eds), Vol. 3, pp. 3–107. New York: Academic Press.

Oldham, J. D., and Sutton, J. D. (1979). Milk composition in the high yielding cow. In *Feeding Strategy for the High Yielding Cow* (W. H. Broster and H. Swan, eds), pp. 114–147. London: Granada.

Smith, S., and Abraham, S. (1975). The composition and biosynthesis of milk fat. *Adv. Lipid Res.*, **13**, 195–239.

Effects of the diet on milk production

Books

Broster, W. H., and Swan, H. (eds) (1979). *Feeding Strategy for the High Yielding Dairy Cow.* London: Granada.

McCullough, E. N. (1973). *Optimum Feeding of Dairy Animals.* Athens: University of Georgia Press.

Miller, W. J. (1979). *Dairy Cattle Feeding and Nutrition.* New York: Academic Press.

Articles

Baldwin, R. L., and Smith, N. E. (1983). Milk production. In *Dynamic Biochemistry of Animal Production* (P. M. Riis, ed.), pp. 359–388. Amsterdam: Elsevier.

Broster, W. H., Broster, V. J., and Smith, T. (1969). Experiments on the nutrition of the dairy heifer. Effect on milk production of level of feeding at two stages of the lactation. *J. Agr. Sci.*, **72**, 229–245.

Broster, W. H., and Thomas, C. (1981). The influence of level and pattern of concentrate input on milk output. In *Recent Advances in Animal Nutrition—1981*, pp. 49–70. London and Boston: Butterworths.

Clark, J. H., and Davis, C. L. (1980). Some aspects of feeding high producing dairy cows. *J. Dairy Sci.*, **63**, 873–885.

Clapperton, J. L. *et al.* (1980). The production of milk rich in protein and low in fat, the fat having a high polyunsaturated fatty acid content. *J. Sci. Food Agr.*, **31**, 1295–1302.

Klopfenstein, T., and Owen, F. J. (1981). Value and potential use of crop residues and byproducts in dairy rations. *J. Dairy Sci.*, **64**, 1250–1268.

Palmquist, D. L. (1984). Use of fats in diets for lactating cows. In *Fats in Animal Nutrition* (J. Wiseman, ed.), pp. 357–382. London: Butterworths.

Palmquist, D. L., and Jenkins, T. C. (1980). Fat in lactation rations. *J. Dairy Sci.*, **63**, 1–14.

Rook, J. A. F. (1976). Nutritional influences on milk quality. In *Principles of Cattle Production* (H. Swan and W. H. Broster, eds), pp. 221–236. London: Butterworths.

Steele, W., Noble, R. C., and Moore, J. H. (1971). The effects of 2 methods of incorporating soybean oil into the diet on milk yield and composition in the cow. *J. Dairy, Res.*, **38**, 43–64.

Storry, J. E. (1981). The effect of dietary fat on milk composition. In *Recent Advances in Animal Nutrition—1981*, pp. 3–34. London and Boston: Butterworths.

Thomas, P. C., and Rook, J. A. F. (1983). Milk production in the cow, ewe and sow. In *Nutritional Physiology of Farm Animals* (J. A. F. Rook and P. C. Thomas, eds), pp. 558–622. London and New York: Longman.

Waldo, D. R., and Jorgensen, N. A. (1981). Forages for high animal production. *J. Dairy Sci.*, **64**, 1207–1229.

Energy requirements and efficiency of milk production

(See also Bibliography to Chapters 14 and 16)

Bines, J. A., and Hart, I. C. (1978). Hormonal regulation of the partition of energy between milk and body tissues in adult cattle. *Proc. Nutr. Soc.*, **37**, 281–287.

Broster, W. H. (1976). Plane of nutrition for the dairy cow. In *Principles of Cattle Production* (H. Swan and W. H. Broster, eds), p. 271–285. London: Butterworths.

Flatt, W. P., Moe, P. W. *et al.* (1969). Energy utilization by high producing dairy cows. In *Energy Metabolism of Farm Animals.* 4th Symposium, Poland (C. L. Blaxter *et al.*. eds), pp. 221–252. Newcastle: Oriel Press.

Flatt, W. P., Moe, P. W. *et al.* (1972). Energy requirements of cows for lactation. In *Handbook of Animal Nutrition* (W. Lenkeit and K. Breirem, eds), Vol. 2, pp. 341–392. Hamburg and Berlin: Paul Parey.

Moe, P. W. (1981). Energy metabolism of dairy cattle. *J. Dairy Sci.*, **64**, 1120–1139.

Moe, P. W., and Tyrrell, H. F. (1974). See Bibliography for Chapter 16.

Moe, P. W., and Tyrrell, H. F. (1975). Efficiency of conversion of digested energy to milk. *J. Dairy Sci.*, **58**, 602–610.

Tyrrell, H. F. (1980). Limits to milk production efficiency by the dairy cow. *J. Dairy Sci.*, **51**, 1441–1447.

Van Es, A. J. H., and van der Honing, Y. (1979). Energy utilization. In *Feeding Strategy for the High Yielding Dairy Cow* (W. H. Broster and H. Swan, eds), pp. 68–89. London: Granada.

Protein requirements for milk production

Book

Buttery, P. J. (1979). *Protein Metabolism in the Ruminant.* London: Agricultural Research Council.

Articles

Broderick, G. A., Satter, L. D., and Harper, A. E. (1974). Use of plasma amino acid concentration to identify limiting amino acids for milk production. **J. Dairy Sci.**, **57**, 1015–1023.

Burroughs, W., Nelson, D. K., and Mertens, D. R. (1975). Protein physiology and its application in the lactating cow. The metabolizable protein feeding standard. *J. Anim. Sci.*, **41**, 933–944.

Chandler, P. T. *et al.* (1976). Protein and methionine hydroxy analog for lactating cows. *J. Dairy Sci.*, *59*, 1897–1909.

Hibbitt, K. G. (1984). Effect of protein on the health of dairy cows. In *Recent Advances in Animal Nutrition—1984*, pp. 189–200. London: Butterworths.

Huber, J. T., and Kung, L. (1981). Protein and nonprotein nitrogen utilization in dairy cattle. *J. Dairy Sci.*, **64**, 1170–1195.

Kaufman, W. (1979). Protein utilization. In *Feeding Strategy for the High Yielding Dairy Cow* (W. H. Broster and H. Swan, eds), pp. 90–113. London: Granada.

Oldham, J. D. (1980). Amino acid requirements for lactation in high-yielding dairy cows. *Recent Advances in Animal Nutrition—1980*, pp. 33–66. London: Butterworths.

Satter, L. D., and Roffler, R. E. (1975). Nitrogen requirement and utilization in dairy cattle. *J. Dairy Sci.*, **58**, 1219–1237.

Mineral requirements for milk production

Fettman, M. J. *et al.* (1980). Chloride nutrition of lactating dairy cows. *Proceedings of the Cornell Nutrition Conference, 1980*, p. 117.

Miller, W. J. (1975). New concepts and developments in metabolism and homeostasis of inorganic elements in dairy cattle. *J. Dairy Sci.*, **58**, 1549–1560.

Miller, W. J. (1981). Mineral and vitamin nutrition of dairy cattle. *J. Dairy Sci.*, **64**, 1196–1206.

CHAPTER 22

The Nutrition of the Horse and the Production of Work

Although mechanical devices have replaced the work horse in the developed countries, horses in increasing numbers are used for racing and other recreational purposes. In horse breeding the emphasis has shifted accordingly, because the physical characteristics and the temperament of the saddle horse differ from those of the work horse. However, the nutritional requirements of horses have not changed; physical performance is the major output of work and saddle horses, and the principles of feeding work horses for optimal performance apply equally to saddle horses. In fact, the specific requirements for work, i.e. physical activity, apply to the mule, the ox, or any other species used for work and to animals that engage in spontaneous physical activity.

In this chapter, the nutrition of the horse, energy conversion in muscle, the efficiency of conversion of dietary energy into work and the nutritional requirements for work are briefly discussed.

THE NUTRITION OF THE HORSE

The horse is a monogastric herbivore. Its stomach is relatively small, even in comparison to other monogastric species, but it has a voluminous caecum and large intestine (colon). The latter compartments of the digestive tract of the horse have functions similar to those of the rumen of cattle and sheep (see Chapter 3). However, the horse does not ruminate and the gases produced by fermentation in the caecum and the colon escape from the gastrointestinal tract through the rectum. In addition, the digestion of fibre in the horse occurs in posterior rather than anterior compartments of the tract. Consequently, all the dietary fibre consumed by the horse must pass practically intact through the stomach and the small intestine; therefore, the rate of passage of digesta from the stomach to the distal end of the small intestine is about three times greater than in ruminants. Although the microflora and the biochemical activity in the caecum and colon are similar to those in the ruminant, fibre digestion in the horse is not as efficient as it is in farm ruminants. It has been estimated that diets containing less than 15% fibre are digested in the horse with an efficiency similar to that of ruminants; diets containing more fibre are less efficiently digested. Nevertheless, the horse digests fibre more efficiently than the rabbit.

Other distinctive features of the digestive tract of the horse are related to the general characteristics described above. Salivary flow and its urea content are lower in the horse than in the ruminant. Although, as already mentioned, gastric emptying is rapid, some bacterial fermentation occurs in the stomach; lactic acid is produced, probably from readily fermentable carbohydrates in the diet. This is consistent with the fact that the pH in the stomach is too low to permit cellulolytic fermentation. The lactic acid produced in the stomach appears to be partly absorbed from the small intestine and partly fermented in the caecum and colon to volatile fatty acids. Although the rate of passage of digesta in the small intestine is also rapid, all the common digestive enzymes have been identified in this part of the tract. Pancreatic trypsin and amylase activities are high; so are the activities of intestinal lactase and maltase. The activities of the intestinal proteolytic enzymes, peptidases and sucrase are moderate. Fat is well digested by the horse; the site of digestion has not been reported. Up to 16% of dietary fat has been fed to horses with no untoward results; this level is higher than that which may be safely fed to ruminants in an unprotected form. The greater ability of the horse to tolerate fat is not surprising. In ruminants, fat which escapes the rumen is also well absorbed in the small intestine, but in the rumen it may severely interfere with bacterial activity, especially fibre digestion. In the horse, fat is evidently digested and absorbed before the digesta reach the caecum and the colon; therefore, in the horse fat does not interfere with the microflora.

In the distal end of the ileum, chloride is absorbed and bicarbonate is excreted. This appears to be an adaptation mechanism which increases the pH of the digesta before they reach the caecum and the colon, where the cellulolytic bacteria and the protozoa require a pH close to neutral. Bicarbonate excretion in the ileum of the horse may be compared to bicarbonate excretion in ruminant saliva.

In addition to microbial cellulolytic activity, bacterial protein degradation also occurs in the caecum and colon of the horse. From these compartments, volatile fatty acids, amino acids and electrolytes are absorbed. Urea added to a low-protein diet increases nitrogen retention in the horse; however, it is unlikely that dietary urea reaches the caecum and colon with the digesta. It is more reasonable to assume that urea diffuses into these compartments from the circulation. Water-soluble vitamins of the B complex are also synthesized by the caecal and colonic microflora; however, the rate of synthesis of these is not sufficient to supply the requirement of the horse. On semisynthetic diets low in these vitamins, horses develop deficiency symptoms of these vitamins.

Generally, the gastrointestinal tract of the horse is adaptable to a wide range of diets. On diets high in concentrates it functions more like a monogastric tract; on diets high in roughage content it is more akin to the ruminant tract.

Horses are reputed to suffer from a variety of mild to severe digestive disturbances. These are probably related to the fact that horses, like ruminants, rely partly on bacterial fermentation for digesting their diet but are unable to eructate. Perhaps more so than cattle, horses must be frequently fed in order to avoid overeating, and the necessary changes in their diets must be very gradually introduced.

ENERGY CONVERSION IN MUSCLE

Muscle is a chemodynamic system, i.e. it converts chemical energy directly into dynamic (mechanical) energy. Muscle differs from the steam engine and the internal combustion engine, because these convert chemical energy first into thermal energy (heat) and the latter into dynamic energy. These engines are thermodynamic rather than chemodynamic systems.

Muscle contraction is an incompletely understood process. However, the energetics of muscle contraction are reasonably well understood. The following is a simplified account of contraction and the conversion of chemical energy into work in muscle.

Muscle consists of fibres which have a diameter of the order to 50 to 100 μ. The fibres in turn are made up of myofibrils with a diameter of 1–2 μ. In the myofibrils, parallel filaments of two types are seen: thin ones consisting of a protein called actin and thick ones consisting of the protein myosin. The diameters are of the order of 0.005 and 0.010 μ. Based on electron microscope studies, during contraction bundles of actin and of myosin filaments slide into each other. Contraction is set off by nervous impulses. The energy required for the contraction of muscle is supplied directly only by the enzymatic hydrolysis of ATP to ADP and phosphate. Therefore, muscle has been referred to as an *actomyosin–ATP machine.*

$$\text{ATP} \longrightarrow \text{ADP} + \text{P}_\text{i} + \text{mechanical energy} + \text{heat}$$

Myosin, in association with actin, has ATPase activity. Both calcium and magnesium ions are present in muscle and are required for the hydrolysis of ATP. In solution ATP relaxes the resting muscle fibres; this effect of ATP is regulated by a special mechanism.

ATP in muscle cells is usually produced from glucose. Muscle contains about 0.4% glycogen; this is the immediate source of glucose in muscle. In the contracting muscle the amount of glycogen diminishes; it is restored from glucose which enters the muscle cells from the circulation. In skeletal muscle, under predominantly anaerobic conditions, ATP is generated by the hydrolysis of glucose to lactic acid (see Chapter 4).

$$\text{Glucose} + 2\,\text{ADP} + 2\,\text{P}_i \longrightarrow 2\,\text{lactic acid} + 2\,\text{ATP} + \text{heat}$$

In the above process, of the total chemical energy available in glucose (2871 KJ/mol) less than 3% is stored in ATP, 5% is lost as heat and 92% is retained in lactic acid. In cardiac muscle more oxygen is available and glucose is more completely oxidized to carbon dioxide and water:

$$C_6H_{12}O_6 + 6\,O_2 + 38\,\text{ADP} + 38\,\text{P}_i \longrightarrow 6\,CO_2 + 6\,H_2O + 38\,\text{ATP} + \text{heat}$$

In this process 69% of the free energy of glucose is converted into ATP. In the aerobic hydrolysis of glucose, lactic acid (or pyruvic acid) is an intermediate. In skeletal muscle only 20% of the lactic acid is completely oxidized; the rest is transported by the circulation to the liver and resynthesized into glycogen, which in turn serves as a source of glucose for all purposes (see Chapter 4).

Although muscular contraction specifically requires ATP and the concentration of ATP in muscle is higher than in the liver, creatine phosphate, produced by the hydrolysis of ATP, is the major high-energy phosphate stored in muscle.

$$\text{ATP} + \text{creatine} \rightleftharpoons \text{ADP} + \text{creatine phosphate}$$

Very little free-energy change occurs in this reaction and no phosphate is set free. When ATP is used by muscle the reaction proceeds from right to left, and when ATP is produced from left to right. Creatine phosphate does not support muscle contraction, but the content of creatine phosphate in muscle (19 mmol/kg) exceeds that of ATP (4.5 mmol/kg) by a factor of about four to five and serves as a reservoir of energy for the production of ATP. The chemical formula of creatine phosphate is:

$$\begin{array}{c} HN \sim PO_3H_2 \\ | \\ C{=}NH \\ | \\ CH_3{-}N{-}CH_2.COOH \end{array}$$

Muscle may contract in the absence of a direct oxygen supply. This is possible because glycogen is present, the hydrolysis of glucose may be anaerobic, creatine phosphate restores ATP and myoglobin supplies some oxygen. Myoglobin is a red pigment similar to haemoglobin but has a greater oxygen-binding capacity (see Chapter 10). In the horse, for several seconds muscle may contract anaerobically and incur an *oxygen debt* which exceeds the amount reaching it five-fold. The ability to incur an oxygen debt, i.e. to produce a deficit of oxygen, is increased by training. However, with the depletion of ATP, fatigue sets in and after prolonged activity the accumulation of lactic acid causes muscle cramps.

To produce ATP, glucose is preferentially oxidized by muscle. However, fatty acids and ketone bodies (see Chapter 6) may also be used. Cardiac muscle, in particular, uses many substrates including fatty acids. The versatility of myocardium in the use of its fuel supply may be regarded as an important factor. In negative energy balance liver glycogen is depleted, the supply of glucose through the circulation decreases and more fat is used as a source of energy. A dietary excess of protein produces both glucogenic and ketogenic amino acid residues which may be used as indirect sources of energy. On the other hand, body protein is a source of energy only in starvation, after body fat is used up.

Some of the biochemical processes which relate to muscle contraction are summarized in Figure 22.1.

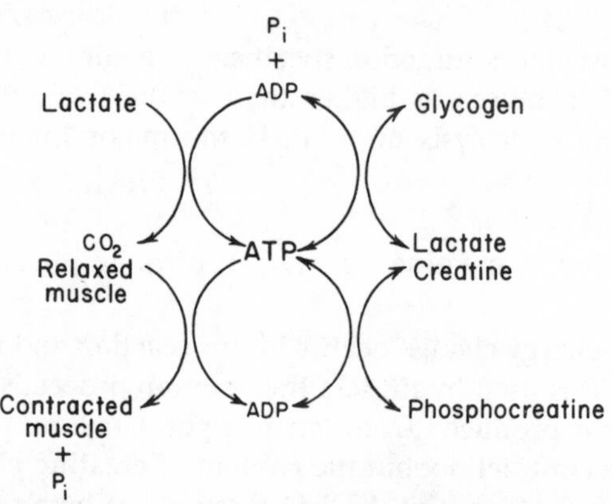

Figure 22.1. The energy for muscle contraction. *By permission from* Principles of Biochemistry, *5th edition,* p. 954, *by A. White* et al. *Copyright 1973 McGraw-Hill Book Co.* The immediate energy source is ATP which may be regenerated from creatine phosphate or from glycolytic or aerobic phosphorylations. (see Chapter 4).

THE NUTRITIONAL REQUIREMENTS FOR WORK

Qualitative Requirements for Work

An animal must be healthy in order to work efficiently. Its nervous, skeletal and muscular systems must function properly. The same holds for the circulation, respiratory system and the thermoregulatory mechanisms. A working animal spends more energy than one that is resting but it does not require more protein, minerals and vitamins. Although vitamin E is needed for the maintenance of the structural integrity of skeletal, cardiac and smooth muscle, and other vitamins and certain minerals are involved in muscular contraction, there is no evidence to support the contention that muscular work *per se* increases the required level of any specific nutrient in the diet. Often claims are made that manipulating the source of energy or increasing the quality or level of protein, or the level of certain vitamins and minerals in the diet, may increase the efficiency of the conversion of feed energy into work or the temporary physical performance of animals; such claims have not been confirmed in controlled experiments. The old belief that muscular work requires an increased level of protein in the diet has been disproved many times in the last 150 years, but appears to persist. In fact, in animals fed a balanced diet ad libitum, a large expenditure of energy caused by work, or by exposure to severe cold, or a combination of both, permits the *reduction* of the level of protein in the diet even in growing animals; the absolute requirement for protein of an animal that is working or exposed to cold weather, or both, is no greater than that of an animal that is resting in the comfort zone (see Chapter 17).

Only if work induces a *secondary* drain on a specific nutrient is the addition of that nutrient to the diet justified. This occurs under two sets of circumstances. (1) When an animal that has not worked before, starts working, its muscle mass increases; the deposition of protein in muscle requires some added dietary protein and energy. However, after the muscles have developed, the dietary requirement for protein reverts to the respective maintenance level. (2) When the performance of work, or intense heat, or the combination of these, causes profuse sweating, the nutrients lost in sweat must be replaced in the diet. In sweat, large amounts of water, sodium and chloride are lost and very small amounts of potassium, magnesium, calcium and phosphorus. If work increases sweating, water and sodium chloride indeed must be supplemented in a continuous and balanced way. This may be done in the form of saline containing 0.1–0.3% sodium chloride, offered at frequent intervals. Adding the other ions to the saline has been recommended but there is no evidence that this is useful. Dehydration is the major danger under conditions of hard work or intense heat. Dehydration in an otherwise healthy animal is caused by an insufficient supply of water, or salt, or both. A horse doing hard work may need up to 60 litres of water and 100 g of sodium chloride per day to avoid dehydration and eventual collapse and death (see Chapter 9). An unbalanced supply, i.e. offering salt and water separately at different times, may aggravate rather than prevent electrolyte disturbances.

Therefore, added energy to an adequate maintenance diet is the only dietary factor required for muscular work proper. Even so, if an animal is in fair body condition, in the short term, i.e. for a day or so, body stores can supply the energy required for hard work. Although the maintenance of hydration and electrolyte balance is important in the short term, the maintenance of energy balance is not. In fact, studies have shown that after a strenuous effort animals and man tend to postpone re-alimentation though not rehydration.

Quantitative Requirements for Work

As indicated above, work *per se* increases only the energy requirement of an animal. The concomitant increase in the requirement for water and sodium chloride is a function of thermoregulatory needs; it may be large in hot weather and insignificant in cold.

The added dietary requirement for work in terms of ME depends on the amount of work done and on the *partial* efficiency of utilization of ME for work. The upper limit on the partial efficiency of the conversion of ME into work is set by the efficiency of conversion of the energy in glucose (or other substrates) into ATP. As mentioned above, the latter is about 69%. However, the working animal which pulls a carriage or a plough—performing useful work—must walk and move its own mass too. Walking, of course, requires energy. Furthermore, unlike an engine, the horse also spends energy on maintenance, whether it is working or not.

The *partial* efficiency of animal work is the ratio between the amount of *useful work* done and the *difference* between the metabolic rate (ME oxidized by the animal) (1) working and (2) walking without working. This ratio* has been measured and is about 33%. In practice, of course, to perform traction the horse must walk and useful work represents only about 25% of the difference between the working and standing metabolic rate. The partial efficiency of animal work varies with the load. It decreases with an increasing load and measurements indicate that it may vary from 37% to 30%. If the efficiency of conversion of glucose to ATP is 69%, the maximum partial efficiency of conversion of ATP into work would be 0.37/0.69 = 0.54, i.e. 54%. In order to illustrate the relationship between the ME requirements for useful work, walking and maintenance, an example follows. It is taken from actual measurements performed by Zuntz.

A horse weighing 500 kg worked 8 hours daily pulling a plough at the speed of 4 km/hour with a force of 67 kg. To keep its body mass constant the horse needed 135 208 kJ of ME per day. The energy value of the work done in 8 hours was 67 kg × 8 h × 4000 m/h = 2144 000 kg m. Since 1 kg m = 9.807×10^{-3} kJ, the amount of useful work done was 21 026 kJ/day. The maintenance requirement of the horse was 47 301 kJ/day. Its requirement for walking was 624 kJ/km, i.e. for a distance of 32 km, a total of 19 968 kJ/day. Therefore, its requirement for work may be calculated:

---

* = Efficiency of animal work

A = total daily requirement		=	135 208 KJ
B = for maintenance		=	47 301 KJ
C = for walking		=	19 968 KJ
D = for work	$= A-(B + C)$	=	67 939 KJ

Thus, the partial efficiency of the production of work was:

$$\frac{21\,026}{67\,968} = 30.9\%$$

and its efficiency for work above maintenance was:

$$\frac{21\,026}{67\,939 + 19\,968} = 23.9\%$$

In the above example the total daily requirement of the horse was nearly three times higher than its maintenance requirement and heat production, i.e. the difference between the total daily requirement and the amount of useful work (135 208 − 21 026 = 114 128 kJ) was about 2.4 times greater than the maintenance requirement. This illustrates the point that work greatly increases the energy requirement and the heat load of the animal.

The specific maintenance requirement of the horse may be calculated from the above example. The metabolic body weight (see Chapter 17) of the horse was 105.7 kg ($500^{0.75}$ kg). Therefore, the specific maintenance requirement of the horse would be:

$$\frac{47\,301\ (\text{kJ})}{105.7\ (\text{kJ})} = 448\ \text{kJ/kg of metabolic body weight}$$

This value is near the average of several estimates reported for the mature horse and is rather higher than that of other farm animals; it merely suggests that the horse spends more energy on spontaneous activity than other species (see Table 17.5).

BIBLIOGRAPHY

Books

Cunha, T. J. (1980). *Horse Feeding and Nutrition.* New York: Academic Press.

National Research Council (1978). *Nutrient Requirements of Horses.* Washington, DC: National Academy of Sciences.

Articles

Hintz, H. F. (1984). Energy and Protein Requirements of Horses. In *Recent Advances in Animal Nutrition—1984* (W. Haresign and D. J. A. Cole, eds), pp. 177–186. London: Butterworths.

Hintz, H. F., and Squires, E. L. (1983). Equine reproduction and nutrition. *J. Anim. Sci.*, **57**, Suppl. 2, 58–74.

Meyer, H. (1983). Protein metabolism and protein requirement in horses. In *Protein Metabolism and Nutrition*, 4th International Symposium, France (as above) (M. Arnal *et al.*, eds), pp. 343–369. Paris: Institut National de la Agronomique Recherche.

Pearce, G. R. (1975). The nutrition of racehorses. *Aust. Vet. J.*, **51**, 14–21.

Robinson, D. W., and Slade, L. M. (1974). The current status of knowledge on the nutrition of equines. *J. Anim. Sci.*, **39**, 1045–1066.

CHAPTER 23

Voluntary Intake of Food

Research on the physiology of the control of food intake is related to two applied problems: the productivity of farm animals and human obesity. Animal productivity, i.e. body-weight gain, lactation and egg laying, depend largely on the intake of food. Therefore, the stockman strives to increase the voluntary intake of farm animals. On the other hand, in the developed countries, where food is abundant, man is mainly concerned with the limitation of food intake in order to control obesity. Gross overweight is associated with several pathological conditions and a reduced life span. High blood pressure and heart disease are the most notorious examples of such conditions.

In the evolutionary process, many species of animals have acquired the ability to select some of the foods available in their environment and to reject others. The classification of animals into carnivores, herbivores and omnivores implies such a characteristic. Obviously, in order to survive, the food selected by any species must satisfy its requirements for specific nutrients. To some extent all animals adapt the amount of food they consume to their requirements. Gross under-consumption would cause death from starvation and gross over-consumption would lead to immobilization due to obesity.

The fact that the food intake of animals is adapted to survival in the wild does not imply that this is the case in farm animals. In the past, the selection of fat animals was popular and produced breeds which consumed more food than was required for protein deposition. This was the case in certain breeds of poultry, swine, cattle and sheep. Obese pigs, unable to carry their own weight before slaughter, were common 50 years ago. For reasons of economy and the reduced demand for animal fat, such breeds have been replaced by leaner ones. Nevertheless, even now, farm animals of heavy breeds, especially old males kept for breeding purposes, may lose their ability to regulate their food intake and become very obese.

QUALITATIVE SELECTION OF FOODS

The qualitative selection of food is based on sensory cues and varies greatly among species. The chemical senses—taste and smell—play an important role in identifying suitable sources of food; they are decisive in man and in the rat. Sight and touch are also used to select foods. The chemical senses play a greater role in the food selection of mammals, and sight is more important in birds. The sense of touch registers the physical structure of the food, its texture, shape and size of particles, hardness, dustiness, etc. These characteristics serve to identify food and may also have physiological effects in the gastrointestinal tract and affect quantitative food intake. The temperature and moisture of food may also affect food selection. In newborn mammals, it appears that the chemical senses and sight have no role in food selection; feeding depends largely on the sense of touch. Newborn mammals eagerly suck on any object that can be held in their mouth, even if it yields no food. They also tend to seek warmth. Furthermore, lambs and chicks often select food by merely imitating the behaviour of other individuals in the flock.

Within species, food selection may vary among individuals depending on genetic traits and on previous experience. The latter also creates an association between the sensory cues presented by food and the expectation of satiety in the animal. Thus animals become conditioned to certain types of foods. It is well known that hunger, whatever its physiological equivalent, induces animals to be less selective. Poisonous plants in the pasture are usually consumed by hungry animals.

In nature, the excrement of one species may be the diet or part of the diet of another. Sparrows feed on the faeces of horses and pigs and poultry select grains from the faeces of cattle. Higher animals have an aversion to consuming the excreta of their own species, except for the coprophagic ones, e.g. the rabbit and the rat, which seize part of their own faeces excreted at night and consume them forthwith. Most animals also refuse to consume the faeces of other species, unless they are part of their natural diets. Nevertheless, in practice, poultry excreta have become a rather common ingredient of the diet of certain classes of ruminants.

Food selection is most common among higher animals. Examples can be given of species able to select an adequate diet, or even an optimal one, for their specific requirements. Such observations have led to the mistaken belief that animals invariably select the best diet and avoid the consumption of useless or harmful material. Herbivores in the pasture prefer forage of a high quality and relatively low is fibre to coarse forage and to poisonous plants.

Also, on diets moderately deficient in amino acids, poultry may slightly increase their food intake and compensate for the deficiency (see Chapters 7 and 18). Laying hens may adapt the quantity of food consumed to egg formation on days on which eggs are formed, if offered a diet high in calcium (3%) (see Chapter 20). Nevertheless, the assumption that animals have nutritional wisdom has been repeatedly proven to be erroneous.

The only nutrients for which higher animals have a specific appetite and which they will actively seek, are water and salt (sodium chloride), though the fowl shows no preference for salt even when it is severely deficient. On the other hand, offering salt ad libitum to animals may increase its voluntary consumption to levels which greatly exceed the nutritional optimum.

QUANTITATIVE CONTROL OF FOOD INTAKE

On empirical grounds it has become customary to distinguish between the long-term and short-term regulation of food intake. The long-term control of food intake was suggested by the fact that mature animals tend to maintain a constant body mass over long periods of time. Short-term control was inferred from observations on hunger and satiety, i.e. the physiological mechanisms which initiate and terminate feeding behaviour in animals and determine the amount of food consumed in one meal or during one day.

In the past, apparently conflicting theories have been put forward to explain the quantitative control of food intake. It appears now that the feeding behaviour of higher animals is controlled by one set of structures in the central nervous system, but many different variables produce peripheral and central stimuli which affect food intake. Although certain statements may be made about the relative importance of certain stimuli and pathways in one species or another, and concerning the shifting emphasis with changing conditions in the same species, in none of the higher animals is food intake wholly controlled by one type of stimulus, or even by a small number of stimuli. The quantitative control of food intake is complex and *multifactorial* in all the higher animals.

The interaction between qualitative and quantitative control of intake is an important one. In man, a person's state of mind affects food intake. In addition, palatability, a difficult parameter to measure but nevertheless a valid one, may greatly influence the intake of food. These factors may override the known mechanisms of the control of food intake. In animals it is difficult to evaluate state of mind though it probably plays a role in controlling food intake. Until recently, the effect of palatability on food intake in animals has been largely dismissed. Studies in laboratory rats and in domestic animals, with diets to which aromatic substances have been added, have produced only minor and transient changes in food intake. However, more sophisticated experiments in both monogastric and ruminant species indicate that palatability may greatly modify intake under certain conditions.

In the following sections the function of the central nervous system in the control of feeding behaviour and some of the many variables affecting food intake are briefly discussed.

Central Nervous Control of Food Intake

The exact location and role of all the central nervous structures which control food intake are not known. Nevertheless, there is conclusive evidence that the visceral brain centre of vertebrates is the *hypothalamus*, and well-defined regions in this part of the brain play a major role in the quantitative control of food intake in birds and in mammals.

The hypothalamus develops from the floor of the embryonic diencephalon and is located underneath the cerebrum, above the pituitary gland. Other structures in the midbrain, derived from the embryonic mesencephalon, appear to be closely associated with the control of food intake. Two well-defined bilateral areas in the hypothalamus control food intake: the *ventromedial hypothalamic nuclei* (VMN) and the *lateral hypothalamic area* (LHA). Bilateral destruction of the VMN induces hyperphagia, i.e. gross overeating and obesity, in at least 11 species of mammals and birds. In humans, injury to the VMN due to pituitary tumour produces the same result and led 40 years ago to the discovery of the role of VMN. Conversely, chemical or electrical stimulation of the VMN, as opposed to destruction, induces aphagia, i.e. it inhibits feeding. Therefore the VMN has been termed the *satiety centre.* On the other hand, bilateral destruction of the LHA induces aphagia until death in rats and in cats, whereas its stimulation induces hyperphagia. Consequently, the term *hunger centre* describes the LHA. The terms satiety centre and hunger centre, for the VMN and LHA, respectively, are convenient but must not be interpreted to mean that they act independently of other areas in the hypothalamus or in the brain to control food intake.

As indicated above, the destruction of the hunger centre inhibits feeding. The added destruction of the satiety centre produces no change. Lesions placed *between* these two structures produce hyperphagia. Therefore, it would appear that the hunger centre acts continuously to induce feeding, as long as it is not inhibited by the satiety centre. The spontaneous activity recorded in these two centres also substantiates their role in the control of food intake.

Long-term Control of Food Intake

Thirty years ago it was proposed that the total energy reserves of the animal are controlled. Later is was suggested that a substance associated with the storage of energy as fat is active in the control of food intake in the long term. Recent observations in partially lipectomized rats suggest that the size of individual fat cells rather than total body fat is controlled. Long-term control is suggested by the fact that mature animals tend to maintain their body weight over long periods of time. The reduction of the amount of body fat by either starvation, reduced food intake, prolonged exercise, lactation, or lipectomy in rats and mice, is followed by a restoration of the fat lost. This is mainly the result of an increase in food intake though other mechanisms have not been excluded. Conversely, forced feeding induces obesity and a reduction of the voluntary intake of food, occasionally to the point of aphagia. In dairy cattle, an inverse relationship has been observed between fatness before calving and food intake after calving (see Chapter 21).

In ruminants the normal volume of the gastrointestinal tract is relatively large; the reduction in the food intake of obese animals is confounded by the expansion of abdominal fat deposits which physically limit gastrointestinal capacity and indirectly food intake. However, sophisticated studies have shown that body fat in ruminants reduces voluntary food intake even on diets supplying a large proportion of concentrates, which minimize the effect of limited gastrointestinal capacity.

The way in which signals from adipose tissue reach the hypothalamus is not known. The simplest hypothesis is that adipose tissue liberates energy-supplying metabolites in proportion to its mass and these affect the hypothalamic centres to reduce the size of meals or their frequency. A steroid hormone was also postulated to be the signal produced by adipose tissue. However, the long-term control of intake is probably more complex. Destruction of the satiety centre, which causes hyperphagia, is associated with a reduction in the circulating levels of growth hormone, thyroxine, glucagon and adrenaline and an increase in the secretion of insulin and prolactin. Cause-and-effect relationships in this picture are far from being clear; all that can be stated with some certainty is that complex reciprocal effects exist between the hypothalamic centres, several endocrine glands and adipose tissue.

The hypothesis that stored energy interacts with food intake has received some support from recent studies with rats. Increased neonatal food intake in rats, caused by reducing litter size and increasing the milk supply, promotes the hyperplasia of adipose tissue. After weaning, the number of fat cells tends to level off though it may continue to increase in overfed animals; the number of fat cells produced does not diminish throughout life and their mass appears to determine subsequent adiposity and food intake.

Short-term Control of Food Intake

The peripheral stimuli which affect and control the food intake of animals are of a physical (or mechanical) and chemical nature. The distention of the gastrointestinal tract is finite and would limit intake even if no receptors innervated it. However, stretch receptors in the digestive tract have been demonstrated. In addition, receptors sensitive to different products of digestion transmit signals to the hypothalamus and control food intake. The hypothalamic satiety and hunger centres also appear to monitor directly the concentrations of certain metabolites in the blood.

The relative importance of the physical and the chemical control of food intake varies among species and with the diet. A useful generalization is that low-energy, high-bulk diets tend to affect food intake physically, whereas on high-energy, low-bulk diets the control of food intake is mainly chemical. This generalization has been graphically presented in Figure 23.1. It has been studied mainly in ruminants because, unlike monogastric species, these animals can subsist on diets ranging from a very low energy content to a very high one. In ruminants it is well established that, up to a certain level of digestibility or metabolizable energy content in the diet, food consumption is controlled by gastrointestinal fill and food intake increases with the energy of the diet. This level of digestibility is approximately 66%, which is the equivalent of about 9.2 MJ of ME per kg dry matter. In studies with forages, a rather constant amount of dry matter has been found in the rumen of sheep consuming diets differing widely in energy content. At a certain level of energy concentration in the diet, the physical limitation of intake ceases; above this level animals tend to maintain a constant rate of total energy intake and voluntarily reduce their dry matter intake below the physically limiting level. As indicated in Figure 23.1, monogastric species usually do not consume diets of a low energy content. Therefore, the physical control of food intake in monogastric species including man, has been underemphasized, even though the intake-limiting effect of inert substances added to the diet of rats and chicks (see Chapter 18) was studied long before the

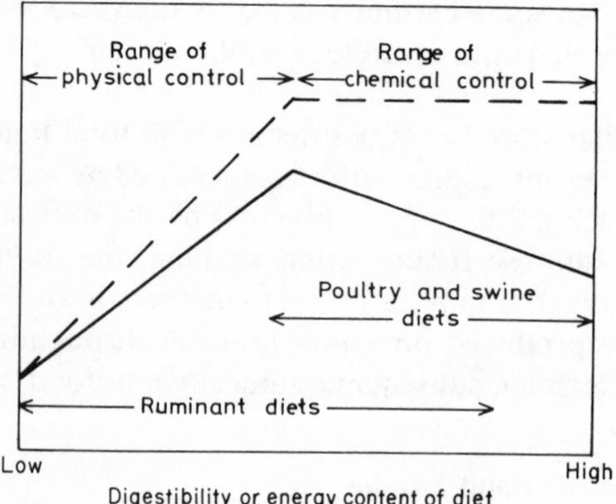

Figure 23.1. Ideal dry matter intake (——) and energy intake (- - - -) of animals assuming perfect physical and chemical control in two distinct ranges of energy concentration (adapted from Baumgardt, 1974).

relationship between energy concentration and intake in ruminants had been established. From the general relationship between physical and chemical control presented above, one must not infer that there is *exclusive* physical or chemical control in any roughage of energy concentration. Within limits, variation in intake may occur due to chemical and physical effects at any point throughout the whole range of energy concentration.

Physical (Mechanical) Control of Intake

The physical control of food intake was first shown in man by passing an inflatable balloon into the stomach. Pressure in the stomach inhibited hunger pangs (powerful contractions of the stomach) and reduced the desire for food, suggesting the existence of stretch receptors in the stomach. Later, rats fed diets diluted with inert materials were shown to maintain a rather constant level of energy intake up to a certain point. With further dilution, the volume of food became the limiting factor and energy intake declined. Similar evidence in ruminants has also been obtained with forages high and low in energy content. Stretch receptors have been found in the oesophagus, in the rumen, in the stomach of monogastric species and in the abomasum of ruminants, in the crop of the fowl and in the intestines of several species. Therefore, the short-term control of food intake by the physical volume of food and digesta is an established fact.

Subsistence on diets low in energy concentration, common in herbivores fed fibrous vegetation which physically limits food intake, may not lead to a high level of production. The productivity of swine and poultry also depends on the energy content of the diet. Only animals fed diets which do not limit food intake physically most of the time may be expected to be high producers.

When food intake is physically limited, most often in ruminants, grinding and pelleting may greatly increase it. The diminution of food particle size reduces or eliminates rumination, increases the rate of passage of digesta through the tract and reduces its retention in the rumen enabling the animal to eat more, provided the food is not dusty. Pelleting further increases intake. In ruminants the apparent digestibility of pelleted foods decreases but their metabolizability does not, because the reduced digestion of fibre in the rumen is compensated by diminished losses of methane (see Chapter 14). The relative effects of pelleting are greater the higher the fibre content of the food and pelleting is an efficient method of maximizing the intake of poor roughage.

Chemical Control of Food Intake

The long-term control of food intake described previously, depends on so far unidentified chemical stimuli presumably produced by adipose tissue. In contrast, the short-term chemical mechanisms which control hunger and satiety depend on substances present in the ingesta throughout the digestive tract, which include food and the digestive secretions, and on metabolites absorbed from the tract. These may affect the central nervous control of food intake directly through receptors in the digestive tract or elsewhere in the body, and indirectly by causing changes in endocrine secretions and in the amount of stored energy, which in turn may affect the central control.

A short list of the variables which have been implicated in the production of stimuli affecting feeding behaviour is glucose, amino acids, fatty acids, mineral ions, osmotic pressure and pH. Of these, glucose has received more attention than others but is not necessarily the principal substance which controls food intake.

Evidence from monogastric animals indicates that circulating glucose may control short-term food intake. Oral, hepatic and central nervous glucose receptors have been shown to exist in several species. The role of glucose was first inferred from the association of postprandial hyperglycaemia with satiety in man, in the dog and in other monogastric species. Later, it was suggested that the arteriovenous differences of glucose is more closely associated with satiety than the absolute level of glucose in blood. Arteriovenous differences of glucose are an indication of its rate of uptake by organs and cells. Glucoreceptors have indeed been found in the hypothalamus, and both the hunger centre and the satiety centre have been found to be involved in the indirect regulation of the level of blood glucose. The satiety centre inhibits insulin secretion and the hunger centre enhances it; in the simplest terms, insulin promotes glucose utilization by cells. Furthermore, the satiety centre is sensitive to glucose and increases the inhibition of the hunger centre as the availability of glucose increases. Thus, the controlling role of glucose is fairly well established in certain monogastric mammals.

In ruminants, the level of blood glucose, arteriovenous differences and glucose utilization decrease rather than increase with feeding, even in animals adjusted to one daily meal of a high-energy diet. Hypoglycaemia in lactating dairy cows and in pregnant ewes is a metabolic disturbance (see Chapter 24) due to an excessive drainage of glucose from the system and is accompanied by reduced food intake, i.e. no hunger, though mild acidosis confounds the picture. More striking, however, is that the intravenous infusion of glucose restores normal appetite. Thus, glucose in ruminants does not regulate food intake as it does in monogastric species.

In ruminants, considerable attention has been paid to the possible role of volatile fatty acids—the major products or rumen fermentation—in the short-term control of food intake. It appears that acetic and propionic acid affect food intake directly or indirectly, whereas the importance of butyric acid is only minor. However, the interpretation of the effects of volatile fatty acids on intake is difficult because they affect the pH and osmotic pressure of rumen fluid. Receptors of pH and of osmotic pressure have been shown in the rumen. In addition, pH and osmotic pressure modify rumen motility and blood flow and the effect of volatile fatty acids in vivo may be confounded by that of a very large number of products of bacterial fermentation. One can only state with assurance that a high rate of fermentation in the rumen is associated with the cessation of food intake and the initiation of rumination. Since rumen fermentation produces large amounts of gases which must be eructated and rumination coincides with satiety, it is reasonable to assume that many chemical and some physical stimuli activate the feedback mechanisms which control food intake in ruminants.

Some evidence suggests that amino acids play an active role in the control of food intake. As mentioned above, on diets slightly deficient in an amino acid, chickens and rats may slightly increase their food intake and compensate for the deficiency. However, on severely imbalanced diets, food intake is sharply decreased. Any large shortage or excess of an amino acid in the diet may severely reduce intake (see Chapter 7). A decrease in the concentration of a limiting amino acid may be the signal to the central nervous system. There is evidence that the brain contains receptors for amino acids or their metabolites; in rats and in chicks, injection of the limiting amino acid into the carotid artery in minute amounts prevented the depression of food intake although its injection into the jugular vein did not. Also, lesions in certain regions of the brain of rats prevented the depression of food intake due to amino acid imbalance. Tyrosine, phenylalanine and tryptophan serve as precursors of specific neurotransmitters and diets deficient in these amino acids appear to reduce food intake through their effect on the availability of the neurotransmitters in the brain. On normal diets, it appears that the role of amino acids in the regulation of food intake is a minor one.

Diets high in balanced protein reduce intake in most species. The effect has been attributed to the high heat increment of protein (see following section) and to a reduced rate of stomach emptying in monogastric animals. Animals fed diets high in protein also tend to be leaner than controls fed a normal diet.

Diets deficient in protein also reduce food intake. In all the species a protein deficiency reduces the production of digestive enzymes, and in ruminants it may inhibit bacterial fermentation in the rumen; furthermore, it causes tissue protein depletion. However, the signals which induce the hypothalamic centres to reduce intake are not known.

Fat supplies considerably more energy per weight than carbohydrates and proteins (see Chapter 14). Ruminants greatly reduce their food intake as the level of fat in the diet increases and they tolerate only about 10% of fat in their diets because fat interferes with normal rumen function. Monogastric animals tolerate large amounts of fat but on high-fat diets they also greatly reduce their intake in terms of dry matter, apparently monitoring energy content. Except for the glycerol moiety, the fat molecule does not produce glucose and no receptors for long-chain fatty acids are known. In fact, there is no more information on the mechanism by which a high level of dietary fat affects food intake than there is on the mechanism of the long-term regulatory effect of adipose tissue.

An elevated level of calcium ions in the brain produces feeding in satiated rats. The hen shows a specific appetite for calcium that is associated with the laying cycle (see Chapter 20), but in mammalian species blood calcium is rather constant and the role of calcium in the regulation of food intake is unexplored.

Increasing the osmotic pressure of ingesta in the duodenum has been shown to reduce food intake in pigs. In ruminants increasing the osmotic pressure in the rumen is also associated with inhibition of feeding. One would expect osmotic pressure to be equally important in the regulation of food intake in other species.

Receptors of pH have been found in the rumen of cattle. It appears that volatile fatty acids and lactic acid inhibit rumen motility and food intake through their effect on pH receptors.

Temperature and the Voluntary Intake of Food

Feeding and the absorption of nutrients rapidly increase the heat production of animals. The increased heat production associated with food consumption is referred to as heat increment (see Chapter 14). Obviously, there is a limited correlation between body temperature and satiety. It has been suggested that 'animals eat to keep warm or to prevent hypothermia' and this teleological dictum has been referred to as the 'thermostatic theory of the regulation of food intake'. Receptors sensitive to temperature are distributed throughout the body, including internal organs and the hypothalamus. In certain species, cooling the hypothalamus induces feeding, whereas warming inhibits it. This indicates that temperature plays a role in the control of food intake, though its effects are only one component of a multifactorial system.

An association also exists between ambient temperature, food intake and body fat. When ambient temperature increases, food intake tends to decrease and vice versa. At temperatures below the comfort zone (see Chapter 17), food intake increases, possibly for two reasons. Heat loss to the environment activates thermoregulatory reflexes (shivering or voluntary activity) which require energy and the energy balance of the animal is disturbed. Long-term control is activated and food intake increases. At the same time, the peripheral temperature of the animal decreases sharply and the short-term control of food intake through the hypothalamic thermoreceptors may be activated. When ambient temperature rises above the comfort zone, food intake is reduced even though energy expenditure increases due to the activation of panting, a thermoregulatory mechanism. If hypothermia occurs, food intake is reduced, probably through the thermoreceptors of the hypothalamus. In the long run, on ad libitum food, exposure to cold weather increases the amount of adipose tissue in animals and exposure to heat decreases it. Therefore, the empirical effect of ambient temperature on the control of food intake is well established. Recently, it has been shown that insulin secretion is inversely related to ambient temperature, suggesting that the reaction to ambient temperature is prompt.

Hormones in the Control of Food Intake

As indicated above, many stimuli produced by feeding are transmitted directly to the central nervous system by receptors or indirectly by changes in endocrine secretion. Several hormones, e.g. secretin and cholecystokinin, are produced by the digestive tract proper; the clarification of their role in the control of food intake has only begun. The most important hormones affecting food intake appear to be insulin, thyroxine, testosterone, progesterone and oestrogen. Insulin, which enhances glucose utilization and lipogenesis, is produced by the endocrine pancreas and increases food intake in monogastric species. Thyroxine, produced by the thyroid gland, increases the rate of metabolism. Exogeneous thyroxine increases the metabolic rate of the animal and indirectly its food intake. In long-term studies with dairy cattle only transient increases in food intake and in milk production have been produced by thyroxine-like compounds. It is noteworthy that all the natural sex hormones affect food intake, possibly acting on the hypothalamus. Testosterone, the male hormone produced in the testes, is anabolic and tends to increase food intake. In females, progesterone, produced by the corpus luteum during pregnancy, is also anabolic and tends to increase food intake. Oestrogen, the hormone of ovulation, depresses food intake. A large number of commercial synthetic steroids with some hormonal activity are used as growth promoters; they may act by being anabolic, by reducing energy expenditure and by increasing food intake (see Chapter 18).

Species and Breed Differences

Many differences observed among species in the control of food intake have been mentioned in this chapter. In this section some of the differences are further emphasized.

Ruminants

The energy and dry matter intake of ruminants as a function of metabolizable energy (ME) is presented in Figure 23.2. It appears that ruminants conform rather well to the model presented in Figure 23.1. At very high levels of ME, total energy intake declines. This is due to the increased intensity and number of negative feedback stimuli due to intensive rumen fermentation. Chemical variables implicated in the decline of food intake are pH, osmotic pressure, the organic acids formic, acetic, propionic and lactic acids, and amines. The dry matter intake of silage is usually lower than that of fresh or dried material from the same crop. This is probably due to the fact that all the variables listed above which depress food intake, accumulate in silage during fermentation. Lactic acid, ammonium salts and amines are thought to depress the intake of silages more than other substances. The intake of very acid silage may often be increased by neutralization, e.g. with sodium bicarbonate. On the other hand, preservatives added before ensiling, e.g. mineral acids, formaldehyde, formic acid or ammonia, considerably limit fermentation and may thus increase the intake of the ensiled material.

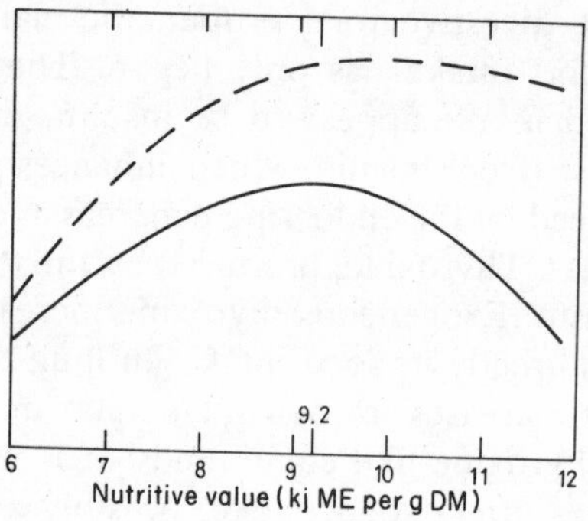

Figure 23.2. Schematic relationship between nutritive value, dry matter intake (——) and energy intake (– – – –) in ruminants (after Ulyatt, 1973, p. 141). *Reproduced by permission of Academic Press and the author.*

Fowl

At increasing levels of ME in the diet poultry reduce their dry matter intake but total energy intake may continue to increase. The relationship between ME, on the one hand, and dry matter and energy intake on the other, is presented in Figure 23.3. The slopes of the lines vary between light and heavy breeds of poultry. An increase of 10% in the ME content of the diet may cause an increase of 5% in the energy intake of heavy breeds and of 2–3% in light breeds. Accordingly, heavy breeds deposit more fat than light breeds. The effect of adipose tissue on food intake in the fowl is probably less important than in other monogastric species, except certain breeds of swine which tend to deposit excessive amounts of fat. To prevent undesirable fattening, different methods of restricting food intake are widely practised in poultry. The role of glucose in the control of food intake in poultry appears to be a minor one

Swine

Very few studies have been devoted to factors controlling food intake

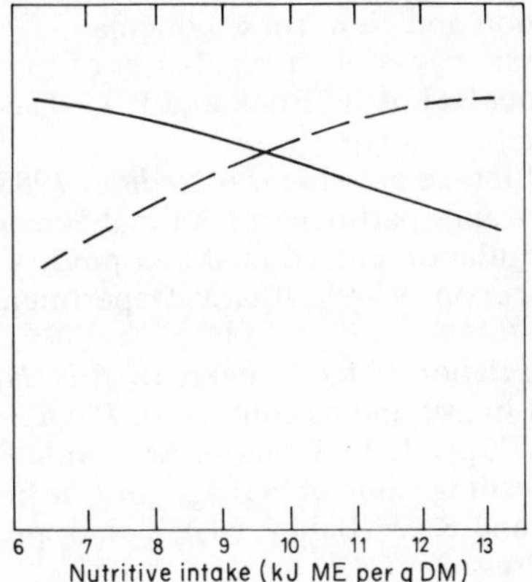

Figure 23.3. Schematic relationship between nutritive value, dry matter intake (——) and energy intake (----) in poultry.

in pigs. The effects of gastrointestinal distention and osmotic pressure are well documented. It also appears that, in the pig, cholecystokinin, a peptide hormone released from the intestinal mucosa after feeding, acts as a negative feedback signal in the control of food intake.

BIBLIOGRAPHY

Book

Boorman, K. N., and Freeman, B. M. (1978). *Food Intake Regulation in Poultry.* Edinburgh: British Poultry Science Ltd.

Articles

Symposium on 'Food intake, and its control by farm animals. (1985).' *Proc. Nutr. Soc.*, **44**, 303–362.

Baile, C. A., and Della-Fera, A. D. (1981). Nature of hunger and satiety control systems in ruminants. *J. Dairy Sci.*, **64**, 1140–1152.

Baile, C. A. (1979). Regulation of energy balance and control of food intake. In *Digestive Physiology and Nutrition of Ruminants* (D. C. Church, ed.), Vol. 2, pp. 291–320. Corvallis: O. and B. Books

Baile, C. A., and Forbes, J. M. (1974). Control of feed intake in ruminants. *Physiol. Rev.*, **54**, 160–214.

Baumgardt, B. R. (1970). Control of feed intake in the regulation of energy balance. In *Physiology and Digestion of the Ruminant* (A. T. Phillipson, ed.), pp. 233–253.

Baumgardt, B. R. (1974). Food intake, energy balance and homeostasis. In *The Control of Metabolism* (J. D. Sink, ed.), pp. 88–112. Pennsylvania State University Press.

Bines, J. A. (1979). Voluntary food intake. In *Feeding Strategy for the High Yielding Dairy Cow* (W. H. Broster and H. Swan, eds), pp. 23–48. London: Granada.

Campling, R. C., and Lean, I. J. (1983). Food characteristics that limit voluntary intake. In *Nutritional Physiology of Farm Animals* (J. A. F. Rook and P. C. Thomas, eds), pp. 457–475. London and New York: Longman.

Forbes, J. M. (1983). Physiology of food regulation of food intake. In *Nutritional Physiology of Farm Animals* (J. A. F. Rook and P. C. Thomas, eds), pp. 177–202. London: Longman.

Haupt, T. R. (1983). Feed intake of swine. *Proceedings 1983 Cornell Nutrition Conference*, pp. 99–107. Ithaca: Departments of Animal Sciences, Cornell University.

Kuenzel, W. J. (1983). Regulation of feed intake in poultry. *Proceedings 1983 Cornell Nutrition Conference*, pp. 94–98. Ithaca: Departments of Animal Sciences, Cornell University.

Lepkovsky, S. (1975). Regulation of food intake. In *Adv. Food Res.*, **21**, 3–21.

Sykes, A. H. (1983). Food intake and its control. In *Physiology and Biochemistry of the Domestic Fowl*, Vol. 4, pp. 1–30. London: Academic Press.

Ulyatt, U. J. (1973). The feeding value of herbage. In *Chemistry and Biochemistry of Herbage* (G. W. Butler and R. W. Bailey, eds), Vol. 3, pp. 131–178. London and New York: Academic Press.

CHAPTER 24

Metabolic Diseases in Ruminants

Strictly speaking, animal diseases are the domain of veterinary science. However, because of their complex gastrointestinal anatomy and physiology and the pressure of high production, ruminants suffer from diseases which are preventable by dietary means. Most, if not all, are not simple deficiency diseases. They are caused by interactions between the nutritional history of the animal, its body condition, production potential, diet and a certain genetic predisposition. Ruminant metabolic diseases represent several greatly different entities. Their common denominators are that none is infectious or degenerative though their incidence may increase with age, and, although breed differences exist, none is due to highly specific inborn errors of metabolism. All are preventable by nutritional management.

KETOSIS AND TOXAEMIA OF PREGNANCY

Ketosis is a disease of dairy cows in early lactation and toxaemia of pregnancy is one of ewes carrying more than one foetus. Both are caused by hypoglycaemia* (see Table 24.1) due to an excessive output of glucose and are common when the animal is in negative energy balance. Since energy is stored as fat, in periods of negative energy balance tissue mobilization supplies few glucose precursors (glucogenic substances). Body fat yields 90% fatty acids and only 10% glycerol; only the latter is glucogenic. Protein yields about 60% glucogenic amino acids but its mobilization is too limited. This is true in monogastric and in ruminant species but the latter have a more precarious glucose economy because in the rumen carbohydrates are fermented rather than merely hydrolysed to hexoses (see Chapter 5). Among

* It has been suggested that 'ketosis' be renamed 'hypoglycaemia'.

the major fermentation products, only propionic acid is glucogenic; acetate and butyrate are not. In the lactating cow glucose may be depleted mainly by its conversion to lactose in the udder, and in the pregnant ewe by the foetuses. In both diseases the timely and sustained intravenous infusion of glucose restores health. Feeding glucose is useless because it is fermented in the rumen and little or none is absorbed. However, glucogenic compounds like glycerol, propylene glycol and propionate may be added to the diet for prevention or therapy, though the infusion of glucose is more efficient.

In hypoglycaemic animals, compounds commonly referred to as *ketone bodies* are present in blood in elevated concentrations and spill over into the urine and into the milk of lactating animals (see Chapter 6). The ketone bodies are acetoacetate, 3-hydroxybutyrate (not a ketone) and acetone. The pathways of ketone body formation are presented in Figure 24.1.

Table 24.1 Levels of metabolites in the blood and liver of normal and hypoglycaemic ruminants

	Cows		Ewes	
	Normal	Hypoglycaemic	Normal	Hypoglycaemic
Blood (mmol/litre)				
Glucose	2.8	<2.2	2.8	<1.7
Ketones	1.7	>5.2	0.7	>3.5
Free fatty acids	0.3	0.6–2.0	0.3	1.0–2.0
Liver (% of wet wt)				
Glycogen	3.0	<0.8	3.0	0.3
Fat	3.0	>10.0	3.0	>20.0

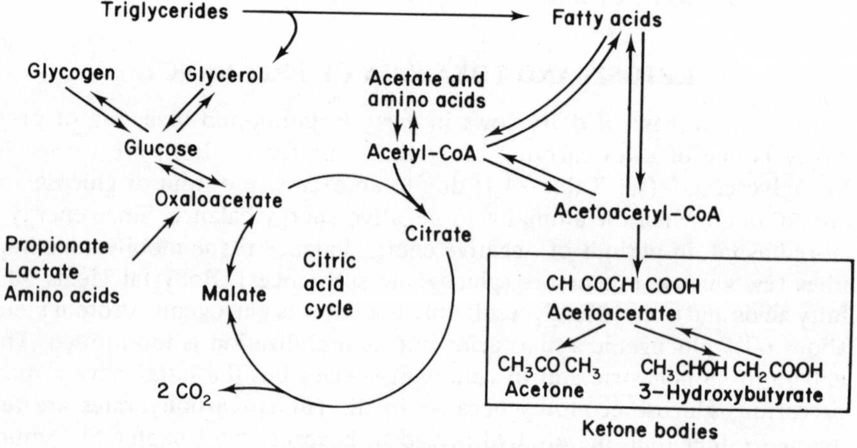

Figure 24.1. Major pathways of ketone body and free fatty acid metabolism.

Hypoglycaemia reduces the glycogen content of liver and the supply of oxaloacetic acid. The shortage of the latter limits the oxidation of acetate, the end-product of lipolysis*, in the citric acid cycle. In hypoglycaemia, insulin release is depressed, and since lipogenesis in the body depends on insulin, body fat is mobilized rather than synthesized; the liver is infiltrated by fat and the excess acetate is converted into acetoacetate. In ketotic animals the proportion of carbon dioxide derived from glucose drops from a normal value of 12% to 5%, and that derived from ketone bodies increase from a normal value of 2% to 20%.

Ketone bodies are normal, oxidizable metabolites when present at relatively low levels in blood; in ketosis and in pregnancy toxaemia, there is a general shift in metabolism from glucose to fat and an increase in the blood level of ketone bodies. Acetoacetate and 3-hydroxybutyrate are relatively strong acids; in ketosis they are incompletely oxidized, produce systemic acidosis (low blood pH), are excreted in urine and waste energy. Acetone, being volatile, may be detected by its odour in expired air.

Ketotic cows and ewes suffering from no infectious disease have no fever but are depressed; their food intake is low, especially that of concentrates. The milk production of cows drops and its fat content increases. Occasionally, ketotic females, especially ewes become excitable, probably because of the specific effect of isopropanol on the central nervous system; this compound is a decarboxylation product of 3-hydroxybutyrate. Primary ketosis and toxaemia of pregnancy occur when the output of glucose exceeds its availability; this is not likely when the intake of energy is curtailed due to the underfeeding of concentrates, an unpalatable diet, obesity, digestive upsets or errors of management. A high-producing cow may need 3000 g of glucose per day as a lactose precursor and for the synthesis of milk, but the circulating amount of glucose is less than 30 g and suffices only for about 15 minutes. Thus, the sensitivity of high producers to hypoglycaemia is not surprising.

Many other diseases which reduce food intake may produce *secondary ketosis*; of course, the primary disease must be diagnosed and treated. The infusion of glucose alleviates hypoglycaemia but does not cure other diseases. Accurate diagnosis of the primary cause is important.

Timely treatment of hypoglycaemia is essential in ewes. Liver and kidney damage may occur and mortality is common. The foetuses may die in utero and aggravate the condition of the ewe; abortion or induced lambing may bring about recovery. Treatment with adrenocorticotrophin or corticosteroids has had some success. These induce the mobilization of tissue protein which supplies glucogenic amino acids.

The incidence of hypoglycaemia may be reduced by preventing the overconsumption of energy *before*, and its underconsumption *during*, the critical period.

---

* Hydrolysis of triglycerides and breakdown of long-chain fatty acids.

PARTURIENT HYPOCALCAEMIA

This severe disorder, also known as *parturient paresis* and popularly as *milk fever*, is common in dairy cows at the onset of lactation. Its incidence increases with age. Jersey cows are more susceptible than other breeds. In ewes, a similar disorder, *lambing sickness*, occurs in late pregnancy. Parturient hypocalcaemia is a failure of the endocrine system to maintain the level of calcium in the blood; hypocalcaemia probably blocks neuromuscular transmission and general paralysis (paresis) occurs. Cows lie on their sternum with the head tucked into the flank, the reflexes disappear and there is hypothermia. Untreated animals usually die. Blood calcium depends on its rate of absorption from the gut, resorption from bone and output in milk and deposition in the foetus. Normally, these processes are regulated by at least three hormones: parathormone, calcitonin and 1,25-dihydroxy-*cholecalciferol*, the latter being a derivative of vitamin D_3 (see Chapter 11) produced in the kidney. Hypocalcaemia occurs when either absorption from the gut or resorption from the bone (or both) fails to respond in time to meet the increased demand for calcium. Blood calcium drops from a normal level of 2.5 mmol/litre to about half the normal level. Blood phosphorus also drops though magnesium rises.

The prevention of hypocalcaemia by hormones has a long history. Although parathormone promotes bone resorption, as a prophylactic it is ineffective. Calcitonin promotes bone accretion and may indeed produce the disease when injected. Vitamin D_3 and some similar compounds prevent the disease and are often used. However, their use is problematic because proper timing of their administration is essential but exact calving time is unpredictable. Repeated doses tend to be toxic. Two recently tested compounds seem to have low toxicity and hold promise to be efficient preventives of parturient hypocalcaemia: 25-hydroxycholecalciferol and 1α-hydroxycholecalciferol. It appears that the derivatives of vitamin D_3 act mainly by increasing calcium absorption.

The incidence of parturient hypocalcaemia can be greatly reduced through stimulation of the calcium regulatory mechanism by manipulation of the diet two to three weeks before calving. Two factors are effective: (1) a low level of calcium, and (2) a high level of *mineral* acidity, i.e. an excess of chloride ions in the diet over the sum of sodium and potassium ions. When parturient hypocalcaemia recurs in a herd, shortly before calving the recommendations for calcium must be set aside and the lowest possible level of calcium offered; the level of calcium must be restored to normal upon calving. Among all the calcium salts, calcium chloride is an exception; because of its high chloride content it stimulates the regulatory mechanism of calcium and may be used as a preventive two to three days before calving. It must be diluted and expertly administered by stomach tube. Dietary phosphorus has no consistent effect on the incidence of parturient hypocalcaemia.

Hypocalcaemia is treated by the intravenous injection of calcium borogluconate. The borogluconate is merely a convenient anion; it yields glucose though probably none is required because in parturient hypocalcaemia blood glucose is high. If necessary, injections of calcium must be repeated.

Downner Cow Syndrome

This term describes the inability of the cow after calving to rise in spite of efforts to do so; *alert downer* and *creeper* refer to the same condition. Death or forced disposal of the cow are common. One cause of the recumbency may be muscle or nerve injury occurring during calving due to mechanical factors, including difficult labour. However, authoritative reports suggest that the condition is often a mere secondary effect of severe parturient hypocalcaemia. In parturient hypocalcaemia, insulin, citric acid and potassium in blood are also low and hypothermia may be severe; irreversible muscle and nerve damage may result. The incidence of this disorder is on the increase in many intensive herds; it is very prevalent in obese cows (see *fat-cow syndrome* below) and it emphasizes the importance of the radical prevention and prompt treatment of parturient hypocalcaemia.

HYPOMAGNESAEMIA

Often called *grass tetany*, hypomagnesaemia is a dietary magnesium deficiency, preventable by magnesium added to the diet and curable by intravenous injections of magnesium salts (see Chapter 10). Nevertheless, it is regarded as a metabolic disease. Untreated hypomagnesaemia produces first muscle tremor, then staggering gait, general muscle spasm (tetany) followed by collapse and death within a few hours. These events reflect the hyperexcitability of the nervous system caused by low magnesium levels in all the body fluids and especially in the cerebrospinal fluid. Unlike the level of calcium, that of magnesium in blood is not known to be under efficient endocrine regulation. The deficiency of magnesium may be chronic, latent, and temporarily compensated for by reduced urinary output of magnesium. Calcinosis, i.e. the calcification of soft tissue, especially of the kidney and the heart muscle, is manifest in chronically magnesium-deficient animals. The dramatic events of hypomagnesaemia are set off in deficient animals by a variety of dietary factors: a high sodium/potassium ratio, lush grass high in nitrogen, high calcium, certain long-chain fatty acids, and possibly other organic acids such as citric and *trans*-aconitic. At least two mechanisms have been postulated to explain outbreaks of this disorder: (1) competitive inhibition of magnesium absorption (e.g. by potassium and by calcium), and (2) chelation of magnesium ions in body fluids (e.g. by organic acids).

Monitoring pastures for magnesium and potassium and the use of magnesium supplements probably prevent the occurrences of this often fatal disorder.

DISORDERS DUE TO READILY FERMENTABLE CARBOHYDRATES IN THE DIET

Concentrates high in starch and sugar are widely used to increase the efficiency of production in ruminants. However, an excessive level or a sudden increase in their level in the diet causes a variety of chronic or acute digestive disorders. All these disorders may be traced to two major events occurring in the rumen: (1) an increased rate of fermentation including gas production, and (2) the accumulation of lactic acid.

Rumen Parakeratosis

This disorder is common on high-grain diets, in feedlot cattle and in dairy cows. It is a chronic reversible keratinization of the rumen mucosa with some inflammation and ulceration. The performance of affected animals is only slightly reduced, but penetration of bacteria into the circulation causes liver abscesses, commonly found in feedlot cattle. The penetration of unidentified toxins also causes *laminitis*, an inflammation of the sensitive laminae of the hooves with oedema. Ruminants affected with rumen parakeratosis are more sensitive to mycotoxins in the food.

Lactic Acidosis (Acute Indigestion, Grain Engorgement or Founders)

Acute cases of this disease range in intensity from mild to fatal. Normally, lactic acid is continuously produced and fermented in the rumen and its level is less than 1 mmol/litre. A drop in rumen pH induces a change in the microflora and the overproduction of lactic acid. The level of lactic acid in the rumen of ill animals may range from 20 to 300 mmol/litre and rumen pH may drop to 4.5 or even lower. Lactic acid is absorbed and the acidosis becomes systemic, i.e. the blood pH drops. Depression sets in and the motility of the gastrointestinal tract is reduced. The high level of lactic acid in the rumen increases osmotic pressure and causes haemoconcentration and dehydration. High proportions of the D-isomer of lactic acid aggravate the disease because mammalian tissue metabolizes it less efficiently than the L-isomer (which is the natural one of higher animals). Therefore, the disorder has been termed *isomer* disease. The disorder is treated with fluid therapy and antacids, mainly magnesium oxide and sodium bicarbonate. Adequate fibre in the diet and gradual changes prevent it. Recently, new antibiotics have been effective in preventing the accumulation of lactic acid in experiments.

Displaced Abomasum

Atony (lack of motility) and gas production in the abomasum may cause its displacement and torsion, usually to the left, and sometimes to the right, side of the rumen. The attachment of the abomasum to the greater omentum is invariably ruptured. Evidence suggests that the overproduction of volatile fatty acids in the rumen and their passage to the abomasum may provoke the occurrence of displaced abomasum. In this disorder there is no interference with blood supply but the displacement interferes with the movement of the digesta. Chronic inanition and secondary ketosis follow. Displaced abomasum occurs more often in large cows fed high-concentrate diets in the dry period. Surgery may be required to correct the disorder; the abomasum is sutured into its proper position. The incidence of displaced abomasum is on the increase and it causes great economic losses.

Bloat

Accumulation of gas in the rumen, mainly because of the formation of stable foam, leads to bloat (ruminal tympany). Foam blocks the cardia and prevents eructation. The distention of the rumen may rupture the diaphragm and cause asphyxiation, but the actual cause of death in bloat is not always clear.

Two distinct dietary causes of frothy bloat are known: (1) legume pasture, and (2) high-grain diets (feedlot bloat). The foaming agent in legume bloat may be protein or chloroplasts in leaf cytoplasm. In feedlot bloat, the foaming agent is probably produced by bacteria and is a viscous carbohydrate polymer. Legume bloat may kill ruminants hours after feeding begins; feedlot bloat develops slowly and may become chronic. Legume bloat may be prevented by spraying the pasture with antifoaming agents: oils, detergents and silicones. Poloxalene is such a synthetic antifoaming agent. Feedlot bloat may be prevented by fibre in the diet. In severe cases it may be necessary to relieve an animal of bloat by the simple device of puncturing the rumen through the left flank with a trocar.

FAT-COW SYNDROME

Cows grossly obese at calving time due to low previous production, long dry periods and overfeeding, suffer from an increased incidence and severity of most of the non-infectious periparturient diseases, like ketosis, parturient hypocalcaemia and the downer syndrome, retained placenta and displaced abomasum, and some of the infectious ones such as mastitis and metritis. Although the syndrome is not well defined, it is well recognized as it increases the incidence of several diseases two- to ten-fold. In addition to obesity, the common signs of the fat-cow syndrome are lack of appetite, fatty infiltration of the liver and abnormal liver function. Probably the best indicator of liver damage is the appearance of liver-specific enzymes in the plasma due to partial liver necrosis. Research to develop diagnostic methods is now in progress, but it is safe to say that routine blood tests, sometimes referred to as *metabolic profiles*, other than for liver-specific enzymes, have little or no value in the *prediction* of overt disease.

The mortality of cows suffering from liver damage is greater than with any disease; liver damage is greater in cows conceiving late and appears to be cumulative. The incidence of the fat-cow syndrome is on the increase because of the widespread use of group-fed, total mixed diets on a free-choice basis.

BIBLIOGRAPHY

Books

Giesecke, D., Dirksen, G., and Stangassinger, M. (1981). *Metabolic Disorders in Farm Animals.* Munich: Institute for Physiology, Veterinary Faculty of the Univesity of Munich.

Kaneko, I. I. (1978). *Clinical Biochemistry of Domestic Animals,* 3rd edn. New York: Academic Press.

Payne, J. M., Hibbitt, K. G., and Sansom, B. F. (1973). *Production Disease in Farm Animals.* London: Baillière Tindall.

Articles

Symposium on metabolic disorders of the ruminant (1970). In *Symposia on Ruminant Physiology, Digestion and Metabolism,* 3rd Internat. Symposium, Cambridge, England (A. T. Phillips, ed.), pp. 545–629. Newcastle: Oriel Press.

Baird, G. D. (1977). Aspects of ruminant intermediary metabolism in relation to ketosis. *Biochem. Trans.,* **5**, 819–827.

Baird, G. D. (1982). Primary ketosis in the high-producing dairy cow: clinical and subclinical disorders, treatment, prevention and outlook. *J. Dairy Sci.,* **65**, 1–10.

Bergman, E. N. (1984). Disorders of carbohydrate and fat metabolism. In *Duke's Physiology of Domestic Animals,* 10th edn. pp., 412–422. Ithaca and London: Cornell University.

Clark, R. T. J., and Reid, C. S. W. (1974). Foamy bloat of cattle. *J. Dairy Sci.,* **57**, 753–785.

Dunlop, R. H. (1972). Pathogenesis of ruminant lactic acidosis. *Adv. Vet. Sci.,* **16**, 259–302.

Kronfeld, D. S. (1971). Hypoglycemia in ketotic cows. *J. Dairy Sci.,* **54**, 949–961.

Kronfeld, D. S. (1971). Parturient hypocalcemia in dairy cows. *Adv. Vet. Sci.,* **15**, 133–158.

Kronfeld, D. S. (1976). The potential importance of the proportions of glucogenic, lipogenic and aminogenic nutrients in regard to the health and productivity of dairy cows. *Advances in Animal Nutrition and Animal Physiology,* Vol. 7. Hamburg and Berlin: Paul Parey.

Lindsay, D. B., and Pethick, D. W. (1983). Metabolic disorders. In *Dynamic Biochemistry of Animal Production* (P. M. Riis, ed.), pp. 431–480. Amsterdam: Elsevier.

Littledike, E. T., Young, J. W., and Beitz, D. C. (1981). Common metabolic diseases of cattle: ketosis, milk fever, grass tetany and downer cow complex. *J. Dairy Sci.,* **64**, 1465–1482.

Pickard, D. W. (1977). Calcium requirements in relation to milk fever. In *Nutrition and the Climatic Environment* (W. Haresign *et al.,* eds), pp. 113–122. London: Butterworths.

Reid, R. L. (1968). The physiopathology of undernourishment in pregnant sheep, with particular reference to pregnancy toxemia. *Adv. Vet. Sci.,* **12**, 164–239.

Schultz, L. H. (1974). Ketosis. In *Lactation* (B. L. Larson and V. R. Smith, eds), Vol. 2, pp. 318–354. New York and London: Academic Press.

CHAPTER 25

Efficiency of Animal Production

World production of human food from animal sources in not likely to diminish in the future although it is problematic in some of the developing countries. In the developing countries, for the time being, farm animals, monogastric species more so than ruminants, compete with the human population for the available supply of the most valuable plant foodstuffs, including protein and energy. Nevertheless, the superior quality of proteins from meat, milk and eggs, and their palatability, have created pressure to increase their production in all but the poorest countries of the world.

The major handicap in the production of food from animal sources is its inefficiency; in terms of edible energy and protein per area of land, plant production is far more efficient. On an annual basis, the efficient production of milk by dairy cows may reach 12 000 MJ and 150 kg protein per ha, whereas that of a crop like potatoes may attain up to 180 000 MJ of edible energy and 220 kg of protein per ha. Sugarcane, the most efficient crop in terms of energy production, may yield up to 18 tonnes per ha with an energy content of 300 000 MJ. On the other hand, even a typical high-protein, low-energy crop like peas may produce up to 15 000 MJ and 325 kg protein per ha, i.e. 25% more energy and more than twice as much protein than obtained by the production of milk.

Plant production is inherently more efficient than animal production because it is carried out mainly by green plants which capture solar energy to synthesize carbohydrates, fats and proteins from carbon dioxide, water and inorganic nitrogen. Farm animals, on the other hand, depend on the consumption of the very same carbohydrates, fats and proteins produced by green plants and convert them into edible tissue, milk and eggs with an efficiency that does not exceed about 26% for protein and 18% for energy but may be very low for both.

Because of the low efficiency of animal production and the fact that it requires added investment and operating expenses, food from animal sources is considerably more expensive than food from plants. Consequently, the proportion of food from animal sources in the human diet is greatly related to income. Because of the increasing demand for food from animal sources, farm animal research everywhere deals with factors which affect the biological and economic efficiency of animal production. Plant and animal production interact in many ways; however, this chapter deals mainly with the major factors which affect the biological efficiency of meat, milk and egg production in the common farm animal species.

TERMINOLOGY

In dealing with the biological aspects of animal nutrition, it is often convenient to compare the net efficiencies of different functions (milk production, growth, etc.). The latter may be expressed as the ratio of net energy (NE), i.e. the combustible energy of the product (added animal tissue, milk or eggs) to the metabolizable energy (ME), digestible energy (DE) or gross energy (GE) consumed above the maintenance requirement of the animal, i.e. NE/ME, NE/DE or NE/GE. These terms have been defined in Chapter 14. The same is true for protein. Occasionally the ratio of NE, ME or DE to GE consumed, including the maintenance, is considered and is referred to as gross efficiency. In this chapter the emphasis is on the overall efficiency of production, which may be defined as the ratio of energy in edible, rather than total, tissue added, and milk or eggs produced, to the total of gross energy consumed by all the animals required to produce it; the same is true for protein. Thus, in estimating overall efficiency, the denominator includes the combustible energy of all the food consumed by (1) the producing animal from birth until production ceases, and (2) the food consumed by necessary non-producing animals until their disposal, e.g. breeding males and sterile females. The numerator, on the other hand, includes the combustible energy in milk or eggs and in the edible tissue obtained from fattening stock, from dairy females (or layers) and non-producing animals at disposal. Thus, overall efficiency measures the conversion ratio of fodder to edible human food by an entire flock or herd. Depending on circumstances, a large or small proportion of the fodder may well be edible human food. This may greatly influence the economics of animal production. Animal fodder fit for human consumption is usually more expensive. It may be noted that small amounts of food from animal sources (fish, feather and bone meal) may serve as fodder, though their effect on estimates of overall efficiency are very small.

BIOLOGICAL FACTORS AFFECTING THE OVERALL EFFICIENCY OF ANIMAL PRODUCTION

Table 25.1 presents some estimated overall efficiencies of the most common

Table 25.1 Estimated percentage overall efficiency of converting fodder into edible animal products (after Wedin *et al.*, 1975)

	Crude protein	Energy
Non-ruminants		
Broilers	23	11
Turkeys	22	9
Layers (eggs)	26	18
Swine	14	14
Ruminants		
Dairy cattle	25	17
Beef cattle	4	3
Mutton sheep*	4	2.4

* Producing one lamb per year.

classes of farm animals. These values indicate that the efficiency of conversion of fodder to edible products varies from about 2% to 18% for energy and from 4% to 26% for protein. The values in Table 25.1 represent near maximal efficiencies which may be obtained at high rates of production in conventional systems. They also suggest that the efficiency of energy and protein conversion vary greatly with the product (meat versus milk and eggs) and with the species. However, within each species and product, the efficiency of conversion depends primarily on the rate of production and the level of feeding, which are closely related.

Rate of Production

On acceptable diets and in a normal environment, the net efficiencies of production of body tissue, milk and eggs are reasonably constant. However, it is well known that the gross efficiency, and therefore the overall efficiency, of these processes is variable and depends on the rate of production, i.e. rate of weight gain, level of milk production and laying percentage. The reason for this variable gross efficiency is the fact that animals have a maintenance requirement. At an increased rate of production the food intake of the animal increases and the proportion of fodder expended on maintenance decreases (see Chapter 17). Since, by definition, food consumed for maintenance produces no product, it is wasted. Obviously, as the level of food intake increases in terms of multiples of the maintenance requirement from one to two, three, four and so on, the proportion of the ration expended on maintenance and wasted, decreases from 1/1 to 1/2, 1/3, 1/4 and so on. At the same time, the proportion of the ration which supports production with a rather constant net efficiency increases in a curvilinear fashion from zero to 1/2, 2/3, 3/4 and so on. The gross efficiency of production increases proportionally. Table 25.2 presents the food intake of different classes of stock in terms of multiples of the maintenance requirement. The ability of the high-producing dairy cows to consume up to four times her maintenance requirement should be noted; it contributes greatly to the fact that milk production vies with egg production for the highest overall efficiency.

Table 25.2 Food intake of producing animals in terms of multiples of the maintenance requirement in some classes of stock

Class of stock and level of production	Food intake in multiples of maintenance
Dairy cow weighing 550 kg	
producing 13 kg milk per day	2.0
producing 30 kg milk per day	4.0
Steer weighing 300 kg	
gaining 0.4 kg/day	1.4
gaining 0.7 kg/day	1.7
gaining 1.0 kg/day	2.0
Pig weighing 50 kg	
gaining 0.75 kg/day	2.3
Fowl weighing 1.0 kg	
gaining 27 g/day	1.5
Fowl weighing 1.4 kg	
laying 40%	1.33
laying 80%	1.67

Gastrointestinal Physiology

The efficiency of all animal production is limited, among other factors, by the incomplete digestion and transformation of fodder into tissue, milk and eggs. In ruminants, microbial fermentation further limits the efficiency of conversion of the combustible energy of fodder. A great proportion of the diet undergoes an added transformation into microbial matter and into volatile fatty acids by an anaerobic and rather inefficient process preceding the absorption of the volatile fatty acids in the rumen and the digestion of the remaining food and microbial matter by the host proper. Although the volatile fatty acids are efficiently absorbed, their heat increment is relatively high. As a result, the ME or NE content of most diets is lower in ruminants than in monogastric species. Nevertheless, because of the microbial activity in the rumen, ruminants are able to consume and utilize low-cost fibrous forage which monogastric species reject and are able to compete with monogastric animals economically. In fact, in many regions producing only fibrous forage, ruminants are the dominant species on the farm.

The efficiency of protein conversion in ruminants is not necessarily lower than in monogastric species. Although on high-protein diets ruminal digestion may well reduce the quantity and quality of protein in the digesta, on many conventional diets the ruminal microflora converts poor-quality protein and even non-protein nitrogen into high-quality protein. In extensive systems the amount of true protein leaving the rumen may be equal to or greater than the amount entering it.

The dietary vitamin requirements of ruminants are almost entirely limited to the fat-soluble ones, since the microbes of the rumen synthesize the others. Both proteins and vitamin metabolism in the rumen probably contribute to the ability of ruminants to compete economically with monogastric animals.

Meat versus Milk and Egg Production

Meat production is considerably less efficient than milk and egg production. There are two reasons for this difference. The first is rather simple: only part of the carcass of animals is edible (see Chapter 18). A variable but large proportion of offal (skin, hair, bone, feathers, blood, intestines, body fat, etc.) reduces the amount of edible meat, although in developed countries all the offal may be converted into useful by-products. On the other hand, milk and eggs (except the shell which contains very little protein and energy) are edible products. For milk this statement must be qualified; when cheese is manufactured from milk, the whey, containing all the lactose and some soluble protein, becomes a rather low- value by-product often fed to stock.

Since most of the body fat in meat animals may be offal and because of the fact that the amount of body fat increases with age, the overall efficiency of meat production decreases when meat animals are raised longer. Many studies deal with the determination of the economically optimal age and size at slaughter of pigs, broilers, cattle and sheep. Optimal slaughter time varies with the genetics of the breed and with the environment, including diet, climate and husbandry.

The other reason for the inferior efficiency of meat production is related to the rate of production discussed above. Animals raised exclusively for meat consume a lower multiple of their maintenance requirement than lactating ruminants or egg-laying hens.

Sex

It is well known that the males of all the farm species are more efficient meat producers than the females. Males grow faster than females and consume a greater multiple of their maintenance requirement. Males are also leaner than females and produce less offal fat.

Fertility

Fertility and fecundity contribute greatly to increasing the overall efficiency of animal production (see Chapter 19). Regular conception and normal pregnancies in dairy cows, ewes and goats minimize the periods of low production, especially the duration of the dry period. Ewes, which have a gestation period of 150 days, are now bred in intensive systems three times in two years to maximize both milk production and the lamb crop.

Prolificacy, i.e. high fecundity in pigs, greatly increases their overall efficiency because it reduces the number of brood sows and boars required to maintain a herd. At the other end of the spectrum, the beef cow at best produces one offspring a year and this considerably reduces the relative overall efficiency of beef cattle husbandry (see Table 25.1).

Sheep and goats are intermediate in prolificacy among farm mammals. Twins and triplets are common in goats and in some breeds of sheep; these breeds of sheep are now becoming more popular. The values for the overall efficiency of lamb production in Table 25.1 are for single-bearing ewes. It has been estimated that twins and triplets increase the overall efficiency of energy conversion of mutton sheep from 2.4% to 3.4% and 4.2%, respectively.

Needless to say, in the domestic fowl, prolificacy surpasses the level required for maximizing the efficiency of meat production and is the condition which permits egg production to be a highly efficient system of animal production.

ECONOMIC EFFICIENCY OF ANIMAL PRODUCTION

The overall efficiency of animal production is only one of many factors affecting its economics. The availability and price of different types of fodder, which in turn depend on the amount of available land and climate in every region, the variable demand for different animal products and many other factors affect the economic efficiency of animal production. However, in intensive systems food cost is the major input.

Because of differences in available land and climate, systems of intensive milk production differ greatly even in the developed countries. In the temperate zones, in countries with ample land, e.g. New Zealand, milk production (and beef production) are based mainly on pasture, whereas in a relatively hot and arid country like Israel high-producing dairy cows are fed mainly concentrates on the farm. In the latter, the level of milk production is higher, and the overall efficiency of conversion is probably greater, but economics favour the former system. Most systems of milk production in the world are intermediate between these two.

In many developing tropical and subtropical countries, beef cattle, mutton sheep and goats are the major classes of stock, though their overall efficiency is very low. In these regions most available fodder is rather fibrous pasture and the major input in animal production is investment and labour rather than food. Many of these regions suffer from periodic droughts and their efficiency could be improved by the judicious use of preserved food or concentrates which, however, are relatively unavailable because of cost. The seasonal nature of the food supply in some of these regions causes a greatly fluctuating growth curve and reduces the overall efficiency of meat production mainly because animals are marketed only after years of slow growth.

Pig and poultry husbandry depends largely on foods which are edible by man and is more intensive in developed countries with a high level of income.

The efficiency of animal production in many parts of the world is far below that which could be realized. Research and development continue to improve the genetic characteristics of farm animals and their environmental conditions. Genetic progress in milk and egg production has been considerable and has produced breeds which are able to express their full genetic potential only in the best environmental conditions. In dairy cattle, artificial insemination and progeny testing have produced these advances; in egg production progress was rapid because the population is large and the generation interval is short. Genetic progress has also been achieved in developing lean breeds of pigs and poultry. In sheep, the emphasis in recent years has been on increasing prolificacy by both genetic and environmental means. Genetically, prolificacy may be increased by the introduction of breeds with superior fecundity, i.e. the ability to produce twins, triplets, etc., rather than singles. Manipulation of the environment (early weaning, hormonal treatment and improved nutrition) in certain breeds of sheep may increase the frequency of lambing from once a year to three times in two years. In beef cattle the major trend is different. Although the traditional breeds are hardy under trying environmental conditions, they are slowly being replaced by faster growing cattle mainly from dual-purpose breeds. In every class of stock and in every system, genetic progress must be accompanied by improved environmental conditions and more sophisticated husbandry.

Improved nutrition, including optimally balanced diets adequately supplied with vitamins and minerals, feed processing, e.g. pelleting to increase food intake, and the adoption of more economic sources of food, the use of growth-promoting agents of different types including hormone-like substances, antibiotics and ionophores, freedom from disease and the alleviation of heat stress in warm climates, are the major environmental factors which contribute to the increased overall and economic efficiency of animal production.

Genetic engineering is the most recent technique employed in the improvement of the efficiency of animal production. Although direct intervention in the genetic traits of farm animals has not yet been achieved, the genomes of laboratory animals have been successfully modified to increase growth rate and mature weight. In farm animals, specifically in sheep, embryo splitting—a form of cloning—has been performed. This is a promising method for accelerating the production of farm animals with desirable characteristics. Another promising achievement is the large-scale production of growth hormone by bacteria which have undergone genetic engineering. The daily injection of growth hormone from this source into high-producing dairy cows has produced increases of up to 40% in the production of milk over a period of 188 days. The same hormone has also enhanced mammary development in cattle and has produced a shift in carcass composition towards more protein and less fat.

BIBLIOGRAPHY

Bauman, D. E. (1984). Effects of growth hormone on growth rates and mammary development of ruminants. *Proceedings 1984 Cornell Nutrition Conference*, pp. 13–17. Ithaca: Departments of Animal Science.

Bondi, A. (1982). Nutrition and animal productivity. In *Handbook of Agricutural Productivity* (M. Recheigl, Jr., ed.), Vol. 2, pp. 195–212. Boca Raton, Florida: CRC Press.

Cravens, W. W. (1981). Plants and animals as protein source. *J. Anim. Sci.*, **53**, 817–826.

Pimental, D. *et al.* (1975). Energy and land contraints in food production. *Science*, **190**, 754–761.

Reid, J. T., and White, O. D. (1978). Energy cost of food production by animals. In *New Protein Foods* (A. M. Altschul and H. L. Wilcke, eds), Vol. 3, Part A, pp. 117–144. New York: Academic Press.

Reid, J. T., White, O. D., Anrique, R., and Fortin, A. (1980). Nutrition energetics of livestock. *J. Anim. Sci.*, **51**, 1391–1416.

Spedding, C. R. W. (1984). New horizons in animal production. In *Development of Animal Production Systems* (B. Nestel, ed.), pp. 399–418. Amsterdam: Elsevier.

Trenkle, A., and Willham, R. L. (1977). Beef production efficiency. *Science*, **198**, 1009–1015.

Wedin, W. F., Hodgson, H. J., and Jacobson, N. L. (1975). Utilizing plant and animal resources in producing human food. *J. Anim. Sci.*, **41**, 667–686.

APPENDIX

Ration Formulation and Production Optimization

By David Sklan

The objective of practical nutrition is to formulate a regime that will supply the requirements of the animal with respect to energy, and all other essential components in the amounts and relationships required. This should be done in such a way as to produce each unit of product (meat, milk, eggs, etc.) at minimal expense. Thus in ration formulation, knowledge of nutrients, feedstuffs and animals is combined to optimize the required level if production. This obviously calls for a blend of information on different levels, and may call for mathematical solutions which are complex. In this Appendix some of the methods will be presented.

INFORMATION REQUIRED

Nutritional Requirements

In order to design a feeding regime we need to know the requirements of all critical nutrients under our particular circumstances. For this, publications such as the National Research Council (NRC) in the USA or the Agricultural Research Council (ARC) in England are usually used. Similar feeding standards have also been developed in other countries (see preceding chapters).

Feedstuffs

The feedstuffs available, including supplements and by-products, should be listed together with the energy and content of all the critical nutrients.

Analytical data from local laboratories is preferable; however, data published in the various tables (for instance, NRC) may be used, with appropriate care if local data are not available. Additional considerations at this stage should include the type of processing used and whether this will adversely affect any of the ingredients. In addition, restrictions on the minimum or maximum amount of feedstuffs for nutritional or technical reasons should be noted.

Feed Costs

Feed costs at the site of calculation (feed mill or farm) should also be obtained.

With this data in hand it is possible to set up a table containing all the required information. An example is given in Table A.1. This information is required for whatever procedure is to be followed for formulating the feed. It is possible to formulate a feed manually; however, as the number of feed ingredients and the nutritional requirements increase it becomes increasingly complex to achieve *any* solution, let alone an optimal solution, as will be seen.

Table A-1 Example of data required for ration formulation

	Milo (M)	Corn (C)	Soya (S)	Fishmeal (F)	Type	Quantity
Amount (kg)	1	1	1	1	=	1000
Metabolic Energy (MJ/kg)	13.4	13.9	9.2	11.3	=	11550
Protein (kg/kg)	0.09	0.08	0.45	0.62	>=	170
Cost	0.9	1.0	1.8	3.0		

The relationships involved in formulating a ration are as follows, in a situation producing 1000 kg of some feed mix. Letters refer to materials as shown in Table A.1. It is helpful (and customary) to list data as shown, with the nutritional requirements as rows, their levels on the right and the ingredients as columns. This type of nutritional requirement can be an equality, minimum or maximum or any combination of these. The following equations can be written:

Amount Equation

$$\text{Amount (kg)} = S + F + M + C = 1000$$

The coefficients stating the amount of kg/kg are 1 in all cases, and the letters the feed ingredients.

Energy Equation

$$\text{Metabolic energy (MJ/kg)} = 9.2\,S + 11.3\,F + 13.4\,M + 13.9\,C = 11550$$

The coefficients represent the metabolizable energy in each ingredient.

Protein Equation

$$\text{Protein (kg)} = 0.45\,S + 0.62\,F + 0.09\,M + 0.08\,C > 170$$

The coefficients are the protein contents of each feed ingredient.

Cost Equation

$$\text{Cost} = 1.85\,S + 3.0\,F + 0.9\,M + 1.0\,C = \text{MINIMUM}$$

the coefficients represent the cost of each material.

Of course, there is no reason why the problem should be limited to the nutritional requirements or feeds shown in the table; this represents a simple example. The problem has thus been set up to make 1000 kg of a feed containing exactly 11550 MJ and at least 170 kg protein. One could also, for example, define an upper limit to protein, and for this one would add another equation with the right-hand side showing a maximum of, say, 220 kg protein.

The formulation is thus a question of solving the above equations. However, it can be seen that such equations do not necessarily have a single unique solution. Several linear programming algorithms (Simplex, revised Simplex) exist to solve, using digital computers, such a series of equations efficiently, yielding in the first instance a feasible solution, followed by the least-cost formulation. Additional data which are calculated by most of the commercial programs available include the actual amount of each nutritional requirement (right-hand side levels), as well as the effect on the total cost of changing the requirement by the first unit. In addition the economic feasibility of feeds that were not selected in the solution are shown. In more sophisticated programs the range over which the solution, right-hand side and costs are valid is also indicated.

Such programs can be run on home computers and any larger. It should be mentioned here that the precision of calculation is very high; however, this only produces a number which is dependent on the accuracy of the data entered.

The units used in any formulation are those which the formulator himself chooses and may be any of the units for which data are available. For energy for example, these can range from gross energy, through digestible energy to metabolizable and net energies; other units such as total digestible nutrients, feed units or any other measure desired can also be used.

Some questions arising in formulating rations for different animals will be discussed.

RATIONS FOR MONOGASTRIC ANIMALS

Knowledge of the nutritional requirements of monogastric animals is extensive, and includes accurate values for requirements of all the essential amino acids, in addition to minerals and vitamins. This results in a relatively large number of 'rows' or equations. Furthermore, feed consumption in birds is usually dependent on the caloric consumption (see previous chapters). The direct result of this is that birds consuming a high-energy density diet will consume less feed. Thus we are required to increase the concentration of other dietary constituents. This can be allowed for in formulating the feed by calculating the nutritional requirement per calorie or joule rather than per gram feed. This can readily be accomplished by setting up the formulation appropriately. Considerable data are found in the literature concerning interactions between nutritional components such as calcium and phosphorus or between lysine and arginine. Where such interactions can be quantified and the relationship expressed in mathematical terms, this can also be incorporated in the formulation. This is relatively easy if the relationship is linear. In addition it should be noted that it is often desirable for both technical and nutritional reasons to restrict the amounts of nutrients such as fat or fibre in the ration.

RUMINANT RATIONS

In comparison with birds the number of nutrients for which requirements are accurately known for ruminants is small. This is, in part, due to the fact that the activity of the microbial population in the rumen alters the form of the feed before it is taken up by the animal. On the other hand, the number of potential feedstuffs is very large and includes many roughages and by-products. Energy units are a little more complex in ruminant nutrition, and the use of NE_g together with NE_m in formulating rations for growing cattle requires a rather different definition of the problem, as the same feed appears at two different energy values. Requirements for fibre in cattle are also more difficult to define as the type of fibre is also important and several systems have been used including crude fibre, acid detergent fibre or crude fibre from roughage. Many nutritionists prefer to formulate fibre requirements (or protein) as a proportion of the total dry matter intake, and this can again be accomplished by appropriate set-up. Interactions such as non-protein nitrogen/available carbohydrates, or calcium/phosphorus can be incorporated where the quantitative relationships are known (Black and Klubick, 1980).

PRODUCTION OPTIMIZATION

The above discussion concentrates on producing the least-cost ration under a defined set of conditions. However, the conditions can very often be varied, and although the ration may be the least-cost this may not represent the optimal production for the farmer. For example, rapid growth of animals may require a more expensive feed, but it will also result in a more rapid turnover of animals, which will be housed for a shorter time on the farm. Such considerations involve more complex examinations, and while using similar formulation calculations also incorporate calculation of cumulative expenses and profit.

In order to attempt to calculate costs of production under different conditions, it is necessary to know the relationships between the different variables. Thus for poultry one would need to determine relationships such as the connection between energy density of the diet and rate of growth. In dairy cattle the relationships between dietary fibre and milk and milk-fat production and between dietary energy and milk production are required.

One such example is the Californian System for dairy cattle profit maximization (Bath *et al.*, 1978). Here, the type and level of the ration are taken into account by means of several appropriate linear relationships, and their effect on the amount and fat percentage of milk produced are calculated. The optimal ration is the one producing the highest profit for the farmer. A more complex example calculates the optimal nutritional trajectory for the whole lactation maximizing profits per lactation, where the relationships are no longer linear and the ration formulation is a subroutine on the main program, which is run many times under an optimizing algorithm (Talpaz *et al.*, 1983).

Production optimization for broilers has also been developed and includes considerations of the dietary energy density and composition, rate of growth, ambient temperature, and cumulative costs on the farm to determine marketing age for optimal profits (Hurwitz and Talpaz, 1985).

OVERALL OPTIMIZATION

An alternative approach is overall optimization of resources. One example would be the distribution of a limited amount of a particular feedstuff between several different animals. For instance, if the amount of fish meal or corn is restricted, it is not at all clear which is the optimal animal to feed it to. Definition of the problem as a linear program with the formulation of each feed simultaneously subject to the restriction of the particular feedstuff will yield the optimal solution for the system defined. A similar approach may be applied to limited resources on any level. However, it should be noted that optimizing resources does not optimize profit for any particular animal crop.

SUMMARY

Formulation of rations for animals can best be done using linear programs which produce least-cost diets under defined conditions. More complex methods are available for optimizing profits or resource distribution. All these methods require detailed knowledge of nutritional requirements and relationships between nutrients and production.

BIBLIOGRAPHY

Bath, D. L., Dickinson, F. N., Tucker, H. A., and Appleman, R. D. (1978). *Dairy Cattle: Principles, Practices and Profits*, 2nd edn. Philadelphia, Pa: Lea and Febiger.

Black, I. R., and Klubick, I. (1980). Basics of computerized linear programs for ration formulation. *J. Dairy Sci.*, **63**, 1366–1378.

Hurwitz, S., and Talpaz, H. (1985). The use of simulation in the evaluation of economics and management of turkey production: dietary nutrient density, marketing age and environmental temperature. *Poult. Sci.*, **64**, 1415–1423.

Talpaz, H., Seligman, N., Goldman, A., Sklan, D., and Hurwitz, S. (1980). Optimal trajectory of lactation and nutrition for the dairy cow. In *Operations Research and Water Resources* (D. Yaron and C. Tapeiro, eds), pp. 285–299. North Holland Publishing Company.

Further Reading

Animal Nutrition

Church, D. C., and Pond, W. G. (1981). *Basic Animal Nutrition*, 2nd edn. New York: John Wiley.

Cole, H. H., and Ronning, M. (1974). *Animal Agriculture.* San Francisco: W. H. Freeman.

Crampton, E. W., and Harris, L. E. (1969). *Applied Animal Nutrition*, 2nd edn. San Francisco: W. H. Freeman.

Cuthbertson, D. (1969). *The Science of Nutrition of Farm Livestock*, 2 vols. Oxford: Pergamon Press.

Haresign, W., and Lewis, D. (eds) (1977–1985). *Recent Advances in Animal Nutrition*, 9 vols. London: Butterworths.

Lenkeit, W., Breirem, K., and Crasemann, E. (1969–1972). *Handbook of Animal Nutrition*. 2 vols. Hamburg and Berlin: Paul Parey.

Maynard, L. A., Loosli, J. K., Hintz, H. F., and Warner, R. G. (1979). *Animal Nutrition.* 7th. ed.New York: McGraw-Hill.

McDonald, P., Edwards, R. A., and Greenhalgh, I. F. D. (1981). *Animal Nutrition*, 3rd edn. London: Longman.

Recheigl, M. (1976–1981). *Comparative Animal Nutrition*, 4 vols. Basel: Karger.

Riis, P. M. (ed.) (1983). *Dynamic Biochemistry of Animal Production.* Amsterdam: Elsevier.

Rook, J. A. F., and Thomas, P. C. (eds.) (1983). *Nutritional Physiology of Farm Animals.* London: Longman.

Ruminant nutrition

Church, D. C. (ed.) (1976–1980). *Digestive Physiology and Nutrition of Ruminants*, 3 vols. Corvallis: D. C. Church.

Jarrige, R. (1987). *Alimentation des Ruminants.* Versailles: Institute National de la Recherche.

Poultry nutrition

Scott, M. L., Nesheim, M. C., and Young, R. I. (1982). *Nutrition of the Chicken*, 3rd edn. Ithaca: M. L. Scott.

General and human nutrition

Alfin-Slater, R., and Kritchevsky, D. (1980). *Human Nutrition. A Comprehensive Treatise*, 4 vols. New York: Plenum.

Darby, W. J. *et al.* (eds) (1981–1985). *Annual Review of Nutrition*, 5 vols. Palo Alto: Annual Reviews Inc.

Davidson, S., Passmore, R., Brock, J. F., and Truswell, A. S. (1975). *Human Nutrition and Dietetics*, 6th edn. London: Churchill.

Draper, H. H. (ed.) (1978–1985). *Advances in Nutrition Research*, 7 vols. New York: Plenum.

Freedland, R. A., and Briggs, S. (1977). *A Biochemical Approach to Nutrition*. London: Chapman and Hall.

Goodhart, R. S., and Shils, M. E. (1978). *Modern Nutrition in Health and Disease*, 6th edn. Philadelphia: Lea and Febiger.

Lloyd, L. E., McDonald, B. E., and Crampton, E. W. (1978). *Fundamentals of Nutrition*, 2nd edn. San Francisco: W. H. Freeman.

Pike, R. L., and Brown, M. L. (1975). *Nutrition. An Integrated Approach*, 2nd edn. New York: John Wiley.

Animal physiology

Gordon, M. S. (1977). *Animal Physiology*, 3rd edn. New York: Macmillan.

Phillis, I. W. (1976). *Veterinary Physiology*. Philadelphia and Toronto: W. B. Saunders.

Swenson, M. J. (ed.) (1984). *Dukes' Physiology of Domestic Animals*, 10th edn. Ithaca and London: Cornell University Press.

Poultry physiology

Bell, D. I., and Freeman, B. M. (1971–1983). *Physiology and Biochemistry of the Domestic Fowl*, 4 vols. London: Academic Press.

Sturkie, P. D. (1976). *Avian Physiology*, 3rd edn. New York: Springer-Verlag.

Index